RIVERSIDE COMMUNITY COLLEGE
1916

Fifth Edition

Functional Anatomy of the LIMBS AND BACK

W. HENRY HOLLINSHEAD, Ph.D.

DAVID B. JENKINS, Ph.D.

The School of Medicine
Department of Anatomy
University of North Carolina
Chapel Hill, North Carolina

W. B. SAUNDERS COMPANY
Philadelphia London Toronto Mexico City Rio de Janeiro Sydney Tokyo

quare
.9105

ussex BN21 3UN, England

1 Goldthorne Avenue
Toronto, Ontario M8Z 5T9, Canada

Apartado 26370 – Cedro 512
Mexico 4, D.F., Mexico

Rua Coronel Cabrita, 8
Sao Cristovao Caixa Postal 21176
Rio de Janeiro, Brazil

9 Waltham Street
Artarmon, N.S.W. 2064, Australia

Ichibancho, Central Bldg., 22-1 Ichibancho
Chiyoda-Ku, Tokyo 102, Japan

Library of Congress Cataloging in Publication Data

Hollinshead, William Henry, 1906–

Functional anatomy of the limbs and back.

Includes index.

1. Extremities (Anatomy) 2. Back – Anatomy. I. Jenkins, David B., joint author. II. Title. [DNLM: 1. Movement. 2. Musculoskeletal system – Anatomy and histology. 3. Musculoskeletal system – Physiology. WE101 H741f]

QM531.H7 1981 611'.98 80–51323

ISBN 0–7216–4755–3

Listed here is the latest translated edition of this book together with the language of the translation and the publisher.

Italian (*4th Edition*) – Aulo Gaggi Editore, Bologna, Italy
Japanese (*2nd Edition*) – Kyodo Isho Shuppan Sha, Tokyo, Japan

Functional Anatomy of the Limbs and Back ISBN 0-7216-4755-3

Last digit is the print number: 9 8 7 6 5 4

Preface to the Fifth Edition

In this edition of *Functional Anatomy of the Limbs and Back* the aging but not yet senile original author welcomes a colleague as a co-author.

We have not changed the format: various facts and concepts that need to be understood before any serious study is undertaken are presented in the first section, and thereafter the emphasis is primarily on the musculoskeletal system. Following this are brief discussions of some of the more important features of the head, neck, and trunk.

Electromyographic findings that were not available at the time of the previous edition have been incorporated in the present one. In response to welcome suggestions, material has been added on levers and their application to the forces bearing upon joints; on the strength of bone, muscle and tendon; and on the courses of nerves and arteries. More practical and clinical applications have been added, especially in regard to the effect of various lesions of nerves. A number of illustrations have been enlarged, and there are twelve new figures.

The senior author is entirely responsible for any errors of facts or concepts that may appear in this book. However, beginning students should be aware that even in the old science of anatomy there are still differences of opinion and gaps in our knowledge. They should therefore not be surprised if their instructor sometimes disagrees with statements found here.

Our thanks are due to the Medical Department, Harper and Row, for allowing us to use figures from the senior author's *Anatomy for Surgeons* and *Textbook of Anatomy*; to our secretary, Mrs. Brenda Kendig, for her cheerful cooperation; and to our publisher, the W. B. Saunders Company,

W. Henry Hollinshead
David B. Jenkins
Chapel Hill, North Carolina

Contents

III THE BACK

IV THE LOWER LIMB

V THE HEAD, NECK, AND TRUNK

Anatomic Terminology 1

The beginning student of anatomy is confronted with the necessity of mastering a largely new, cumbersome and complicated vocabulary. The difficulties are increased by the fact that, in common with most scientific terminology, anatomic names are given a Latin form. It is now agreed that each language group may use the vernacular as it deems proper, therefore we depart from the strict Latin by anglicizing many expressions or using a direct English translation; however, the Latin terminology is the foundation upon which our scientific vocabulary is built. The problem is further complicated for most of us through the fact that, regardless of their endings, most anatomic terms have Greek or Latin roots. As most of them convey very definite meanings it is well worth the effort to consult a medical dictionary, when necessary, to discover the original meaning of the word and thus translate it in one's mind from a term that must be merely memorized to one that is understood. If the student will make a conscientious effort to understand the terminology of anatomy, he will find it much easier to learn the facts and concepts of anatomy.

A further difficulty in anatomic terminology is the abundance of synonyms that have accumulated for generations. The international anatomic terminology (NA, or Nomina Anatomica, adopted in 1955) recognizes practically no synonyms, but even anatomists, because they are used to an older terminology, have difficulty in eliminating synonyms from their talking and writing. Certainly, many of the clinicians with whom the therapist will come in contact will use the synonyms with which they are most familiar, and it is impossible to eliminate synonyms from older texts and figures that the student may consult. In this book the use of synonyms has been reduced to an approximate minimum. Only where it seemed obvious that the student might be handicapped through ignorance of another commonly employed although officially outdated term has that term been given as a synonym. Some of the more common synonyms are listed beginning on page 384.

The major subdivisions of the body are the head, neck, trunk, and limbs. Although it is perfectly proper to use these English names, they also have Latin names that are used in many terms that the student will meet. Thus the head is "caput," and "capitis" therefore means "of the head." The neck is "collum"; cervix also means neck, especially its anterior part, and nucha means its posterior part. Thus "colli" means "of the neck," and we also find the terms "cervical" and "nuchal" or "nuchae" used in referring to structures in the neck.

Our word "trunk" is obviously the same as the Latin "truncus," but we have no particular reason to use the latter word. However, the subdivisions of the trunk need to be understood. The Latin word for chest is thorax, and this word will be met often both in this form, in its possessive "thoracis," which means "of the chest," and in its adjectival form "thoracic." The abdomen is the part of the trunk with muscular walls that lies below the thorax. It is most easily translated into our word "belly," but since this is considered inelegant we are left with no acceptable translation, and therefore use the words "abdomen" and "abdominal." ("Stomach," commonly used and understood to mean the abdomen, means no such thing; the stomach is one of many organs in the abdominal cavity.) The lowest part of the trunk is the pelvis (meaning "basin"), and the Latin term is always used for this; "pelvic" is the adjective pertaining to the pelvis.

The Latin for the limbs is "membra" (member), but has no common usage. "Appendage" is a term long used by zoologists to describe limbs in general, and sometimes appears in human anatomy in the adjectival form "appendicular." Names of smaller subdivisions of the limbs are best reserved for the times when the limbs are studied.

Since anatomy is a descriptive science, many anatomic terms are in themselves descriptive, referring to shape, size, location or function of a part, or its fancied resemblance to some nonanatomic structure. Muscles, especially, are usually so named, but since many muscles may have a given shape or a given function it is usual to use more than one descriptive adjective in naming a muscle. Thus there are two muscles called biceps, or muscles with two heads of origin, so we distinguish one as the biceps brachii, or two-headed muscle of the arm, the other as the biceps femoris, or two-headed muscle of the thigh. Similarly, the pronator quadratus is a quadrilateral muscle that pronates or turns the palm of the horizontally held hand downward, the quadratus lumborum is a quadrilateral muscle located in the lumbar region; the rectus abdominis is a muscle running vertically (rectus means straight) in the abdominal wall, the rectus capitis anterior is an anteriorly situated muscle running vertically to attach to the caput, or head, and so forth.

In addition to the various technical names of structures in the body there are also certain general terms describing surfaces of the body, planes through the body, relative positions of one structure to another, and so forth, that must be understood from the very beginning. The surfaces of the trunk may be conveniently described as the dorsum, or back, the ventral or belly surface, and the two sides, or lateral surfaces. The cranium is the skull, and cephalon is the Greek word for head, so cranial and cephalic both mean toward the head; similarly, caudal means toward the tail, or in man toward where the tail would be had it persisted from embryonic life.

The above-defined terms are all understandable regardless of the position of the body, but there are others that require agreement as to what position of the body we are referring to before they can be understood. For instance, superior, meaning up or upward, implies a relation to gravity, and therefore might differ entirely in meaning according to whether one was erect, lying upon one's back, or standing upon one's head. For this reason, anatomists have agreed that such terms of relative position should always be used in relation to a fixed position of the body termed the anatomic position; the anatomic position is the erect one, with the heels together and the feet pointing somewhat outward, the arms by the sides and the palms facing forward. With reference, then, to the anatomic position, superior always means toward the head, and is therefore used interchangeably with cephalic or cranial; similarly, inferior means toward the feet, and is usually synonymous with caudal. Anterior, referring to the part of the body habitually carried forward in progression, is thus synonymous with ventral in the human being, and posterior and dorsal are also synonymous. Even in the new anatomic terminology there is some inconsistency in the use of the terms "anterior" and "ventral," and similarly in "posterior" and "dorsal"; in essence, however, dorsal and ventral are used in human anatomy only in referring to parts of spinal nerves and of the hand and foot (compare our English "back of the hand"). Elsewhere anterior and posterior are preferred, so what was once, for example, the dorsal interosseous artery is now the posterior interosseous artery. The term "ventral" was rarely used in regard to the limbs (except in developmental stages), the more common term here being "volar" (vola = the palm or sole); this has now been abandoned as the antithesis of posterior, so what was once the volar interosseous artery is now called the anterior interosseous artery.

Other terms of relative position are medial and lateral, that is, toward the midline or toward a side. In the case of the limbs, however, some confusion might exist as to whether these terms refer to the body as a whole or to the limb itself; thus the little finger is medial to the other fingers in regard to the midline of the hand, but is lateral, as is the thumb, in regard to the midline of the body. It is best in this case, therefore, to use medial and lateral only when referring to the limb as a whole, and its relation to the body in the anatomic position; if relative mediolateral relationships of structures within the limb are to be described, radial and ulnar, and tibial and fibular are the terms best employed. These terms refer to the paired bones of the limbs. In the upper limb, the radius is on the thumb side, the ulna on the little finger side, and in the lower limb the tibia is on the side of the big toe while the fibula is on that of the little toe; the sides of the limbs are thus named from the positions of these bones. Additional terms especially useful in regard to the limbs are proximal, meaning toward the attachment of the limb to the trunk; distal, or away from the base of the limb; palmar, referring to the palm of the hand; and plantar, referring to the sole of the foot.

In regard to the planes of the body (fig. 1-1), the sagittal plane is either one passing through the midline of the body so as to divide it into right and left halves or one (often called parasagittal) parallel to but to one side of it. A coronal or frontal plane is one dividing the body into an anterior and a posterior (ventral and dorsal) portion, thus running roughly parallel with the front of the body and with a suture of

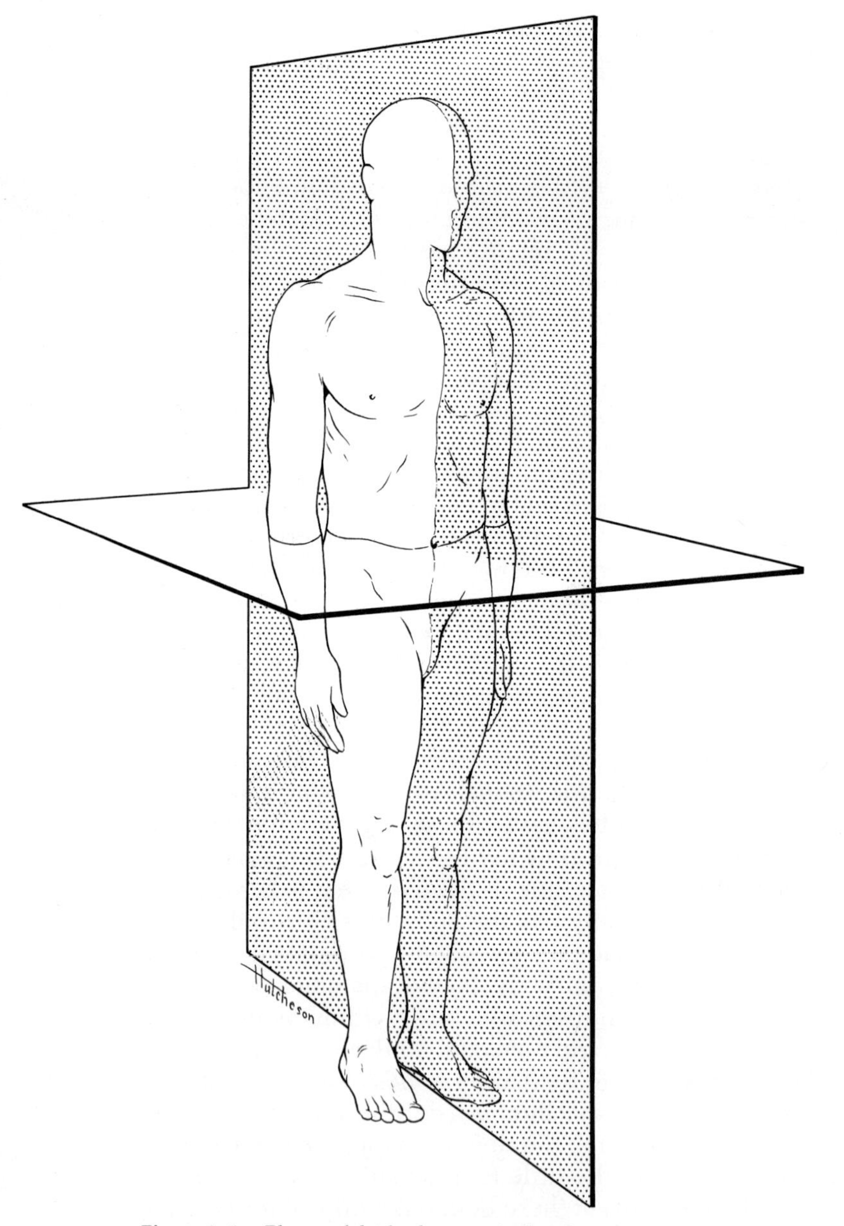

Figure 1–1. Planes of the body — sagittal and transverse.

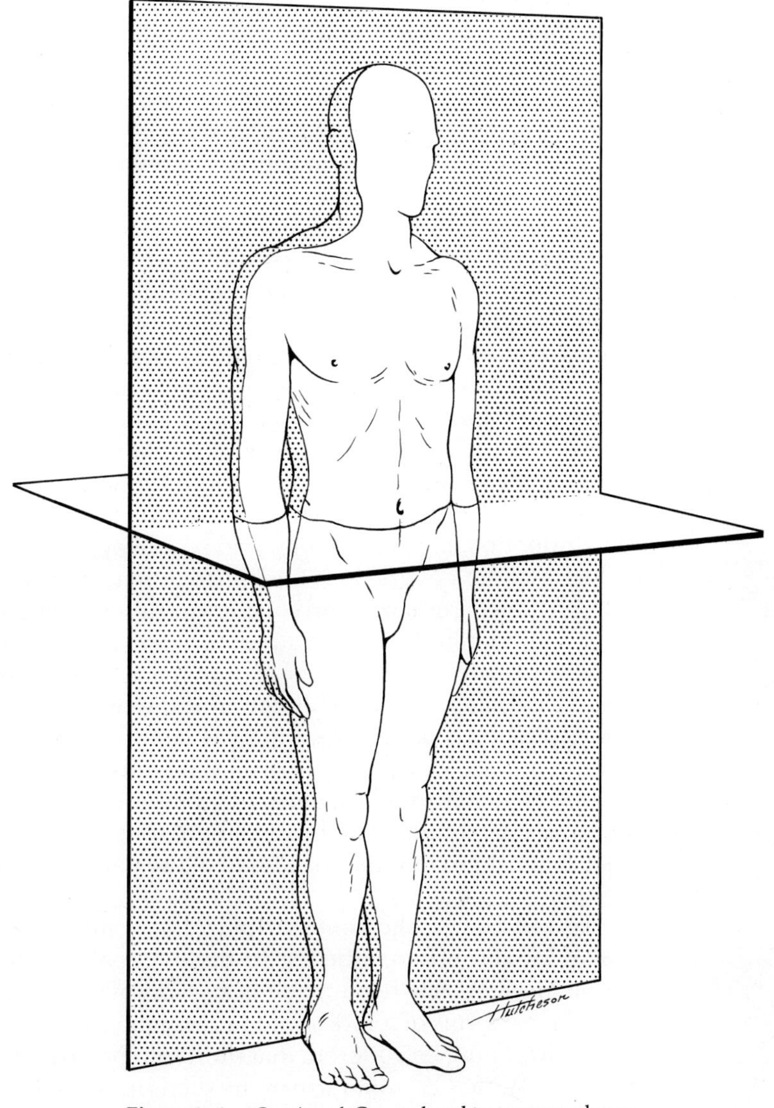

Figure 1–1. *Continued.* Coronal and transverse planes.

the skull called the coronal suture. A horizontal or transverse plane divides the body or limbs into upper and lower parts, in relation to gravity and the anatomic position.

In describing movements we may refer to them as being toward or away from a given plane, or we may find it convenient to speak of the axis of motion. For any given movement the axis of motion is an imaginary line about which a part describes a rotatory movement. These axes of motion typically make angles with one of the chief planes of the body. For instance, the axis for bringing the forearm up against the arm ("bending" the arm) is a line passing through the elbow region from lateral to medial sides, and is therefore at approximate right angles to the sagittal plane, and correspondingly approximately parallel to the frontal plane.

In addition to the terms of position, we must also agree in general upon the meaning of terms describing movement. As a rule, bending of the trunk or the limbs occurs most freely toward the original ventral surfaces, therefore flexion, which means bending, is usually a bending in the ventral direction; if we want to designate a bending as being in the dorsal direction, it is proper to say dorsiflexion. Extension, the straightening out of a bent part, is a movement opposed to that of flexion, and therefore occurs typically in the dorsal direction; if, at the joints where this is possible, we continue the movement of extension beyond that necessary to straighten the part, we can designate it as hyperextension. Dorsiflexion and hyperextension are plainly synonymous, as they both designate movement in the same direction. From these terms of movement we have acquired terms of position that are frequently used colloquially, although not a part of official terminology. Thus we may speak of the flexor or extensor surfaces of a limb, and these correspond to the original ventral and dorsal surfaces of the developing limb.

Abduction means moving apart, or away from the midline, and adduction, the reverse, is moving together, or toward the midline; both of these are particularly useful in describing movements of the limbs. Protraction is moving a part forward, retraction moving it back; elevation is lifting a part, that is, moving it superiorly, while depression is moving it inferiorly. Rotation is the twisting of a part around its longitudinal axis, and is described as lateral or external rotation if the anterior surface of the part is turned laterally, and medial or internal rotation if it is turned medially. Circumduction is a combination of successive movements of flexion, abduction, extension and adduction in such a fashion that the distal end of the part being moved describes a circle.

In relation to the limbs, especially, some of the terms defined above are apt to be confusing in their specific applications. For this reason an effort has been made in the following chapters of this book to so define terms of relation and movement, as they apply specifically to the part being considered, that there can be no ambiguity as to what is meant. With the proper use of the dictionary, and with due regard to the correct usage of anatomic terms as displayed by your instructor and in textbooks, the bugaboo of anatomic terminology can be largely overcome, and what may seem at first an almost incomprehensible jargon will become a useful part of your scientific language.

2 The Tissues of the Body

The human body, like most of the better-organized forms of animal life, consists of various types of specialized cells and a varying amount of intercellular substance; much of the actual weight of the body is water, both within the cells and without. The cells represent the living portion of the organism, while the intercellular substance, regardless of its nature, represents nonliving material that owes its existence to the activities of the cells.

Most types of cells tend to occur in groups in which the constituents are somewhat similar both in appearance and in function; such organized groups of cells are known as tissues. The tissues of the body are in turn not independent of one another but, rather, various types of tissues are interwoven to form more complex anatomic and functional units known as organs or organ systems.

According to their general appearance and functions, the various tissues of the body are usually classified into a few great groups: epithelial tissue, connective tissue, muscular tissue, nervous tissue, and blood.

EPITHELIA

Epithelial tissue (fig. 2-1) occurs most commonly in sheets, and is adapted especially for covering other tissues. It serves the general functions of protection, absorption and secretion. An epithelium is characterized by the fact that its cells are closely packed together with a minimum of inert intercellular cement substance between them. The cells may vary in shape from extremely flat ones resembling paving stones (called squamous cells; squama means a scale, such as that

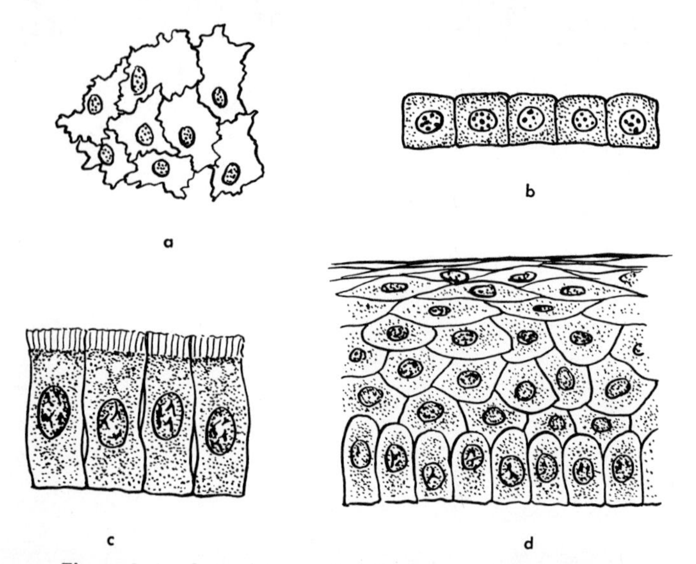

Figure 2–1. Several types of epithelium. *a.* Simple squamous, surface view. *b.* Simple cuboidal. *c.* Ciliated columnar. *d.* Stratified squamous. *b, c,* and *d* represent views of sections.

of a fish), to cuboidal ones shaped somewhat like children's blocks, to tall columnar ones. There are many subvarieties in shape, general appearance and function.

An epithelium may take the form of a single-layered sheet of cells, a multilayered sheet, or of essentially tubular outgrowths (glands) from such sheets. One type of epithelium covers the external surface of the body where, as the outer layer of the skin, a stratified or multilayered epithelium with dead outer cells protects the more delicate deeper-lying cells and serves as a membrane to seal off intercellular spaces from contact with the outside. Another type of epithelium, adapted for absorption and secretion, lines the digestive tract, and outgrowths from this epithelium form the digestive glands, including also the characteristic cells of such large organs as the liver and pancreas. Other types of epithelium line the tubules of the kidneys, the ureters, and the urinary bladder, and of course continue along the urethra (the tube leading from the bladder) to unite with the epithelium of the skin. Thus epithelium occurs primarily either on the outside of the body or as a lining of those cavities of the body that communicate with the exterior.

Two specialized types of epithelium are also found lining closed cavities within the body. Mesothelium, a single-layered very flattened epithelium of the pavement or squamous type, lines the four great cavities of the trunk, namely, the two pleural cavities surrounding the lungs, the pericardial cavity surrounding the heart, and the peritoneal cavity surrounding the abdominal viscera. Endothelium, essentially similar to mesothelium in appearance, forms the inner lining of the heart, of all blood vessels, and of the lymphatics.

CONNECTIVE TISSUE

Connective tissue, in sharp contrast to epithelial tissue, has cells that are more or less widely dispersed and separated from each other by nonliving intercellular material; it is the presence and character of this intercellular material that give connective tissue its specific characteristics.

The most pervasive type of connective tissue in the body is **fibrous connective tissue** (fig. 2-2). In this type the spaces between the cells are occupied by numerous fibers that make the tissues tough and capable of withstanding distortions and strains. The fibers between the cells may be of several types, and may occur either in the form of a loosely woven net with large quantities of fluid in the interstices of the net, or as an apparently solid structure such as a tendon, with closely packed fibers and very little interfibrillar space. The most common type of fiber found in connective tissues is the collagenous fiber. These fibers are essentially nonelastic; therefore, when they occur in places in which some deformation must be possible they are arranged in wavy bundles that allow movement until the slack of these bundles is taken up. Elastic fibers are the other important type of intercellular fibers; these fibers are actually elastic as their name implies: they may be stretched, and when the tension upon them is relaxed, they will shorten again. Their cut ends often curl. They are frequently mixed with more numerous collagenous fibers, but in certain locations great bundles of almost pure elastic tissue are found. At intervals, in the interstices between connective tissue fibers, connective tissue cells occur. Some of these, known as fibroblasts, are responsible for the formation and repair of the connective tissue fibers. Others possess the property of ingesting formed material; in this duty they may be aided by cells from the blood, some of which pass freely into the fibrous connective tissues as a part of the reaction of inflammation.

Collagenous tissue, with or without the admixture of elastic fibers, is the most ubiquitous of all tissues. Taking various forms, it permeates and surrounds practically all the tissues of the body, serving as a binding agent for these tissues. Indeed, it has been aptly said that if it were possible to dissolve out all the tissues of the body so as to

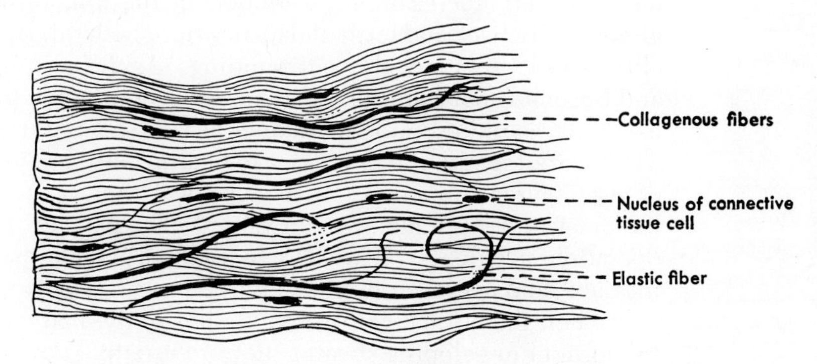

Figure 2–2. Connective tissue consisting largely of collagenous fibers, but with some admixture of elastic fibers.

leave only the fibrous connective tissues, the essential organization of the body would still be represented and recognizable through the arrangement of this fibrous tissue.

Where fibrous connective tissue forms the deep layer (the dermis or corium) of the skin it consists of densely matted fibers running in all directions. Most of these are collagenous, but there are also elastic fibers to lend resiliency to the skin. The dermis of animals is the source of leather. Deep to the skin, elastic and collagenous fibers are more loosely woven to form a subcutaneous layer, the tela subcutanea (subcutaneous network) that allows movement of the skin over the deeper structures. To a varying extent in different parts of the body and in different individuals this subcutaneous connective tissue contains modified tissue cells that are filled with fat. If these fat cells are sufficiently numerous the tissue is known as adipose tissue or fat. Varying amounts of loose connective tissue, often containing fat, occur elsewhere throughout the body, where this type of tissue forms padding between various organs, about blood vessels, and so forth. Special accumulations of connective tissue form the outer wall of blood vessels, and surround and permeate nerves to bind their nerve fibers together. Epithelia are rather regularly supported by connective tissue; muscles are surrounded and their cells are held together by connective tissue; and bone contains large quantities of connective tissue fibers. Thus fibrous tissue permeates practically all the organs of the body.

Because of its ubiquity, fibrous connective tissue is almost always involved in any injury to the body. It normally plays an important part in the healing process, for new connective tissue fibers form in the injured area and reunite the parts that have been separated by the injury. Connective tissue formed in an attempt to repair an injury is known as scar tissue. If the injury has been unduly severe or of long duration, more scar tissue is formed than is needed to repair the defect (nature's way of insuring at least enough is to overdo it somewhat). As this newly formed tissue grows older the fibers shorten and become more densely packed together, and may thus form a hard mass of considerable size which may—on a finger, for instance—interfere with movement of the part. Moreover, if the scar tissue is attached to a movable part, for example, a tendon in the finger which normally glides freely back and forth, it may interfere with this by binding it too closely to its less movable surroundings. As the scar tissue contracts and becomes more dense, it may in turn so pull upon the tendon that the finger is pulled into and maintains a flexed and useless position. Thus scar tissue, while necessary to healing, also has its dangers. One of the common functions of the physical therapist or occupational therapist is to minimize the unwanted effects of scar formation after operation, accidents, or disease, by the use of such methods as heat, massage, and exercise.

When the normal connective tissues of the body are arranged in the form of enveloping sheaths they are usually known as **fascias** (fascia means a bandage or band, and thus connotes a layer enveloping or binding together other structures). Thus the subcutaneous tissue or tela subcutanea is frequently called the superficial fascia; numerous examples of well-developed, tough, deep fascias occur, especially in the limbs where fascia forms heavy membranes surrounding the limb as a whole. Individual muscles are also surrounded by thin fascia, called perimysium, and separated from each other by looser connective tissue. This is especially well developed where two adjacent muscles cross each other rather than running parallel, and the fluid between the fibers of the tissue then acts as a lubricant to allow free movement of one muscle upon the other. In some locations between muscles or between muscles or tendon and bone, or even beneath the skin over bony prominences, connective tissue spaces coalesce to form pocket-like accumulations of fluid known as **bursae** (bursa = a purse). From the fascia surrounding a muscle, connective tissue septa pass into the muscle and subdivide it into bundles; these septa in turn divide until delicate connective tissue fibers surround each muscle fiber within a muscle.

The connective tissue fibers of a fascia, although arranged in approximately the same plane to form membranes, run in various directions within this plane so that they appear interwoven, with no main direction of fibers predominating as a rule. In tendons and ligaments, in contrast, connective tissue fibers are arranged roughly parallel to one another, and are closely packed to form definite cords or bands that are especially adapted to resist movement in one direction. **Tendons** are formed of heavy collagenous bundles and delicate cross fibers; a tendon is defined as such a bundle that attaches muscle to bone or, occasionally, to some other structure. In the abdomen the tendons of the lateral abdominal muscles form broad flattened membranes known as aponeuroses. The tendons of most muscles are, however, more narrow bands, or frequently, as is true of many of the tendons of the limbs, rounded cords. The fibers of which tendons are composed are attached firmly to the muscle cells at one end, while at the other end they enter the bone and blend both with the connective tissue surrounding the bone (periosteum) and with the fibers within the bone itself.

Although most of the collagenous fibers composing a tendon run in the same direction, they are not strictly parallel, but intertwine to form small bundles that in turn intertwine to form the larger parallel bundles that give tendons their distinctive appearance. As the tendon nears its attachment to the bone the larger tendon bundles also intertwine with each other. The end result, therefore, is that the pull of any part of the muscle, instead of being limited to a tendon bundle originating in that part, is widely spread through the tendon; also, although in a broad tendon different fibers are in the direct line of pull as the bone is moved, a large part of the muscle constantly acts on these fibers.

Ligaments represent another type of dense connective tissue, frequently similar to tendons in appearance but uniting bone to bone rather than muscle to bone. Most ligaments are composed of dense collagenous tissue, but a very few are almost pure elastic tissue.

A type of connective tissue that at first sight appears to have little in common with fibrous connective tissue is **cartilage**. Like fibrous connective tissue, however, cartilage consists largely of intercellular material with cells scattered only at intervals through this. Although it is frequently not apparent, the intercellular material of cartilage has as its groundwork a feltlike mass of fibrous tissue. This fibrous tissue is in turn impregnated by a matrix that renders it harder, tougher, and more homogeneous than ordinary fibrous tissue. The fibers within cartilage are usually collagenous in nature, but in cases in which brittleness would be an especial disadvantage, as in the cartilages of the external ear and the tip of the nose, the cartilages contain elastic fibers.

Cartilage serves as a supporting framework for softer tissues, being more resistant to deformation than is fibrous connective tissue but less resistant, and therefore more resilient, than bone. In the embryo and fetus the early skeleton consists almost entirely of cartilage, but in the adult most of this cartilage has been replaced by bone, and cartilage is found in only a relatively few locations. The thyroid cartilage (Adam's apple) in the neck, and the cartilaginous rings that support the trachea (windpipe) and its branches represent a supportive type of cartilage that is sufficiently strong for the duty of keeping open the airway to the lungs and yet is less brittle than similar-sized bones would be. Since the most common type of cartilage, called hyaline cartilage because of its glassy appearance, presents a much smoother surface than does bone, this type of cartilage is found covering the ends of bones at joints that are freely movable; thus hyaline cartilage typically forms the bearing surfaces between two adjacent bones as they move one upon the other. In other locations, where a cartilage must support great crushing force, the collagenous fibers within the cartilage are exceedingly heavy and prominent and the cartilage is then known as a fibrocartilage. Outstanding examples of fibrocartilages are the intervertebral disks, heavy pads that form a part of the backbone and thus must withstand the weight of the body and yet allow some movement between bones at the same time.

Bone is the hardest of the connective tissues and forms most of the skeleton of the adult human body. Like cartilage, bone consists of a fibrous connective tissue imbedded in more solid matrix. The matrix of bone contains a large amount of minerals, primarily in the form of tiny crystals of a complex compound of calcium and phosphorus, which are responsible for the hardness of bone. In the young child, in whom the deposition of calcium has not gone on to completion, the fibrous tissue of bone overbalances the mineral content and the bone therefore has toughness without adequate hardness. Thus the bones of a young child are relatively easily deformed by weight-bearing, and when young bones are broken they tend to break irregularly and splinter just as does a green stick. The disproportion between crystalline minerals and fibers leads to this splintering, which is described as a "green stick" fracture because of its appearance. In a young adult the balance between calcium deposit and fibrous content of the bone is usually well maintained, so that the bone possesses both maximum hardness and resistance to stress. In old age calcium continues to replace water in the bone, and the balance is thus in favor of the calcium compound. The bone then remains hard but is no longer tough, so that strains thrown upon it may easily result in fracture.

Bone occurs in two typical forms, spongy and compact. Compact or cortical bone forms the outer surface of all bones (cortex means "bark," and thus implies an outer surface); spongy bone lies within compact bone. Compact bone varies in hardness and thickness, but is distinguished by the fact that it is laid down in concentric layers and appears solid. Spongy bone actually appears spongy in texture, for it is composed of very thin plates of bone that meet other plates at various angles, and the interstices between these plates or trabeculae (trabecula = beam) are relatively large. In a typical long bone such as one of those in the limbs (fig. 3-1) cortical bone forms the entire thickness of the bone, from its outside to its hollow interior or marrow cavity, except at the ends of the bone. Here the cortical bone is thinner, and inside it the spongy bone subdivides the marrow cavity into the numerous interstices typical of spongy bone.

In a thin section through a piece of typical cortical bone (fig. 2-3) the bone can be seen to be laid down mostly in layers around a series of branching tubes (shown in cross section rather than from the side in figure 2-3) that contained the blood vessels of the cortical bone, and are known as Haversian systems or osteons. The cavity of the osteon is officially called the canal of the osteon but is more frequently known as the Haversian canal. The concentric layers of bone around a Haversian canal are marked by the layers of rather prickly-looking bone cells, or the spaces that these cells occupied during life. The layers belonging to one osteon are bound to those of adjacent ones by layers resembling parts of osteons, and the whole is enclosed on its outer surface by layers that encircle the entire bone, and on its inner surface by layers that encircle the marrow cavity.

The blood vessels entering the bone are so distributed through the Haversian canals that none of the cells that lie between the layers of bone is too far removed from a blood vessel. The living cells within the bone, although separated by the layers of bone matrix, communicate with each other and finally with the Haversian canal

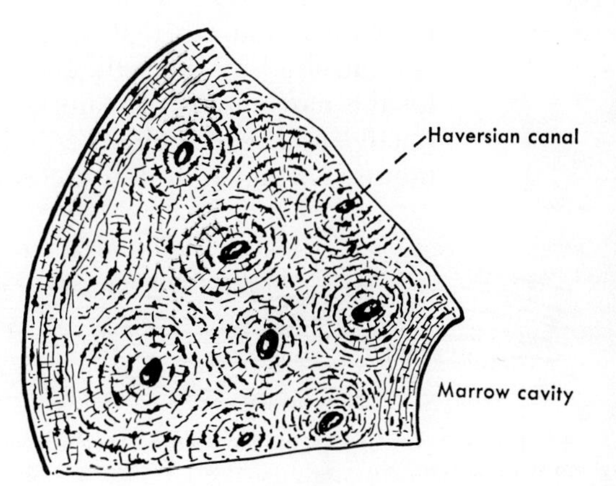

Figure 2–3. Typical structure of cortical bone. The bone cells mark off the layers of bony matrix, and are arranged concentrically about the marrow cavity and on the outer surface of the bone; between these inner and outer layers the bone is arranged in concentric layers around numerous Haversian canals.

by means of tiny threadlike processes. Through these communications substances from the blood, especially calcium salts, may be passed out into the bone, or calcium from the bone may be passed back into the blood stream. Even the bone of an adult, in which the growth in length and diameter have both ceased, is therefore not an inert unresponsive mass of tissue. Rather, the living cells in and about the bone are capable of bringing about modifications within this tissue, and bones therefore constantly adapt themselves to changes in the body as a whole.

Modification of the calcium deposit within the bone is especially striking in connection with tumors of the parathyroid glands; in persons with such tumors calcium may be so withdrawn from the bones that even turning over in bed may fracture a rib or a limb. Similarly, modifications of the entire bony structure may occur when a bone is fractured and improperly set, or when the forces exerted upon a bone, in the form of weight-bearing and muscle pull, are markedly changed. In normal spongy bone, for instance, the trabeculae are so arranged as to support the stresses normally placed upon that bone. If the direction of these stresses is changed much of the bone may undergo reorganization, resulting in an entire rearrangement of the trabeculae with disappearance of those no longer useful and formation of new trabeculae to withstand the new forces acting upon the bone.

MUSCLE

Muscle is a type of tissue that is especially adapted for shortening, or contraction; therefore, it consists of rather long cells. There are three distinct types of muscle in the human body, known respectively as smooth muscle, cardiac muscle, and voluntary striated muscle.

Smooth muscle (fig. 2-4) typically occurs in sheets surrounding hollow viscera, as for instance in the walls of the digestive tract or the walls of blood vessels. The individual smooth muscle cell is elongated, with tapering ends, and contains within its cytoplasm delicate muscle fibrils. Smooth muscle cells are usually firmly interlocked with each other, and contraction occurs regionally rather than involving individual cells. Smooth muscle forms one of the two types of involuntary muscle, that is, muscle whose contraction cannot be directly controlled by the will. Involuntary smooth muscle is responsible for the movements of material along the digestive tract, for the contractive ability of such other hollow viscera as the urinary bladder and uterus, for the control of the very small arteries whose diameter is in

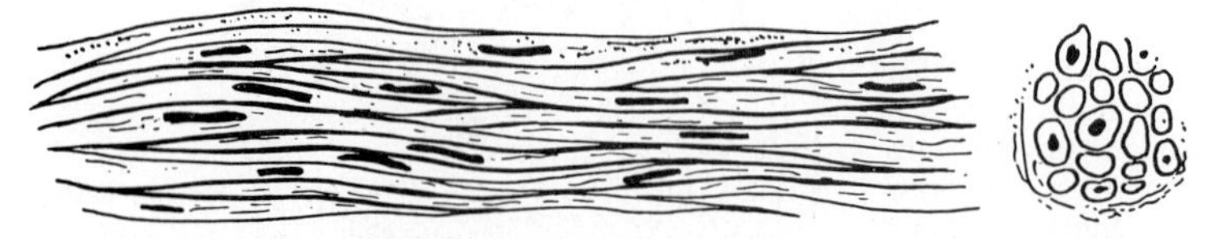

Figure 2–4. Groups of smooth muscle cells, seen from the surface and in cross section.

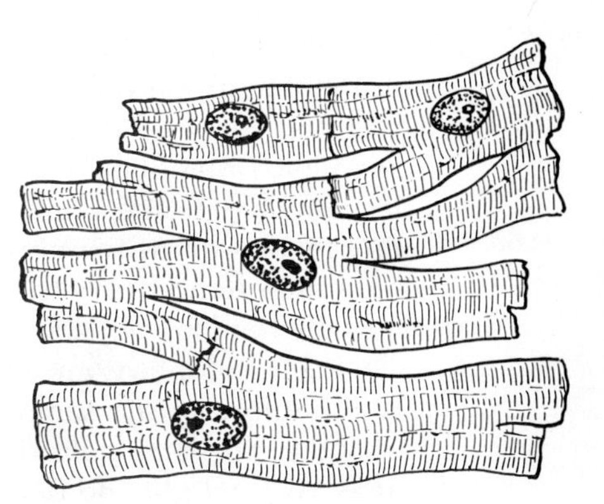

Figure 2–5. A small piece of cardiac muscle. Note the branching and anastomosing of the fibers and their cross striations.

turn so important in affecting the blood pressure, and in various other activities including even the formation of "goose flesh" by small smooth muscle bundles connected with hair follicles.

Cardiac muscle (fig. 2-5) is, as one might suspect, confined to the heart (cor) and the bases of the great vessels immediately adjacent to the heart. Physiologically, this muscle resembles smooth muscle in that it also is involuntary; anatomically it appears to be somewhat intermediate between smooth muscle and the striated voluntary muscles, for its cells, as do those of voluntary muscle, present a striated appearance when viewed under the microscope. Cardiac muscle differs sharply from voluntary muscle in one respect, however: its cells branch and are closely united to each other, so that contraction starting within one localized region of cardiac muscle spreads widely over the heart through the close contact of the cardiac muscle cells with one another. Essentially, the cardiac muscle of the atria of the heart contracts as a unit, and that of the ventricles also contracts as a unit; while various muscles or muscle layers in the heart are described, these consist of only partially separable sheets of fibers, the whole, therefore, forming interconnecting layers by means of which an impulse for contraction may travel over the entire cardiac muscle of the atria or of the ventricles.

Voluntary or **striated muscle** (fig. 2-6) constitutes by far the greater mass of muscle of the body and is the tissue that in domestic animals we usually recognize as meat. The individual cells or fibers of volun-

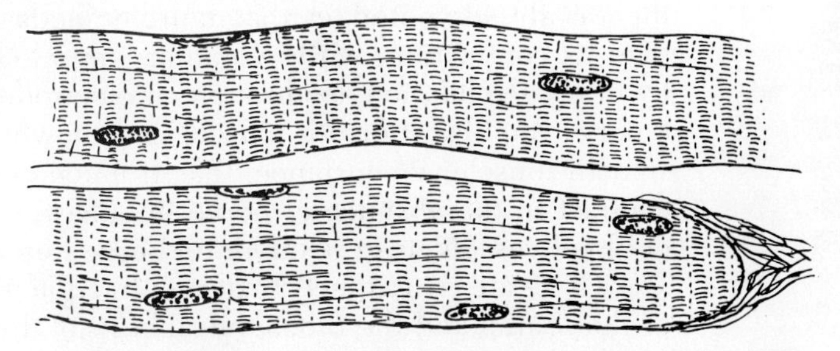

Figure 2–6. Portions of two adjacent voluntary muscle fibers. The fibers, very much longer than shown here, neither branch nor anastomose. Cross striations are prominent. On the lower fiber the junction of muscle and tendon is indicated on the right.

tary muscle are extremely threadlike, being only a small fraction of a millimeter in diameter but extending as much as 2 inches (about 5 cm.) or more in length. The surface of the fiber is called the sarcolemma. Immediately outside the sarcolemma is a very delicate layer of connective tissue, continuous with the perimysium surrounding the muscle, that binds the muscle fiber loosely to other muscle fibers and, more important, binds the end of the fiber to the end of another fiber or to the tendon.

Each voluntary muscle cell or fiber contains numerous nuclei, usually close to the sarcolemma, and closely packed, longitudinally arranged fibrils (myofibrils) that present alternating light and dark areas. As the light and dark areas of each fibril are approximately adjacent to the similar areas of other fibrils, these closely packed areas present the appearance of alternating light and dark stripes running transversely across the muscle fiber and give to this type of muscle its name of striated muscle. This type of striated muscle differs sharply, however, from cardiac muscle, in that the fibers run roughly parallel to one another and have no anastomoses. Each cell in a voluntary muscle has associated with it a nerve ending that deeply indents the sarcolemma; under normal conditions, the muscle fiber contracts only as a result of impulses received through this nerve ending.

Myofibrils are the contractile elements of muscle fibers and therefore occur not only in voluntary striated muscle but in smooth and cardiac muscle. Each myofibril, in turn, contains still smaller fibrils, visible only with the electron microscope and designated as myofilaments. The myofilaments of voluntary muscle are clearer than those of smooth and cardiac muscle and have been more extensively studied, but myofilaments of cardiac muscle are similar to those of voluntary muscle, and those of smooth muscle are assumed to be. Two types of myofilaments are described: thick ones, that seem to be composed primarily of the protein myosin, and thin ones of which the chief constituent is thought to be another protein, actin.

Contraction of Muscle

Contraction of cardiac muscle cells and of many smooth muscle cells spreads from one cell to the next, while the contraction of one voluntary muscle fiber has no effect on adjacent muscle fibers. Further, as already noted, cardiac muscle needs no nerve impulse to initiate its contraction, while voluntary muscle cannot contract (except by direct stimulation, as through an electrode) unless it does receive a nerve impulse. Smooth muscle is somewhat in between, for some smooth muscle, for instance, that of much of the digestive tract, will contract in the absence of nerve impulses, while other, such as that of blood vessels, is dependent on nerve impulses for contraction.

There are two aspects to the contraction of muscle: the mechanics of shortening and the biochemical basis of this shortening. In the uncontracted or resting voluntary muscle the thick (myosin) myofilaments, which form the dark band of the fiber as a whole, are only partially overlapped by the thin (actin) myofilaments; these project beyond the ends of the thick filaments into the light band in the muscle fiber and are the only filamentous occupants of that band. It is known that during contraction the light band shortens and finally disappears; this is thought to be due to a sliding of the thin filaments toward each other, between the thick filaments, until they meet and are completely overlapped by the latter.

In contrast to this relatively simple mode of change in length, the biochemical changes responsible for and associated with contraction are very complicated, and are as yet not completely understood. They will be only briefly summarized here.

In the case of voluntary muscle, the nerve impulse initiates the contraction by releasing at the nerve endings on the muscle a substance known as acetylcholine. The acetylcholine so released is believed to change the permeability of the sarcolemma so as to allow an influx of sodium ions, and this influx in turn creates an electrical potential or muscle impulse which travels extremely rapidly along the length of the muscle fiber. The muscle impulse then causes release of calcium ions which permit interaction of actin and myosin filaments to produce contraction. The immediate source of energy for contraction is adenosine triphosphate (ATP). Glucose, derived from glycogen stored in the muscles and in the liver, is the chief original source of energy for muscle contraction; when sufficient oxygen is available, the glucose is oxidized to CO_2 and water, and the energy thus released is used in part to form additional ATP (some, of course, is wasted in heat). When the respiratory and vascular systems cannot supply sufficient oxygen, as during vigorous exercise, the glucose is converted to lactic acid, but the lesser energy liberated by that reaction also helps form additional ATP; since lactic acid is essentially a poison to the muscle, and oxygen is necessary to remove it, the muscle is said to have accumulated an "oxygen debt." The resting muscle, now receiving sufficient oxygen, uses that oxygen in part to re-form glucose and glycogen from lactic acid and in part to oxidize the lactic acid to CO_2 and water.

NERVOUS TISSUE

Nervous tissue is specialized for conduction. The essential part of nervous tissue is the nerve cell (fig. 2-7), which has a somewhat rounded cell body distorted by fibrous processes that extend outward from this cell body. The body and its processes are known as a neuron. Every neuron or nerve cell has at least one process, and most neurons have many processes. One process of the nerve cell is threadlike and rarely branches until close to its ending; it is known as the axon. Most nerve cells have other processes that are relatively short and branch abundantly in the manner of a tree, and are therefore called dendrites. As a rule, the dendrites of a cell are limited in their distribution to the immediate region of the cell body, while the axon may be short or long. The axons of some cells extend only to closely adjacent cells and may be only a fraction of a millimeter in length; in contrast, there are

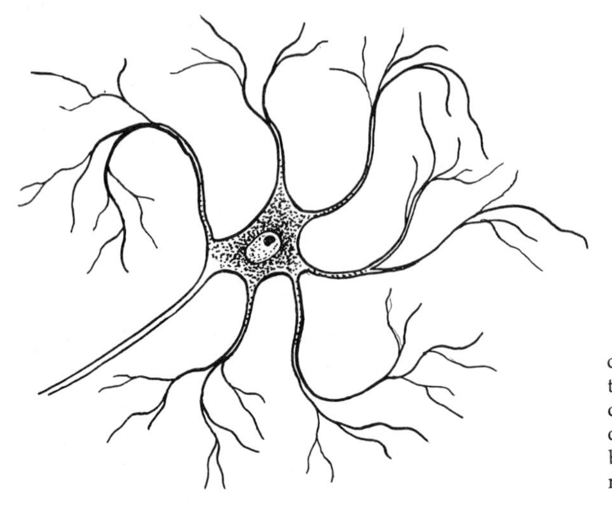

Figure 2–7. Diagram of a nerve cell from inside the central nervous system. Numerous dendrites branch frequently and end in the vicinity of the cell body, while the axon is so long before it branches that only a short segment of it is shown here.

cells in the brain that have axons measuring up to 20 inches (about 500 mm.) or more in length, and the nerve cells that supply the muscles of the foot have axons that extend the whole length of the lower limb and thus may be a yard or more (say 9000 mm.) long—this in spite of the fact that the axon may be in the neighborhood of 10 microns (one hundredth of a millimeter) or less in diameter. The axon is the fiber that takes the nerve impulses away from the cell body; dendrites, not as specialized in structure as axons, conduct nerve impulses toward the cell body and probably serve also in the cell's metabolic functions.

Most nerve cell bodies lie within the central nervous system; that is, they form a part of the brain or spinal cord (p. 46). In these locations they and their fibers are held in place by a special connective tissue, peculiar to the nervous system, known as the neuroglia. Other nerve cells routinely lie outside the central nervous system and form groups of cell bodies that are known as ganglia. Ganglion means swelling, and this term is sometimes applied to any swelling but is more often limited to a swelling produced by an accumulation of nerve cell bodies outside the central nervous system.

Nerve cell bodies vary greatly in size and shape, but the largest ones rarely exceed 100 microns (0.1 mm.) in diameter. Their shapes depend primarily upon the number of processes to which they give rise. Since the process represents a non-nucleated extension of the cytoplasm of the cell body (the nucleus being located in this cell body), nerve fibers cannot survive after they have been detached from their connections with the cell bodies. Thus when a nerve fiber is cut in two that part of it which lies distal to the cut dies, since it no longer has a connection with the nucleated part of the cell. Nerve fibers that have been interrupted outside the central nervous system can, under the proper circumstances, grow back and form connections that replace the old degenerated ones. On the contrary, when nerve fibers within the central nervous system are interrupted there is no functional restitution of the degenerated fibers, although the reasons for this difference are not clearly understood. Once nerve cells are formed, they are incapable of replacing themselves, and therefore when a given number of nerve cells is destroyed the individual has lost this number of cells forever.

Nerve impulses typically travel through the cell body and out along the axon; these nerve impulses are initiated by an ionic change in the cytoplasm of the cell, essentially similar to that initiating contraction of muscle: an adequate stimulus allows the influx of sodium ions into the cytoplasm, producing a reversal of polarity so that the inside of the cell very briefly becomes positive in regard to the outside; the electrical change in turn triggers a similar change in polarity in the immediately adjacent part of the cell, so the impulse travels through the cell or along the fiber. It is through the electrical change that the speed and progress of the nerve impulse can be followed. The speed of the impulse varies according to the diameter of the fiber along which it is traveling, being faster in large axons and slower in small ones. It is nevertheless very fast in all types of nerve fibers, being of the order of 120 meters per second in the faster fibers. Nerve fibers are capable of conducting a nerve impulse in either direction, but nerve impulses proceeding in the wrong direction are kept from being propagated farther by the synapse, or junction between two nerve cells. This synapse, usually formed by the close apposition of the terminal branches of the axon of one cell to the dendrites or cell body of another cell, allows the nerve impulse to pass across it only in one direction. This is because conduction across the synapse involves chemical rather than electrical transmission, and neither dendrites nor the cell body can release the chemical substance; it can be released only by axons. The chemical substances (neurotransmitters) released by axons vary, but the best known are acetylcholine and noradrenaline (norepinephrine). Transmission across the synapse (and from axons to an effector organ, such as muscle) is therefore brought about when the electrical nerve impulse reaches the axonal ending where it causes the release of the chemical transmitter. Since nerve cells and nerve fibers usually conduct only in one direction, we can distinguish some as motor (efferent) cells and fibers, conducting impulses away from the central nervous system and to some effector organ such as a gland or muscle; others are afferent or sensory cells and fibers, conducting impulses to the central nervous system from the skin, muscles, joints, viscera, and so forth. Within the central nervous system many nerve cells have such numerous connections that it is difficult to classify them as motor or sensory. Instead, we usually describe these cells as sending their fibers primarily up the central nervous system, that is, toward the brain, down the central nervous system, or making relatively local connections. We thus speak of ascending fibers within the central nervous system, descending fibers, and intercalary or connecting fibers and neurons.

BLOOD

Blood is sometimes classified as a separate primary tissue of the body, and sometimes regarded as a special type of connective tissue

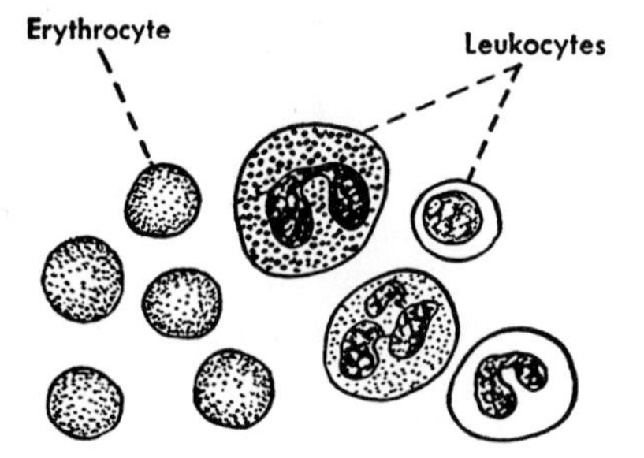

Figure 2–8. Some typical blood cells. The red blood cells are non-nucleated; among the leukocytes may be seen both granular and nongranular forms.

since it consists of cells (fig. 2-8) widely separated by a nonliving, in this case fluid, intercellular substance. The cells of the blood are of two types, red cells (erythrocytes) and white cells (leukocytes). The normal red blood cell in the human being is a flattened, non-nucleated, biconcave disk. Its red color is due to the presence of hemoglobin, which possesses the special property of readily taking up oxygen and also readily releasing this oxygen under certain circumstances. The red blood cells serve, therefore, for the transport of oxygen from the lungs to the other tissues of the body.

The so-called white cells of the blood, really not white but transparent and colorless, are nucleated cells of several different types. Depending upon the size and appearance of the cell, the shape of the nucleus, the presence or absence of granules within the cytoplasm, and the type of granules present, these white blood cells or leukocytes are further subdivided. The nongranular leukocytes include the lymphocytes and the monocytes; the granulocytes (cells with granules) are polymorphonuclear cells (that is, have variously shaped instead of the usual spherical nuclei) and are divided, on the basis of the manner in which their granules are stained by certain dyes, into neutrophils, eosinophils and basophils.

Since the efficiency of the transport of oxygen by the blood depends both upon the concentration of red cells present in the blood and upon the amount of hemoglobin in each red cell, information concerning these is often useful to the clinician. The number of red blood cells per cubic millimeter of blood can be calculated by counting the number in a thin smear of known volume, and the amount of hemoglobin can be determined by colorimetric methods. Blood counts usually include both red and white cells; the normal number of red blood cells per cubic millimeter is approximately 5,000,000, being usually greater in the male than in the female. White blood cells are very much scarcer, averaging about 7,000 to 8,000 per cubic millimeter. In making cell counts, it is also possible to observe whether immature forms of either the red or the white cells are appearing in the blood stream. The appearance of such forms in appreciable numbers indicates some derangement of body mechanisms. Similarly, an increased leukocyte count typically is a sign of infection within the body, as the leukocytes are one of the defenses of the body against infections.

3 The Organs and Organ Systems

An organ is a combination of several different tissues that work together to perform a given function, while the organ systems are groups of organs of somewhat similar make-up and with somewhat similar functions. The stomach, for instance, an organ of the digestive system, is composed largely of epithelial tissue, connective tissue and smooth muscle, with smaller amounts of vascular and nervous tissues; all these tissues are necessary to the proper functioning of the stomach. In turn the digestive organs as a whole are built upon the same fundamental plan as is the stomach, and each of the various organs contributes something toward the total digestive process. In the same way, a single muscle is an organ, while all the skeletal muscles together constitute the muscular system; the various individual bones (organs) together form the major portion of the skeletal system, and so forth. In this chapter we shall consider only general features of the various organs and organ systems necessary to an understanding of more detailed features of the form and functions of the individual organs and their relationships to each other. It is these more detailed features that form most of the lore of gross anatomy, some phases of which are of particular importance to the student of physical therapy.

THE SKELETON

The skeleton of the body consists largely of bones, with cartilage of one type or another located at strategic points. However, bone is usually preceded by a cartilaginous model. While the cartilage is still growing, a blood vessel erodes the cartilage and grows into it at about its middle, but also brings with it bone-forming cells that begin to lay down bone. As the cartilage is eroded bone is also laid down on the inside and outside, so that a hollow bone replaces the solid cartilage in the middle of the structure. In order to grow in diameter, this bone must be eroded on the inside, and added to on the outside; thus growth of a bone from its first appearance as cartilage to its definitive size and shape involves a simultaneous destruction of previously formed cartilage or bone and addition of new bone, all the while providing the necessary support for surrounding tissues—a process roughly comparable to enlarging the exterior of a house by advancing its outer walls while at the same time enlarging and changing the number of rooms, and keeping the roof intact.

Bones can be classified according to their shape, and we frequently talk of flat bones and long bones, although there are gradients between these two, and many bones belong to neither category. Sesamoid bones are a special type, developed from sesamoid cartilages in connection with ligaments or tendons; most of them are tiny nodules as implied by their name, which refers to sesame seed.

A typical long bone (fig. 3-1), such as most of those of the limbs, consists of a body (also called shaft) and two extremities. The body of the bone is formed of compact or cortical bone (p. 13) that surrounds a large medullary or marrow cavity. Within the medullary cavity during a period of fetal life the red cells and many of the white cells of the blood are formed. During adult life the marrow of many of the bones of the body ceases its function of producing blood cells, this function being restricted then primarily to the flat bones, but the long bones may reassume this function if there is an excessive demand for newly formed blood elements. In those bones in which the marrow is not actively forming blood cells, the connective tissue of the marrow develops numerous fat cells and therefore assumes a yellowish-white appearance; marrow active in the formation of red blood cells is known as red marrow, from its color. The marrow cavity is supplied by a nutrient artery that pierces the body of the bone to reach and branch within the marrow cavity. This artery is also the chief supply of the bone forming the body, for many of its branches enter the bone to run in Haversian canals.

The ends of long bones are provided with a thin outer shell of compact bone, but are largely filled by spongy bone (p. 13). Here, therefore, the marrow cavity is subdivided by the bony trabeculae.

The major middle part and the ends of adult bones are firmly united. In a newborn infant, however, parts of both ends of the larger long bones of the limbs, and a part of one end of each of the small long bones of the hand, foot, fingers, and toes are largely or entirely made of cartilage. Blood vessels invade this cartilage to allow its replacement by bone, but during childhood and early adolescence this new bone remains separated by cartilaginous plates from the major part of the bone, called the diaphysis. The separate ends of bones are called epiphyses, and the cartilaginous plates between them and the diaphysis are called epiphyseal cartilages. There may be other epiphyses and epiphyseal cartilages connected with a bone, but the cartilages near the ends of the bones are the most important ones, for they are

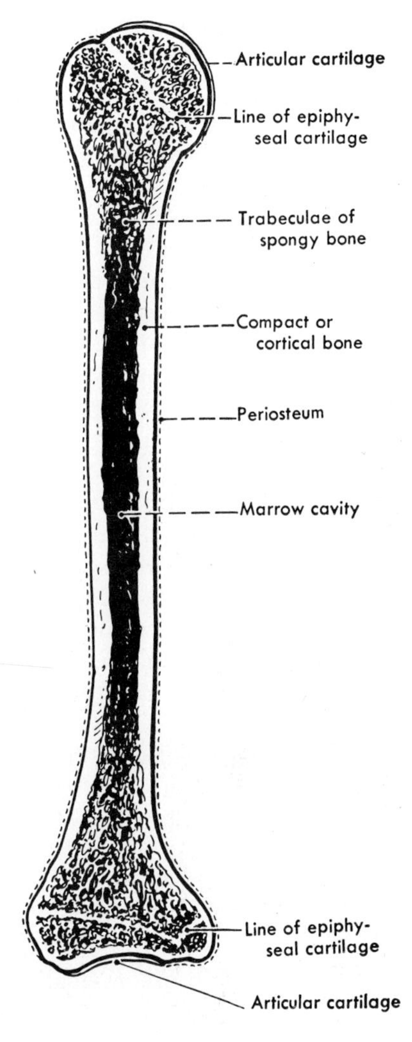

Figure 3–1. A typical long bone in longitudinal section.

entirely responsible for the growth in length of the diaphyses of the long bones. As the epiphyseal cartilage grows, the part of it nearest the diaphysis is being constantly transformed into bone continuous with the diaphysis. (Diaphysis and body are often, but not always, used as synonyms: the diaphysis of a developing bone is all of that bone except the epiphysis or epiphyses; "body" or "shaft" usually does not include the wider ends, called metaphyses, adjacent to the epiphyses.)

As long as the epiphyseal cartilages are growing and are not being replaced by bone faster than new cartilage is formed, growth in a long bone continues. When, however, the destruction of the epiphyseal cartilage and its replacement by bone proceeds faster than the cartilage can grow, the cartilage will soon disappear and no further growth in length of the bone is possible. For a time after the epiphyseal cartilage has disappeared, however, its former position in the bone is often fairly clear.

The importance of the epiphyseal cartilages in growth in length of the long bones is dramatically illustrated by a condition, achondroplasia, in which for unknown reasons the epiphyseal cartilages cease their growth early in life. However, since growth in diameter of a bone is not dependent upon the presence of cartilage, but occurs as a result of deposition of successively additional layers of bone on the outside of the bone already present, the bones of the limbs continue to grow in diameter even though they have ceased to grow in length. In consequence, the limbs remain short, little longer than those of an infant, although they attain a diameter approaching that of an adult, and the trunk and head also usually reach normal size. The adult so affected has therefore extremely short limbs attached to a more normal trunk, and is known as an achondroplastic dwarf.

A number of factors may affect the growth of the epiphyseal cartilages and their transformation into bone. An important mechanical one is pressure, which must be, however, much greater than that exerted by the weight of the body before it has any effect. It is common for a limb paralyzed by poliomyelitis to grow more slowly than does the normal limb, and one of the factors in this slower growth may be the pressure exerted when the bone grows and paralyzed muscles fail to grow likewise. Regardless of this, it is possible to insert staples across an epiphyseal cartilage so that they hold the body and epiphysis of a bone together, and as the growing cartilage builds up pressure, growth ceases. This stapling technic has been used to retard or halt growth of a normal limb so that there will not be too great disparity in length between the otherwise normal limb and a paralyzed one. Another method of accomplishing the same result is to remove one or more epiphyseal cartilages.

Hormones (p. 63) that affect growth of the body as a whole, particularly thyroxine secreted by the thyroid gland in the neck, and the growth hormone of the hypophysis (pituitary gland) lying in the skull just below the brain, also affect growth of bone. Too little secretion of either hormone leads to dwarfism.

In contrast to undersecretion, oversecretion of the growth hormone leads to growth which may go on far beyond the age at which the epiphyseal cartilages normally disappear; such an overgrowth may produce marked gigantism, such as is found in circus giants. Sex hormones also affect epiphyseal cartilages, but in a different manner from the growth hormone. They hasten the replacement of cartilage by bone, and thus lead to total disappearance of the cartilage. The earlier sexual maturity of the female is thus the cause of the earlier cessation of growth of girls as compared to boys.

The ages at which epiphyseal cartilages disappear and growth in length at that end stops have been carefully recorded. They vary much for different bones, and even for the two ends of a single bone. There are also variations among individuals, but epiphyseal fusion tends to follow a general pattern in which there is usually a range of only a year or two among individuals of the same sex; girls, however, typically have epiphyseal fusion as much as three years before boys.

Still other bones that are first formed in cartilage—the ribs, most of the bones of the wrist and ankle, and many bones of the skull—do not have epiphyses. In these cases, once the growing cartilage has been destroyed growth occurs as does growth in diameter of long bones, by addition of bone to the outside.

While most bones are first formed in cartilage, some of the flat bones of the skull never go through a cartilaginous stage. Instead, they are formed directly within a cellular membrane, and are referred to as membrane bones. The membranes connecting the bones of the roof of the skull have not, at birth, been completely transformed into bone, and the bones of the skull of an infant can therefore overlap somewhat during childbirth.

Bones are covered by a dense fibrous connective tissue membrane called the periosteum (that is, the "around the bone"). This tough membrane is usually firmly united to the bony tissue through some of its fibers which penetrate the bone to mingle with the collagenous tissues there. The tendons of muscles insert into the periosteum and blend with it, and also send many of their fibers into the bone.

In addition to its fibrous, relatively vascular, outer layer, periosteum also has a more delicate and more cellular inner layer, lying against the outer surface of the cortical bone. The cells of this layer are capable of forming bone, and it is by these in the fetus and child that new bone is laid down on the outside of the old bone, thus producing growth in the diameter of the bone. Other bone-forming cells lie against the inner surface of the cortical bone, thus lining the marrow cavity, to form an endosteum. While these endosteal cells can and do form bone, many of them perform the more useful function in this location of destroying bone, to allow growth in diameter of the marrow cavity and thus prevent the bone from becoming too solid and heavy as it increases in size.

While the cells within the cortical bone are living, it is the potential bone-forming cells of the endosteum and periosteum that are especially capable of new bone formation in the adult. When a fracture occurs these bone-forming cells begin to lay down bone across the break; they usually overdo the process of repair and thus form an enlargement or callus where the fracture occurred. In removing a piece of bone, the endosteal cells are necessarily removed, but the periosteum can be saved if so desired; therefore, if it becomes necessary to remove a piece of bone, and subsequent replacement of this bone is wanted, the surgeon carefully strips the periosteum from the bone and removes the bone itself, but leaves the periosteum so that it may form new bone.

While cortical bone varies much in strength, both its tensile strength—resistance to being pulled apart—and its compressive strength—resistance to being crumbled—exceed those of granite and of white oak, although they do not approach that of medium steel. Bone is reported to have a tensile strength along its long axis of from 13,200 to about 17,700 pounds per square inch, a compressive one of 18,000 to 24,700 pounds per square inch. The corresponding figures for granite are a tensile strength of 1500 pounds, a compressive one of 15,000; for white oak along the grain, a tensile strength of 12,500 pounds, a compressive one of 7000 pounds; and for medium steel a tensile strength of 65,000 pounds, compressive strength of 60,000 pounds.

Joints

Joints (articulations) are defined as a union between one bone and another, and therefore range the gamut from absolutely immovable joints to very freely movable ones. They are classified into three great groups according to the method of union between the bones participating in forming the joint. These groups are fibrous joints, cartilaginous joints, and synovial joints, and there are subclassifications of each group.

In fibrous joints the bones are united by connective tissue fibers. Many of these joints are immovable because of the shapes of the articulating surfaces and the shortness of the collagenous fibers that bind them together. An example of such a joint is the type known as a suture, occurring between many bones of the skull (fig. 21-1). Here two adjacent bones have serrated edges that interlock and are held firmly together by a small amount of fibrous tissue. At the other extreme, the two bones entering into a fibrous joint may be some distance from each other and if the connecting ligaments are of elastic tissue, as they are between posterior parts (laminae) of the vertebrae where they are known as ligamenta flava (fig. 13-3), a good deal of movement is allowed.

The most common type of cartilaginous joint in children is that formed by the epiphyseal plate uniting the bodies and ends of long bones, and these also are immovable. In most cartilaginous joints of the adult, union is by fibrocartilage instead of hyaline cartilage, and a small amount of movement is permitted. Such a joint is called a symphysis (for example, pubic symphysis, p. 235). Belonging to the class of symphyses is the union between the bodies of the vertebrae, where heavy fibrocartilaginous disks, the intervertebral disks (fig. 13-1), unite the bones and allow the limited movement between any two that is necessary for movements of the back.

Synovial joints, in which there is a cavity between the articular surfaces, permit more movement than the preceding types. Synovial joints may be simple, between two bones, or composite, between several bones, but are more completely classified according to the shapes of the articulating surfaces (fig. 3-2). These shapes in turn determine the type of movement allowed at the joints. In a plane joint the two articulating surfaces are almost flat, and allow only a gliding movement; such joints are found, for instance, between some of the bones of the wrist. In spheroid joints, commonly called ball-and-socket joints, one of the articular surfaces is rounded, the other concave, as at the shoulder joint. (Cotyloid is another name for these joints, but was once restricted to those with a particularly deep socket; the word refers to a cup.) Unless movement is restricted by ligaments or muscles, this joint allows the greatest freedom of movement—forward, backward, away from or toward the body, and even rotation around the

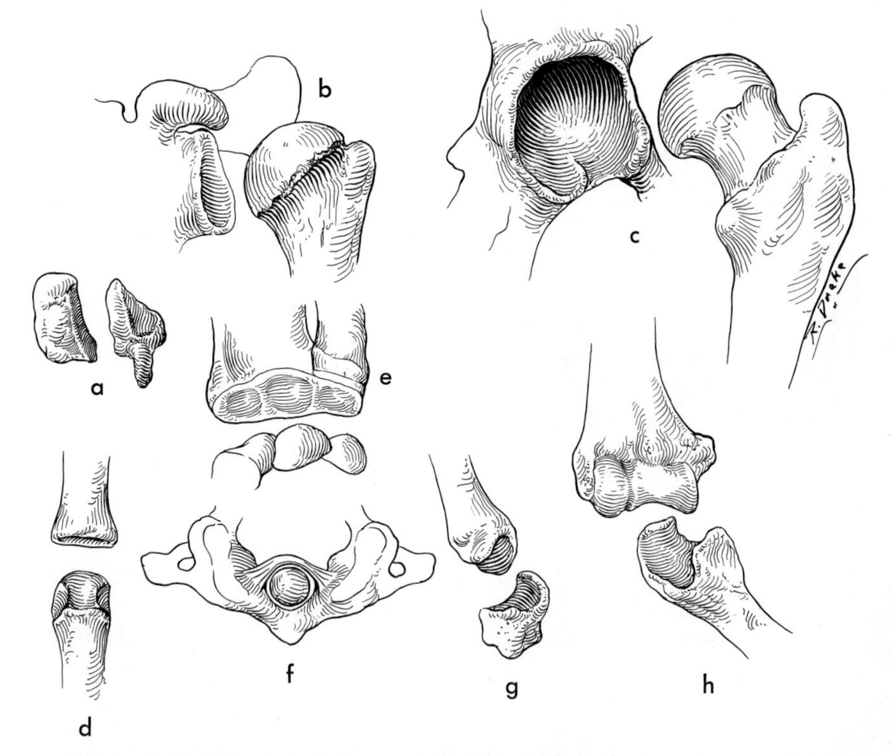

Figure 3–2. Major types of synovial joints. *a.* Plane, between two carpal (wrist) bones; *b* and *c.* Spheroid or cotyloid, the shoulder and hip joints; *d.* Condylar, a metacarpophalangeal or knuckle joint; *e.* Ellipsoid, the radiocarpal or major wrist joint; *f.* Trochoid, the middle atlantoaxial joint (between the first two cervical vertebrae); *g.* Sellar, between the trapezium, a carpal bone, and the metacarpal of the thumb; and *h.* Ginglymus or hinge, between the humerus and the ulna at the elbow.

long axis of the bone. A ginglymus or hinge joint allows primarily back and forth movement, such as occurs at the elbow or the knee. A condylar joint is a modified spheroid one in which ligaments or muscles severely limit rotation (as in the joints at the bases of the fingers); an ellipsoidal joint is also a modified spheroid one, but again allows little or no rotation because the concavity and convexity of the two bones are ellipsoid rather than rounded. In a trochoid joint one element resembles a peg and is so held against the second element that rotation is the primary movement allowed; an example of this type of joint is that between the first and second cervical vertebrae. Finally, a sellar joint is one in which both surfaces are saddle-shaped, concave in one direction and convex in the other, the concave surface of one fitting onto the convex surface of the other, and vice versa. An example of this type is the joint at the base of the thumb (carpometacarpal joint). This joint allows a limited amount of all movements, including rotation.

In general, therefore, where two surfaces come together to form a synovial joint they are reciprocally curved. The two curves are usually not identical, however; a certain amount of discrepancy allows better lubrication of the joint.

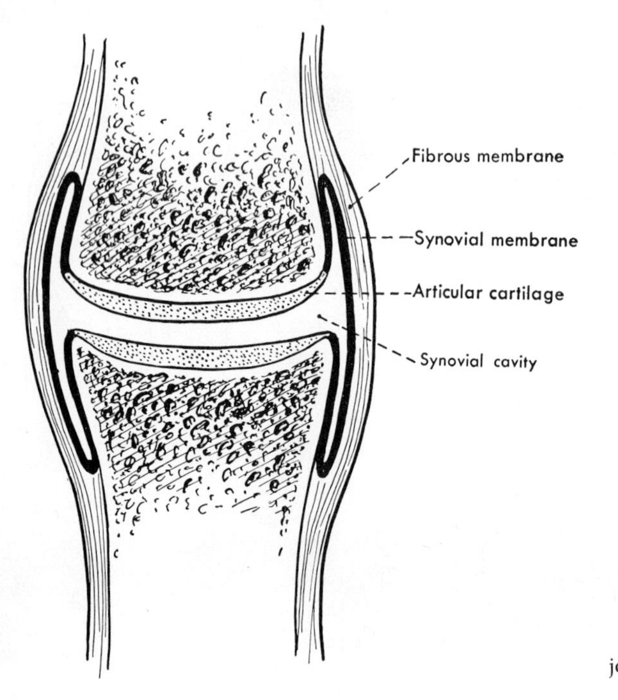

Figure 3–3. A typical synovial joint.

A typical synovial joint (fig. 3-3), regardless of the shape of the articulating surfaces, has a constant structure. The portions of the bones that are in contact with and move upon each other, and therefore constitute the articulating surfaces, are covered with cartilage. This cartilage, almost always of the hyaline variety, offers a much smoother surface than can be obtained from the bone itself; it may therefore be compared to the special tempering or plating of bearing surfaces in machinery. These surfaces may be subject to considerable pressure, even in non–weight-bearing joints. For instance, when flexion of the extended forearm is attempted, the line of pull of the flexor muscles is almost parallel to the bones, and therefore much of the muscle's force is exerted on the elbow joint, forcing the articular surfaces against each other; because of the lack of complete congruency, this pressure is concentrated into an area less than that of the apposed articular surfaces. If there is a weight in the hand, the muscles must contract still more strongly in order to flex the forearm, thus exerting more pressure on the joint. In weight-bearing joints, the pressure on the joint may be very much increased by the pull of supporting muscles. If all the weight of a 200 lb. man is supported on one limb, for instance, the hip joint is subject not only to that weight less the weight of one limb, but also to the pull of the muscles necessary to maintain the weight on one limb. One limb should be about 15 per cent of the body weight, or 30 lbs., leaving 170 lbs. to be supported. Using the formula for calculating the force necessary to balance the body on one limb (see Levers and Muscular Action, p. 34), it is found that this amounts to 425 lbs. Thus the hip joint is subjected to a pressure of 425 plus 170 lbs., or a total of 595 lbs. This is during quiet standing. Imagine if you can the total stress on the joint when the person is running instead of standing still.

The ends of the bones entering into a movable joint are united by a membrane, the articular capsule, which stretches from one bone to the other. The outer layer of the articular capsule is the fibrous membrane; it is typically composed of collagenous tissue, completely surrounds the joint, and blends with the periosteum of the bones entering into the joint. In many joints this layer attaches some distance from the articular surfaces of the bones. The inner layer of the articular capsule is the synovial membrane, also made of fibrous tissue but more cellular, especially on its inner surface, and also more vascular. The synovial membrane lines the inner surface of the fibrous one, but is also reflected along the bones to the edges of the articular cartilages. Thus a synovial cavity is lined by the synovial membrane except over the articular cartilages. The synovial membrane produces a viscous substance, the synovia or synovial fluid, that somewhat resembles the white of an egg (synovum means like an egg). The synovial fluid is the lubricant of the joint, and also the source of nourishment to the articular cartilage.

The thin fibrous membranes of synovial joints are thickened in certain locations by ligaments, bands of dense fibrous connective tissue, almost always collagenous, whose constituent bundles run largely in the same direction (for instance, fig. 15-7). Most ligaments blend with the fibrous capsule on their deep surfaces, and are therefore really local thickenings of the capsule. In the case of a hinge joint the anterior and posterior parts of the capsule are usually thin and protected by the muscles passing in front of and behind the joint, while the sides are reinforced by well-developed lateral ligaments.

Ligaments play an important part in the physiology of joints, for while the type of movement allowed at a joint usually depends primarily on the shape of the articular surfaces, ligaments sometimes guide the movement, and they regularly assist muscles in limiting the amount of movement allowed at a joint. They are an important source of strength to the joint, and are typically much stronger than is necessary to resist the forces that ordinarily act upon them. If, however, unusual forces act upon them over a long period they gradually stretch and allow the bones to slide out of their normal positions. An excellent example of this is acquired flatfoot, in which carrying the weight constantly on the inner border of the foot leads to stretching of the supporting ligaments and flattening of the arch. A certain amount of dislocation (termed subluxation) between bones may therefore occur as a result of lax or stretched ligaments. For complete dislocation to occur, however, ligaments must be torn. A sprain is a tearing of ligaments without dislocation.

Any swelling of the capsules and ligaments of joints, whether produced by strain or sprain, infection, or arthritis, is painful because the capsules and ligaments are provided with nerve endings of which some belong to pain fibers. The articular cartilage itself has no nerve fibers in it. It is a general rule that a joint is supplied by all the major nerves that cross it. Thus the elbow joint, for instance, is crossed by four nerves and typically receives one or more branches from each nerve, although any one may fail to supply it. Nerves to joints are usually difficult to observe, because of both their small size and the frequency with which the supply to a joint is a continuation of a nerve into an adjacent muscle.

The swelling and stiffness typical of a joint long immobilized, as by a splint or cast, is due, at least in part, to faulty circulation to the joint. (This may explain the report that immobilization of a joint leads to regressive changes in that joint's ligaments, which become more easily stretched and more easily ruptured; they are said to recover slowly after mobility is restored.) When possible, therefore, joints distal to the immobilized one should be regularly exercised in order to increase the circulation to the affected member.

MUSCLES

Muscles are, of course, composed primarily of voluntary muscle fibers (p. 15), but as organs they contain also a certain amount of connective tissue, and abundant blood vessels and nerves. A typical muscle moves bone upon bone, and is therefore attached to each of two bones across a movable joint. For purposes of description, it is obviously preferable to have terms by which one attachment of the muscle may be distinguished from the other attachment; the terms adopted for this purpose are "origin" and "insertion." We consider the origin of a muscle to be that attachment which is, under usual circumstances, the less movable end of the muscle; similarly, the insertion of a muscle is therefore the attachment to the more movable part. Difficulties sometimes arise in deciding which of the skeletal attachments of a muscle is more likely to move when the muscle shortens, but as a whole it is usually easy to distinguish on the basis of common sense between origin and insertion. In regard to the limbs, it is clear that in general a more distal part of the limb may be moved more easily than a more proximal part; therefore the origins of limb muscles are usually at their proximal ends, their insertions at their distal ends.

The definitions of origin and insertion do not imply that the origin of a muscle may not be moved by contraction of that muscle. As a muscle shortens it necessarily moves its ends closer together but if, due to particular circumstances, the insertion of the muscle is at the moment more fixed than is the origin, the origin of the muscle will then be moved by contraction of the muscle. Thus, for instance, muscles passing across the shoulder joint have in general their origin on the shoulder or the back, insert on the arm, and usually move this arm. If, however, one suspends one's self by one's arms, contraction of the shoulder muscles then moves the body as a whole, for the limbs are then the fixed points. The fact that a muscle may work in the reverse direction does not make us change our concepts of origin and insertion; in the above example we would still speak of the origins of the muscles as being proximal, their insertions distal—unless, perhaps, we were talking about a species which routinely used the arms to suspend and propel the body and only rarely used them as free limbs.

Muscles are regularly attached to bone by dense fibrous connective tissue which at one end attaches to the ends of the fibers and at the other end blends with the periosteum about the bone and with the fibrous connective tissue within the bone itself. If these connective tissue fibers are short the muscle fibers may appear to arise almost directly from the bone and we then speak of a fleshy origin of the muscle. Many muscles arise by longer connective tissue bundles which are aggregated to form a tendon (p. 11), and most muscles insert by tendons. A tendon has several advantages over muscle fibers. In crossing a bone or joint, for instance, a muscle closely applied to bone may be subjected to considerable wear and tear in this location, which may then lead to the injury or death of the muscle fibers. On the contrary, a tendon is composed of nonliving fibers and is much tougher than are living muscle cells, and therefore much more suited to withstand such strain. Still another advantage of a tendinous insertion is that it allows a bulky muscle to insert on a very small area of bone, for tendon is much stronger than muscle and therefore small tendons can withstand the pull of large muscle bellies. For instance, most of the muscles of the forearm attach in the hand; obviously, if these muscles continued as muscles into the hand that organ would have to be very much larger in order to accommodate them. The muscles are therefore replaced by tendons as they near the wrist, and the reduced bulk of these tendons contributes much to the flexibility of the hand.

Tendons are very much stronger than the muscles that act upon them, so that a very large muscle can act through a small tendon or even a small part of a small tendon. The maximal tensile strength of muscle (its resistance to a pull) has been reported to be about 77 pounds per square inch, while tendons have been found to have a tensile strength of from 8600 to 18,000 pounds per square inch. Thus, while there is a great difference in the strength of different tendons, they are all enormously stronger than muscle. This accounts for the fact that normal tendons that are ruptured by sudden force never break in their middles, but instead pull away from one end. If this occurs at the bony attachment, a piece of bone may be avulsed with the tendon; if it is at the other end, the tear comes at the musculotendinous attachment. Certain tendons around the shoulder joint sometimes rupture in or near their middles, but this is a result of repeated damage to the tendon with eventual weakening of it. In one experiment it was found that rupture of a tendon, rather than of its attachments, occurred only when about half of its fibers had been cut.

It has already been mentioned that connective tissue pads between muscles help to facilitate free movement of one muscle upon its neighbor, and that sometimes fluid-filled sacs or bursae develop in such locations. A similar development of connective tissue spaces in the form of bursae or tendon sheaths occurs about many tendons. A tendon sheath may be regarded as a bursa that completely surrounds a tendon (fig. 3-4); it has two walls, of which the inner, closely attached to the tendon, forms a smooth glistening outer surface on the tendon,

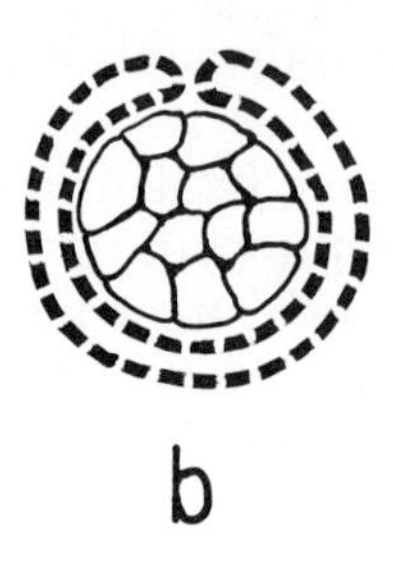

Figure 3–4. Diagram of the similarity between a bursa and a tendon sheath, the synovial membrane being represented by the heavy broken lines. In *a*, a bursa underlies a tendon; in *b*, the synovial membrane has extended around the tendon to form a tendon sheath.

while the outer wall of the sheath forms a closed sac, uniting with the inner layer at the ends of the sheath (fig. 3-5). Within the sheath, a thin membrane (mesotendon) may unite a part of the inner and outer walls, and serve as a point of ingress for tiny blood vessels. The cavity of the tendon sheath, between its inner and outer walls, contains a substance (similar to that occurring in joints) that acts as a lubricant to allow frictionless play of the tendon. Synovial tendon sheaths occur especially at the wrist and ankle, where the tendons pass close to bone and are held down against this by heavy ligaments.

Mechanics of Muscular Action

The strength of a muscle and the range of movement that it can produce at a joint vary with several different factors. Although the strength depends, in the last analysis, upon the number and size of the constituent muscle fibers, the mechanical factors of the arrangements of fibers within muscles and of the varying leverage afforded by the attachments of muscles across the joints make it impossible to compare the effective movements and strengths of different muscles solely upon the basis of their sizes. While individual muscle fibers, regardless of what muscle they are in, apparently can contract maximally to the same percentage of their length (variably reported as from

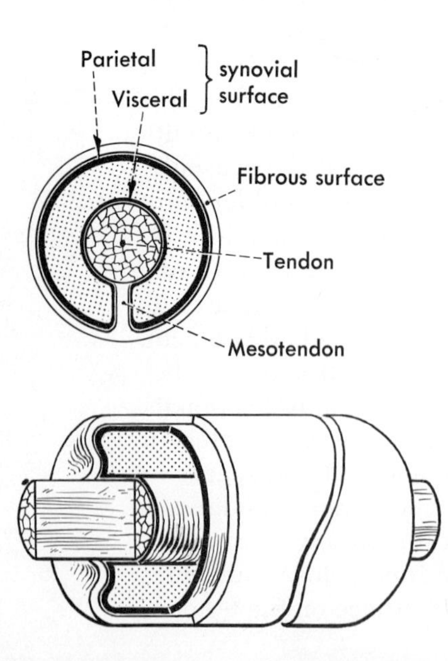

Figure 3–5. Detailed diagrams of a tendon sheath. The cut edges of the synovial membrane are represented by the heavier black lines. Above is a cross section, with a mesotendon included. Below is a view from the side, with a gap to indicate that a segment has been removed, and parts cut away at various levels on the left side.

60 to 43 per cent of the length they have when they are stretched by maximal movement in the opposite direction of the part on which they act), each muscle crossing a joint may contract by a different amount; the extent to which a muscle as a whole can contract is the distance over which it normally shortens as the part to which it is attached moves through its complete range of movement, from the extreme in one direction to the extreme in the other. Each muscle is therefore accurately adapted to the amount of movement it can carry out, and this adaption depends upon both the length of the muscle fibers and their arrangement in the muscle.

In muscles with essentially parallel fibers (fig. 3-6), the fibers either run the length of the muscle or are attached approximately end-to-end, so that the amount of shortening the muscle can undergo is roughly 50 per cent of its length; thus a long muscle with parallel fibers produces a great range of movement. In other muscles, however, the fibers are at an angle to the line of pull and their length is therefore always less than that of the muscle as a whole. Thus for a pennate muscle, in which the fibers insert into a tendon somewhat as the barbs of a feather attach to its quill, the distance over which the muscle can contract bears no fixed relation to the length of the muscle, but is proportional to the length of its muscle fibers. This characteristic varies from muscle to muscle, depending upon the angle at which the fibers approach their insertion, the width of the muscle, and whether the muscle belly is flat or rounded. (Descriptions of various types of muscles are unsatisfactory, since muscles vary much, but they are often designated as unipennate—really semipennate, since they more closely resemble half of a feather; bipennate, or like a feather; fusiform, in which the muscle belly is rounded with tapering ends,

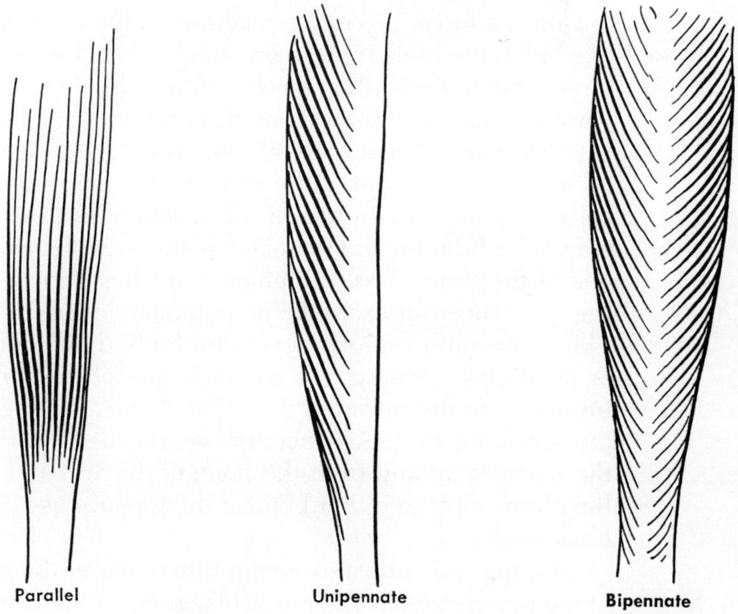

Parallel

Unipennate

Bipennate or fusiform

Figure 3–6. Various arrangements of muscle bundles within a muscle.

and the fibers curve between their origins and insertions; and multipennate, in which there are within the muscle many tendons that give origin and insertion to the fibers.)

Although the distance over which a muscle can contract thus depends on the length of its muscle fibers, the strength of contraction is dependent on the size and number of contracting fibers. Thus the maximal strength of contraction of a muscle is dependent on the total cross-sectional size of all its muscle fibers. In the parallel type, this is also the cross-sectional area of the muscle, but in all other types it is different; obviously, for a given muscle length, there are more muscle fibers in a unipennate muscle than in a parallel one of the same width, and still more in a bipennate one. Thus it is impossible to compare two muscles of different types on the basis of their sizes alone; moreover, of two muscles of the same size in which the fibers are not parallel, the one with the longer fibers necessarily has fewer, so that while it has the greater range of contraction it also has the lesser strength.

Levers and Muscular Action

A consideration of levers will indicate that a similar inverse relationship between range and strength of movement, one being sacrificed to a greater or lesser extent for the other, appears when two muscles of similar size and shape differ appreciably in the distance of their insertion from the joint over which they act. We employ mechanical levers in many of our daily activities to increase strength, as in using a claw hammer to pull a nail, or to increase the range and rapidity of movement, as in swinging a golf club. Similarly, the musculoskeletal system is largely a series of levers.

There are four components of a lever system: the lever itself, typically a rigid bar, in the case of the body a bone or bones; a fulcrum, the joint; a force, the muscle, acting on the lever; and the resistance, or weight, the force must overcome to move the lever. Levers fall into three classes, depending on the relationship between the fulcrum, the point at which the force is applied (effort point), and the resistance. In a lever of the first class, the fulcrum lies between the effort point (in man, the insertion of the muscle) and the resistance (as examples, a see-saw, an oar); in one of the second class, the resistance lies between the fulcrum and the effort point, as with a wheelbarrow; and in one of the third class, the effort point lies between the fulcrum and the resistance (for example, an automatic door closer).

All or almost all the levers in the body belong to either the first or third class. Figure 3–7 shows levers of the third class, since the insertions of the muscles, the effort points, lie in both cases between the joint and the resistance, or weight distal to the effort point. All the muscles arising from the bone of the arm and passing in front of the elbow joint to either bone of the forearm use these bones as third class levers.

Figure 3–7 indicates how a difference in the point of attachment of the muscle affects the range of action, and a simple calculation will show how it also affects the strength the muscle needs to overcome a

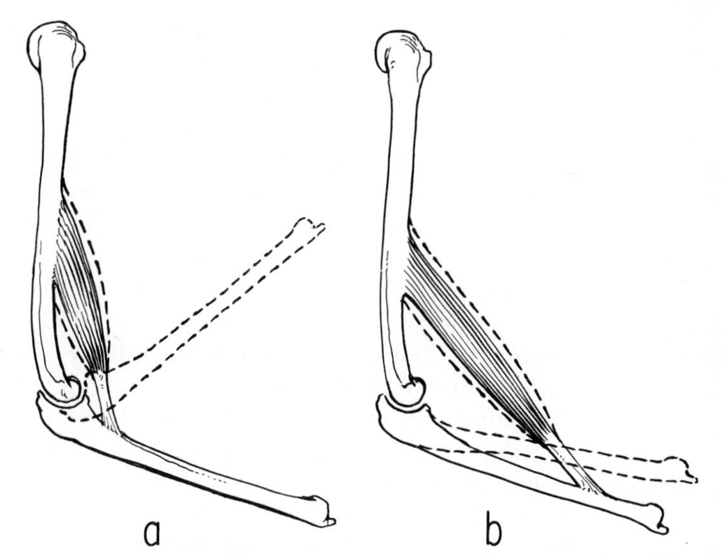

Figure 3–7. Effect of the place of attachment of a muscle on the range of movement. Both muscles are shown shortening the same amount (the difference between the lengths of the solid and broken outlines), but that in *a*, attached closer to the joint, moves the lever much more (the difference between the solid and broken outlines) than does that in *b*.

given resistance. In a lever system, a perpendicular line from the line of force to the fulcrum is called the effort arm (or effort moment arm), while a perpendicular line from the line of resistance to the fulcrum is called the resistance arm (or resistance moment arm). For the lever to be in balance the effort times the length of the effort arm must equal the resistance times the length of the resistance arm, or $E \times EA$ must equal $R \times RA$. In figure 3–7*a* the effort arm is initially about 0.2 inch, the resistance arm about 1.5 inches; in figure 3–7*b* the effort arm is initially about 0.5 inch, the resistance arm again about 1.5 inches. Assuming a weight or resistance of 10 lbs., our equation would read for 3–7*a*: $E \times 0.2 = 15$, $E = 75$ lbs.; for 3–7*b* it would read $E \times 0.5 = 15$, $E = 30$ lbs. Thus the muscle in 3–7*a* would have to contract with $2\frac{1}{2}$ times the strength of the muscle in 3–7*b* in order to produce any movement; through its closer attachment to the joint, it has therefore sacrificed strength in favor of range of movement (and speed, since range and speed parallel each other), while the muscle in 3–7*b* has gained effective strength at the expense of range and speed of movement.

(In the calculation of the stress on the hip joint when standing on one limb — p. 28 — use is made of the fact that the resistance arm has been found to be about $2\frac{1}{2}$ times the effort arm. Thus the equation here reads $E \times 1 = 170 \times 2\frac{1}{2}$, or $E = 425$.)

The muscles shown in figure 3–7 are flexors; therefore a muscle arising from the posterior surface of the upper bone and inserting into the proximal end of the lower bone would be an extensor, and the lower bone would then function as a lever of the first class — the joint being between the insertion of the muscle and the resistance. There is a muscle of the arm, the triceps, that has these attachments. Obviously such a muscle, because it inserts so close to the joint, produces rapid movement over a wide range. However, a similar muscle inserting farther from the joint, an insertion that could be afforded by a longer posterior projection of the lower bone, would have greater effective strength but produce a lesser range of movement. Thus regardless of the type of lever, effective strength of a muscle and range and rapidity of movement vary inversely with each other.

Another important aspect of levers in relation to muscular contraction is the effect of the length of the lever on speed. For instance, if the arm with extended forearm is abducted at the shoulder to 45°, the elbow and the hand, although moving together, travel at different speeds, for the tips of the fingers will be approximately twice as far from the side as is the elbow. The use of multiple levers can much increase this effect. Compare, for instance, throwing a baseball while limiting the movement to the shoulder, and the usual throwing movement in which the lower limb, the trunk, and the upper limb act together to multiply the speed at which the ball leaves the hand.

Types of Contraction

In the foregoing discussion, shortening of a muscle as a result of its contraction has been assumed. In a smooth movement, such shortening may or may not demand any great variation in the strength of contraction, but is called an isotonic (equal tension) contraction, nevertheless. However, muscle may contract and perform work in other ways. Thus if opposing muscles across a joint act with equal strength there will be no movement of the part and neither set of muscles will shorten in spite of their contraction; since they retain the same length, this is called an isometric contraction. Finally, if a movement that can be carried out by gravity, such as bending the knees, is to be controlled, muscles that oppose this movement must first contract and then gradually lengthen. This lengthening reaction is known as an eccentric contraction.

Determination of the Actions of Muscles

The action of a muscle, meaning how it moves a part, was originally deduced from observations on the origin, insertion, and placement of the muscle. In time this was supplemented by electrical stimulation of many of the muscles, and we obtained a considerable body of information on what certain muscles can do when they contract alone. Careful studies of patients with various paralyses have further defined the possible contribution of unparalyzed muscles to movements that they do not necessarily normally carry out, and palpation of superficial muscles during various movements has revealed in part which muscles normally do participate in a given movement.

Now, with the newer technic of electromyography, or recording the electrical impulses generated by muscular contraction, it is possible to determine very precisely not only which muscles, superficial and deep, contract during a given movement, but to record the sequence in which each of several participating muscles contracts, and to estimate the strength of contraction of each.

Through electromyography we have obtained much more accurate information concerning the actions of many muscles, although our knowledge of others is still incomplete and research is continuing. While the method is precise, the results must be interpreted very carefully, for we need to know not only what muscles are contracting but how they are participating in the movement or, really, why they are contracting.

In analyzing the participation of various muscles in a movement it has been customary to categorize them as prime movers, protagonists, or agonists, as synergists, and as antagonists. A prime mover is a muscle that carries out the action with which we are concerned, and when we give the chief action or actions of a muscle it is its function as a prime mover that we are attempting to define. A synergist is a muscle that contracts at the same time as the prime mover; an antagonist has an action more or less directly opposed to that of the prime mover. Depending upon the movements being considered, the same muscle may at one time be classified as a prime mover, at another as an antagonist, and perhaps at another as a synergist.

Although synergists (working together) are muscles that contract at the same time as the prime mover in order to facilitate or potentiate the effect of the prime mover, the term is a loose one. In a broad sense it can include the second of any two muscles that regularly contract together, regardless of the function of the second muscle, and the term is sometimes used to describe the second of two muscles that carry out the same action. A more useful definition of a synergist, however, is to regard it as a fixating or stabilizing muscle, one that contracts at the same time as the prime mover in order to prevent some unwanted movement that would otherwise ensue. The determination of whether a contracting muscle is serving as a second prime mover or as a synergist in the restricted sense may present great difficulty, although it may be obvious, and examples of the latter function have long been known. For instance, clenching the fingers should also flex the wrist since the tendons of the fingers cross the front of the wrist, but if the wrist does flex, the fingers cannot also be tightly clenched because of the too-short range of contraction of the finger flexors. Hence, in clenching the fingers, muscles that cross the back of the wrist contract synergistically to bend the wrist a little dorsally, and thus make the muscles moving the fingers more effective. Similarly, some muscles about the shoulder regularly contract synergistically with other muscles, not to move the arm but to prevent displacement at the shoulder joint through the action of the prime mover.

Antagonists also require further definition. The word is useful in designating a muscle that has approximately the opposite action of the prime mover we are studying, but only in this sense can it be called an antagonist; in the normal individual an antagonist does not fight against the prime mover, but either relaxes completely or cooperates with it, preventing some unwanted effect and therefore acting strictly as a synergist. In other instances, such as in lowering the outstretched arm or in bending over, gravity substitutes for the prime mover while the antagonist, by its lengthening reaction or eccentric contraction, controls the movement. In general, then, once a movement is learned, antagonists contract only when they can in some way aid the movement.

Nerve Supply

Study of a muscle as an organ implies observation of its size and shape, of its origin and insertion, and of its nerve supply. The tendons, if any, of origin and insertion are an intrinsic part of the muscle as a whole, and so also is the nerve supply to the muscle. The blood supply of the muscle also forms an essential portion of the organ, but there is no necessity for learning this. Whereas the nerve supply of a muscle is sharply limited to the one or several nerves specifically destined for that muscle, the blood supply is usually derived from all the blood vessels in the neighborhood. Learning the blood supply to a muscle is therefore simply a question of putting together general knowledge concerning the location of that muscle and knowledge of the several blood vessels in that area. On the other hand, a general knowledge of the locations of various nerves is of little significance in predicting which of these nerves will supply the muscle. Of several nerves in the neighborhood of a muscle, only one as a rule will supply this; therefore, nerve supplies of muscles must be learned.

Every voluntary muscle receives at least one and sometimes two or more nerve branches. These are in turn regularly derived from more than one spinal nerve, so that most muscles have a plurisegmental innervation—that is, are supplied with fibers from two or more of the segmental spinal nerves (p. 54). The activity of the muscle is dependent upon the nerve or nerves reaching it; if the nerve supply to a voluntary muscle is destroyed, the muscle is paralyzed and remains so until a nerve supply is re-established. As a nerve enters a muscle it divides to be distributed within it; the branching is for the most part the separation of smaller bundles of nerve fibers, but eventually individual nerve fibers branch, and every muscle fiber receives a nerve supply.

A typical nerve to a muscle does not consist entirely of motor fibers (that is, fibers that cause the contraction of the muscle), as might perhaps be expected, but contains also a large number of sensory fibers—about 40 to 60 per cent of the nerve fibers entering a muscle are sensory in character. Some of these sensory fibers are concerned with the conduction of pain, as evidenced by the appreciation of soreness in a muscle from overexercise or strain, or the pain arising from tears of the muscles or tendons. These pain fibers are probably associated with the connective tissue and blood vessels of the muscle, rather than with the muscle fibers themselves, and are relatively few in number. Most of the sensory fibers are, however, of the type known as proprioceptive. These fibers are concerned with registering the stretch or contraction of a muscle and the tension within a tendon, and thus carry to the central nervous system impulses concerning the activity of the muscles and the pull upon their tendons.

The majority of the sensory or afferent impulses from muscle do not reach the level of consciousness, but they do play an extremely important part in the subconscious regulation of muscular contraction. Practically all movements require the coordination of a number of individual muscles, and each of these muscles must contract at exactly the proper moment and with exactly the proper force if the movement is to be a smooth one. The afferent fibers from muscles and tendons, together with similar fibers from about the joints themselves, play a determining role in this coordination. Their importance is clearly brought out in such clinical affections as tabes dorsalis, in which disease the larger fibers from muscles and joints are among the first to be affected. Such apparently simple everyday actions as buttoning one's dress or coat, or even walking (really very complicated actions from the standpoint of the muscular coordination required) become difficult for a patient afflicted with tabes. The lack of both conscious and subconscious information as to what the muscles are doing and what is the position of the fingers or limbs at any particular moment results in clumsy and ill-coordinated movements that must be guided primarily by the eyes. Thus a patient with tabes, although suffering no paralysis or weakness of voluntary musculature, walks with a peculiar gait. He is unable to estimate how high he has lifted his foot from the ground, and therefore in order to keep from stumbling he lifts it too high; as he puts his foot down, he again cannot estimate the movement required and therefore tends to sling the foot onto the ground. Progression of the tabetic is better in the light, where he can watch his feet and therefore consciously guide them somewhat, but in the dark it becomes much more difficult or even impossible. The learning of movements, both in infancy and in adulthood, and the acquisition of greater skill in movements, are primarily dependent upon the proprioceptive sensory fibers, and we constantly depend upon them for guidance. Thus these fibers from muscles, tendons and joints are of very great importance for the proper functioning of the muscles.

Endings in tendons, muscles, and about joints are of several different types, and except for the fact that some of those related to joints are particularly important in the conscious appreciation of position and movement, relatively little is known concerning the function of the various types. In muscle, the chief sensory organ is the muscle spindle, a group of two to 10 small muscle fibers. One sensory nerve fiber winds intricately around the center of the muscle spindle (fig. 3-8*b*), and is called an "annulospiral" or primary ending; other nerve fibers form what are called "flower spray" or secondary endings closer to the two extremities of the muscle fibers of the spindle. In addition, the tendons of muscles typically contain afferent end organs ("Golgi" tendon organs) close to the attachment of tendon and muscle fibers. None of the afferent fibers from these endings gives rise to impulses that reach consciousness.

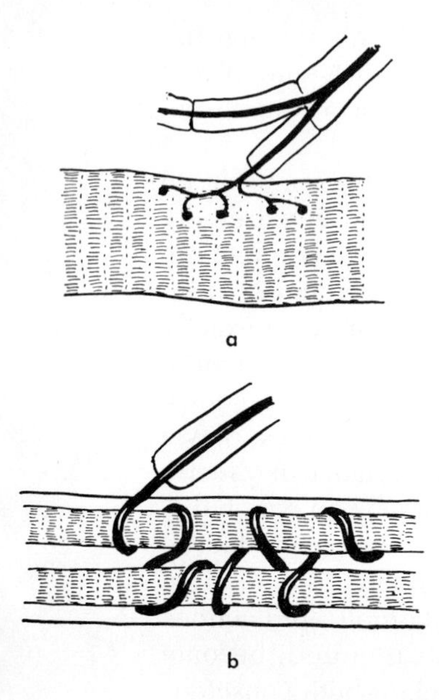

Figure 3–8. Motor and sensory (proprioceptive) nerve endings in voluntary muscle. *a.* Motor ending, in which the nerve fiber or axon loses its fatty sheath and indents the outer wall of the muscle fiber to end on a special part of the fiber. *b.* Sensory ending, in which a nerve fiber penetrates a special connective tissue sheath surrounding several small muscle fibers and winds intricately around these fibers.

Muscle spindles respond to stretching of their annulospiral regions, hence they are commonly stimulated by stretch of the muscle as a whole. The annulospiral afferent fibers from the muscle spindles of a muscle make direct contact with the motor nerve cells in the central nervous system that supply that muscle, and contraction of the muscle in response to this stretch then ensues and relaxes the tension on the annulospiral region. The response of muscle to stretch is one of the basic reflexes (p. 51), and this particular reflex is fundamental in resisting gravity and thus maintaining posture. It is the predominant reflex activity resulting from muscle stretch. Flower spray endings in muscle respond also to stretch, but end differently in the central nervous system; their activation has been said to facilitate flexion, regardless of the muscle stimulated. Finally, the tendon organs also respond to stretch, but since they do not lie within the muscle, this stretch can be imposed by either stretch of muscle and tendon or contraction of the muscle. Their sensitivity to stretch is far less than that of the annulospiral ending, however, and their action, inhibiting contraction of the muscle concerned, becomes prominent only when the stretch is excessive. It is a protective effect, tending to prevent undue strain on muscle and tendon.

The muscle fibers of the spindle, like all voluntary muscle fibers, receive motor nerve fibers. These are distinctly smaller than the motor fibers to the rest of the muscle and arise from a different set of cell bodies in the central nervous system. They and the neurons of which they are a part are called gamma fibers and neurons, while the neurons and fibers to the bulk of the muscle are designated as alpha. The gamma neurons, in contrast to the alpha neurons, receive no impulses from the muscle spindles but do receive impulses from higher centers in the central nervous system. Their activity produces contraction of the two ends of muscle fibers of the spindle, thus stretching the annulospiral region and sensitizing the spindle to stretch of the muscle as a whole. Disturbance of the normal control over the gamma innervation in various disease conditions apparently accounts for the occurrence of the abnormal states of contraction of muscle known as spasticity and rigidity.

The motor nerve fibers to the muscles all end upon muscle fibers, so indenting the cell membrane that they appear to be actually in the fiber (fig. 3-8*a*); the specialized ending and the modified portion of muscle fiber in which it lies are called a motor end-plate. If one counts the number of motor fibers entering a muscle, and calculates the approximate number of muscle fibers within that muscle; it will be found that there is always a disparity between the two numbers. For instance, it may be found that the muscle fibers outnumber the entering motor nerve fibers by about 100 to 1—that is to say, that there are about 100 muscle fibers for every motor fiber in the nerve or nerves entering the muscle. Since, normally, every muscle fiber within a muscle is capable of contraction, and since no voluntary muscle fiber can contract unless it is supplied with a functional nerve ending, this means that each nerve fiber after it enters the muscle must branch repeatedly. While the ratio between muscle fibers and nerve fibers varies greatly from one muscle to another, in the example given an average nerve fiber must give off about 100 branches or sub-branches in order to supply its quota of muscle fibers, each of these terminal branches then ending upon a single muscle fiber.

There is a so-called law of physiology, termed the "all or none law," which states in relation to muscle that, if a given muscle fiber contracts, it contracts with all the force of which it is capable at that particular moment—that is, a stimulated muscle fiber contracts with all its strength or does not contract at all. While the all or none law is a statement of fact concerning the contraction of individual muscle fibers, it is apparent upon even the slightest reflection that it does not apply to a muscle as a whole. For instance, we will use the same muscles in grasping and picking up a heavy lead ball that we would in similarly grasping and picking up a very delicate egg shell. If we exerted the same strength in grasping the shell as we must to lift the lead ball, we would destroy the shell. Thus it is a matter of everyday knowledge that we grade our voluntary movements and exert only the desired strength and speed. So far as any one muscle fiber is concerned, such gradation is impossible. Similarly, it is impossible to send an impulse of contraction to only one muscle fiber of the many innervated by a single nerve fiber, for a nerve impulse, once started along a nerve fiber, is propagated along all the branches of that fiber. There is no known mechanism by which a nerve impulse can be routed along only certain branches of a single fiber.

From these facts it follows that regulation of the strength of a movement depends upon activation of groups of muscle fibers. If we desire a delicate movement, we may use perhaps only 10 per cent of the nerve fibers to a muscle to activate 10 per cent of the muscle fibers in that muscle; whereas if we require the strongest possible movement we will send impulses along all the nerve fibers to the muscle and thus activate all of the muscle fibers. Insofar as any given muscle is concerned, therefore, it is possible to get a smooth gradation of contraction, from a minimal one that produces no movement to a maximal one that produces the strongest movement possible. Since all the muscle fibers that are innervated by a single nerve cell and its branching nerve fiber necessarily contract at the same time, the nerve cell and the group of muscle fibers it innervates constitute a "motor unit." The size of the motor unit is determined by the number of muscle fibers composing it, and varies from muscle to muscle; since it represents, however, the smallest number of fibers in a muscle that can contract at one time, and the smallest increment by which strength of contraction can be increased, it might be expected that its size should vary with the type of movement demanded of the muscle. Thus some of the muscles about the hip and thigh, concerned in general with very coarse movements, have motor units variously quoted as from approximately 150 to perhaps 1600 muscle fibers. Those governing the rather delicate movements of the thumb have much smaller motor units; and the muscles governing movements of the eye, which must be very precise, have the smallest motor units of all, averaging perhaps no more than three muscle fibers per nerve fiber. Thus some muscles have a built-in delicacy of movement that others do not have, and that no amount of training could establish.

Just as all the muscle fibers within a muscle do not have to contract together, so also the various larger portions of a muscle do not necessarily contract together. For instance, the pectoralis major muscle on the chest (thorax) is so arranged that some of its fibers will aid in elevation of the arm while others aid in depressing the raised arm. Obviously, if both upper and lower fibers necessarily acted together they would tend to cancel each other out; they are used together in some movements and this use of the muscle as a whole results simply in pulling the arm toward the side or across the front of the chest. However, either some of the upper fibers or some of the lower fibers may be used alone; if you desire to raise your arm forward, you use such muscles or parts of muscles as will accomplish this action, including also the upper part of the pectoralis major but not the lower part. If, after raising the arm forward, you wish to pull it back toward the side you may then use all muscles or parts of muscles that will carry out this action, including the lower fibers of the pectoralis major. This selection of the proper portion of a muscle to carry out a given action is obviously brought about by selective activation of those nerve fibers going to that part of the muscle.

The Integration of Muscular Action

The dispatching of impulses along only those nerve fibers that end in parts of muscles useful in carrying out a desired movement is an automatic one, taken care of by cellular arrangements within the brain. While it is possible with special training to learn to contract a single motor unit, most of us cannot at will contract only part of a muscle except by carrying out a movement which we have learned by experience involves contraction of the desired portion. Obviously, then, this selection lies largely below the conscious level; we produce a given movement not by deciding what muscles we should use (if that were necessary we would all have to be very expert anatomists before we could learn to move at all!) but rather by the simple expedient of deciding that we want a given movement. Learning is involved here, for we all know that we have had to learn movements, but the important point is that the motor centers, and especially the voluntary movement center in the cerebral cortex (p. 344), are organized not just anatomically, on the basis of muscles, but physiologically, on the basis of movements. Artificial stimulation of the motor cerebral cortex regularly produces integrated movements, and only in appropriate cases isolated contraction of an individual muscle. Similarly, it is movements we learn, consciously and subconsciously, and integrated movements that we get when we volitionally stimulate our own motor centers; depending upon the movement and the strength necessary to carry it out, one or several muscles, or only appropriate parts of one or several muscles, may be involved.

Many movements require, of course, very precise coordination in the timing and strength of contraction of various muscles and their synergists. Learning a movement often requires that we first learn not to use the antagonists to that muscle, which interfere with it and make movement clumsy and difficult; we then acquire greater skill by learning to use more precisely only those muscles that will produce the desired effect.

While, in general, we all use the same muscles in the same way, electromyography has shown that there may be differences among individuals. Of two muscles that produce the same movement, for instance, one may initiate the movement in one person, the other one may do so in another person.

Some Consequences of Muscle Contraction

Mechanisms of muscle action have already been discussed, as have the physicochemical changes involved in the contraction of muscle fibers. In the latter discussion, it was pointed out that under conditions of insufficient oxygen, metabolites from the oxygenation of glucose, particularly lactic acid, accumulate in the muscle. This accumulation of metabolites is thought to be a cause of soreness after excessive exercise. Both massage and heat increase the circulation within the muscle and therefore aid in the destruction or the removal of the metabolites and the consequent relief from soreness.

Muscles, like manmade engines, are not completely efficient in their use of energy, and some of the energy is dissipated in the form of heat. This production of heat as a result of muscular contraction is obvious, and needs little comment. We are all aware that exercise warms the body, and may lead to such increased heat production as to bring forth efforts upon the part of the body to dissipate this heat more quickly, by the production and evaporation of sweat and the dilation of the blood vessels in the skin. Similarly, when we become too chilly we call upon our voluntary muscles to produce more heat, and shivering is the response of the muscles to this demand.

The Neuromuscular Ending

Discussion of functional aspects of muscular contraction would be incomplete without some further reference to the neuromuscular endings or motor end-plates. As already pointed out, the neuromuscular end-plate or junction represents anatomically the point of contact between two different tissues, nerve fibers and muscle fibers, while physiologically it represents the mechanism by which the nerve impulse is transmitted to the muscle fiber and creates the muscle impulse that results in contraction. Since this transmission takes place through a humoral mechanism, and transmission of the nerve impulse and spread of contraction along the muscle fiber are electrical phenomena, the neuromuscular ending presents features that are found neither in the nerve fiber nor in the muscle fiber. It is, instead, essentially similar to the synapse, or junction between two nerve cells—even to the degree that acetylcholine, the transmitter involved at the neuromuscular junction, is also the active agent at many synapses.

Since the acetylcholine stored at the nerve endings cannot be replaced as rapidly as it can be released, it might be expected that repetitive stimulation of the nerve fiber can result in such depletion of the acetylcholine that transmission between nerve and muscle becomes largely ineffective or ceases entirely. This is often referred to as fatigue of the neuromuscular junction. Thus it has long been known that continuous stimulation of a nerve to a muscle results at first in a tetanic (constantly maintained) contraction of the muscle, as the nerve impulses reach the individual muscle fibers so fast that none of the fibers relax. If such stimulation is long continued, however, the muscle begins to relax in spite of it, and eventually becomes completely relaxed because of fatigue at the neuromuscular ending. This fatigue then prevents further contraction of the muscle until recovery has taken place. It can be shown, however, that the muscle fibers themselves are still capable of contraction, as they can be stimulated directly with an electric current, and that the nerve fibers will still conduct impulses. Complete tiring of the muscle really involves, therefore, an inability of the nerve impulses to pass the neuromuscular junction.

Under ordinary circumstances, fatigue at the motor end-plate is minimized through a rotation of contraction among the muscle fibers that will carry out a given movement or maintain a certain posture. We know that if we exert all our strength in carrying out a certain movement we tire very quickly, and the movement soon becomes weaker and weaker; yet the same movement may be repeated for a much longer time if less effort is involved. If the desired strength of a movement requires only 5 per cent of the total number of muscle fibers capable of carrying out that movement, then we could obviously use any one motor unit on an average of only once out of every twenty contractions, thus allowing a considerable rest period before the same unit must be used again.

In addition to its susceptibility to fatigue, the neuromuscular junction is also susceptible to certain chemical agents. Among the best known of these is curare, long used by certain South American Indians as a poison to paralyze game, and now of great clinical importance. Curare has the effect of blocking the neuromuscular junction, and therefore of paralyzing voluntary muscles. In contrast to anesthetics, which affect primarily the nervous system rather than the neuromuscular junction, the carefully controlled clinical use of curare produces relaxation of the voluntary musculature without undue effect upon the nervous system.

Effects of Training and Exercise upon Muscle

The supervision of therapeutic exercise plays an important part in the activities of the physical therapist and occupational therapist, and the physical educator must supervise normal motions. Such workers should therefore have some clear idea of what can and what cannot be accomplished through exercise. In the first place, no amount of exercise will increase the number of muscle fibers in a muscle. Increase in size and strength of a muscle results from increase in the size (hypertrophy) of the muscle fibers already present, and not to an increase in their number. Since there is a maximal size and strength that muscle fibers can reach, the useful effects of exercise in increasing the strength of a muscle are limited by this. There is no way of restoring to normal a muscle in which many fibers have degenerated completely and been replaced by fibrous tissue. All that can be done is to assure the most effective action of the remaining muscle fibers, and to hope that this may prove functionally adequate.

The training of muscle really involves training the nervous system, since the activities of muscle are entirely dependent upon the nervous system. If, in trying to produce a movement, the antagonists are also used, training must include relaxation of the antagonists as well as the most efficient use of the prime movers. Also, since it is patterns of movement that we learn, re-education in a movement may allow a new pattern of muscular contraction to be set up in the central nervous system, so that the weakness of a given muscle or muscle group is at least partially compensated for by the use of other muscles, perhaps not habitually used in the weakened movement, but having functions overlapping with those of the weakened group.

An important factor in muscular imbalance is the fact that muscle fibers tend to so adjust their lengths (under the control of the nervous system) that they are exactly long enough, but no longer than is necessary, to bring about the range of movement ordinarily required of them. In order to retain their original lengths, they need to be stretched and made to contract over the total distance that they normally do. Thus if a part is so bent that the muscle or muscles crossing it need to contract over only half the distance that they ordinarily would have to, these muscles contract enough to take up the slack. The longer the part is kept in such a position, the more "set" the muscle fibers become in this short, partially contracted condition, and the more difficult it will be later to stretch them back to their original lengths.

This shortening of muscle can occur either through a part being kept in a flexed position by splints, or as a result of weakness of an opposing muscle group. Thus in paralyses caused by peripheral nerve injuries, it often occurs that the unparalyzed muscles draw the part toward themselves, and if the regenerative process is lengthy they can become fixed in this shortened position before the paralyzed muscles can recover. This shortening can be prevented by passively carrying the affected part through its complete range of movement, and thus subjecting the unparalyzed muscles to a normal amount of stretch.

Another type of contracture that has nothing directly to do with muscle results from the deposition of collagen in joints, ligaments and tendons, as in the acute phase of poliomyelitis or in a completely paralyzed (flail) limb. Here also the physical therapist, by taking the part through its complete range of normal movement, can stretch the newly formed tissue and prevent deformity.

Closely related to the problem of re-education in cases of loss of muscular power is the possibility of substitute movements, sometimes known as "trick" movements. These depend primarily upon taking advantage of some mechanical disposition of muscles and tendons at joints so that a movement that is otherwise impossible can be carried out. For instance, in paralysis of the extensors of the wrist and fingers the wrist can frequently be extended adequately by clenching the fingers tightly, and some extension of the fingers can be obtained by sharply flexing the wrist. Both these movements of flexion mechanically put tension on the extensor tendons. The importance of such substitute movements can hardly be exaggerated, and the physical therapist must become familiar with many of these in order to aid the patient in overcoming handicaps.

THE NERVOUS SYSTEM

The essential elements of the nervous system are nerve cells and their processes, that is, cells that are especially differentiated for the conduction of impulses. These important cells have been briefly described in the preceding chapter. However, it is the arrangement of these cells, and therefore their functional connections, that are of first importance in an understanding of the nervous system. While an adequate discussion of either the anatomy or the physiology of the nervous system is far beyond the confines of this book, certain fundamentals of the organization and function of this system are necessary to an understanding of the function of muscles and of the body as a whole, and some of these fundamentals will be pointed out here.

Anatomically, the nervous system is partially divisible into two great parts, the central nervous system and the peripheral nervous system. The central nervous system is composed of the brain and spinal cord, the peripheral nervous system of the cranial and spinal nerves and of the autonomic nervous system. Such a division is useful for descriptive purposes, but it must be clearly understood that all parts of the nervous system are dependent upon one another. The peripheral nervous system arises in part within the central nervous system, and it both receives impulses from the central nervous system and sends impulses into this system. The central nervous system, on the other hand, is dependent upon the peripheral nervous system for all the information it receives, and it is only through the peripheral nervous system that it is able to exert its effects. Various details of the distribution of the peripheral nervous system will be found throughout this book. Some further basic features of the anatomy of the spinal cord and brain are described in Section III, The Back, and Section V, The Head, Neck, and Trunk.

Origin

The central nervous system arises as a thickening of the epithelium on the dorsal surface of the embryo; this sinks into the underlying tissue to form a groove and the lips of the groove then roll together to form a tube called the neural tube, which separates from the overlying epithelium. As this separation occurs, epithelium at the junction of the two separates from both, to lie alongside the neural tube. This, the neural crest, subsequently forms the sensory ganglia of the cerebrospinal nerves, and also the ganglia of the autonomic system. Only the cells adjacent to the lumen of the spinal cord and brain retain their epithelial shape. The others, after a period of proliferation, differentiate into connective tissue peculiar to the central nervous system or into neuroblasts. A neuroblast becomes a nerve cell by giving rise to sprouts which grow out as dendrites and axons. Axons in the central nervous system grow for varying distances within that system, or if they are to emerge as motor fibers, leave it as components of a root of the nerve (ventral root for a spinal nerve). Each axon then must continue to grow until it reaches the muscle or other structure in which it is to end. The neuroblasts of the sensory ganglia likewise give rise to sprouts, but to only two, one of which grows centrally into the central nervous system while the other, like the motor axon, grows peripherally to form a sensory ending.

The Synapse

The essential feature of the organization of nerve cells is their functional connection to each other by their processes. Unless they end peripherally on muscle or glands, axons of nerve cells end primarily in connection with the dendrites or cell bodies of other nerve cells. Through these connections the cells of the central nervous system are arranged in innumerable interlocking rings, and it is the function of neuroanatomy to describe some of the more important of these rings, and to analyze their significance. Such an analysis is made possible only because certain of these interconnecting rings are used constantly in a fixed and definite pattern. If there were no "choice" as to the routing of a nerve impulse, it might wander haphazardly from one ring to another, or spread simultaneously in many directions so as to involve eventually the entire nervous system in aimless and uncoordinated activity. Once an impulse is initiated in a nerve cell, there is no choice as to the pathway within that cell—the impulse must follow out the axon and travel along all its branches. When, however, the impulse arrives at the terminations of the axon, other conditions prevail. There exists between one axon and the next nerve cell a slight gap, minute anatomically but very important physiologically (p. 19), that the nerve impulse must cross if it is to affect the next nerve cell. This tiny gap is known as the synapse, and it is due to the character of the synapse that coordinated activity of the nervous system is possible. It is at the synapse that it is determined whether an impulse shall cross to the next nerve cell or be obliterated, and some synapses are much more "resistant" to the passage of a nerve impulse than are others. If resistance at the synapses is generally broken down, as occurs in strychnine poisoning, then nerve impulses do spread through the nervous system without rhyme or reason, and totally uncoordinated activity results. The response of the synapse to the timing and the number of nerve impulses reaching it plays a decisive role in the activity of the nervous system. By making use of our knowledge of the important functional connections existing among nerve cells, and under what conditions certain synapses become usable, we are able to outline some of the more simple features of neural activity, especially in relation to sensation and to activity of voluntary muscles. Knowledge of the pathways and processes involved in more abstruse neural activity, such as learning, memory, abstract thought, and so forth, can hardly be said to exist.

The Spinal Cord

The spinal cord, the lower and least complicated portion of the central nervous system, is continuous above with the brain; it approximates a finger in its diameter, and is in an average adult some 17 or 18 inches long. It is protected by the vertebral (spinal) column, and is connected to voluntary muscles, skin, and other structures by the spinal nerves. In cross section (fig. 3-9) the fresh spinal cord may be seen to contain a center of pinkish gray material, shaped roughly like a butterfly or a distorted letter "H" and known as the gray matter, and a peripheral glistening white area known as the white matter. Both gray and white matter extend throughout the length of the spinal cord and are continued upward into the brain. The horns seen in cross sections of the cord are therefore actually parts of continuous gray columns, and the words "horns" and "columns" are used somewhat interchangeably in referring to the gray matter. The gray matter of the spinal cord consists primarily of nerve cell bodies, and of course of the fibers leaving these cell bodies and of fibers entering the gray matter to end upon them.

The posterior or dorsal projections of the gray matter on either side are known as the posterior or dorsal horns, and are concerned especially with receiving impulses coming in over the spinal nerves, and routing such impulses upward to the brain. The anterior or ventral projections of the gray matter are the anterior or ventral horns, and are concerned primarily with sending impulses out the spinal nerves to the voluntary muscles. Many of the cells of this horn give rise to fibers

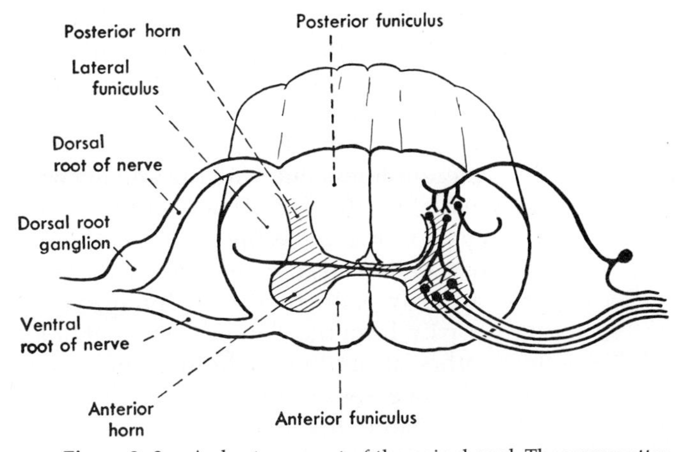

Figure 3–9. A short segment of the spinal cord. The gray matter of the cord is surrounded by white matter. On the right, a sensory fiber is shown coming into the cord over a dorsal root, and is seen to branch extensively; one of its branches turns upward in the posterior funiculus, others end in the posterior horn of the gray matter where they make synaptic connections with the cells located here. These latter cells in turn send their axons upward in the white matter (either crossing to the opposite side of the cord or remaining uncrossed) or make connections with the motor cells in the anterior horn.

that leave the spinal cord as the ventral (anterior, motor) roots of the spinal nerves, to end eventually upon voluntary muscles. Associated with the voluntary motor fibers as they emerge to form the ventral roots of the nerves are motor fibers controlling the activity of smooth muscles and glands. These are known as autonomic, sympathetic, or involuntary motor fibers, and arise from cells placed laterally in the posterior parts of the ventral horns of some, but not all, parts of the cord. There are also many cells in both horns that receive impulses from higher centers or from incoming nerve fibers and distribute them to other cells of the cord.

The white matter is composed of nerve fibers. As one looks at a cross section of the spinal cord it is obvious that the cord is partially divided into right and left halves by a posterior septum and an anterior fissure, and that the projecting anterior and posterior horns of the gray matter tend in turn to divide the white matter of one side into three great bundles, the funiculi or white columns—a posterior, a lateral, and an anterior. Each of the funiculi is in turn composed of several more or less definite, but often overlapping, smaller bundles of fibers. These smaller fiber bundles (fasciculi) are also known as the tracts of the spinal cord. The tracts represent groupings of fibers of a similar function, and by tedious processes of analysis we now know the locations and some of the functions of many of those in the cord. The tracts of the spinal cord are discussed briefly in Section III. For purposes of our present discussion, it need only be noted that they include groups of nerve fibers which connect different parts of the spinal cord, ascend to various parts of the brain, or descend from the brain to the cord. Fibers in the lateral and anterior funiculi consist of all three types, and are derived almost entirely from cells that lie in the central nervous system; those of the posterior funiculus are mostly ascending fibers, and are the central processes of the nerve cells that form the dorsal root ganglia and send their peripheral processes, as sensory fibers, into the spinal nerves.

Afferent fibers entering the spinal cord arise from cell bodies that are located on the dorsal roots of the spinal nerves, where they form the dorsal root ganglia (fig. 3-9). The peripheral processes from these ganglion cells end in connection with many tissues in the body, and are variously concerned with impulses of touch, pressure, temperature or pain from the skin or from deeper structures, or with impulses from other structures including muscles, joints and tendons. As the central processes of these nerve cells enter the spinal cord, each fiber branches. Some branches end in the gray matter of the spinal cord at the level at which the nerve fiber enters the cord, while others run upward or downward within the white matter of the cord for varying distances. Through these branches of a single entering nerve fiber it is possible for an incoming sensory impulse to travel in several directions within the spinal cord at one time. Many of the branches of the sensory fiber end about cells of the posterior horn, through which their impulses are propagated both within the cord and, by way of the long ascending tracts, upward to various parts of the brain; some end also on the large cells in the anterior horn that control voluntary muscle.

The long descending tracts of the cord originate from neurons in many parts of the brain; the majority of their fibers end either on the anterior horn cells supplying voluntary muscle or on cells that in turn send fibers to these anterior horn cells, so most of these tracts are known as motor tracts, and the groups of cells from which they originate in the brain constitute motor centers. The pathway by which the motor centers of the brain influence voluntary muscle therefore always involves at least two neurons, one in the brain and one in the anterior horn of the spinal cord. Since both are motor neurons, they are distinguished by calling the cell in the anterior horn the lower motor neuron, the one in the brain the upper motor neuron.

The large lower motor neurons of the anterior horn not only receive connections from all the motor tracts of the cord that affect voluntary muscle, but are also influenced by incoming sensory fibers of various types. Thus nerve impulses from many different sources converge on the lower motor neuron. On the other hand, the anterior horn cells and their fibers constitute the sole connection between the spinal cord and voluntary muscles. It does not matter whether the muscle cell is to be affected by a nerve impulse originally derived from stimulation of the skin, from stretching of a muscle or movement of a joint, or from any of those numerous parts of the brain having to do with the action of voluntary muscle—the lower motor neuron of the anterior horn, with its fiber (axon) ending upon the muscle, constitutes the only pathway to this muscle. This cell and its axon have thus been called the "final common path" by Sherrington, a famous neurophysiologist. Destruction of the lower motor neuron, the final common path, means that no impulses, whatever their original sources, can reach the voluntary muscle. Therefore, such destruction leads to complete paralysis of the muscle concerned. If the paralysis of the muscle is due to a peripheral nerve injury, that is, to damage to the axons of the lower motor neurons, the body will attempt to repair the nerve injury. However, the complete lack of activity in the paralyzed muscle leads to atrophy and degeneration of the muscle fibers; so, while the nerve repair is being awaited, the physical therapist attempts to check the atrophy as much as possible by stimulating the circulation to the muscle, and perhaps stimulating the muscle fibers to contraction through an electric current.

It is the cell body of the lower motor neuron that is especially affected in poliomyelitis. If all the lower motor neurons to a given muscle are destroyed, that muscle will always remain paralyzed, since new nerve cells cannot be formed to replace those destroyed. On the other hand, lower motor neurons may be so damaged that for a while they cease to function, and the muscle is therefore paralyzed, yet the neurons may later recover with consequent recovery of function on the part of the muscle. It is impossible to distinguish between paralysis caused by total destruction of the nerve cells supplying the muscle, and paralysis caused by their temporary injury. Therefore, hope of recovery following muscular paralysis should not be abandoned until it seems obvious that the nerve cells innervating the muscles have actually been destroyed.

The Reflex Arc

The least complicated of the controls over the lower motor neuron is that of incoming sensory fibers at the spinal level. These fibers, their connections with the lower motor neuron, and this neuron itself can be regarded as forming an anatomic and functional unit which is called a "simple spinal reflex arc" (fig. 3-10). A particular characteristic of a reflex (a reflex being the activity resulting from the functioning of a reflex arc) is that the response following the application of a given stimulus is stereotyped—that is to say, the nerve connections are so relatively simple that there is little choice as to the routing of the nerve impulse, and therefore little opportunity for deviation from one particularly appropriate action. It is upon reflex arcs of varying complexity that all our activities are built up. The simple spinal reflexes therefore constitute the basis of all muscular activity, even though in normal life spinal reflexes are much affected, modified, and sometimes rendered almost unrecognizable by impulses from numerous other reflex arcs and from volitional centers.

A pure spinal reflex, free from inhibiting or inciting influences from higher centers, can be seen only when the spinal cord is cut in two so as to isolate a lower portion of the cord. An animal or a man with the cord so cut is known as a "spinal" animal or man; in such an individual the muscles supplied from this lower segment of the cord are paralyzed, inasmuch as they can no longer respond to the efforts of the individual to move them. However, they are not paralyzed in the sense that they cannot contract, for indeed a great deal of activity is

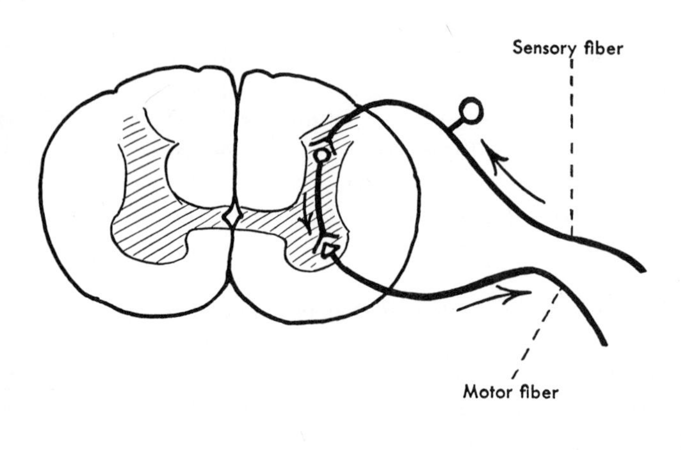

Figure 3–10. A simple spinal reflex arc. This particular arc is diagrammed as involving only three elements: a sensory neuron brings into the nervous system an impulse originating in the sense organ; an intercalated cell located in the posterior horn transmits this impulse to the motor cell in the anterior horn; the motor cell itself (lower motor neuron) then transmits the impulse to the effector organ, the voluntary muscle. Other connections of the sensory fiber are not shown, nor are there indicated the many other fibers that end in connection with the lower motor neuron, the final common pathway to the muscle.

maintained in them through the activity of the reflex arc. If the muscles supplied from an isolated segment of the cord are palpated, it will be found that they are relaxed even less than in the normal individual; the limbs may be maintained in abnormal postures by this excessive contraction, and may offer very great resistance to passive movement by the observer. This is because various sensory impulses arising both from the skin and from the muscles, tendons, and joints are being constantly poured into the spinal cord where they act upon the lower motor neuron. This lower motor neuron, being now freed of all impulses from centers in the brain, responds more actively than normal to the incoming impulses at its own level. If all sensory impulses coming into the cord are abolished, the activity of the muscles then ceases. If, on the other hand, sensory impulses are increased by stimulation, then the activity of a certain muscle or muscle group is likewise increased and may produce a definite movement or reflex action. In physiologic terms, a simple spinal reflex involves stimulation of receptors; transmission, of the nerve impulses thus set up, along the sensory fibers of the spinal nerve to the cord; and relay of these impulses to the motor neurons with subsequent transmission to an effector organ.

A simple type of spinal reflex may be elicited in a "spinal" animal by pricking the skin of the foot. Regardless of how many times this is done, if proper attention is paid to the conditions of the experiment the result will always be the same: the animal will withdraw his foot as if to escape from the source of the pain—an obviously useful movement. Another simple spinal reflex is the muscle stretch reflex; if a muscle is stretched, nerve endings within that muscle (p. 40) are stimulated, and this stimulation, being transmitted to the central nervous system, brings about the activity of the lower motor neuron and contraction of the muscle. A very familiar example of this stretch reflex is the patellar reflex, or knee jerk, in which the tendon below the knee cap is hit smartly and the blow results in a contraction of the extensor muscle of the knee, therefore in a sudden kick. While this reflex, as elicited, may seem purposeless, it is actually the basis of the support of the body weight—stretching of the large muscle on the front of the thigh usually means that the knee is being bent, and therefore giving beneath its load. Further contraction of the extensor of the knee is a logical response to control this movement, to prevent tumbling to the ground.

All spinal reflexes are not so simple as those just mentioned. For instance, if the foot of a "spinal" dog is pricked with a pin he will withdraw that foot (in spite of the fact that he feels nothing, because the pain impulse cannot reach the brain), and this is a simple spinal withdrawal reflex. At the same time, however, that he withdraws the pricked foot the extensor muscles of the other limbs will increase their contraction in order to support the shift in weight of the body. Spinal reflexes, therefore, whether simple or complicated, are coordinated reactions that seem to be carefully planned to bring about an appropriate response to a given stimulus. They differ from more complicated reactions in the nervous system through their relative simplicity and the inevitability of their results. An intact dog, when pricked upon the foot, will most certainly lift that foot, but may also turn to bite the individual or attempt to run away—therefore, he can choose among several courses of action. The "spinal" dog, on the other hand, will simply continue to lift the pricked paw, making no effort either to bite the individual or to run away. In fact, such a dog does not know his foot is being injured or that he is lifting it, since section of the cord prevents sensory impulses of any type from reaching the brain.

As evidenced by the above examples, some of the more important spinal reflexes involving voluntary muscle are concerned with the protection of the body from harmful influences, or with the maintenance of posture. Since harmful stimuli usually result in pain, reflexes initiated by pain are of common occurrence. Thus we "double up" with an abdominal pain because this pain causes increased tension in the abdominal muscles, which contract in order to protect the abdominal contents from outside pressure. The contraction of the muscles in its turn increases intra-abdominal pressure, and this is relieved by flexing the trunk and thus approximating the origins and insertions of the muscles concerned. The physician, in examining the abdomen, notes the occurrence and the location of muscle spasm as one clue to localization of the origin of the pain. In a similar fashion, meningitis, or inflammation of the coverings of the brain and spinal cord, may result in the head and back being arched backward by the contraction of the back muscles. This posture somewhat relieves the tension of the coverings over the convexity of the brain. Another example of the protective reflex action of muscle is found in the fixation of a painful joint; all the muscles about such a joint may go into spasm so as to prevent that joint from being further moved.

Since flexion, representing a withdrawal movement, is a basic reflex response to pain, the flexor muscles frequently contract more than do the extensors about a painful joint, thus maintaining the joint in flexion in spite of the fact that the extensors (usually the antigravity muscles) are typically the stronger. Thus in the very painful rheumatoid arthritis, for instance, the limb is characteristically maintained in flexion spasm.

Supraspinal Influences

While the activity of the lower motor neuron is thus obviously maintained in part by spinal reflexes, this cell is also, in the intact individual, under the control of the various descending motor pathways from the brain; its final action is therefore a summation of all of the various excitatory and inhibitory impulses reaching it. The physician examines spinal reflexes because he knows that these can be affected not only by damage to the components of the spinal reflex arc but also by damage to various higher centers or pathways. Indications of a change from normal in a reflex may give important information concerning the functional activities of these higher centers or their descending tracts.

The motor pathways descending into the cord are responsible for all of the voluntary and much of the automatic or higher reflex control over the lower motor neuron. Thus a typical result of an upper motor neuron lesion is paralysis of voluntary movement, as in the familiar "stroke" in which a whole side of the body is paralyzed. The same injury that prevents impulses for voluntary movement from reaching the lower motor neurons also cuts off a variable number of other impulses, especially inhibitory ones, so that the lower motor neuron is simultaneously deprived of its ability to carry out voluntary movements and yet excited more than normally by various reflex arcs converging on it. Thus many peripheral stimuli may produce a contraction of the affected muscles. Upper motor neuron paralysis, therefore, is typically a "spastic" paralysis, in which the muscles cannot be used voluntarily and at the same time their contraction and their resistance to passive movement are increased markedly above normal. This type of paralysis is in sharp contrast to lower motor neuron paralysis, in which again the muscles are not subject to voluntary influence, but are in addition completely relaxed or flaccid, since destruction of the lower motor neurons prevents any nerve impulses from reaching the muscles.

Other motor centers in the brain are concerned not with initiating a voluntary movement, but with coordinating the action of the numerous lower motor neurons involved so that a movement is carried out smoothly and accurately. Disease of some of these centers results in abnormal distribution of impulses to the lower motor neuron; depending upon the center involved there may be almost constant contraction of all muscles, called rigidity because the part is then hard to move; there may be trembling whenever a part is moved, and it may be moved inaccurately and weakly; or there may be trembling at rest, and perhaps uncontrollable movement of a part, such as flinging a limb around in a purposeless fashion. All these give evidence of the intricacy of control needed to assure proper function of the lower motor neurons and through them the muscles.

The Nerves

The nerves arising from the brain and spinal cord are essentially cables, or bundles of the axons of nerve cells, that connect the central nervous system to the rest of the body. The voluntary motor fibers of a nerve arise from lower motor neurons and extend to the voluntary muscle, however remote that may be from the central nervous system (p. 18). The sensory fibers of a nerve have their cell bodies located in ganglia (collections of nerve cell bodies outside the central nervous system) that are located very close to the central nervous system (see p. 50), being protected by the skull and vertebral column (backbone) which also protect the central nervous system. A sensory neuron in a ganglion has two processes that arise together but soon separate: one, usually very long, runs peripherally in the nerve to reach skin, muscle, or other tissue with a sensory innervation; the other enters the central nervous system and then branches to end upon cells located here.

The nerves as they attach to the central nervous system are divisible into two groups: spinal nerves (see also p. 221), which attach to the spinal cord and are therefore associated with the vertebral column at their attachments; and cranial nerves, which attach to the brain and make their exits through the skull. All the smaller nerves found in the body are branches of either the spinal or the cranial nerves.

There are 31 pairs of spinal nerves, and although they differ much in size they are similar in composition. Each spinal nerve is attached to the spinal cord (see fig. 3-9) by a dorsal root (also called posterior root) which is sensory and a ventral root (also called anterior root) which is motor. On the dorsal root there is a swelling, the dorsal root ganglion, produced by the accumulation of the cell bodies of the sensory fibers. The ganglion lies on the dorsal root at the point at which the nerve is about to leave the shelter of the vertebral column. Just distal to the ganglion, as the nerve emerges from between the adjacent vertebrae (bones of the vertebral column, p. 197) the dorsal and ventral roots join and their fibers become mixed together. In consequence of this mixing, almost all the branches of a spinal nerve contain both sensory and motor fibers. While injury to or surgical section of the dorsal or ventral roots separately will injure only sensory or motor fibers, injury to the spinal nerves after their roots have joined will typically involve both types of fibers.

After the mixed (sensory and motor) spinal nerve leaves the vertebral column it divides into two branches, a dorsal (posterior) branch that turns backward to supply muscle and skin of the back, and a ventral (anterior) branch that runs laterally and forward (fig. 13-7). The ventral branches supply the muscles and skin of all parts of the body except the back. Except in the thoracic (chest) region, where they run between the ribs and are separated by these, the ventral branches run close together and exchange branches with each other, such an exchange being known as a nerve plexus (for instance, fig. 5-3). The nerves leaving a plexus are never the same as the nerves entering it. Instead of containing fibers of only one spinal nerve, as the spinal nerves entering the plexus obviously do, the nerves leaving the plexus contain fibers from more than one spinal nerve.

Because the nerves supplying the limbs form plexuses, determination of the exact distribution of spinal nerves to the muscles and skin of the limbs is difficult; it is the peripheral nerves, containing fibers from several spinal nerves, that can be traced by dissection. However, the clinician must distinguish sharply between peripheral nerve distribution and segmental (spinal) nerve distribution, and be familiar with both. The area of distribution of a spinal nerve to skin is known as a dermatome, and there are charts that show the dermatomes of the body as they have been determined by clinical means. Many textbooks also contain tables showing the approximate segmental innervation of the voluntary muscles. While there is disagreement on details, probably resulting both from incomplete knowledge and from variations among persons, a general rule is that most muscles are innervated through two or more spinal nerves, and that any one spinal nerve to the limbs helps to supply a large number of muscles.

In speaking or writing of the spinal nerves it is often necessary to differentiate among them, therefore they are all named and numbered. The eight that leave the vertebral column in the neck are called cervical nerves, and are distinguished from each other by number—the first cervical nerve is the highest, the eighth the lowest. Similarly, there are 12 thoracic nerves, or nerves associated with the chest; five lumbar nerves, leaving the mobile part of the vertebral column below the ribs; and five sacral and one coccygeal nerve leaving the lowest part of the vertebral column. In identifying the nerves a sort of shorthand is sometimes used: thus the third cervical nerve, for instance, is C3 (also written C-3), the letter "C" standing for cervical; in the same way, T1 through T12 (also written T-1, etc., and in older texts Th1 or even D1—D meaning dorsal or thoracic) identifies thoracic nerves; L1 through L5, lumbar; S1 through S5, sacral; and Co, the coccygeal.

The Autonomic Nervous System

The autonomic nervous system is that portion of the motor nervous system that controls the activities of smooth muscle, cardiac muscle, and certain glands. It has, therefore, no direct effect upon voluntary muscles, but its activities so influence the activities of the body as a whole that brief mention must be made of it. The autonomic nervous system is classified as a part of the peripheral nervous system because a large part of it lies outside the central nervous system and this is the most obvious part. However, there are important centers and pathways within the brain and spinal cord that influence the activity of the autonomic system, and are therefore referred to as autonomic centers and pathways. Moreover, the autonomic nervous system is like the rest of the peripheral nervous system in that its activities are dependent upon the activities of the central nervous system as a whole.

The autonomic nervous system can in its peripheral portion be conveniently divided into two subsystems that differ from each other, in part, both in their anatomy and their physiology. These subsystems are the sympathetic and parasympathetic systems. They are similar in that they both control involuntary structures, and in that it takes *two* neurons to transmit impulses to these involuntary structures from the central nervous system—in sharp contrast to the single neuron that transmits impulses from the central nervous system to voluntary muscle.

The first neuron of the autonomic system proper sends its fiber outside the central nervous system but has its cell body located in the brain or spinal cord. Its axon leaves the brain along with other fibers forming certain of the cranial nerves, or leaves the spinal cord with the voluntary motor fibers in the ventral roots of the spinal nerves. The cell bodies and axons of the elements leaving the central nervous system are known as preganglionic neurons. Only a few cranial nerves contain preganglionic fibers as they leave the brain; similarly, preganglionic fibers do not occur in the ventral roots of all spinal nerves. Spinal preganglionic fibers usually occur only in the ventral roots of the twelve thoracic and of the upper two lumbar nerves, and in those of the second and third or third and fourth sacral nerves. Thus the preganglionic fibers occur in three groups: a cranial outflow is associated with the cranial nerves and is separated from the thoracolumbar outflow by the cervical nerves, in which there is no preganglionic outflow; the thoracolumbar outflow then forms a second group and is in turn separated from the third or sacral outflow by the lower lumbar and one or more upper sacral nerves, which also have no preganglionic outflow. The cranial and sacral outflows of the autonomic nervous system closely resemble each other in their anatomy and physiology, and they and their further distribution are grouped together as the parasympathetic nervous system. The thoracolumbar outflow and its distribution is called the sympathetic nervous system. Figure 3-11 illustrates the essential structure of this latter system.

Both sympathetic and parasympathetic nervous systems are associated with collections of nerve cell bodies that lie outside the central nervous system and are therefore known as ganglia. The autonomic fibers leaving the central nervous system end in these ganglia to form synapses with their cells, and the axons of these second neurons, the postganglionic neurons, then run farther to end upon involuntary muscle or glands. The ganglia of the sympathetic nervous system form paired trunks or chains lying on the sides or front of the vertebral column and extending for the whole length of the column. In addition, other sympathetic ganglia are located around the origins of the great vessels going to the abdominal viscera where, with the tangled fibers entering and leaving them, they form the celiac or "solar" plexus (fig. 23-5). The ganglia of the parasympathetic system are small and are scattered, being located in or very close to the organ that they innervate. Except for a few parasympathetic ganglia in the head, most of the parasympathetic ganglia are too small to be seen during dissection.

Postganglionic fibers from both the sympathetic and parasympathetic ganglia usually run with other nerves or along blood vessels to the organs that they innervate. For instance, the postganglionic sympathetic fibers to the limbs join the spinal nerves close to the spinal cord and are thereafter distributed by these nerves along with other fibers to the various parts of the limb, while the postganglionic sympathetic fibers to the head and to the abdominal viscera reach these parts

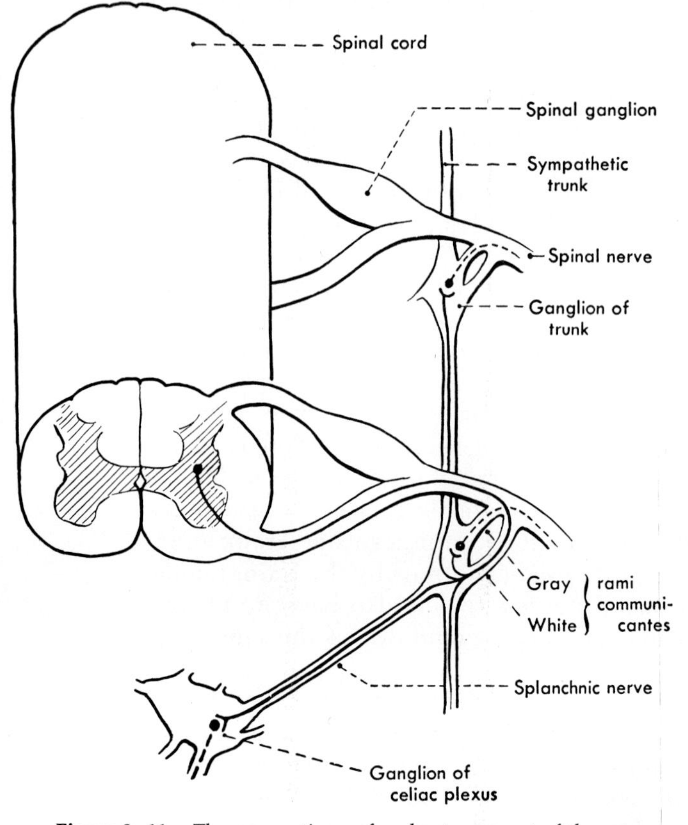

Figure 3–11. The connections of a short segment of the sympathetic trunk. Preganglionic fibers (solid lines) are seen emerging in the ventral root of a spinal nerve, leaving the nerve to reach the sympathetic trunk, and there having one of several courses: they may synapse with the ganglion cells in the first ganglion they reach, they may run up or down the trunk to synapse in ganglia above or below the level at which they enter the trunk, or they may leave the trunk without synapsing, to end in ganglia of the celiac or associated plexuses. Postganglionic fibers (broken lines) arise from cells of the trunk ganglia and return to the spinal nerves to be distributed with them, or arise from the ganglia of the celiac plexus, and so forth, and are distributed along the blood vessels to the viscera.

by following blood vessels. Both sympathetic and parasympathetic fibers frequently go to the same organ, so that many involuntary structures have a double innervation. Generally speaking, when both systems innervate an organ they have opposite effects upon it. For instance, the sympathetic innervation to the pupil of the eye dilates the pupil; the parasympathetic innervation to the pupil constricts the pupil. Not all organs, however, are supplied by both sets of fibers; the smooth muscle of the blood vessels in the limbs, for instance, is innervated only through sympathetic fibers. It must not be thought, however, that the two systems actively oppose each other in those organs that have a double innervation; rather they cooperate in very much the fashion that prime movers and antagonists of voluntary muscles do.

The functions of the two parts of the autonomic nervous system can be broadly compared as follows: the sympathetic nervous system is concerned with preparing the body for emergency actions, and functions especially in times of fright or anger; the parasympathetic nervous system is concerned with maintaining the everyday activities of the body and minimizing some of the strains put upon its parts. Sympathetic stimulation tends to stop the digestive functions and constricts the blood vessels to the digestive tract and to the skin, so that more blood can be available to go to the limb muscles where it may be needed. It increases the cardiac output so that the blood circulates faster, and it dilates the air passages in the lungs so that the blood may be fully aerated. These things being done, the body is better prepared to react to danger by either fleeing or fighting, as may seem wiser at the moment. On the other hand, the parasympathetic nervous system promotes the orderly activity of the digestive tract, slows down the heart and aids in emptying the rectum and bladder.

Since the autonomic system is, by definition, a motor system, the sensory fibers that accompany it to visceral structures are not, strictly speaking, a part of this system. Although it is not unusual to hear references to "sympathetic sensory fibers," such fibers are better called "visceral sensory fibers"; they are concerned especially with pain and with reflex activity initiated in the viscera. In their anatomy they are similar to other sensory fibers, as their cell bodies lie in the sensory ganglia (not in autonomic ganglia), and their processes extend to the sensory endings. However, in order to reach the viscera they travel with autonomic fibers going to the same location; therefore most autonomic nerves and plexuses are really a mixture of autonomic and sensory fibers.

THE VASCULAR SYSTEM

The vascular system consists of the heart, blood vessels, and blood-forming organs with their contained blood, and of the lymphatic vessels, lymph nodes, and the lymph itself. The anatomy of the heart is briefly described with the thoracic viscera (p. 369), and the general courses of many blood vessels are described in other parts of this text. Here we will consider only a few general aspects of this system.

The heart is a muscular pump that propels the fluid blood. The tubes along which the blood is pumped as it leaves the heart are known as arteries. Since the blood enters the arteries under pressure, the arteries must have strong walls to withstand this pressure. The walls are also somewhat elastic, being composed especially of varying mixtures of elastic tissue and smooth muscle. This elasticity in turn helps to force the blood along, and when it is lost through arteriosclerosis, or hardening of the arteries, the heart must beat harder in order to move the same amount of blood. As the arteries branch and become smaller they eventually give rise to very small branches that are known as arterioles. The chief component of the arteriolar wall is smooth muscle, and the contraction of this smooth muscle under the control of the sympathetic nervous system determines the ease with which the blood can pass through these smaller vessels. In simplest terms, the blood pressure depends upon the amount of blood pumped into the arteries by the heart, and the size of the vascular bed into which this blood can pass. If the arterioles are constricted, the size of the vascular bed is decreased and blood flows more slowly from the arteries, so that on the next beat the heart will have to contract more forcibly in order to push its contained blood through the arterial bed. Therefore, if the arterioles are contracted the blood pressure rises, while if they are relaxed the peripheral resistance and hence the blood pressure are lowered. The arterioles also control the distribution of blood through various parts of the body. If the arterioles of one part of the body or one organ contract more than usual, then that part will receive less blood than normal and some other part will then receive more than normal.

The arterioles open into capillaries, tiny vessels whose walls consist only of endothelium, the thin cellular lining found throughout the entire vascular system. Through these very thin walls exchange between materials in the blood and materials in the fluid outside the vascular system can take place, and the capillaries are therefore responsible for supplying the tissue cells with their required food, oxygen, and hormones, and for removing from the fluid about these cells the carbon dioxide and other metabolic products that should be eliminated. The capillaries are in turn continuous with the veins, which join together to form larger and larger veins and eventually return the blood to the heart. The vascular system therefore forms a closed circuit, the blood circulating from the heart through the arteries through the capillaries through the veins and thus back to the heart.

The lymphatic system forms an accessory drainage for the tissue spaces. In contrast to the blood vascular system, the lymphatic system begins in blind capillaries among the cells; these capillaries unite to form larger vessels and eventually one major and several minor chief lymphatics, which empty their contents into the blood stream by way of the great veins in the base of the neck. Most lymphatics, and the veins of the limbs but not of many other parts of the body, are supplied with valves that allow the flow of fluid in one direction, toward the heart, but resist a backward flow.

Associated with lymphatic vessels are structures known as lymph nodes or lymph glands. These are, in general, bean-shaped structures varying in size from that of a pinhead to that of a lima bean, interposed along the pathways of the lymphatics. Lymph passes through one or more of these lymph nodes as it is returned to the blood stream. The lymphatics frequently act as the pathway for migration of infections and cancer, and the lymph nodes act as filters along the lymphatic path, tending to catch and hold for a time such bacteria or cancer cells as may reach them. The physician may look for red streaks up the arm or for hard swollen lymph nodes in the armpit or elsewhere as a sign of infection, and the pathologist may examine lymph nodes for cancer cells or for other pathologic processes that may involve the nodes. The surgeon likewise is extremely conscious of the importance of the lymph nodes; in removing a cancerous part he also removes insofar as possible the lymph nodes to which the cancer is most apt to have spread, hoping that in so doing he will have removed all of the cancer cells within the body and thus cured the disease by eliminating it. If the cancer has not yet spread beyond those nodes that can be removed, cure should result.

The fluid blood, composed of plasma in which the blood cells (fig. 2-8) are suspended, is of course the sole means by which most of the tissues of the body can obtain the elements they need for life and through which they can dispose of harmful products of metabolism. The red cells of the blood transport oxygen, picking it up as they circulate through the capillaries of the lung and in turn giving it up as they circulate in the capillaries of the body tissues in general. The white blood cells or leukocytes serve primarily to repel invasions of the body by noxious agents or cells that are foreign to that body (for instance, tissues introduced by transplants of skin, kidney, or heart). While most of the white blood cells are carried passively in the blood, they are living cells and are capable of movement. Moreover, they can leave the blood stream by crawling between the interlocking edges of the capillary endothelial cells and thus congregate in any tissue where they are needed. Large numbers so leave the blood stream in an area of infection, and act to overcome this infection both by engulfing the bacteria concerned and by the production of chemical substances that interfere with the growth or further life of the bacteria. In any localized area of severe infection, leukocytes form a prominent element of the pus produced by the infection. In regard to transplants, the leukocytes that are responsible for their rejection respond to immunologic differences between host and transplant. Thus a skin graft derived from the person on whom the skin is grafted, or from an identical twin, provokes no immunologic reaction. In grafts between persons of less or no consanguinity, however, the major problem is usually how best to minimize, typically by the infusion into the blood stream of various chemical substances, the immunologic reaction.

The blood is normally protected from contact with other tissues by the endothelial walls that line the blood vessels. If the blood vessels are broken, however, blood may obviously spread among the tissue cells or perhaps be lost to the body through an opening to the outside. In either case, there is a tendency for the blood to check its own flow by clotting as it comes in contact with a strange environment. This clotting involves certain chemical reactions in the plasma, and in the case of small vessels results in the formation of a fibrous plug at the break in the wall of the vessel. The addition of heat hastens the chemical reaction leading to clotting. On the other hand, heat dilates the smaller blood vessels, whereas cold will contract them. Thus there is room for argument as to the efficacy of hot packs versus cold packs for the control of minor hemorrhages.

Even a slight injury to the endothelial lining results in the formation of a clot at the point of injury. If the injury is extensive, a large clot may be formed; if this clot subsequently breaks away from the vessel wall, it will be carried along by the blood stream until it reaches a vessel too small for it to traverse. It then occludes this vessel, with varying results: there will be little effect if other vessels supply the same tissue as the obstructed vessel, while if there is no other blood supply, death of that tissue, and perhaps even death of the individual, will result.

There is normally a constant interchange between the fluid in the blood capillaries, that in the tissue spaces (living cells have to be surrounded by fluid), and that in lymphatic capillaries. If there is obstruction to either the venous or the lymphatic drainage of a part, fluid is likely to accumulate in the tissues faster than it can be removed, leading to swelling of the tissues by the fluid (edema). Another cause of local edema is infection.

THE DIGESTIVE SYSTEM

The digestive system includes the mouth, pharynx, esophagus, stomach and intestines, with their associated glands. The mouth is provided with three pairs of large salivary glands, whose secretion both moistens the food and starts the process of digestion. All three are closely associated with the lower jaw (mandible); the one between the jaw and the ear that is involved in mumps is the parotid gland. The mouth opens into the pharynx, as does the nose, but digestive system and respiratory system separate at the level of the larynx, or voice box, where the esophagus begins. The esophagus traverses the lower part of the neck and the entire thorax to reach the abdomen and end in the stomach. The gross anatomy of these parts, beginning with the stomach, is described with the other abdominal viscera.

Both the pharynx and the upper part of the esophagus are provided with voluntary muscle, but once swallowing is started further propulsion through the essentially tubular digestive tract is carried out by smooth muscle. Within the smooth muscle is a highly glandular mucosa, which adds digestive enzymes and other substances to the ingested material, and also absorbs the products of digestion and, finally, much of the water. The activity of the digestive tract is largely automatic, but is in part under the control of the autonomic nervous system.

THE RESPIRATORY SYSTEM

The respiratory system begins with the nose, where the air passages are bilateral and provided with projections (conchae) from their walls that bring the air into close contact with the nasal mucosa, by which it is warmed and humidified as it passes toward the lungs. Since nasal and oral cavities both empty into the pharynx, the mouth can also be used for respiration, and is, of course, so used when the nasal passages cannot accommodate the desired air flow. The pharynx is held open by its attachments to the skeleton of the head and neck, and thus allows a free flow of air to the point of origin of the trachea or windpipe. The trachea begins at the larynx and is supported by a series of C-shaped cartilages that keep this part of the air passage open. In the upper part of the thorax the trachea divides into two large bronchi, one for each lung; within the lung the bronchi divide and subdivide to end finally in tiny, very thin-walled air sacs, the alveoli. The alveoli are surrounded by networks of blood capillaries; the air in the alveoli is thus in close contact with the blood within the capillaries, and interchange between the gaseous contents of blood and air takes place freely.

UROGENITAL SYSTEM

The urogenital system consists of the kidneys, ureters, urinary bladder, and urethra, composing the urinary system, and the sex glands or gonads and their associated reproductive organs. The paired kidneys, through which the blood is filtered free of certain impurities that are dissolved in water to form the urine, lie in the abdomen (fig. 23-3). The urine formed in the kidneys is transported by the peristaltic action of the ureters to the urinary bladder, where it accumulates. The tube leading from the bladder to the exterior is the urethra; in the male, the urethra is also a part of the genital system.

Much of the urogenital system lies in the pelvis, and is briefly mentioned beginning with page 380, and depicted in figure 23–4. The difference in position of the gonads, the testes in the male and the ovaries in the female, accounts for the greater incidence of inguinal hernia in the male. While the ovary remains in the pelvis, the testis migrates into the scrotum, carrying its duct and blood vessels (constituting the spermatic cord) with it; the testis and cord thus create a larger defect in the lower abdominal wall than does the small ligament that passes through the wall in the female.

DUCTLESS GLANDS

The ductless or endocrine glands are a scattered rather than a united system such as those already discussed. They lie in the head, the neck, and the trunk, and are dissimilar to each other in practically all details of their anatomy and physiology except that unlike most glands they have no ducts, and their secretions, hormones, are discharged into the blood stream; the latter accounts for their characteristically great vascularity.

Hormones, chemical agents that in very tiny quantities affect the activity of cells and tissues, may be released by tissues that are not truly endocrine in structure—as, for instance, the cellular lining of the duodenum, the first part of the intestine, releases into the blood stream a hormone that causes the gallbladder (p. 378) to contract—but the majority of hormones that have been identified are produced by the endocrine glands. The structures that are usually listed as endocrine glands are the hypophysis, the thyroid gland, the parathyroid glands, the suprarenal glands, and certain parts of the pancreas and of the ovaries and testes.

The **hypophysis** (pituitary gland) is located at the base of the skull immediately below the brain, to a part of which it is attached (fig. 21-5). It is actually two different glands, although they are closely bound together. One part (posterior lobe) is primarily concerned with regulating the amount of water excreted by the kidneys. The other part (anterior lobe) is sometimes spoken of as the "master gland," for in addition to producing a hormone necessary for growth (p. 24) it produces hormones that largely govern the secretion of the other most important endocrine glands: the thyroid, part of the suprarenals, and the ovaries and testes.

The **thyroid gland** lies in the neck, largely on the sides of the trachea (windpipe) but with its two large lateral parts connected across the front of this (fig. 21-12). Its hormone primarily governs the rate of oxidative processes in the body, and its activity is measured by the basal metabolism.

The **parathyroid glands,** typically four, are about the size of very small peas and usually lie on the posterior surface of the thyroid gland. The fact that they govern the level of calcium in the blood has already been noted (p. 14).

The paired **suprarenal** or **adrenal glands** get their names from the fact that they lie on the kidneys (renes). Each is really two glands. An inner part, the medulla ("marrow"), liberates into the blood stream a hormone called epinephrine or adrenaline that produces the same general effect that stimulation of the sympathetic nervous system does. For instance, it increases the strength of the heart beat and at the same time causes many arterioles to contract, so that it raises the blood pressure. Epinephrine is released particularly when one is frightened or angry. The outer portion of the gland, the cortex (again "bark"), apparently liberates a number of complex substances; the best known of these is cortisone, particularly useful in treating certain crippling conditions of joints.

Most of the cells of the pancreas produce digestive enzymes that reach the intestine through a duct, but scattered through the pancreatic tissue are small groups of different cells, the **pancreatic islets** (islands of Langerhans), that are endocrine glands. They produce insulin, which has to do primarily with the level of blood sugar (glucose) and its storage as glycogen. An insufficient secretion of insulin produces the common form of diabetes, in which the high level of sugar in the blood leads to its appearance in the urine.

In addition to the sex cells that they form, both the **testis** and **ovary** also contain cell groups that release the sex hormones (hormones affecting sexual characteristics may also come from other places, as the suprarenal cortex or the placenta). In general, these sex hormones, of which there are a particularly large number in the female, govern especially the growth and activity of the other parts of the reproductive system. For instance, the uterus of an immature animal will not grow if the ovaries are removed, and shifting balances of sex hormones are responsible for the menstrual cycle, which therefore ceases when the ovaries atrophy. They are also responsible for the development of the secondary sex characteristics such as the distribution of hair and the development of the female breast.

THE SKIN

The skin, composed of an outer layer of stratified epithelium, the epidermis, and of a densely matted deeper layer of connective tissue, the dermis or corium, obviously varies much from person to person in such characteristics as pigmentation and texture, and varies also in the same person according to differing circumstances: it may reflect health or illness; it may reflect the emotions, as in sweating from nervousness or blushing from embarrassment; and it gives evidence of the circulation to a part, becoming flushed when the arterioles and capillaries are dilated, as they are by heat, blanched or even blue when exposed to cold, or when there is some other interference with the arterial supply. Because the nails are translucent, they also reflect these changes in blood supply.

The skin has many functions. A primary one is to seal off the body fluids, which living cells require if they are to survive, from the surrounding air or water; thus one of the problems in burns is the loss of body fluids that may ensue. It also protects against infection; and it offers first resistance to physical forces such as friction. For the latter reason, the skin of the dorsal parts of the body is usually thicker than that of the ventral, less exposed, parts, but the reverse is true of the palm and sole; here the epidermis is particularly thick, and if subjected to more than the usual friction will thicken still more to form calluses; and the dermis is also thick and tightly bound down to deeper structures. Nails, hair, and glands, developed from the epidermis, also have protective functions: nails protect against mechanical trauma, hair against cold, and sweat glands against heat. The skin thus has thermoregulatory properties, largely confined in man to dissipation of heat through radiation, increased by the evaporation of sweat; and its various glands, which include the breast and the sebaceous glands connected with hairs, make it also a secretory organ.

The skin is also a particularly important sense organ containing, especially in the dermis immediately adjacent to the epidermis, nerve endings that respond to touch, pressure, heat, cold, and painful stimuli. These sensations are mediated through sensory fibers of the cranial and spinal nerves and testing cutaneous sensation is therefore a routine part of examining the nervous system.

In considering the innervation of skin, the same problem exists as in the innervation of muscle: in the limbs, the peripheral nerves have derived their fibers from several spinal nerves, and the innervation of skin can thus be described in two ways, in terms of its peripheral nerve innervation or of its segmental nerve innervation. Cutaneous peripheral innervation can be reasonably accurately determined by dissection, although because of overlap between adjacent nerves the area of sensory loss after section of any one nerve is never as large as might be expected. The segmental or spinal nerve distribution to skin is known as a **dermatome,** and can be determined only by clinical methods. Because there is overlap between adjacent spinal nerves, the most common method of determining dermatomes has been to record the area in which sensation remains when an intact spinal nerve is bordered above and below by nerves that have been interrupted. On the trunk—because the nerves do not traverse a plexus—dermatomes and peripheral nerve distribution are identical, but on the limbs the dermatomes bear no apparent relation to the distribution of the various peripheral nerves. While dermatomal charts vary somewhat, one which is widely used is shown in figure 3-12.

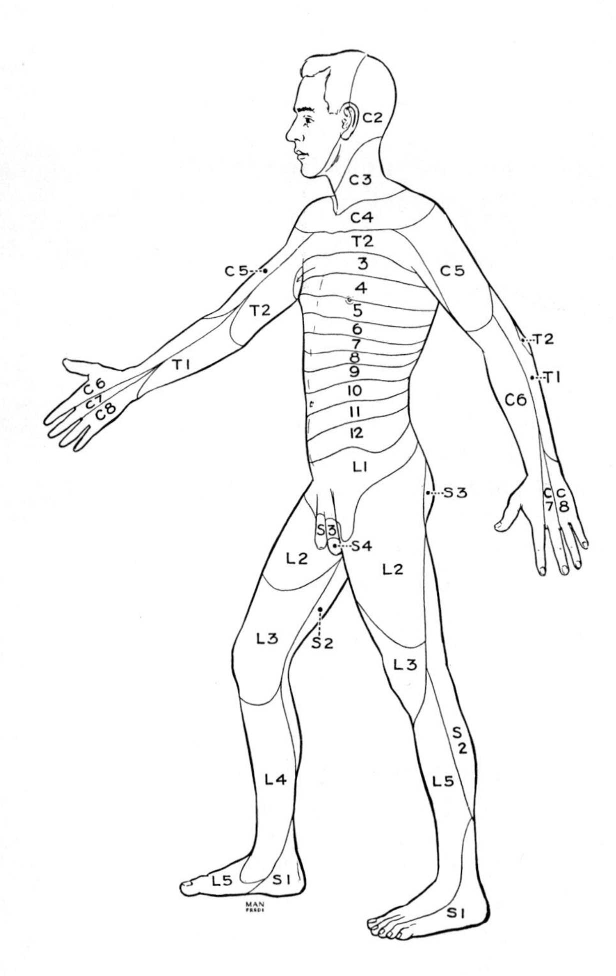

Figure 3–12. An anterolateral view of the dermatomes. Compare the regular arrangement of distribution of the spinal nerves on the trunk with the irregular arrangement on the limbs. (After Foerster, from Haymaker, W., and Woodhall, B.: *Peripheral Nerve Injuries* [ed. 2]. Philadelphia, W. B. Saunders, 1953.)

4 General Survey of the Upper Limb

The upper limb (the Latin is membrum superius, "superior member") may be divided into the shoulder region, the arm or brachium (above the elbow), the forearm or antebrachium, the wrist or carpus, and the hand (manus). The hand ends in digits, a neutral name that applies equally well to the thumb (pollex) or to the fingers, and also to the toes. As the upper limb of man has been freed from weight-bearing function it has been possible to sacrifice the greater stability necessary for this purpose and thus gain the mobility that has added so much to the development of man. This mobility is especially marked in the hand and digits, but to a lesser degree extends throughout the whole limb.

The upper limb first appears as a swelling on the side of the embryo; as it rapidly grows outward, it projects first laterally and then ventrally and slightly caudally. The distal end becomes a flattened plate for the hand, flexures indicating the elbow and wrist appear, and ridges on the hand plate differentiate into digits. This growth, and a medial rotation of the growing limb, distorts the relations between the surfaces of the limbs and those of the trunk, but the original relations can be restored in the adult by simply raising the limb to a horizontal position with the palm forward: there the back of the hand, arm, and forearm face dorsally or posteriorly, the thumb or radial side is directed cranially, the palm and the flexor side of the forearm and arm form the anterior or ventral surfaces, and the little finger is directed caudally.

The limb bud consists initially of a core of mesenchyme (embryonic connective tissue) within a thin sheet of epithelium. The latter will form only the outer layer of the skin, while the mesenchyme of the bud forms the remaining tissues except for the blood vessels and the nerves, which grow into the bud. While the base of the bud is relatively broad, extending from about the fifth cervical to the first thoracic segment, nerves from these segments grow into the developing limb. Condensations of mesenchyme (which will be transformed into cartilage and then bone) indicate the positions of the skeleton, and condensations around the primitive skeletal elements gradually differentiate into muscle groups (for example, extensor on the dorsal side, flexor on the ventral side) and then into individual muscles. Some of those at the base of the limb grow back into the trunk, to attach to the ribs, sternum, or vertebral column.

A convenient subdivision of the skeleton is into axial skeleton (skull, ribs, sternum, and vertebral column) and appendicular skeleton (the skeleton of the limbs, a more appropriate term when the limbs were known as "appendages" rather than "members"). For both limbs, the skeleton is divisible into a girdle (cingulum) and the skeleton of the free limb. The girdle of the upper limb (fig. 4-1), also called shoulder girdle, or pectoral girdle because it lies in the region of the chest (pectoral means both breast and chest), consists of the clavicle or collar bone and the scapula or shoulder blade. The scapula is largely attached to the body by muscles; the clavicle is attached at one end to the sternum (breastbone), at the other to the scapula (fig. 5-2). The clavicle serves the purpose of a strut to keep the upper limb away from the body wall. It is provided with movable joints at both its sternal and scapular ends; because of these joints it limits movement of the scapula much less than might be supposed.

The scapula has upon it a shallow oval fossa that receives the upper rounded end or head of the humerus, the single bone of the arm. The shoulder joint is thus a shallow ball-and-socket joint, relatively freely movable in most directions. The forearm contains two bones, the radius on the thumb side and the ulna on the little-finger side. The ulna is so articulated with the lower end of the humerus as to allow only movements of flexion and extension between the two bones; the radius participates in flexion and extension, of course, but also can rotate upon its long axis, thus allowing the palm of the hand to be turned downward and upward (pronation and supination) when the forearm is horizontal.

While the ulna forms the chief articulation at the elbow, the distal expanded end of the radius is the chief forearm element entering into the wrist joint. At the wrist there are two rows of small bones, carpals, a total of eight in all. Movements of flexion, extension, and so forth, go on here, aided by movements between the carpals, and especially between the proximal and distal rows of these elements. The long bones of the palm of the hand are the metacarpals. For the most part they are much limited in their movements, but as you can observe, the first or thumb metacarpal is freely movable to allow opposition of the thumb (touching it to the tips of the fingers or to the palm); the fifth and fourth metacarpals are also more movable than the second and third (which are essentially immovable), this movement allowing a firmer grasp with the ulnar (little finger) side of the hand. The digits are composed of phalanges; the thumb has only two of these, the other digits three. The phalanges articulate with each other by hinge joints.

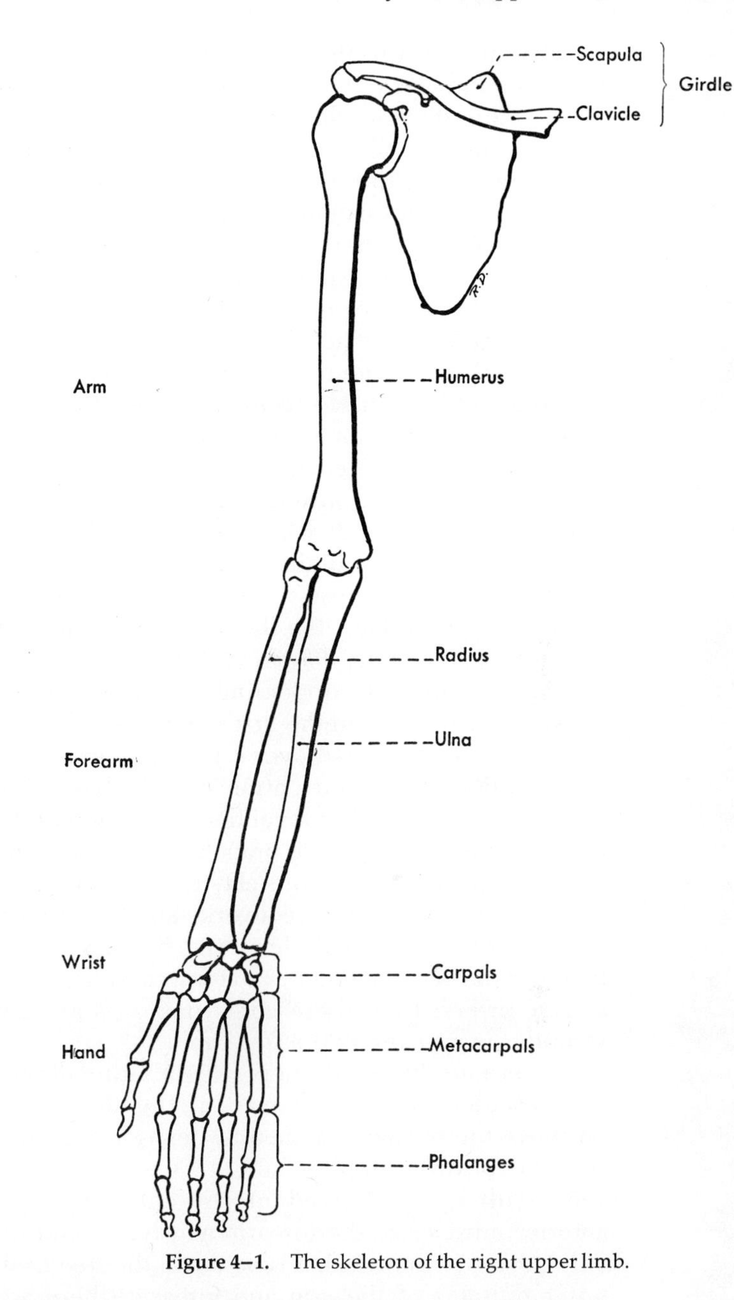

Figure 4–1. The skeleton of the right upper limb.

Some of the muscles of the upper limb lie for the most part on the thorax (chest) in the pectoral (breast) region or on its side and are therefore from their positions muscles of the thorax, although they act upon the shoulder girdle and the free limb. Other muscles of the limb have spread over the back to attain an origin from the vertebral column (backbone), and are therefore, in position, muscles of the back although their primary actions are upon the girdle or the free limb. Other muscles of the shoulder arise from the girdle and insert on the humerus, the uppermost bone of the free limb; these are listed as muscles of the upper limb proper. These muscles as a whole move the scapula or move the arm on the scapula.

The muscles in the arm form fleshy masses on the anterior and posterior surfaces of the humerus and act primarily across the elbow joint, where they are flexors or extensors of this joint, or, in addition, rotate (supinate) the radius. Some of these muscles arise from the girdle, however; these can act also at the shoulder joint, and one of them acts only at this joint.

The muscles in the forearm, divided conveniently into flexor (anteromedial) and extensor (posterolateral) muscle masses, act primarily at the wrist or upon the digits, but some of them have an accessory or their chief action at the elbow. Among the forearm muscles are those that pronate or supinate the forearm.

Many of the muscles connected with the fingers and thumb have their bellies in the forearm, forming parts of the flexor and extensor masses here, and therefore act upon both wrist joint and digits. These longer muscles are continued into the hand by relatively narrow tendons (leaders) many of which can be easily felt or seen at the wrist. Other muscles of the digits, situated in the hand itself, act only upon the digits. Some of these shorter muscles form two prominent groups on the palmar surface, the thenar (thumb) and hypothenar (little finger) groups; others lie deeper.

The nerves of the upper limb are derived almost entirely from the lower four cervical and the first thoracic spinal nerves. At their origins these nerves are spread over a territory considerably wider than the space available for their entrance at the base of the arm, and for the most part are at a higher level than the origin of the free limb. As they run into the arm they give branches to some of the shoulder muscles and then pass between the clavicle and first rib to enter the axilla (armpit). In so doing they converge and then branch in a complex pattern to form the brachial plexus (fig. 5-3). Essentially, the nerves come together in such a manner as to form three main cords arranged about a large artery; from these cords branches are given off both to the shoulder region and to the free limb.

There are four main nerves continuing down into the free limb: the musculocutaneous, the median, the ulnar, and the radial (fig. 4-2). Of these the musculocutaneous, derived from an anterior part of the brachial plexus, supplies anterior muscles of the arm; the median and the ulnar, also derived from the anterior part of the plexus, supply anterior muscles of the forearm and hand; and the radial nerve, the only posterior branch that runs down the free limb, supplies the posterior muscles of the arm and forearm (there are no true posterior muscles in the hand). These nerves also help to supply skin of the limb. Their distributions are best considered later, but that of the musculocutaneous is shown diagrammatically in figure 6-1; that of the median in figure 7-1; that of the ulnar in figure 7-2; and that of the radial in figure 7-3.

The chief artery to the upper limb is the subclavian artery, which is a large direct or almost direct branch from the aorta (p. 370) close to the heart. While it is in the base of the neck the subclavian gives rise to branches that help supply shoulder muscles, but it passes into the free limb with no diminution in size. As it crosses the first rib its name is changed to axillary artery, which gives off branches (most of which

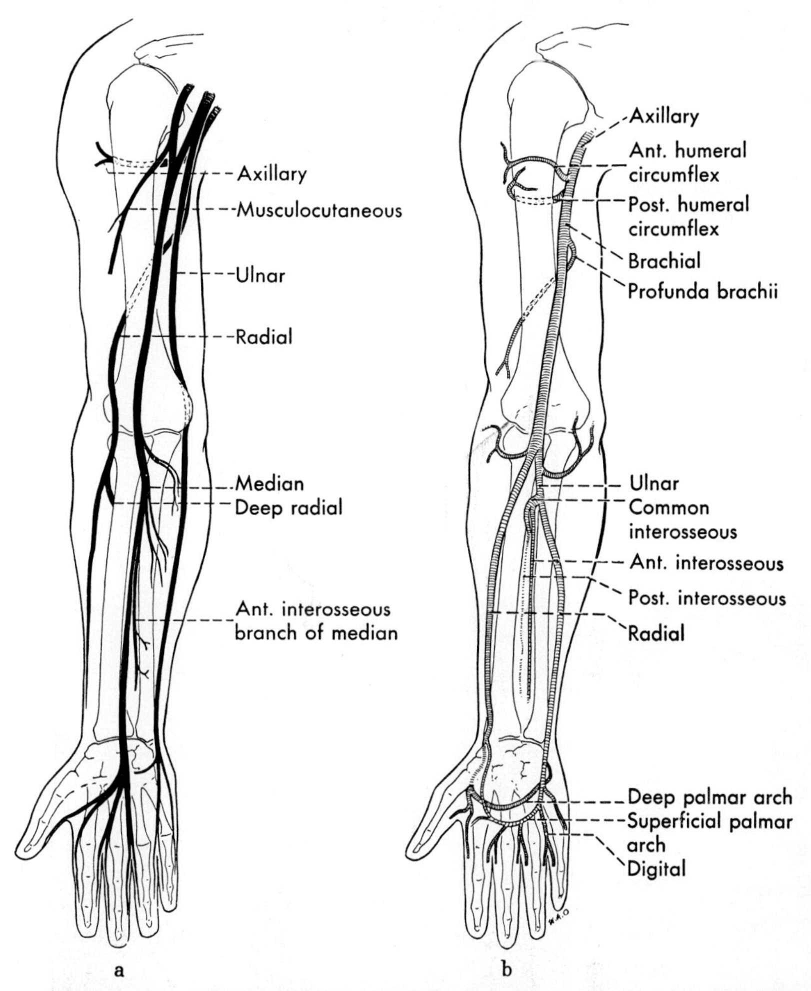

Figure 4–2. *a*. The principal nerves of the upper limb. The deep branch of the radial nerve is distributed on the back of the forearm to extensor muscles. *b*. The chief arteries of the limb.

are not shown in figure 4-2) to muscles of the shoulder and to the wall of the chest, and then continues into the arm where its name is changed to brachial artery. Branches of the brachial artery supply the arm, and in the upper part of the forearm the brachial artery ends by dividing into radial and ulnar arteries. These run down the front of the forearm, so there is no large posterior artery, but the ulnar gives off a branch that does run deeply on the posterior aspect. Both ulnar and radial arteries go mainly to the palm of the hand.

Subcutaneous bursae similar to other bursae (p. 11) are found in a few locations on the upper limb. The more constant of these are the acromial subcutaneous bursa, over the acromion (shoulder tip); the olecranal subcutaneous bursa, between the olecranon process (back of the elbow) and skin; and dorsally placed bursae lying over the knuckles or the proximal joints of the fingers. These bursae have essentially the same functions as bursae elsewhere in the body; in these particular cases they enable the skin to slide freely over a projecting bony surface.

5 The Shoulder

GENERAL CONSIDERATIONS

The shoulder region in human anatomy includes in a broad sense not only the rounded contour between the arm and the body, but also the pectoral (breast) region, the region of the back around the shoulder blade, and the axilla or armpit. The shoulder muscles cover the upper part of the chest and spread posteriorly so that they almost completely cover the true back muscles. Thus an inspection of the shoulder must necessarily include much of the trunk, and an upper part of the arm.

The muscles of the shoulder attach not only to the skeleton of the shoulder, the clavicle and the scapula, which will be studied in more detail later (p. 78), but also to the skeleton of the anterior thoracic wall (fig. 5-1) and to the vertebral column. These parts should therefore be briefly inspected. The following account calls attention to those parts that are visible and palpable, and are therefore useful for general orientation.

Perhaps the most familiar landmark of all is the clavicle or collar bone (fig. 5-2) at the base of the neck. The midline depression between the ends of the two clavicles and the prominent muscles of the neck (sternocleidomastoids) that attach here is the jugular fossa. The clavicle can be traced laterally to where it joins the bone forming the tip of the shoulder, which is the acromion of the triangular scapula or shoulder blade (also fig. 5-2).

The skeleton of the chest or thorax is partly covered anteriorly and almost entirely covered posteriorly by shoulder muscles; some of the anterior shoulder muscles form the anterior fold of the axilla, while some of the posterior muscles form the posterior fold. However, many of the ribs can be felt or seen both anteriorly and laterally, and in the anterior midline the sternum or breastbone can easily be felt. The

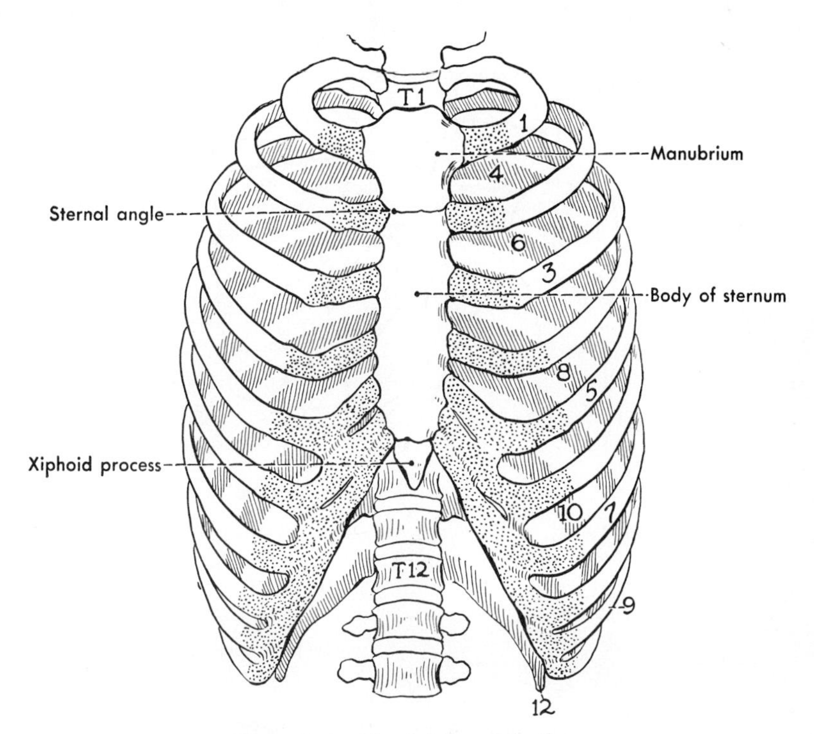

Figure 5–1. The skeleton of the thorax.

uppermost rib that can be felt attaching to the sternum is the second rib, for the first one is covered by the clavicle. The point of attachment of the second rib to the sternum is regularly marked by the sternal angle, which can be felt by running a finger lightly down the upper portion of the sternum. The sternal angle is formed by the junction of the upper segment of the sternum, the manubrium sterni, and the main portion or body, which lie in slightly different planes. The small third portion of the sternum, the xiphoid process, lies in the infrasternal angle, formed by the cartilages of some of the lower ribs of each side as they ascend to attach to the sternum.

On the back, the posterior tips or spinous processes (fig. 13-1) of many of the vertebrae can be palpated or seen in the midline (fig. 5-11), their prominence being increased by flexion (forward bending) of the trunk. Most of the cervical vertebrae (cervix means neck) lie so deeply buried in the muscles of the neck that they cannot be felt distinctly, and usually the first distinct spinous process is that of the seventh cervical. This usually forms a marked projection at the base of the neck, and the vertebra is sometimes known, therefore, as the vertebra prominens. Occasionally the sixth vertebra may be easily felt above the seventh, and sometimes the spinous process of the first thoracic vertebra is more prominent than that of the seventh cervical. The spinous processes of the thoracic vertebrae (twelve in number) are in general long and pointed, and overlap each other; those of the five lumbar vertebrae are broad and blunt (fig. 13–2). Below the lumbar spinous processes the sacrum can be palpated in the midline, and laterally the attachment of the hip bones to the sacrum can be felt.

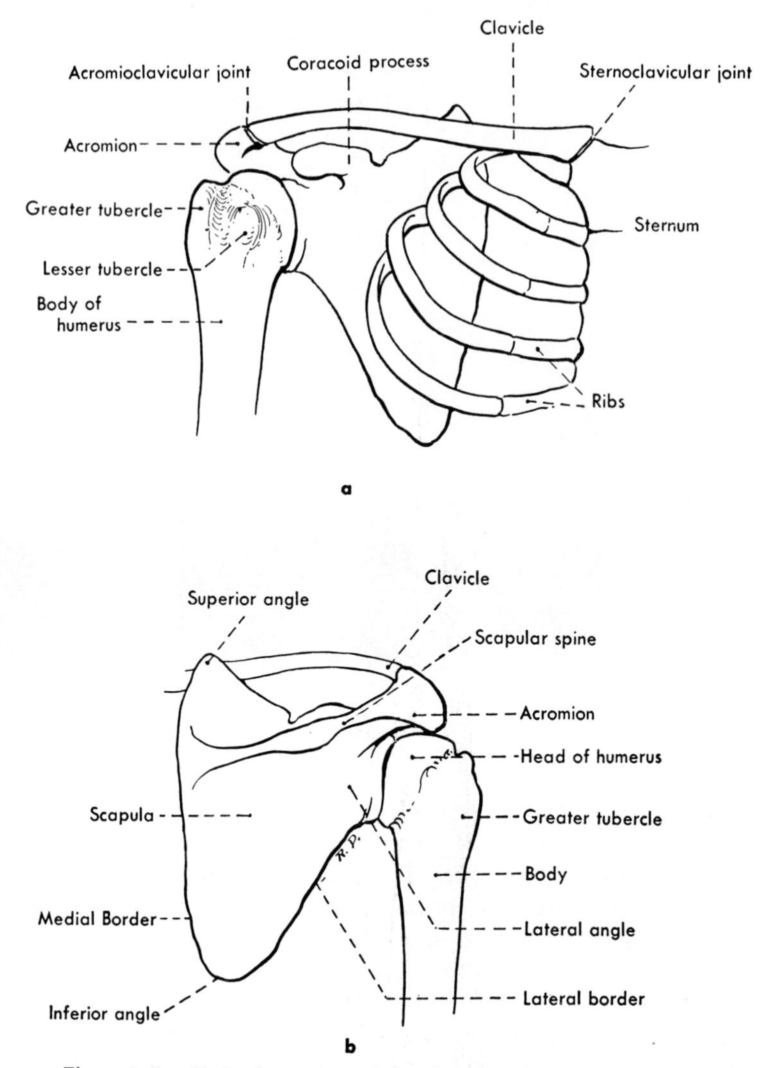

Figure 5–2. Skeletal anatomy of the shoulder region. *a.* Anterior view. *b.* Posterior view.

On the posterior aspect of the head there are also markings that are of importance for the study of the limb muscles. These are the external occipital protuberance, which is the prominent posterior projection of the skull in the midline, and the mastoid processes, the bony enlargements behind the ears.

As already stated, the girdle of the upper limb consists of the clavicle and scapula (fig. 5-2). The scapula is largely suspended by muscles, and is therefore rather freely movable. The lateral end of the clavicle is articulated to the scapula and therefore moves primarily with it, while its medial end, articulated to the sternum, moves on that bone. The upper end of the humerus, the bone of the arm, articulates with the scapula to form the shoulder joint.

The muscles of the shoulder arise in part from the axial skeleton, and in part from the girdle. Those arising from the girdle (intrinsic muscles of the limb, since they both arise and insert on bones of the limb) insert on the humerus and thus act at the shoulder joint; some of those arising from the axial skeleton (extrinsic muscles, since they are not confined to the limb) also attach to the humerus and act primarily at the shoulder joint, but others attach to the scapula and clavicle and move these bones.

Most of the nerves to the muscles of the shoulder are derived from the upper part of the brachial plexus (see the following section), but they are quite numerous and must be described individually. The blood supply of these muscles is chiefly from branches of the subclavian and axillary arteries, thus from great vessels at the base of the neck and the base of the arm.

The movements of the scapula are defined as elevation (raising it, toward the head), depression (lowering it), protraction (moving it forward) and retraction ("straightening" the shoulders). Rotation of the scapula also occurs, and may be either upward or downward. In upward rotation the inferior angle is moved laterally and forward around the thoracic wall, and the lateral angle, articulating with the humerus, is tilted upward. In downward rotation the inferior angle is moved toward the vertebral column and the lateral angle is lowered. These movements of the scapula accompany movements of the arm—for instance, when the arm is reached forward, the scapula slips forward on the chest wall in protraction; when the arm is raised above the head, the accompanying upward rotation of the scapula tilts the lateral angle upward also. As the scapula moves upward, downward, forward, and backward the lateral end of the clavicle follows it, pivoting at the sternal end.

Movements of the arm at the shoulder are particularly free, and include flexion, extension, abduction, adduction, circumduction, internal rotation, and external rotation. These movements should probably be defined in terms of their relation to the scapula, but are usually referred rather to the body as a whole. Thus flexion of the arm (also called flexion at the shoulder) is a forward movement of the arm. Extension, the reverse of this, is backward movement of the arm. Abduction is the movement of raising the arm laterally away from the body; adduction, the opposite of this, is then bringing the arm toward the side. Circumduction is a combination of all four of the above-defined movements, so that the hand describes a circle. Internal rotation (also called medial rotation) is a rotation of the arm about its long axis, so that the usual anterior surface is turned inward toward the body; external rotation (also called lateral rotation) is the opposite of this. It might be noted that during normal movement with the forearm straight the apparent effects of rotation of the arm are increased by the somewhat similar movements of pronation and supination (p. 112) occurring in the forearm. Therefore, if the amount of rotation of the arm itself is to be observed the forearm should be held in flexion while this movement is being tested; this allows a dissociation between rotation and pronation-supination.

All movements of the arm at the shoulder can be described by the terms used above, although usually movements of the arm are combinations of two or more of the above-defined movements. Thus, in bringing the arm forward across the chest we flex and adduct it and usually also internally rotate it at the same time; in scratching the lower part of one's back, one extends, internally rotates, and abducts or adducts.

THE BRACHIAL PLEXUS

This lies primarily in the lower part of the neck, being situated deeply here, and extends between the clavicle and the first rib into the axilla. On the lateral wall of the axilla the lower end of the brachial plexus may be rolled between a finger and the humerus.

The brachial plexus cannot be studied in the dissecting room until the axilla and the lower part of the neck have been rather completely dissected, but a general knowledge of its composition and branches, such as may be obtained from diagrams or pictures, is of great aid in understanding the nerve supply to the muscles of the upper limb. The following description may be compared now with available diagrams, such as figure 5-3, and checked later on the body.

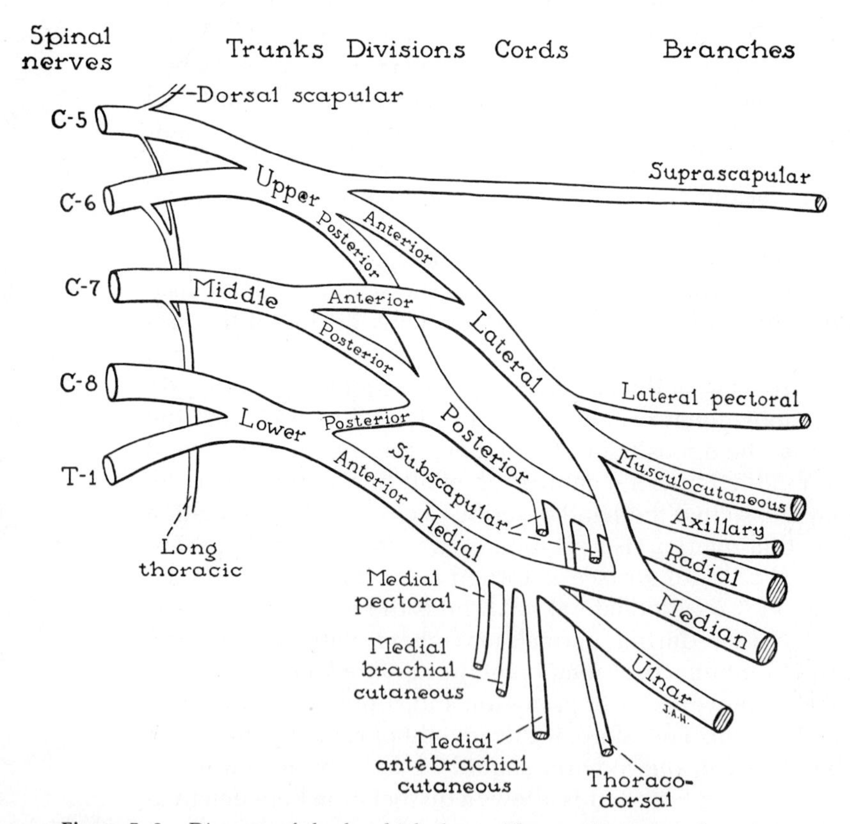

Figure 5–3. Diagram of the brachial plexus. The small nerve to the subclavius, from the upper trunk, is omitted.

The ventral branches (p. 55) of the fifth, sixth, seventh and eighth cervical nerves and of the first thoracic nerve unite to form the brachial plexus (the dorsal branches of the spinal nerves turn sharply backward around the vertebral column to supply muscles and skin of the back). To these five ventral branches there may be added small communications from the ventral branches of C4 or T2. (Spinal nerves are often designated by letters and numerals, the letter designating the region to which the nerve belongs, the numeral its place in the series – p. 56.) The ventral branches of C5 and C6 (with any contribution there is from C4) join to form an upper trunk; C7 comes out alone as a middle trunk; and C8 and T1 unite to form a lower trunk. Each of these three trunks then divides into an anterior and a posterior division. The posterior divisions of all three trunks unite to form the posterior cord; the anterior divisions of upper and middle trunks unite to form the lateral cord; and the anterior division of the lower trunk is continued as the medial cord. These cords are named from their relations to the axillary artery, about which they are arranged.

The plexus is clearly divided into two fundamental parts, an anterior (flexor) and a posterior (extensor) portion; this division of the plexus corresponds to the division of the musculature of the limb into flexor and extensor groups. The medial and lateral cords of the brachial plexus together form the anterior portion of the plexus and through their branches supply the muscles of the pectoral region and all the muscles on the anterior aspects of the arm, forearm and hand, that is, all muscles originally arising on the ventral or flexor surface of the limb. The posterior cord represents the posterior element of the brachial plexus; it supplies most of the muscles of the shoulder proper, and all the posterior muscles in the arm and forearm, that is, muscles originally associated with the dorsal or extensor surface of the embryonic limb.

The lateral cord, carrying fibers primarily from C5, 6 and 7, gives off the lateral pectoral nerve to the pectoralis major and the larger musculocutaneous nerve to the anterior muscles of the arm; the remainder of the lateral cord is then joined across the front of the axillary artery by a branch of the medial cord to form the median nerve. The medial cord, carrying fibers from C8 and T1, gives off the medial pectoral nerve; two cutaneous nerves in succession, one to the medial side of the arm and one to the forearm; the much larger ulnar nerve; and then continues as the medial root of the median nerve to join the lateral root from the lateral cord. The posterior cord, containing fibers from most of the elements of the brachial plexus, but usually relatively few fibers from C8 and sometimes none from T1, gives off an upper subscapular nerve to the subscapularis muscle, the thoracodorsal nerve to the latissimus dorsi, a lower subscapular nerve to the subscapularis and teres major muscles, and then divides into axillary and radial nerves. The axillary nerve runs posteriorly around an upper part of the humerus while the larger radial nerve, the obvious continuation of the posterior cord, takes a more gradual course around the humerus.

In addition to the branches from the three cords of the brachial plexus certain important nerves arise higher, that is, closer to the origins of the nerves contributing to the plexus. The tiny nerve to the subclavius (not shown in fig. 5-3) and the larger suprascapular nerve arise from the upper trunk of the brachial plexus; the dorsal scapular nerve to the rhomboids arises from the fifth cervical nerve before this joins the upper trunk; and the long thoracic nerve to the serratus anterior arises from the ventral branches of C5, 6, and 7, especially C6. Further, the ventral branches, before their union, contribute nerves to muscles of the neck, and the fifth cervical nerve also regularly sends fibers into the phrenic nerve (to the diaphragm).

BONES AND JOINTS OF THE SHOULDER

The **scapula** (fig. 5-4) is a triangular bone possessing three borders and three angles. The borders are named superior, medial, and lateral. The angles are the superior, inferior, and lateral, the latter being an expanded end on which is set the smooth glenoid cavity, the articular surface receiving the head of the humerus. In the fresh condition the glenoid cavity has attached around its edge a narrow rim of fibrocartilage, the glenoidal labrum, that slightly widens and deepens the cavity. The hooked coracoid process projects forward from the narrowed neck, close to the glenoid cavity. The costal (related to the ribs) surface of the scapula is relatively smooth, while the dorsal surface is divided into two parts by the projecting spine. The area above the spine is the supraspinatous fossa; that below the spine is the infraspinatous fossa. The costal surface is called, incorrect-

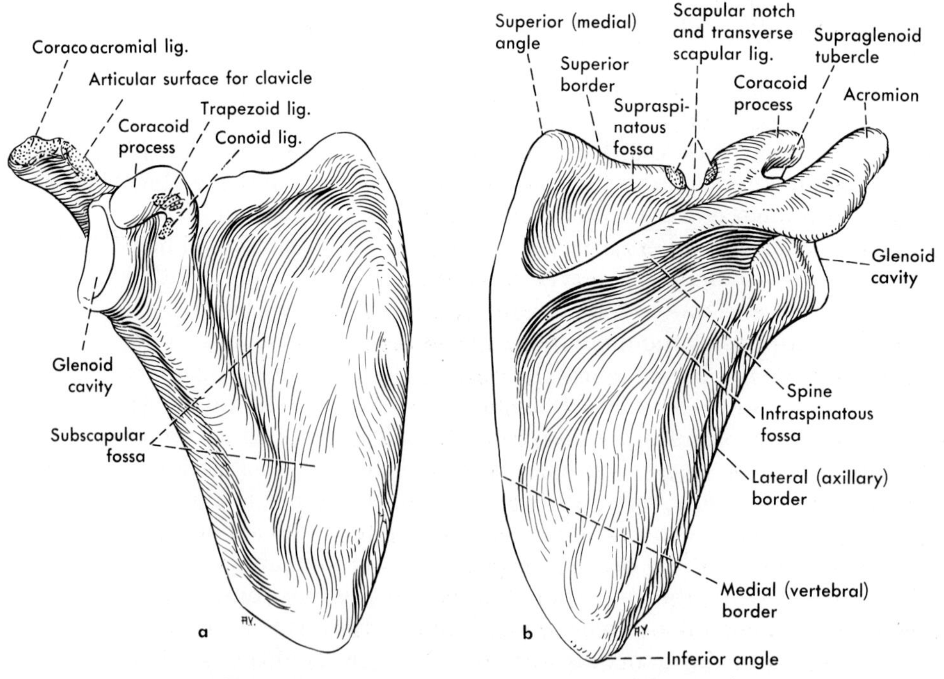

Figure 5–4. Costal (*a*) and dorsal (*b*) views of the scapula.

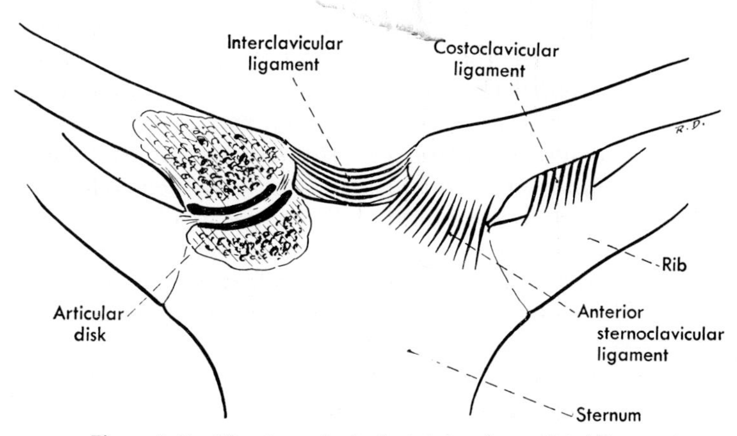

Figure 5–5. The sternoclavicular joint and associated ligaments.

ly in man, the subscapular fossa. These fossae are occupied by muscles bearing the same names: supraspinatus, infraspinatus, and subscapularis. The spine of the scapula is prolonged as a free projection, the acromion, which forms the point of the shoulder and articulates with the clavicle. The acromion is connected to the coracoid process by the coracoacromial ligament (fig. 5-6); acromion and ligament thus form an arch above the shoulder joint. The scapula is largely suspended by muscles; its only articulations are with the humerus and clavicle.

The **clavicle** is a long bone, therefore roughly cylindrical, and has a slight S-shaped curve. Its sternal (medial) end is somewhat expanded and fits poorly into the notch on the manubrium sterni, while its acromial end is expanded and flattened and articulates with the acromion of the scapula. Since the clavicle acts primarily to keep the free limb free, or away from the body, it must be attached firmly at both ends; since it must allow movement of the scapula, it must also possess joints at both ends.

The **sternoclavicular joint** (fig. 5-5) contains two synovial cavities which are separated by a fibrocartilaginous articular disk. While the adjacent surfaces of the clavicle and sternum do not fit too well together, this articular cartilage allows the joint to move somewhat like one of the ball-and-socket type, for up-and-down, forward-and-backward, and rotatory movements are all allowed. In up-and-down movement, the freest, the clavicle moves on the disk as on a hinge, while in the other movements the disk moves with the clavicle. The joint includes also a small portion of the first rib as it attaches to the sternum. The sternoclavicular joint slants in such a way that medial thrust on the clavicle would tend to displace its sternal end upward and medially; also, downward movement of the shoulder tends to bring the clavicle against the first rib and, if continued, to raise the sternal head of the clavicle from its bed, using the first rib as a fulcrum. These movements of dislocation are resisted by several ligaments. The anterior and posterior sternoclavicular ligaments reinforce the capsule, the posterior being the stronger, and are directed downward and slightly medially from clavicle to sternum, so they help prevent both upward displacement and the lateral displacement which tends to occur if the arm is pulled. A similar function is served by the costoclavicular ligament, which runs downward and medially between the clavicle and the first rib, and by the articular disk. This is attached below to the first rib and above to the clavicle, so that any upward and medial movement of the sternal head of the clavicle produces tension on the disk. It tears away from the rib, however, if the posterosuperior part of the capsule, which offers most of the resistance against upward dislocation, is cut. Lateral displacement is prevented not only by the sternoclavicular and costoclavicular ligaments, as already noted, but also, to some extent, by the interclavicular ligament, which extends from one clavicle to the other across both joints and has also some attachment to the sternum.

The **acromioclavicular joint** is small and its gliding surfaces are so sloped as to favor over-riding of the acromion by the clavicle. The joint capsule itself is of little strength, and the scapula could be easily displaced medially beneath the clavicle if it were not for the coracoclavicular ligament (fig. 5-6). This strong ligament is divided into two parts, the more medial and posterior part being the conoid ligament and the more lateral and anterior the trapezoid ligament. Both parts of this ligament prevent medial displacement of the scapula; the conoid

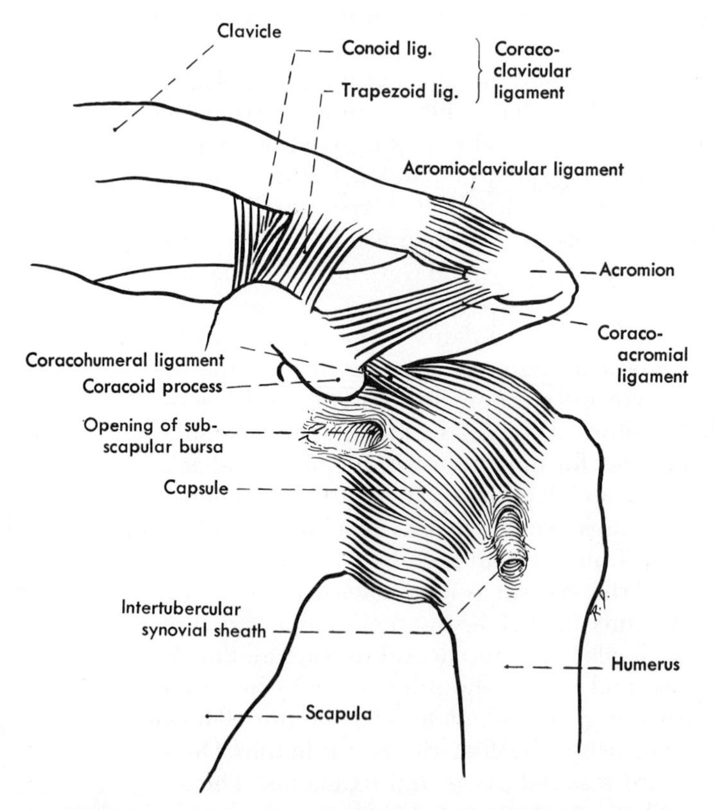

Figure 5–6. Ligaments of the shoulder joint and distal end of the clavicle.

passes upward and slightly backward from the coracoid process to the clavicle and therefore resists also forward movement of the scapula without corresponding movements of the clavicle, while the trapezoid resists independent backward movement of the scapula. In forcible dislocation at this joint ("shoulder separation") the coracoclavicular ligament is usually torn, as are the muscles (trapezius and deltoid) attaching across the joint.

The importance of the synovial joints at both ends of the clavicle can be easily demonstrated by noting the manner in which the shoulder blade and clavicle move together as the shoulders are raised or lowered, or thrust forward or backward. In all these movements the clavicle has to move rather freely at the sternoclavicular joint. The small gliding movement allowed at the acromioclavicular joint is necessary because the lateral angle of the scapula necessarily follows the clavicle, thus describing an arc of which the clavicle is the radius, while the medial border of the scapula follows the different curve of the thoracic wall, to which it is closely held by muscles. Thus constant adjustment at both ends of the clavicle is necessary for the scapula to move smoothly. These movements of the scapula, in turn, greatly increase the mobility of the shoulder joint, as they result in alterations of the position of the glenoid cavity on which the head of the humerus moves.

To understand the shoulder joint and the muscles acting across it, the upper end of the **humerus** (figs. 5-2 and 6-2) must be studied. The humerus is the long bone of the arm, and consists of a body (shaft) and two expanded ends. The smooth articular part of the upper end of the humerus is the head; a marked prominence on the anterior surface of the upper end is the lesser tubercle, which is clearly separated from the more lateral greater tubercle by the intertubercular groove. Below the tubercles, this groove is bordered by crests that extend downward from each tubercle. The anatomic neck of the humerus is at the point of junction of the head with the body, and lies in part between the head and the tubercles. The surgical neck, so called because of the more frequent occurrence of fractures here, lies below both the head and tubercles and is a narrow, not clearly delimited, portion of the upper part of the body. (The remainder of the humerus, and its participation in the elbow joint, is described on page 113.)

The **shoulder joint** is formed by the articular surfaces of the glenoid cavity and the head of the humerus. It is surrounded by a thin articular capsule of relatively little strength, except where this capsule is somewhat strengthened above by a thickening known as the coracohumeral ligament; internally there are two or three very slightly thickened bands on its anterior wall. The fibrous capsule of the shoulder joint is attached proximally to the glenoidal labrum (p. 78), and by this to the edge of the glenoid cavity, but is very often deficient anteriorly close to the labrum, where the synovial cavity of the joint may communicate with a subscapular bursa lying on the costal surface of the scapula (p. 97). Distally the capsule is attached to the anatomic neck of the humerus, but between the tubercles it extends downward as a thin-walled tube, the intertubercular synovial sheath, that surrounds the tendon of the long head of the biceps brachii muscle (p. 118). Through this sheath the tendon enters the shoulder joint and thereafter runs through its cavity to an origin on the upper edge of the glenoid cavity.

The glenoidal labrum is often torn partially away from the edge of the glenoid cavity when there has been repeated dislocation of the shoulder, but it is also frequently torn in older persons who have never had dislocations. While the coracohumeral ligament will support the weight of the arm hanging by the side, additional weight evokes muscular action; moreover, even slight abduction of the arm releases the ligament, rendering it useless during almost all movements. Thus the chief strength of the shoulder joint lies in certain muscles and tendons that are closely applied to the capsule anteriorly, above, and posteriorly and that together are called the musculotendinous or rotator cuff of the shoulder (p. 95).

The relatively free movements of the ball-and-socket or spheroid joint of the shoulder are limited by the short muscles about the shoulder joint, as well as by the tubercles and the overhanging acromion process. On the other hand, the apparent range of movement at the shoulder is greatly increased by movements of the scapula, as is also the strength of the movements of the arm.

SUPERFICIAL NERVES AND VESSELS; FASCIA

The skin over the shoulder is supplied by a number of nerves. Anteriorly and laterally the supraclavicular nerves, branches of the cervical plexus (mostly C3 and C4), pass downward over the clavicle to supply the skin over the upper part of the chest and the top of the shoulder. The skin of the pectoral region is supplied by branches of intercostal nerves (ventral branches of thoracic nerves). The skin over the shoulder muscles on the back is supplied partly also by branches of intercostal nerves, partly by the dorsal branches of cervical, thoracic and lumbar spinal nerves. The skin over the lateral side of the upper part of the arm is supplied by a cutaneous branch of the axillary nerve, while that of the floor of the axilla and the medial side of the upper part of the arm is supplied by a branch (intercostobrachial nerve) from the second or the second and third intercostal nerves, and by a small branch from the medial cord of the brachial plexus.

The only superficial vessel of any size in the shoulder region is the upper end of the cephalic vein; this lies between the deltoid and pectoralis major (muscles shown in figure 5-7) and passes deeply between the two in the infraclavicular fossa.

The subcutaneous tissue or superficial fascia in the pectoral region contains a variable amount of fat, and also encloses the glandular tissue of the breast, a gland of the skin which has expanded into the subcutaneous tissue. Elsewhere about the shoulder the superficial fascia is less well developed and is fused with the deep fascia.

The deep fascia of the shoulder presents few features worthy of particular description. Essentially, it simply splits to surround each

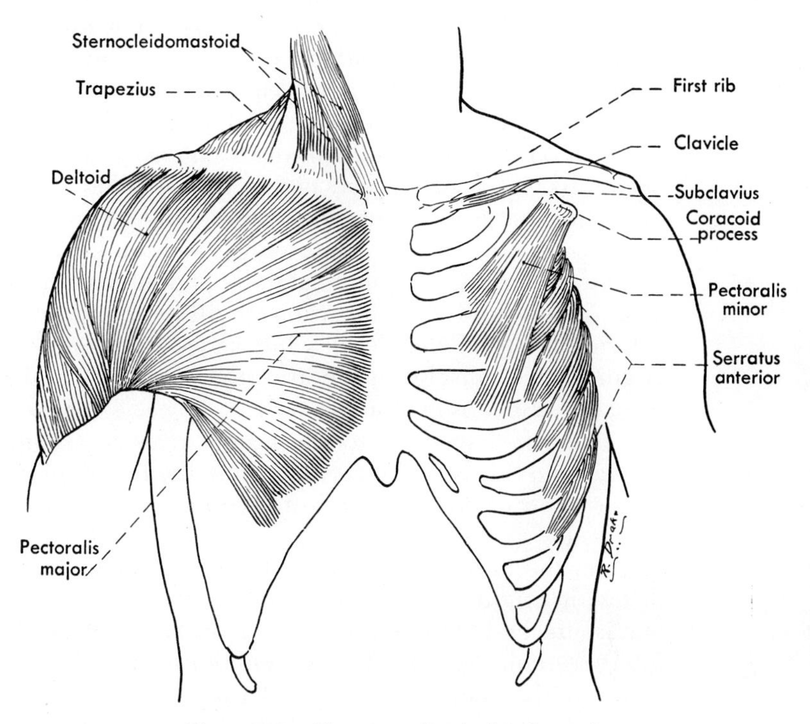

Figure 5–7. The pectoral and related muscles.

structure it encounters, and then unites again into a single layer on the other side of that structure. It is attached to various bony prominences and in certain regions – for instance, the supraspinatous and infraspinatous fossae – gives attachment to some of the fibers of the underlying muscles. The fascia on the deep surface of the pectoralis major muscle contains the larger nerves and vessels to the muscle. A special layer of fascia surrounding the pectoralis minor muscle, under cover of the major, and extending up to the clavicle is the clavipectoral fascia (p. 85).

MUSCLES, NERVES AND VESSELS OF THE SHOULDER

The Pectoral Region

The two **pectoralis major** muscles are the large muscles covering most of the upper part of the chest; each forms the anterior wall of the axilla (armpit) as it extends across to attach to the humerus. After the skin and fascia over the thorax and anterior aspect of the upper part of the arm are removed, the relations of the pectoralis major can be observed (fig. 5-7). At its attachment to the clavicle the muscle is covered by the thin platysma ("flat") muscle, lying mostly in the fascia of the neck but extending downward over the clavicle (not shown in fig. 5-7). Attached to the upper border of the clavicle above the clavicular origin of the pectoralis major is a more important muscle of the neck, the sternocleidomastoid. The pectoralis major lies medial to the anterior portion of the deltoid muscle, the large muscle below the point of the shoulder. Usually there is a distinct groove between these two muscles, occupied by a superficial vein of the arm, the cephalic, which serves as a landmark for the boundary line between them. The pectoralis minor may or may not be visible at the lower lateral edge of the major; the external oblique muscle and its aponeurosis (fig. 23-1) and the serratus anterior muscle (fig. 5-7) can be identified if the dissection is carried far enough laterally and inferiorly. As the pectoralis major is followed to its insertion it will be found to pass deep to the anterior fibers of the deltoid but across the front of the conjoined origin of two muscles of the arm (the coracobrachialis and the short head of the biceps brachii, fig. 6-5).

The fan-shaped pectoralis major arises from approximately the medial two thirds of the clavicle, from the length of the sternum and, under cover of this portion, from the upper six costal cartilages, and, finally, by a small slip from the aponeurosis of the external oblique muscle, the most superficial of the lateral abdominal muscles. From this wide origin the muscle bundles converge to a tendon of insertion that is attached to the crest of the greater tubercle, or lateral lip of the intertubercular groove. The tendon of insertion is bilaminar; the tendon from the fibers of the clavicular portion of the muscle blends with the tendon from the upper part of the sternocostal portion to form an anterior lamina; the fibers of the lower sternocostal and the abdominal parts pass upward behind the insertion of the upper portion to form the posterior layer of the pectoralis tendon, the lowest fibers being inserted highest on the humerus.

The pectoralis major is supplied by the medial and lateral pectoral nerves. The medial pectoral nerve, innervating the lower part of the muscle, arises from the medial cord of the brachial plexus (hence its name of medial) and runs around the lateral border of the pectoralis minor (the muscle deep to the major) or pierces that muscle to enter the lateral part of the pectoralis major. It brings into the muscle nerve fibers from the eighth cervical and first thoracic nerves. The lateral pectoral nerve arises from the lateral cord of the plexus, and contains fibers derived from the fifth through seventh cervical nerves. It runs forward into the upper part of the pectoralis major with pectoral branches of the thoracoacromial artery (the first major branch of the axillary artery, figs. 5-9 and 5-10), passing above the pectoralis minor and therefore perforating the fascia (clavipectoral) stretching from this muscle to the clavicle. In addition to the branches that the thoracoacromial artery sends into the pectoralis major, other branches emerge above the insertion end of the muscle to run toward the tip of the shoulder and also downward with the cephalic vein between the deltoid and pectoralis major muscles, supplying both. Deep to the pectoralis major a twig of the thoracoacromial artery runs toward the sternoclavicular joint; and on the thoracic wall, approximately along the lateral border of the pectoralis minor muscle, are the lateral thoracic vessels. (The artery arises also from the axillary, usually below the origin of the thoracoacromial, but sometimes with that vessel.)

The muscle as a whole adducts the arm, and brings it forward and medially across the chest. The clavicular fibers, working alone, add an upward movement of flexion, such as is involved in touching the lobe of the opposite ear. The lower fibers of the sternocostal portion depress the arm and may therefore through it depress the shoulder. The sternocostal portion of the muscle, acting alone, will extend the arm—bring it downward—if the arm is already flexed, but cannot hyperextend it (carry it backward beyond its normal position at the side). Also, since the muscle as a whole crosses the front of the arm to insert lateral to the intertubercular groove, it will, upon contraction, tend to move the groove medially, thus producing medial rotation of the humerus. The pectoralis major is occasionally congenitally absent, and all or most of it is removed in radical amputation of the breast. The approximately normal movement of the arm across the chest in the absence of the pectoralis major indicates the extent to which other muscles, especially the anterior portion of the deltoid and the coracobrachialis, can substitute for the functions of the missing muscle.

The **pectoralis minor** (also fig. 5-7) is a small rather triangular muscle lying under cover of the major. It is surrounded by a thin fascial sheath, the clavipectoral fascia, whose anterior and posterior layers come together at the upper and lower edges of the muscle. Below and lateral to the pectoralis minor, the clavipectoral fascia is continuous with the fascia covering the superficial muscle (serratus anterior) of the anterolateral thoracic wall; above and laterally it joins fascia that forms the floor of the axilla. From the upper and medial border of the pectoralis minor the clavipectoral fascia extends to the clavicle, being especially tough laterally. It is pierced by the lateral pectoral nerve and the large vessels to the pectoralis major. As it reaches the clavicle it divides to go on both sides of the small subclavius muscle, which lies between the clavicle and the first rib.

The muscle arises from about the third to the fifth ribs, the origin varying somewhat, crosses the front of the axilla where it is in close contact with the vessels and nerves here, and inserts upon the coracoid process of the scapula behind its tip (which gives rise to muscles of the arm). The pectoralis minor is supplied by fibers from the medial pectoral nerve, and is usually pierced by the portion of this nerve that continues to the pectoralis major. The pectoralis minor depresses the shoulder and since it acts close to the lateral angle may aid in downward rotation of the scapula.

The pectoralis minor is supplied by branches of the various arteries appearing on the thoracic wall, notably the thoracoacromial and lateral thoracic arteries from the axillary.

The Axilla

The contents of the axilla, or armpit, can be more easily approached after reflection of the pectoralis major and minor. The axilla itself varies somewhat in shape and size depending upon the position of the arm. With the arm slightly abducted it can be regarded as a space in the form of a somewhat misshapen truncated pyramid, the base of which is formed by the skin and fascia extending from the arm to the thoracic wall. The pectoralis major and minor form the anterior wall of this space, and the serratus anterior, a muscle on the lateral thoracic wall (fig. 5-8) forms the medial wall. The posterior wall is formed by shoulder muscles, the latissimus dorsi and teres major below and the subscapularis above, while the narrow lateral wall is the intertubercular groove of the humerus. The misshapen apex of this pyramid lies between the first rib, the clavicle and the upper edge of the subscapularis muscle, and through this apex pass the great nerves and vessels of the upper limb.

Arranged around the chief arterial stem of the upper limb, termed the axillary artery during its course in the axilla, is the lower end of the brachial plexus (see also p. 76 and fig. 5-3). The lateral cord of the brachial plexus can be identified lateral to the artery, and the medial and posterior cords also have the relations to the artery indicated by their names (fig. 5-9) except that as they emerge from behind the clavicle the posterior cord is at first lateral to the artery, while the medial cord, unlabeled in figure 5-9, lies at first behind the artery. The formation of these cords must be examined in a dissection of the neck, but many of their branches can be identified in the axilla. After giving off the lateral pectoral nerve at about the level of the clavicle, the **lateral cord** gives off the musculocutaneous nerve, which penetrates the coracobrachialis muscle; then, as the lateral root of the

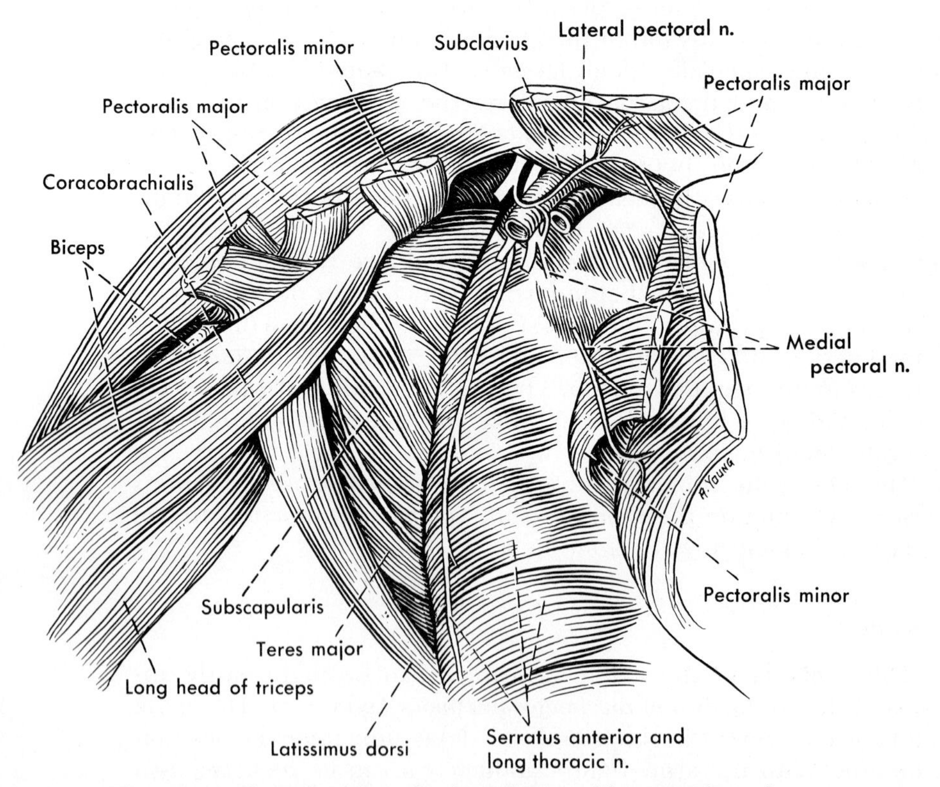

Figure 5–8. The walls of the axilla after reflection of the pectoral muscles.

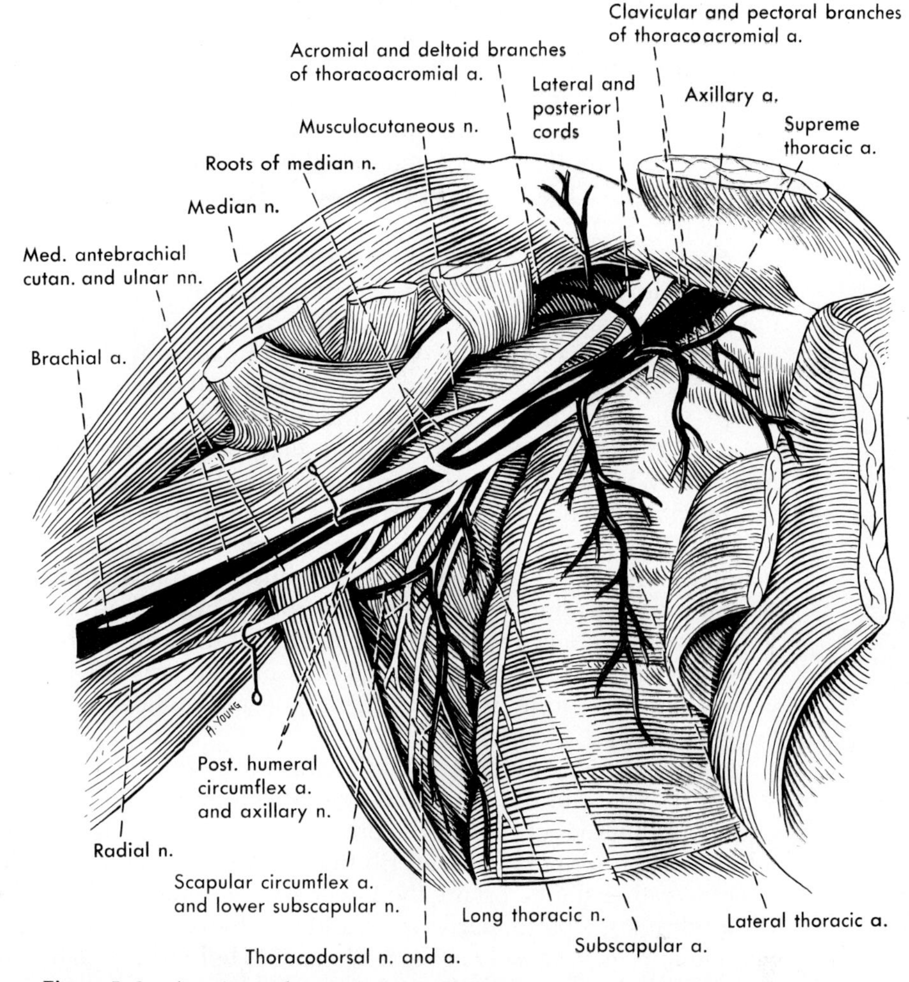

Figure 5–9. Arteries and nerves of the axilla. The muscles shown here can be identified from Figure 5–8.

median nerve, the remainder of the lateral cord joins with a similar contribution from the medial cord to form the median nerve.

The **medial cord** gives off the medial pectoral nerve, then two cutaneous branches, a tiny upper one that supplies skin on the medial side of the arm and a second larger one that runs downward to supply skin of the forearm. The medial cord then ends by dividing into the ulnar nerve, which passes down the arm slightly posterior to the artery, and the medial root of the median nerve, which crosses in front of the axillary artery to join the lateral root of the median nerve. The median nerve is thus formed anterolateral to the axillary artery, and its two roots form a loop embracing this vessel.

The **posterior cord** lies on the surface of the subscapularis muscle behind the axillary artery and gives off in succession an upper subscapular nerve, a nerve to the latissimus dorsi, and a lower subscapular nerve, after which it divides into radial and axillary nerves. The axillary nerve immediately passes posteriorly around the surgical neck of the humerus; the radial nerve gives off small motor and cutaneous branches in the axilla and then disappears posteriorly deep to the triceps muscle (the muscle of the back of the arm, p. 121). In addition to these branches of the brachial plexus in the axilla, the long thoracic nerve can be found running downward on the surface of the serratus anterior (fig. 5-8) to supply this muscle. This nerve arises high in the neck from some of the nerves giving origin to the brachial plexus (fig. 5-3).

Because of its position between the first rib and the clavicle, the brachial plexus may be brought against these or other structures about the axilla and stretched over them, with consequent damage to nerve function. A familiar example of temporary impairment of the function of the fibers of the brachial plexus is that involved in the arm "going to sleep" when one lies in bed with the arm above the head. Here the brachial plexus is stretched over the clavicle and the head of the humerus. While the discomfort ordinarily resulting from such abuse to the brachial plexus is mild, and there are no permanent sensory or motor effects, exaggerated positions of the arm under some conditions may lead to more permanent damage. Thus, the brachial plexus may be injured by undue stresses exerted upon it while an individual is under anesthesia, or by abnormal postures of the arm maintained by the faulty application of casts or splints. Constant pressure in the armpit, as from a splint or a crutch, may also produce injury to the brachial plexus.

Chronic injury to the brachial plexus and the subclavian-axillary stem at the base of the neck as they pass into the axilla has been ascribed to a number of causes: the presence on the seventh cervical vertebra of an abnormal (cervical) rib that the neurovascular bundle has to cross, rather than crossing the lower-lying normal first rib (cervical rib syndrome); a particularly broad and tendinous or a spastic anterior scalene muscle, a muscle in the neck (p. 355) behind which the plexus and the subclavian artery lie (scalenus anticus syndrome); pinching of the plexus between the clavicle and the first rib (costoclavicular syndrome); and various other conditions. These are sometimes grouped together as the "thoracic outlet (or inlet) syndrome," that is, the symptoms resulting from injury at the upper border of the thorax; a term often preferred now is "neurovascular compression." Most of them have one thing in common, namely, that symptoms are brought on or increased by habitually carrying the shoulder lower than normal, so that the plexus is subjected to abnormal stretch and pressure. In many such cases the patients have been reported to be relieved or cured by appropriate physical therapy, a toning up of the elevators of the scapula so that the patient carries his shoulder higher. This avoids the surgical treatment that is otherwise necessary.

The **axillary artery** (fig. 5-10) is the direct continuation of the subclavian and continues into the arm as the brachial artery. Thus what is actually the same artery receives different names according to where it lies. The upper boundary of the axilla is defined as the first rib, the lower boundary as the lower border of the teres major; the axillary artery therefore extends between these two parts. The axillary artery

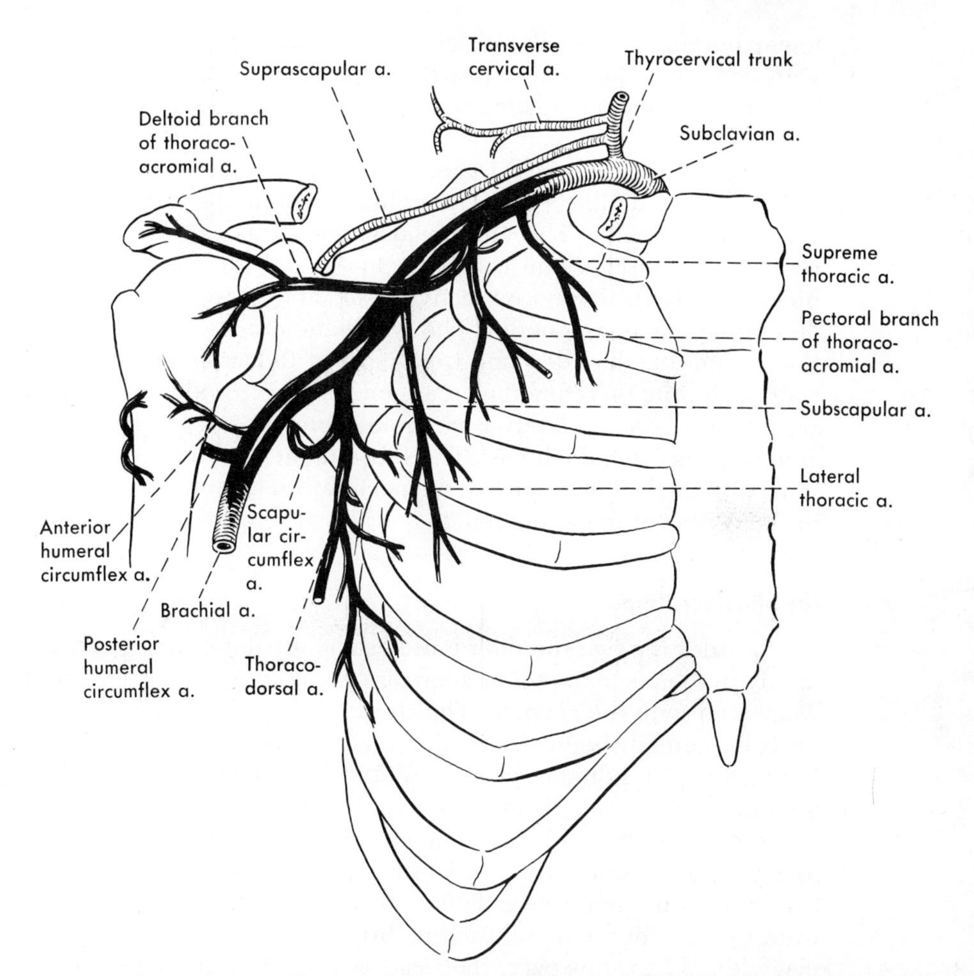

Figure 5–10. Arteries of the pectoral region and shoulder. The axillary artery and its branches are black, other vessels are shaded. The suprascapular and transverse cervical arteries may arise directly from the subclavian artery, or the branches of the transverse cervical may arise separately, one from the subclavian and the other from the thyrocervical trunk.

gives off branches to the thoracic wall and its covering muscles, to the shoulder, and to the uppermost part of the arm. It has six named branches. Of these, the supreme thoracic artery is a small artery to the upper thoracic wall; the thoracoacromial supplies primarily the pectoral muscles, the anterior part of the deltoid, and the joints at both ends of the clavicle; and the lateral thoracic artery supplies both the thoracic wall and the pectoral muscles, especially the pectoralis minor. The subscapular artery is the largest branch of the axillary and through its scapular circumflex branch (fig. 5-13) supplies muscles on the dorsal surface of the scapula. Its continuation downward on the thoracic wall is the thoracodorsal artery (fig. 5-10). The last two branches of the axillary are the anterior and posterior humeral circumflex arteries of which the anterior is small, while the larger posterior one encircles the surgical neck of the humerus with the axillary nerve (fig. 5-13).

The **veins** in the axilla are somewhat variable. A large superficial vein, the basilic, has its name changed to axillary vein as it crosses the lower border of the teres major to enter the axilla. Into the axillary vein come two brachial veins, the cephalic vein, and various deep branches corresponding approximately to arterial branches of this region.

Most of the complex nerves and vessels thus far described are surrounded by a tube of fascia brought down from the neck and termed the axillary sheath. The additional space in the axilla is occupied by parts of muscles (the coracobrachialis and the short head of the biceps – p. 118), and by connective tissue and fat in which are imbedded numerous lymph nodes. The lymphatic drainage from the upper limb, from the shoulder, and from most of the anterolateral thoracic wall, including therefore the breast, ends in these nodes. These nodes are removed as completely as possible in one type of operation for carcinoma (cancer) of the breast. The subsequent swelling (edema) of the limb, and the fact that much of the pectoralis major muscle is removed in the course of the operation, may make physical therapy necessary.

The Shoulder Proper

Besides the two shoulder muscles already described as lying primarily in the pectoral region, there are two anteriorly located muscles connected with the clavicle. The **sternocleidomastoid,** a muscle in the neck, has already been observed as it attaches to the clavicle (fig. 5-7). It has also a tendinous head of origin from the sternum, and runs obliquely upward and backward across the neck to attach to the prominent mastoid process behind the ear (fig. 21-10). If this muscle takes its fixed point from below it will so pull upon the skull as to turn the face toward the opposite side and at the same time flex the neck toward the side of the muscle acting (bringing the ear down toward the clavicle). When, however, the head is fixed by contraction of other muscles attaching to the skull, the sternocleidomastoid acts upon the clavicle and sternum to raise them. Through this action it becomes an accessory respiratory muscle.

The sternocleidomastoid is innervated by the accessory or eleventh cranial nerve. This nerve runs obliquely downward and backward to reach the muscle only an inch or so below the mastoid process; it supplies the muscle as it passes either through it (most common) or deep to it, and then continues its oblique course toward the trapezius muscle. Fibers of spinal nerves, usually from C2, also enter the muscle either separately or after joining the accessory nerve. The accessory nerve is motor only, while the cervical nerve fibers to the sternocleidomastoid are probably all sensory.

The **subclavius** muscle (fig. 5-7) arises by a short tendon from the first rib and passes laterally and upward to a muscular insertion on the lower surface of the clavicle. It can perhaps slightly depress the clavicle, or help in raising the first rib as the clavicle is raised. Since it runs markedly laterally to reach its insertion, it aids in retaining the sternal end of the clavicle in place. Aside from these actions, the muscle also sometimes affords protection to the subclavian artery in fractures of the clavicle, its muscular belly intervening between the artery and the possibly sharp edges of the fractured bone. The subclavius is supplied by a tiny branch from the upper trunk of the brachial plexus, which can be found only by careful dissection of the neck.

Most of the remaining shoulder muscles can best be studied from the back. Two large, flat ones, the trapezius and the latissimus dorsi, between them cover almost the entire back, extending from the skull to the sacrum and crest of the ilium (fig. 5-11). The **trapezius,** the upper of these muscles, has an extensive origin from the midline of the back, including the ligamentum nuchae (the "ligament of the back of the neck," extending from the skull to the prominent vertebral spinous process at the base of the neck), the lowest cervical, and all the thoracic spinous processes. It often also attaches directly to and lateral to the external occipital protuberance of the skull. From this wide origin the muscle converges to a more limited insertion on the spine of the scapula, the acromion, and the lateral third of the clavicle. The acces-

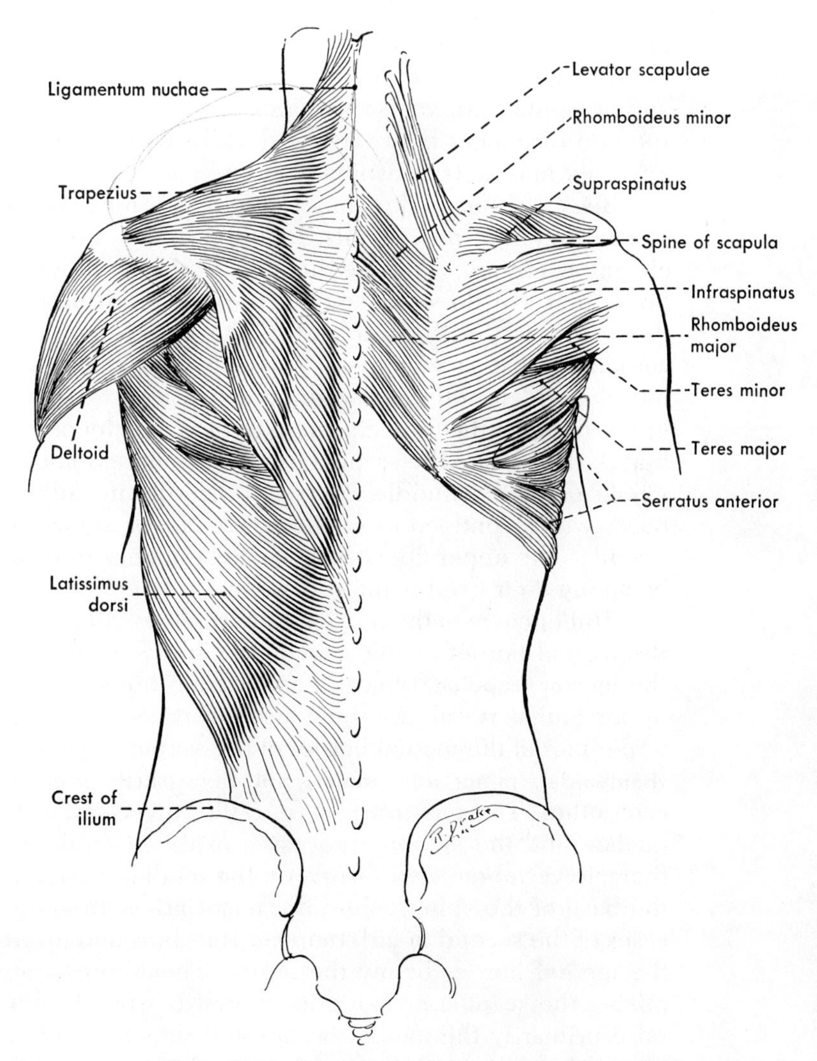

Figure 5–11. Musculature of the shoulder from behind.

sory nerve runs (with a branch of the transverse cervical artery) on the deep surface of, and ends in, this muscle after supplying the sternocleidomastoid, and branches of the third and fourth cervical nerves also go to it. As in the case of the sternocleidomastoid, the fibers of the accessory nerve are known to be motor; while there has been much argument concerning the function of the fibers from the cervical nerves, it seems probable that these are all sensory.

Two of the vessels to the shoulder, the transverse cervical and suprascapular arteries (fig. 5-10), arise in the neck. Whether they both originate from the thyrocervical trunk, or one arises from the subclavian artery directly, they follow a similar course: they run laterally and posteriorly, with the transverse cervical artery above and the suprascapular artery behind the clavicle, to disappear deep to the trapezius. Under cover of this muscle the transverse cervical artery divides into a superficial and a deep branch (or if the branches arise separately, as the superficial cervical and dorsal scapular arteries, they separate from each other). It is the superficial branch that runs on the deep surface of the trapezius, while the deep or dorsal scapular runs on the deep surface of the rhomboidei with the dorsal scapular nerve. The suprascapular artery is accompanied by the corresponding nerve, and the two disappear between the dorsal surface of the scapula and its covering muscle (supraspinatus).

Because of its wide origin the trapezius muscle can help carry out several different movements. Its upper fibers, inserting on the clavicle and acromion, can raise the point of the shoulder and are the only fibers that can do so directly. (The part inserting on the clavicle is thin, so it is the thickest part of the muscle, which inserts on the acromion, that is effective.) Working with the lower fibers of the muscle, which pull downward on the base of the scapular spine, the upper fibers of the trapezius help to turn the glenoid cavity upward—that is, they rotate the scapula upward. Contraction of the muscle as a whole or of the middle fibers only results in pulling the shoulder back, while contraction of the lower fibers alone will depress the scapula. The upper fibers can also flex the neck toward the same side, by taking their fixed point from below.

Under cover of the trapezius are three smaller muscles attached to the medial border of the scapula (fig. 5-11). The upper one of these, the **levator scapulae,** typically arises from transverse processes of the upper four cervical vertebrae and inserts on the superior angle and upper part of the medial border of the scapula. Below this are the two **rhomboidei, minor** and **major,** not necessarily clearly distinct from each other. The minor arises from the lower part of the ligamentum nuchae and the spinous processes of the seventh cervical and first thoracic vertebrae, and inserts on the medial border of the scapula at the base of the spine, while the major arises from the spinous processes of the second to fifth thoracic vertebrae and inserts on the rest of the medial border below the minor. These three muscles all aid in raising the scapula or fixing its medial border. Acting together, they raise primarily the medial border and thus produce downward rotation of the glenoid cavity; the rhomboidei also retract the scapula. The nerve supply to the rhomboidei is through the small dorsal scapular nerve (figs. 5-3 and 5-12) arising from the fifth cervical. This nerve runs transversely across the neck, paralleling the transverse cervical and suprascapular arteries, passes deep to or through the levator scapulae, and then runs with the deep cervical or dorsal scapular artery on the deep surface of the rhomboids very close to the medial border of the scapula. The levator scapulae is sometimes supplied also, in part, by the dorsal scapular, but is supplied chiefly by small twigs from the third and fourth cervical nerves into its anterior surface.

The **latissimus dorsi** (fig. 5-11) arises from about the lower six thoracic and all the lumbar and sacral spinous processes and from a posterior portion of the crest of the ilium by a broad aponeurosis of origin that covers the back muscles. Fleshy slips from the lower four ribs join the deep surface of the muscle. It converges to a relatively narrow, flat tendon of insertion which, passing across the posterior wall of the axilla, runs forward around the medial surface of the humerus to insert on the crest of the lesser tubercle (medial lip of the intertubercular groove), and into the groove. As it does so, it twists so sharply around an underlying muscle (teres major) that its original anterior surface is directed posteriorly. The muscle is an extensor, internal rotator and adductor of the arm, a movement used, for instance, in chopping wood, in the overhand swimming stroke, and so forth. In

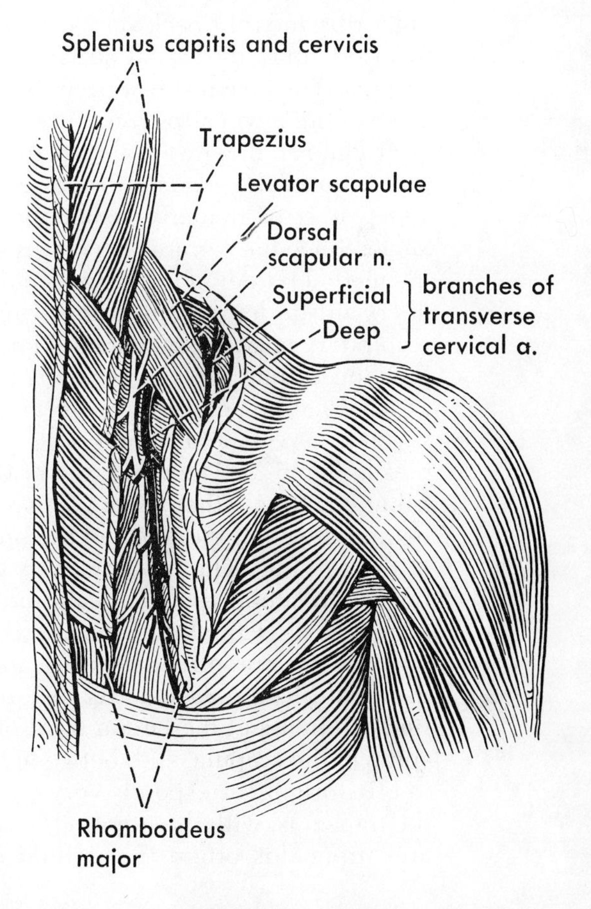

Figure 5–12. The dorsal scapular nerve and transverse cervical artery.

contrast to the lower portion of the pectoralis major, with which it works in the first part of these movements, the latissimus dorsi will carry the arm backward beyond its position at the side, that is, will hyperextend it. Through its action on the arm the latissimus dorsi can also depress the shoulder. While the latissimus and the lower part of the pectoralis major form a sort of anteroposterior sling from the trunk to the free limb, the latissimus is much the more powerful and important component of this sling in extending the arm and depressing the shoulder. For instance, "chinning" one's self is possible when the latissimus dorsi is intact but the lower part of the pectoralis major is damaged, yet becomes impossible if the latissimus alone is gravely weakened. Similarly, one cannot walk on crutches unless the latissimus dorsi is functioning to prevent the shoulder from being pushed up by the weight on the crutch. The nerve supply to the latissimus is through the thoracodorsal nerve (fig. 5-9), a branch from the posterior cord of the brachial plexus, which transmits fibers derived primarily from C7 and C8. It lies at first on the costal surface of the subscapularis muscle, on which it runs downward to the deep or costal surface of the latissimus. The chief artery entering the muscle is the thoracodorsal, a branch of the subscapular that runs on the deep surface of the muscle and also helps to supply the adjacent serratus anterior.

The **serratus anterior** muscle (figs. 5-7, 5-8, 5-9) arises on the anterolateral thoracic wall by slips (digitations) from about the upper eight ribs and runs backward, closely applied to the curve of the chest, to insert upon the costal surface of the entire medial border of the scapula. The heaviest insertion, however, is on the inferior angle, and some of the lower slips run upward to reach this while others, of higher origin, run downward to this insertion. The muscle as a whole is a protractor of the scapula; because it protracts especially the inferior angle, it is particularly important in upward rotation of the lateral angle. Since the serratus curves around the thoracic wall, it also keeps the medial border of the scapula closely applied to this wall. Its lower fibers aid in depression of the scapula; although once regarded as a respiratory muscle, since it could raise the ribs, it apparently does not function as one. It is innervated by the long thoracic nerve, which arises from the anterior primary branches of about the fifth, sixth and seventh nerves before these enter into the formation of the brachial plexus (fig. 5-3). The chief root of the nerve is usually from C6; contributions from either C5 or C7 may be lacking. This nerve runs down behind the other elements of the brachial plexus, on the outer surface of the serratus. The blood vessels supplying the muscle are those of the anterolateral thoracic wall and of the scapular region (fig. 5-9), primarily the lateral thoracic and thoracodorsal arteries.

The most prominent intrinsic muscle of the shoulder (arising from the girdle) is the **deltoid** (figs. 5-7 and 5-11). This muscle arises anteriorly from about the lateral third of the clavicle, posteriorly from the spine of the scapula, and, between these two origins, from the acromion. Its origin corresponds very closely, therefore, to the insertion of the trapezius, with the upper fibers of which it works in abducting the arm; from this origin the deltoid converges to its insertion on the lateral surface of the humerus, about its middle. Since the fibers of the deltoid pass in front of, lateral to, and behind the shoulder joint, this muscle has several actions here. The middle fibers raise the arm away from the side, that is, abduct it. The anterior fibers working alone will flex and internally rotate the humerus, while the posterior fibers will extend and externally rotate it. Finally, the lower fibers of both anterior and posterior parts of the muscle may be brought into play in forcible adduction of the arm, although it has been suggested that this is primarily a protective action against the downward displacement of the humerus that the more powerful adductors, the latissimus dorsi and pectoralis major, tend to produce. The muscle is supplied by the axillary nerve (C5 and 6) and is fed by the posterior humeral circumflex artery (fig. 5-13) as well as by other vessels in the neighborhood. The nerve and artery pass below the subscapularis and teres minor muscles, and above the teres major, and circle anteriorly close against the surgical neck of the humerus.

There are five muscles that arise entirely from the scapula and insert on the humerus. Three of them attach to the greater tubercle and form the upper and posterior parts of the "rotator cuff" or "musculotendinous cuff" about the shoulder joint. The **supraspinatus** arises from the dorsal surface of the scapula above the spine (supraspinatous fossa) and from the fascia covering the muscle; it passes over the top of the shoulder joint to insert on the upper part of the greater tubercle. Between this muscle and the overhanging acromion there is an important subacromial (subdeltoid) bursa. The **infraspinatus** muscle

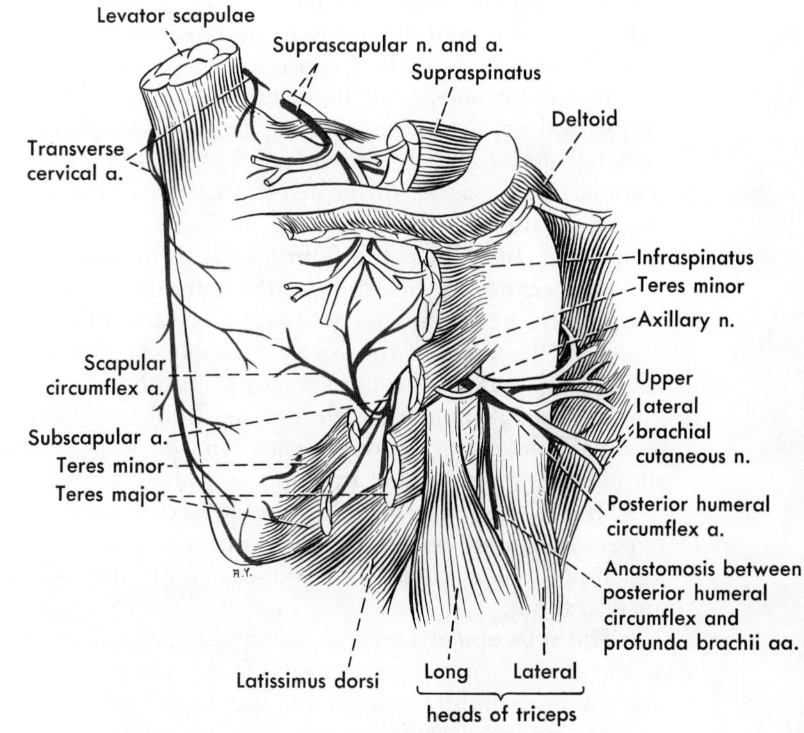

Figure 5–13. Posterior nerves and vessels of the shoulder.

arises from its covering fascia and from the infraspinatous fossa and inserts on the greater tubercle directly below the insertion of the supraspinatus muscle. The **teres minor** arises from about the upper two thirds of the dorsal surface of the lateral border of the scapula and from septa between it and the infraspinatus above, the teres major below, and inserts upon the greater tubercle directly below the insertion of the infraspinatus. These muscles and the following, the teres major, are shown in figure 5-11.

The supraspinatus is primarily an abductor of the arm, assisting the deltoid in this action. The infraspinatus and teres minor are primarily external rotators of the humerus, and are also important in maintaining the head of the humerus in position during other movements of the arm. The supraspinatus and infraspinatus muscles are supplied by the suprascapular nerve (C5 and 6). As this nerve reaches the superior border of the scapula it passes through the scapular notch to run between the bone and the supraspinatus muscle, and then continues deep to the spine of the scapula to reach the infraspinatus muscle. The teres minor receives a branch from the axillary nerve, given off as this nerve passes below the muscle on its way to the deltoid. The suprascapular artery, arising anteriorly at the base of the neck, accompanies the suprascapular nerve to these muscles (fig. 5-13); it passes above a ligament across the scapular notch, but otherwise has a course similar to that of the nerve. The scapular circumflex artery rounds the lateral border of the scapula by passing through the origin of the teres minor, and ramifies in the infraspinatous fossa. The suprascapular, transverse cervical (particularly its deep or dorsal scapular branch), and scapular circumflex arteries anastomose freely with each other and thus provide an alternative route by which blood from the subclavian artery can reach the axillary artery.

The **teres major** is at its origin closely associated with the teres minor and infraspinatus, arising from septa between it and these muscles and from the dorsal surface of the inferior angle of the scapula. As it passes to its insertion it is, however, separated from the teres minor by the long head of the triceps muscle (not shown in figure 5-11, but labeled in figure 5-13). It becomes closely associated with the tendon of insertion of the latissimus dorsi, a bursa usually intervening between the two, and passes with this latter muscle to insert on the crest of the lesser tubercle, or medial lip of the intertubercular groove. The teres major is supplied by the lower subscapular nerve (C5 and 6), which arises from the posterior cord of the brachial plexus, runs downward on the subscapularis muscle to supply the lower part of this and continues into the teres major (fig. 5-9). In its action it resembles the latissimus dorsi, being an extensor, hyperextensor, internal rotator, and adductor of the humerus, but assisting the latissimus in these movements only when there is resistance to them.

The **subscapularis** muscle (fig. 5-25*b*) arises from most of the costal aspect of the scapula (subscapular fossa) and passes across the front of the shoulder joint to insert on the lesser tubercle and its crest. A subscapular bursa, usually opening into the synovial cavity of the shoulder joint, intervenes between the muscle and the neck of the scapula. The insertion of the muscle forms the anterior part of the rotator or musculotendinous cuff of the shoulder. This muscle is supplied by the upper and lower subscapular nerves from the posterior cord of the brachial plexus (fig. 5-3); the upper nerve passes directly into the muscle, while the lower one runs down to supply the lower part and the teres major. The muscle is an internal rotator, and is particularly important in preventing anterior dislocation of the head of the humerus; laxity of it, or even tears, are usually found in operations for recurrent anterior dislocation.

Aside from minor variations of origin and insertion, usually not particularly altering their functions, the muscles of the shoulder are rather constant. Occasional cases of absence of various muscles – pectoralis major, pectoralis minor, trapezius, rhomboids – are reported in the literature, but the very fact that they are deemed worthy of reporting indicates their rarity. The more common supernumerary muscles of this region are the sternalis, small paired or unpaired muscles on the anterior thoracic wall superficial to the pectoralis major, and an axillary arch muscle. This latter consists of muscle fibers that tend to arch across the axilla, hence its name. While it may assume various forms and attachments, one simple type is a bundle of fibers connecting the latissimus dorsi and pectoralis major.

MOVEMENTS OF THE SHOULDER

Scapular Movements

Movements at the shoulder joint proper, that is, between the humerus and scapula, are ordinarily accompanied by movements of the scapula itself. The coordinated movement of both elements is sometimes referred to as the scapulohumeral rhythm, and disturbances of the normal rhythm are typical of certain lesions about the shoulder region. Many of the muscles acting across the shoulder joint are short ones that attach close to the upper end of the humerus and therefore do not have the leverage that could be obtained by a lower insertion. Movements of the scapula increase the force of arm movements, and also, by tilting the glenoid cavity in the desired direction, increase the range of movement of the free limb. As the arm is abducted, for instance, the deltoid and the supraspinatus are obviously the active movers at the shoulder joint. Accompanying this abduction there is also an upward rotation of the glenoid cavity, variably reported as being 1 degree of rotation for each 2 degrees of abduction, and as 2 degrees for each 3 degrees of abduction. This upward rotation is brought about by the lower part of the serratus anterior and the upper and lower parts of the trapezius. In a similar way, movements of extension and flexion at the shoulder typically involve both scapular and humeral movement. Actually, fairly good use of the arm may persist in almost total destruction of the shoulder joint, scapular movements in this case substituting for the normal combined action of both scapula and humerus.

Of the muscles acting on the scapula some, as we have seen, act directly upon it through their attachment here while others act primarily through their attachment to the humerus. There are only four muscles that are capable of elevating the scapula (fig. 5-14). The upper fibers of the trapezius inserting upon the spine and acromion of the scapula, and upon the clavicle, are solely responsible for elevation of the lateral angle of the scapula; the levator scapulae and the two rhomboids are so attached that they can act only upon the medial border. In paralysis of the trapezius muscle (injury to the accessory nerve) the lateral angle of the scapula, having nothing to support it, is dragged downward by the weight of the free limb; the weight of the entire limb is then thrown upon the levator scapulae and the rhomboids which, reflexly increasing their activity in response to this greater stretch, contract to produce excessive elevation of the superior angle of the scapula.

In depression of the shoulder the pectoralis minor, subclavius, and latissimus dorsi, and lower fibers of the trapezius, serratus anterior, and pectoralis major, may all participate (fig. 5-15). The pectoralis minor tends to rotate the scapula downward at the same time, while the serratus anterior tends to rotate it upward. The subclavius,

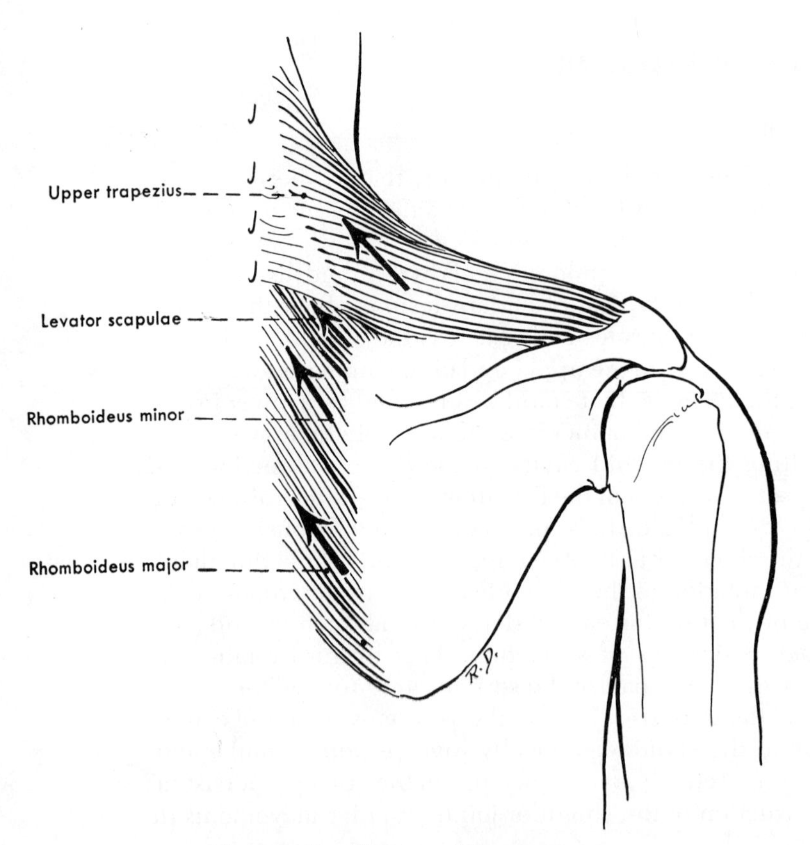

Figure 5–14. Elevators of the scapula.

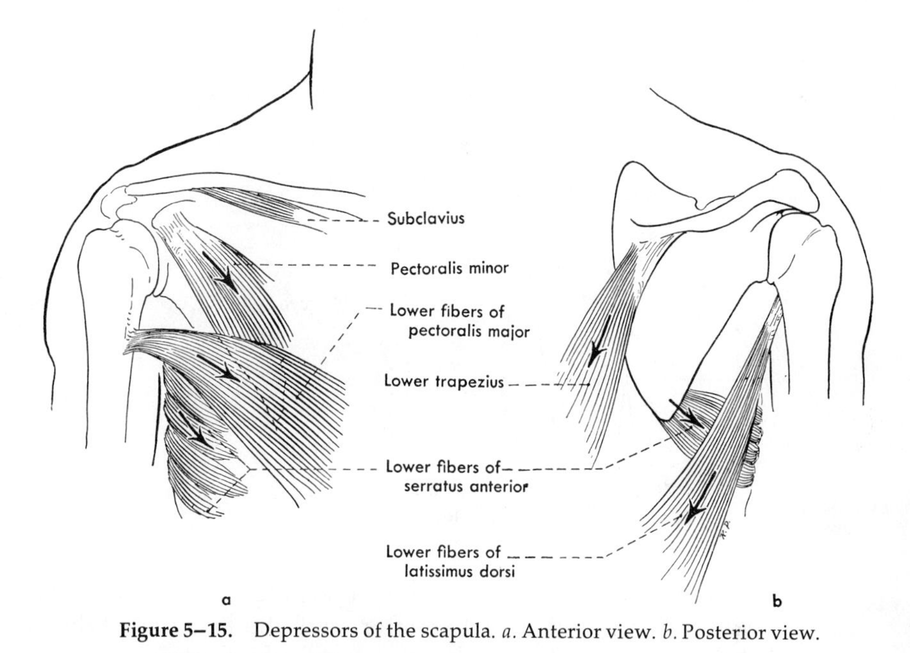

Figure 5–15. Depressors of the scapula. *a*. Anterior view. *b*. Posterior view.

while listed as a depressor, can actually have little of such effect upon the shoulder, both because of its size and its very oblique position. The lower fibers of the trapezius tend to retract the scapula as they depress it; so does the latissimus dorsi, which, like the pectoralis major, exerts its effect on the scapula through the short muscles attaching the humerus to the scapula. The lower fibers of the pectoralis major tend to protract the scapula as they assist in depressing it. An apparently simple movement such as depression of the shoulder may involve most of the muscles of the shoulder, either as prime movers, or as fixators to prevent rotation and maintain the contact between the glenoid cavity and the head of the humerus.

Upward rotation of the scapula, necessary to allow abduction of the arm above the horizontal, is carried out by the combined actions of the trapezius and the serratus anterior (fig. 5-16). The upper fibers of the trapezius pull upward on the clavicle and acromion, the lower fibers pull downward on the base of the spine. Through their strong insertion on the inferior angle, the lower fibers of the serratus anterior pull this portion of the medial border laterally and forward, and are therefore very important in upward rotation of the lateral angle.

While some of these lower fibers of the anterior serratus tend also to depress the scapula, this tendency is overcome, in upward rotation, by the contraction of the upper part of the trapezius. Obviously, this movement also demands a delicate distribution of action among several shoulder muscles, in order to prevent the scapula as a whole from being dragged anteriorly and inferiorly as it is rotated. Usually, upward rotation of the scapula is accompanied by elevation of this bone, rather than depression, thus assisting the arm to reach higher. However, when the trapezius is paralyzed there is first a depression and

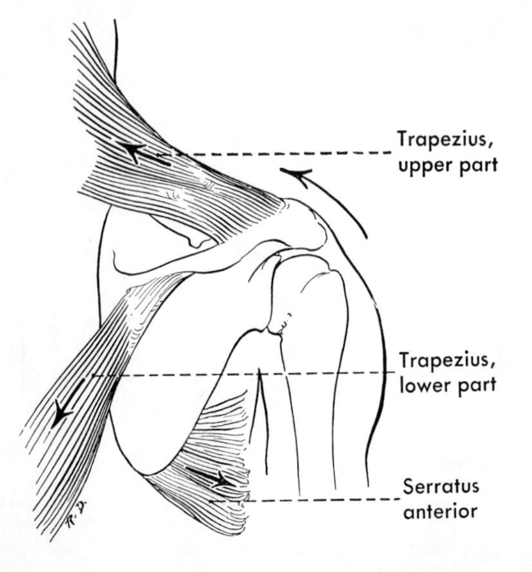

Figure 5–16. Upward rotators of the scapula.

downward rotation produced by the weight of the arm and the pull of the serratus, and only thereafter, with the levator and rhomboids stabilizing the medial border, does upward rotation occur. This reverses, at least in part, the downward rotation but does not actually turn the glenoid cavity upward, so that abduction even to the horizontal is frequently not obtainable.

The opposite movement of downward rotation is brought about through the action of the rhomboids and the levator scapulae in raising the medial border of the scapula while the pectoralis minor, the pectoralis major and the latissimus dorsi, aided also by the effect of gravity on the free limb, pull down the lateral angle (fig. 5-17).

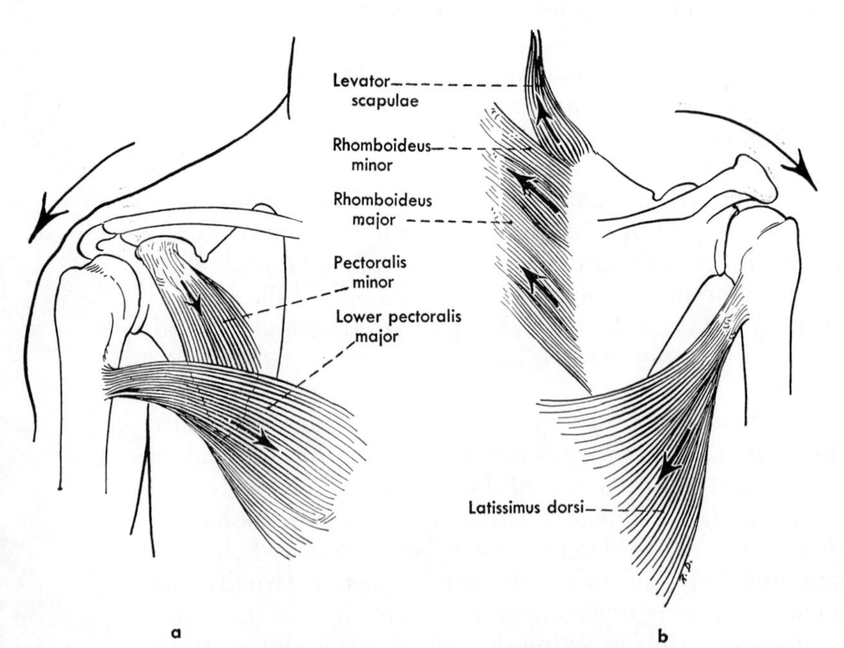

Figure 5–17. Downward rotators of the scapula. *a*. Anterior view. *b*. Posterior view.

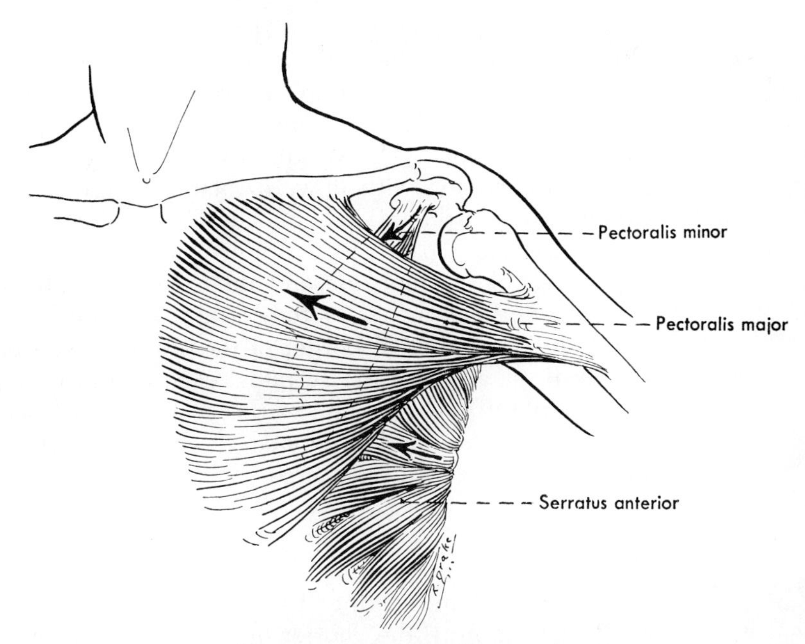

Figure 5. Protractors of the scapula.

Downward rotation of the scapula is usually associated with depression—as, for instance, in reaching down to pick up a suitcase.

Protraction of the scapula is brought about by the serratus anterior and by the pectoralis major and minor (fig. 5-18). Retraction is due to the middle fibers of the trapezius or to the trapezius acting as a whole, to the rhomboids, and to the latissimus dorsi (fig. 5-19).

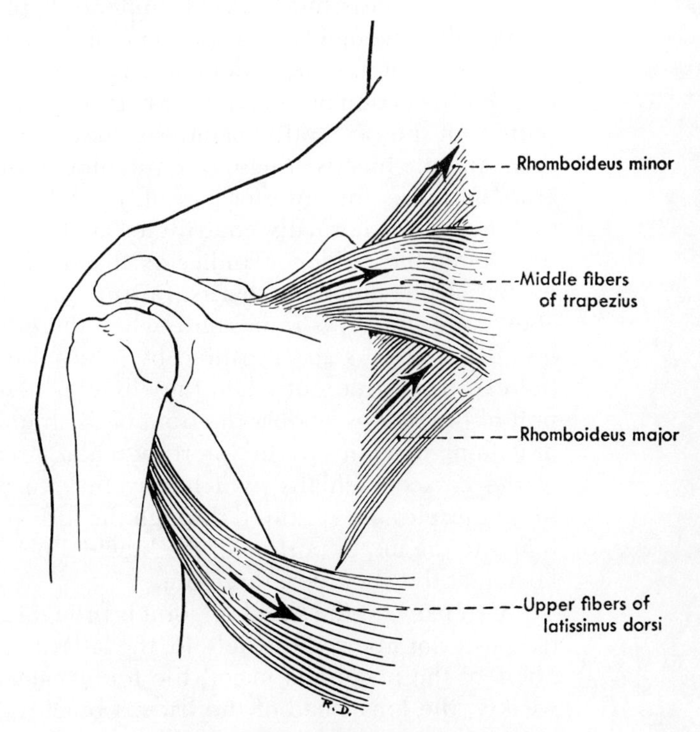

Figure 5–19. Retractors of the scapula.

For most of these movements of the shoulder a number of muscles cooperate. Since they belong to several different muscle groups and have rather widely separated innervations, marked interference with shoulder girdle movements through injury to a single nerve is uncommon. The striking exceptions relate to simple elevation or to elevation with upward rotation. Elevation of the scapula without downward rotation of the lateral angle is impossible when the upper trapezius is paralyzed; similarly, upward rotation is weakened by paralysis of the trapezius, and almost abolished by paralysis of the serratus anterior. Further, as the serratus anterior is also responsible for retaining the medial border of the scapula close against the thoracic wall, paralysis of the serratus leads to a projection of this border, described as "winging" of the scapula. This becomes obvious when the arms are held horizontally forward, and even more so if the person pushes against resistance.

Humeral Movements

The musculature acting at the shoulder joint can be divided into two general groups, the shorter ones that act primarily to retain the humerus in its socket and to rotate it therein, and the longer ones that are responsible for much of the free movement between humerus and glenoid cavity. With the arm by the side, downward displacement of the humerus is resisted by the coracohumeral ligament, assisted if necessary by the supraspinatus and the posterior fibers of the deltoid. During flexion or abduction, however, the ligament is relaxed, and it is the short muscles, the supraspinatus, infraspinatus, teres minor, and subscapularis that prevent humeral displacement; they contract during all movements of flexion and abduction.

Flexion at the shoulder joint can be brought about (fig. 5-20) through the action of the anterior portion of the deltoid, the clavicular portion of the pectoralis major, the coracobrachialis (a muscle of the arm), and the biceps brachii (the prominent muscle on the front of the arm). Of these, the anterior part of the deltoid is the most important; that the biceps normally contributes at all has been denied, but electromyographic evidence indicates that it probably does. Complete flexion at the shoulder, that is, raising one's limb forward until it is above one's head, is impossible when the elbow is kept straight unless the flexion is accompanied by internal rotation of the humerus, but can be carried out when the elbow is bent so as to diminish the pull of the biceps against the front of the humerus, where the tendon of its long head lies in the intertubercular groove. With the exception of the coracobrachialis, which may receive fibers from C7 also, all these muscles are supplied through the fifth and sixth cervical nerves; injury to the upper portion of the brachial plexus may markedly affect flexion at the shoulder, therefore.

Extension at the shoulder joint is brought about (fig. 5-21) through the posterior fibers of the deltoid, the latissimus dorsi, the sternocostal fibers of the pectoralis major, the teres major against resistance and, weakly, the long head of the triceps brachii (the muscle on the pos-

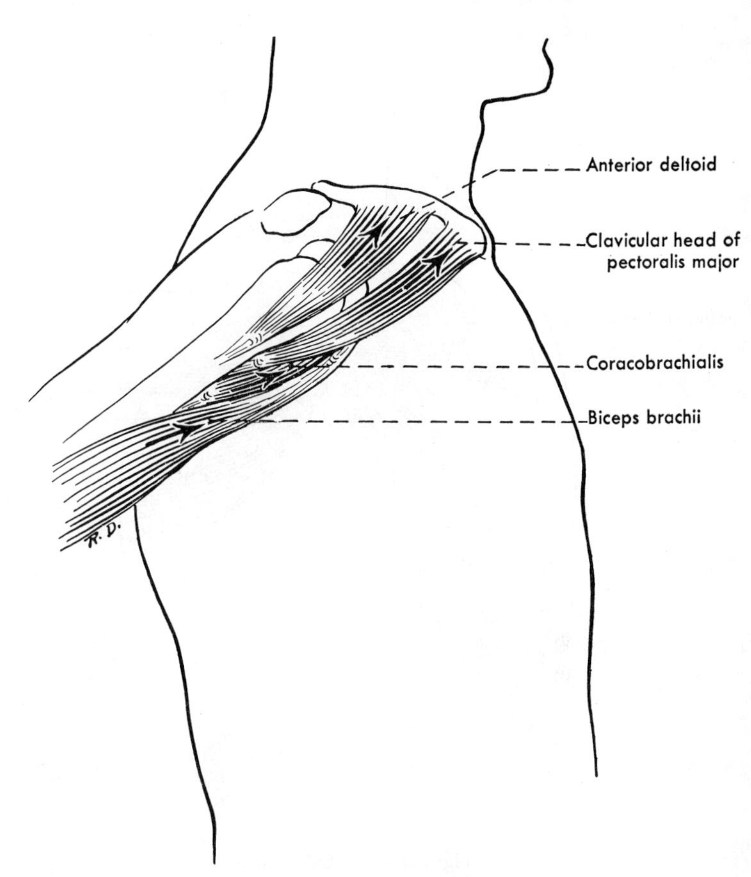

Figure 5–20. Flexors of the arm.

terior aspect of the arm). While the lower fibers of the pectoralis major can assist this movement only to the extent of bringing the flexed arm downward until it reaches the side, they are nevertheless an important contributor to such extensor actions as bringing an axe downward, the pull of a swimming stroke, or "chinning" one's self. The posterior fibers of the deltoid can draw the arm farther back than can any of the other muscles, making possible such movements as placing the hand into a back pocket. The segmental nerves involved in extension of the arm are all those contributing to the brachial plexus.

Abduction is brought about (fig. 5-22) by the simultaneous action of the deltoid, especially its middle, or more lateral, part, and by the supraspinatus. With rotation of the humerus the anterior or posterior parts of the deltoid are brought into a more lateral position so that they abduct more strongly. Abduction in external (lateral) rotation is stronger than it is in internal (medial) rotation; since the movement is weakest from internal rotation, weakness of the deltoid is most easily demonstrated by testing abduction from this position. Of the two muscles, the more powerful deltoid can produce full abduction—to about 90 degrees when there is no accompanying scapular rotation; the supraspinatus sometimes can, but more frequently cannot, carry out good abduction when the deltoid is paralyzed. External rotation of

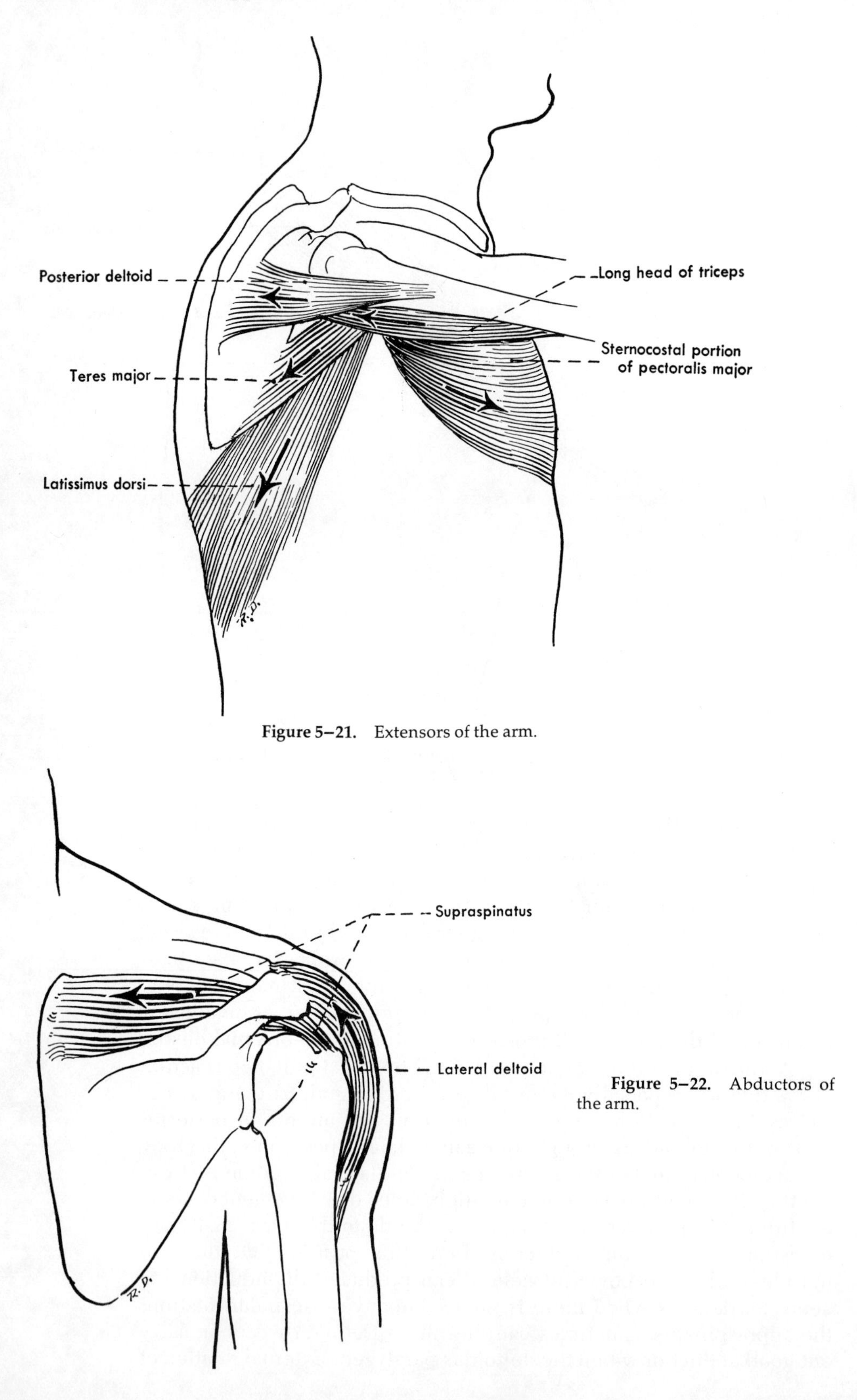

Figure 5–21. Extensors of the arm.

Figure 5–22. Abductors of the arm.

the humerus always accompanies complete abduction of the arm. This, apparently, is necessary to allow the greater tubercle to slide under, rather than hit against, the acromion.

The two abductor muscles are innervated exclusively through the fifth and sixth cervical segments and therefore abduction, like flexion, is easily interfered with by lesions of the upper portion of the brachial plexus.

The pectoralis major, the latissimus dorsi, and the teres major are the chief adductors of the arm; the coracobrachialis, and the long head of the triceps to a small extent, also assist. Further, the deltoid, the chief abductor, can also aid in adduction. Because the posterior fibers are lower, they can assist in adduction while the arm is some 45 degrees from the side, but the anterior fibers cannot help until the arm is fairly close to the side, since only then do they lie below the axis of motion at the shoulder joint. (It has also been said that the anterior fibers act only when there is simultaneous flexion, and that the posterior fibers contract primarily to prevent the pectoralis major and latissimus from internally rotating or depressing the humerus.) The muscles composing the adductor group (fig. 5-23) are innervated through fibers arising from all elements of the brachial plexus.

Medial or internal rotation is brought about (figs. 5-24 and 5-25) primarily by the subscapularis. The pectoralis major and the latissimus dorsi medially rotate as they adduct, or flex and extend, respectively, and the clavicular fibers of the deltoid medially rotate as they flex. The teres major apparently contracts, although somewhat weakly, for pure medial rotation. The medial rotators are innervated through all segments contributing to the brachial plexus. Lateral or external rotation is carried out (figs. 5-24 and 5-26) by the infraspina-

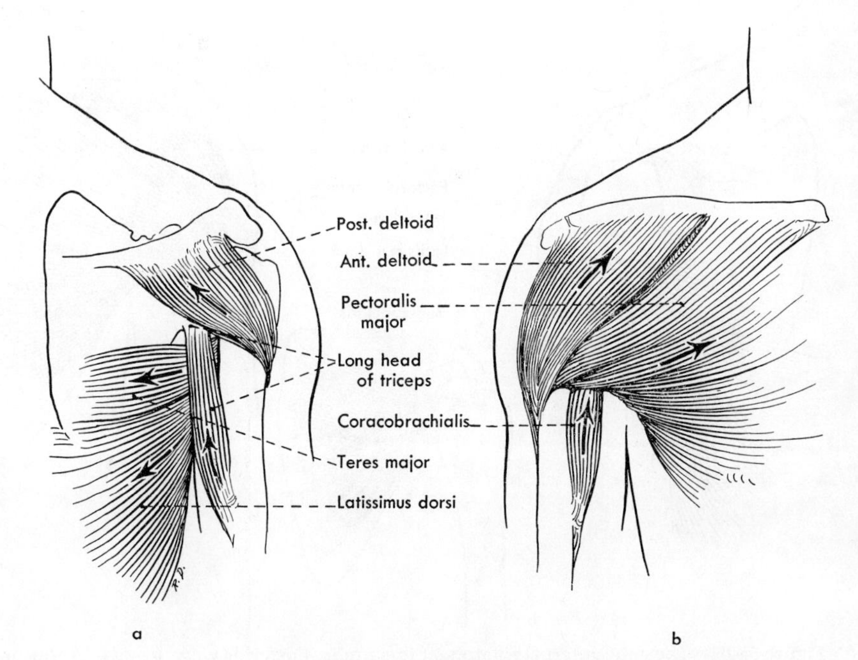

Figure 5–23. Adductors of the arm. *a*. Posterior view. *b*. Anterior view.

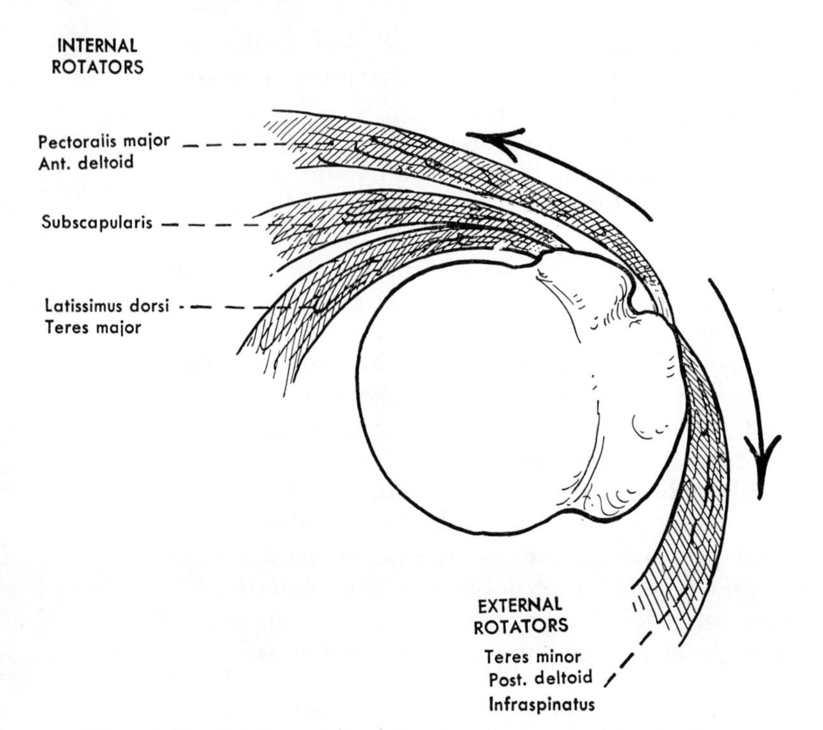

Figure 5–24. Relations of the external and internal rotators to the upper end of the humerus. Right arm, viewed from above.

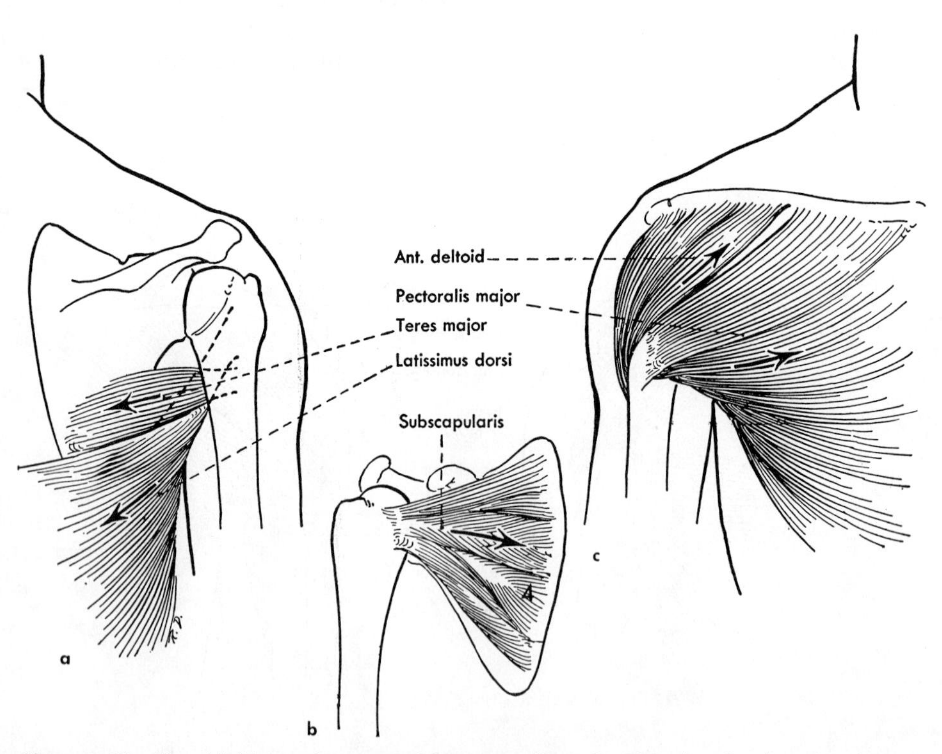

Figure 5–25. The chief internal rotators of the arm. *a*. Posterior view. *b* and *c*. Anterior views.

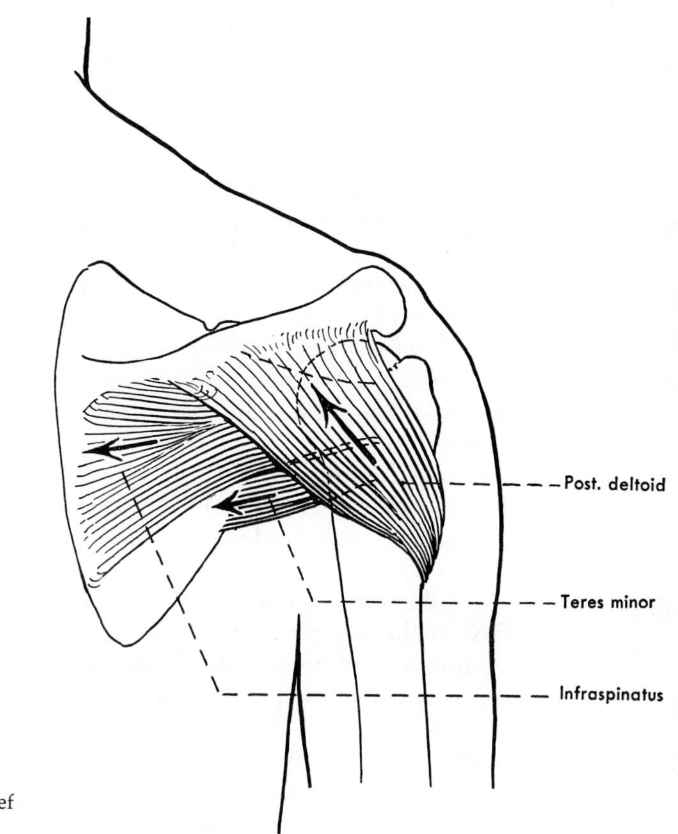

Figure 5–26. The chief external rotators of the arm.

tus and teres minor, and by the posterior fibers of the deltoid if extension and lateral rotation are combined. These three muscles are innervated through the fifth and sixth cervical nerves.

From the comments on the innervation of these various muscle groups it should be clear that injury to the upper elements of the brachial plexus (that is, to the fifth and sixth cervical nerves or the upper trunk, or to the seventh nerve also)—sometimes called "Erb's paralysis"—may involve all the muscles acting as flexors, abductors and lateral rotators, while only some of the muscles involved in extension, adduction and medial rotation can be implicated. The injured limb in such cases will therefore be extended, adducted, and medially rotated.

The manner in which many muscles may cooperate in what at first sight appears to be a very simple movement has already been commented upon, and is well illustrated by an analysis of the movement of abduction of the arm. The primary muscles in this movement are only two, the supraspinatus and the deltoid. However, abduction of the arm demands that the humeral head be held firmly in its socket. Otherwise the action of the deltoid upon the humerus, with the arm by the side, would raise the head of the humerus rather than abduct the limb, and as the arm is abducted its weight would tend to dislocate the head of the humerus downward. Thus the infraspinatus, the subscapularis, and the teres minor all contract to help retain the head of the humerus in the socket. Finally, as has already been pointed out, abduction always involves an upward rotation of the scapula, and therefore the cooperation of the muscles involved in this movement. Abduction at the scapulohumeral joint and rotation of the scapula go on simultaneously in an almost constant ratio (p. 97), with no more than a little irregularity in the scapulohumeral rhythm at the beginning and end of the movements. In upward rotation of the scapula the serratus anterior acts with the upper and lower fibers of the trapezius, while the smoothness of the movement is aided by the levator scapulae, especially, and to a lesser extent by the rhomboids. Thus about ten muscles assist directly in abduction of the arm—and, if you have ever had a strained back, you will probably recall that even the back muscles indirectly participate during movements of the shoulder!

A consideration of shoulder movements would not be complete without reference to the role of the subacromial (subdeltoid) bursa. There may be two, a subacromial and a subdeltoid, but they function as one and frequently are fused; whether fused or not, subacromial bursa is the more common name. This bursa rests upon the upper surface of the supraspinatus muscle, intervening between it and the overlying deltoid muscle, acromion process, and coracoacromial ligament. Any upward movement of the humerus tends to force the head and greater tubercle, with the covering supraspinatus muscle, against the arch of the acromion and coracoacromial ligament. The bursa between the muscle and the arch acts therefore as an accessory joint cavity of the shoulder, allowing for movements of the upper end of the humerus beneath the arch and being especially important, therefore, in abduction.

Calcification or other lesions in tissues adjoining the shoulder joint or the subacromial bursa will lead to painful crippling of humeral movements. Tears of almost any of the muscles about the shoulder may occur, but the two most common lesions of this type here are tears of the musculotendinous cuff (insertions of subscapularis, supraspinatus, infraspinatus, and teres minor), especially of the supraspinatus portion, and rupture of the long head of the biceps. While rupture of shoulder muscles may be largely traumatic, those involving tendons are typically preceded by degenerative changes of these tendons. Degenerative changes in the supraspinatus are apparently often initiated by impingement of its tendon against the edge of the acromion, while those of the long head of the biceps result from wear against irregularities of the bone in the intertubercular groove.

REVIEW OF THE SHOULDER

In review and summary of the shoulder, the bony landmarks (p. 72) should be palpated upon one's self or a classmate, and as many of the muscles as can be recognized by inspection or palpation should be examined while movements designed to produce contraction of each individual muscle and of any part that acts differently from another are being carried out. Reference to the descriptions of the various muscles, and to the text and figures pertaining to specific movements, should be sufficient to indicate what movements will best bring out the various muscles and their parts.

It should be remembered that muscular action is never any stronger than is required to bring about the desired movement, and some muscles that can help bring about the movement will not contract at all unless they are needed. Thus resistance to the movement, most easily produced in many cases by having the subject push against resistance offered by the observer's hand, will bring out muscular contraction both more plainly and more completely. However, it must also be remembered that contraction of a muscle does not necessarily mean it is bringing about the movement, for it may be contracting synergistically to prevent some other, undesired, movement. It is not always possible to determine which muscles are prime movers and which are synergists, but a knowledge of their anatomy will often allow the decision to be made.

Of the **bones,** the external occipital protuberance, the spinous processes of the vertebrae, and the upper borders (iliac crests) of the hip bones should all be identified in examining the back. The spine of the scapula can also be easily identified and traced to the acromion at the tip of the shoulder, although there is no way of locating exactly the junction of these two continuous parts. The medial border of the scapula and its inferior angle can be palpated with no difficulty and are usually visible when the subject pushes with his arm and hand against resistance. The other parts of the scapula are covered by too much muscle to be readily identifiable, except the coracoid process, which is palpable anteriorly just below the clavicle between the thorax and the shoulder. The clavicle can be easily felt and seen. The various parts of the sternum can also be identified with no difficulty—the sternal angle marks the junction of manubrium and body, and the xiphoid process can be felt at the lower end of the sternum, where the arches of the ribs of the two sides come together.

Of the **muscles** of the shoulder, some are easily seen or felt or both, while others are difficult or impossible to identify on the living person. The pectoralis major can be easily identified on the front of the thorax; in the male it is largely responsible for the contour of the pectoral region, and in both sexes it is the chief component of the anterior axillary fold. In muscular individuals the slips of origin of the serratus anterior from the ribs can be easily seen on the anterolateral thoracic wall below the pectoralis major. Both the pectoralis minor and the subclavius lie too deeply (the minor behind the major, the subclavius under cover of the clavicle) to be recognizable, nor is there any way to test specifically their actions.

On the back, the lateral border of the upper part of the trapezius as it runs from the neck to the shoulder can be both seen and felt; atrophy of the muscle is easily recognized because of the change in contour. Because of its flatness, however, other parts of the muscle are difficult to identify. Similarly, only a lateral part of the latissimus dorsi, as it extends toward the posterior axillary fold, is clearly recognizable. Much of the muscle is too flat, and in the axillary fold it is difficult or impossible to distinguish between the latissimus and the teres major, since the muscles are closely applied together and have in general the same actions. The teres major, however, forms the larger bulk of the musculature in the fold.

The deltoid is easily recognizable as it gives shape to the junction of shoulder and arm; it is a particularly favorable muscle in which to demonstrate different actions of various parts. The supraspinatus is sometimes visible as it produces a slight outward bulging of the trapezius immediately above the scapular spine, but is difficult to palpate distinctly because when it contracts in abduction the overlying upper part of the trapezius also contracts. Because the deltoid also abducts, there is no good way of estimating the strength of the supraspinatus.

The other muscles intimately attached to the scapula—the infraspinatus, teres minor, and subscapularis—are not usually identifiable in the living person. However, since the first two are pure external rotators and the subscapularis is a pure internal rotator, the strength of these muscles can be estimated by observing the strength of rotation of the humerus, taking care that neither flexion nor extension is attempted at the same time. Carrying out rotation of the arm with the

TABLE 5–1. Nerves and Muscles of Shoulder

NERVE		MUSCLE	
Name	*Origin**	*Name*	*Chief action*
Accessory	Cranial	Sternocleidomastoid	Lateral flexion and rotation of head
		Trapezius	Elevation of tip of shoulder
Nn. to levator scapulae	C3, 4	Levator scapulae	Elevation of scapula
Dorsal scapular	C5	Both rhomboidei	Retraction of scapula
N. to subclavius	C5, 6	Subclavius	Depression of clavicle?
Axillary	C5, 6	Teres minor Deltoid	External rotation of arm Abduction of arm
Upper subscapular	C5, 6	Subscapularis	Internal rotation of arm
Lower subscapular	C5, 6	Subscapularis Teres major	Internal rotation of arm Extension and internal rotation of arm
Suprascapular	C5, 6	Supraspinatus Infraspinatus	Abduction of arm External rotation of arm
Long thoracic	C5, 6, 7	Serratus anterior	Upward rotation of scapula
Lateral pectoral	C5, 6, 7	Upper pectoralis major	Adduction-flexion of arm
Medial pectoral	C8, T1	Lower pectoralis major	Adduction-extension of arm
		Pectoralis minor	Depression of shoulder
Thoracodorsal	C6, 7, 8	Latissimus dorsi	Extension-adduction of arm

*The common segmental origin. Muscles innervated by a nerve may or may not receive fibers from all the spinal nerves contributing to the peripheral nerve, but when a nerve is distributed to only one or two muscles, as are those in this table, the segmental innervation of the muscle is the same as the segmental composition of the nerve.

elbow flexed will allow one to distinguish clearly between rotation of the arm and pronation and supination (p. 112) of the forearm and hand.

The **vessels** and **nerves** of the shoulder are for the most part not recognizable in the living person. The cephalic vein is frequently visible through the skin as it runs up the anterolateral side of the arm, and may be visible between the deltoid and pectoralis major muscles. Unnamed superficial veins are also frequently visible through the skin of the pectoral region, especially in the female where these veins are larger because they participate in the drainage of the breast.

The subclavian artery can be palpated (its pulse felt) behind the clavicle in the depression at the base of the neck; and the axillary artery, which gives rise to all the other important arteries to the shoulder, can be palpated in the axilla against the humerus.

The only nerves recognizable are those forming the lower part of the brachial plexus; these are not individually recognizable, but can be rolled against the humerus by the thumb. The fact that it is nerves that form at least a part of this mass (the axillary artery is also a part of it) can be recognized by the unpleasant sensation produced by slight pressure upon the nerves.

The distribution of nerves to the muscles of the shoulder, and the chief action of each of these muscles, are indicated in Table 5-1.

6 The Arm

GENERAL CONSIDERATIONS

Movements between the arm and forearm are of two types, flexion-extension and pronation-supination. Flexion of the forearm, or flexion at the elbow, means simply bending the elbow, while extension of the forearm increases the angle between forearm and arm. Pronation and supination are most easily described in reference to a forearm held in the horizontal position, although actually they may occur in any position of the forearm. With the forearm horizontal, pronation is the movement of turning the palm of the hand down; supination is the movement of turning the palm up.

The muscles in the arm are few in number, and are clearly divided into anterior (flexor) and posterior (extensor) muscle masses. The chief action of both groups is at the elbow, but both have also some action at the shoulder joint.

The nerves to the muscles of the arm pass through the axilla with the axillary artery. Neither the median nor the ulnar nerve, the larger components from the anterior portion of the brachial plexus, has any significance for the innervation of the muscles of the arm; they supply muscles of the forearm and hand only. The muscles of the anterior surface of the arm are supplied by the musculocutaneous nerve, from the lateral cord of the brachial plexus; after supplying the coracobrachialis, both heads of the biceps, and the brachialis, this nerve supplies skin on the forearm (fig. 6-1). The radial nerve, from the posterior cord of the plexus, supplies the posterior musculature of the arm (fig. 7-3). The blood supply to the muscles of the arm is largely from branches of the brachial artery, the continuation of the axillary artery into the arm; these are supplemented by branches of the axillary that descend across the shoulder joint, and by small branches that ascend across the elbow joint from the radial, ulnar and interosseous arteries in the forearm.

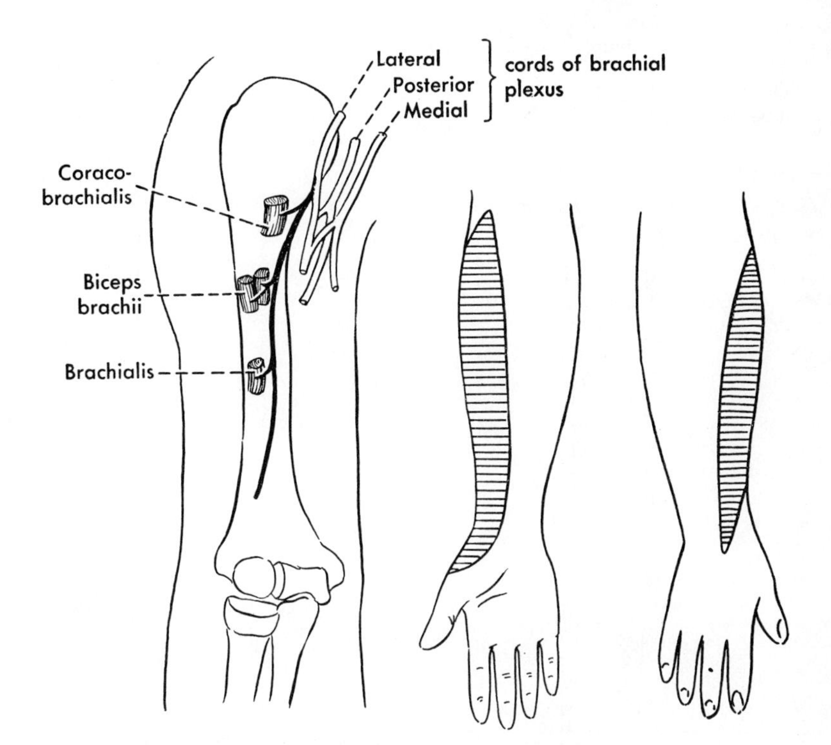

Figure 6–1. Distribution of the musculocutaneous nerve. Its cutaneous branch to the forearm, here unlabeled, is called the lateral antebrachial cutaneous nerve.

THE HUMERUS AND THE ELBOW JOINT

Just as the shoulder girdle and the upper end of the humerus, the sole bone of the arm, had to be studied in connection with the shoulder muscles, so must the humerus (fig. 6-2) and the upper ends of the radius and ulna, the two bones of the forearm, be studied in connection with the arm muscles. (The lower ends of these bones are described beginning with page 129.) Except for a roughened surface laterally, marking the insertion of the deltoid muscle, and a shallow spiral groove posteriorly, marking the course of the radial nerve, the approximately cylindrical body of the humerus presents no features of special interest. At its lower end, however, it expands laterally and medially, and at the same time becomes flattened anteroposteriorly. The sharp medial and lateral borders of the lower end of the humerus give origin to some of the muscles of the forearm and end below in more rounded but prominent medial and lateral epicondyles, which are also projections for the attachment of forearm muscles. These are easily palpable in the living person. On the posteroinferior surface of the medial epicondyle is the groove for the ulnar nerve. The lower end of the humerus bears two articular surfaces, a lateral capitulum for articulation with the head of the radius and a medial trochlea (pulley) for articulation with the ulna. Above the rounded trochlea anteriorly is the coronoid fossa (receiving the coronoid process of the ulna when

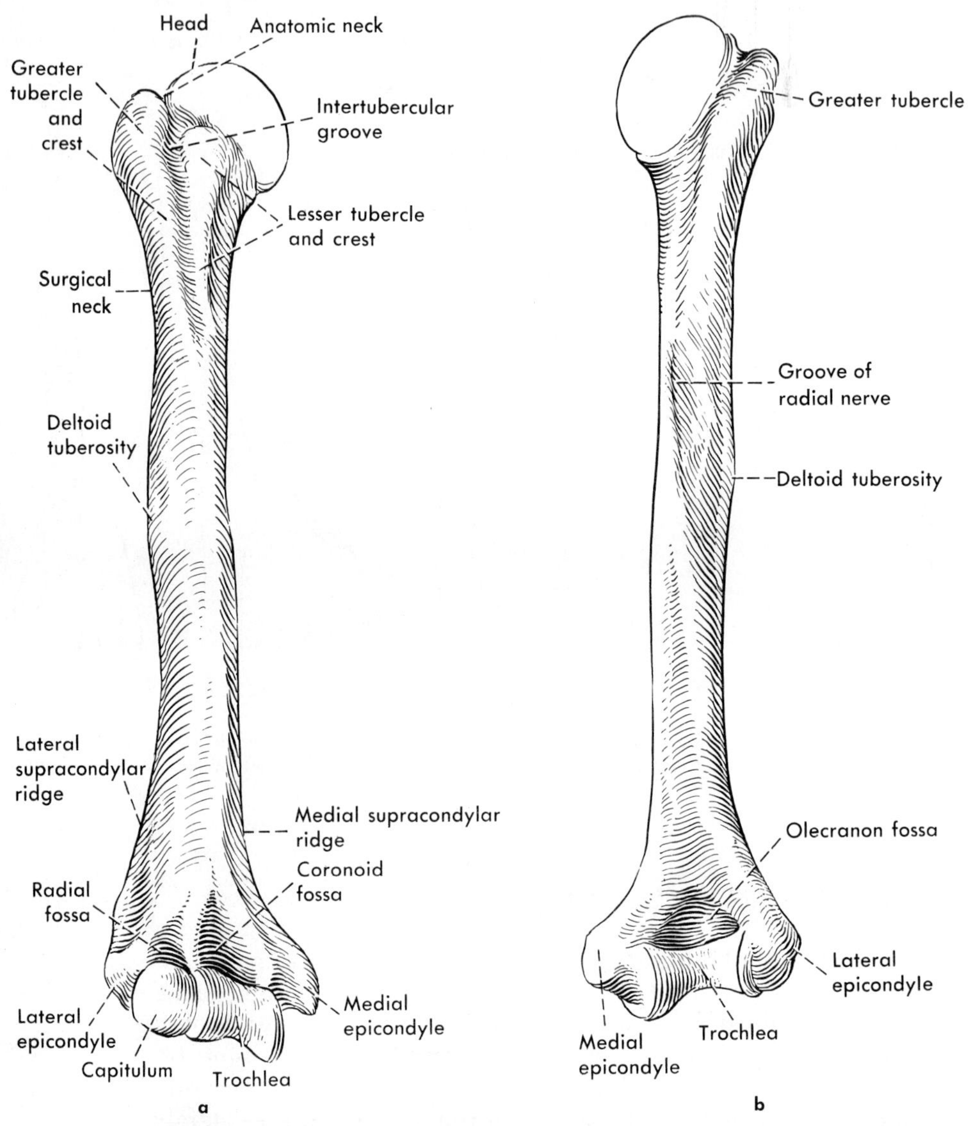

Figure 6–2. Anterior (*a*) and posterior (*b*) views of the humerus.

the forearm is flexed) while posteriorly is the olecranon fossa, receiving the olecranon (the backward-projecting portion, or upper end, of the ulna) when the forearm is extended. Occasionally the bone is completely deficient between these two fossae and a hole appears here in the dried bone, although in life this hole was bridged by a membrane. The concavity on the front of the humerus above the capitulum is the radial fossa, which receives the head of the radius when the forearm is flexed.

The upper end of the ulna (figs. 6-3 and 7-4), the more medial of the two forearm bones, is the olecranon; it is subcutaneous, as is much of the ulna throughout its length. On the anterior surface of the ulna is the deep trochlear notch (incisure) for articulation with the trochlea. The articular surface of the notch is shared by, and limited inferiorly by, the projecting coronoid process. On the lateral side of the coronoid process there is a second articular surface, the radial notch, which receives the head of the radius. Below the coronoid process is the ulnar tuberosity, marking the insertion of the brachialis muscle; distal to this the ulna narrows to become more rounded, and finally even triangular in cross section in the middle of its body.

The proximal end of the radius presents an expanded disklike head, smooth not only on its upper surface but also on its edges. The upper surface is slightly concave and fits against the capitulum of the humerus, while the circumferential part of the articular surface is in contact with the radial notch on the ulna and with a ligament (annular ligament) that holds it against this notch. The proximal end of the radius presents also a slightly constricted neck and a rather well-marked tuberosity for the insertion of the biceps. Beyond this point the radius, like the ulna, becomes rounded and then roughly triangular in cross section.

Although there is a single joint cavity at the elbow, there are two functional types of joints within this cavity. From the standpoint of function it is usually impossible to speak simply of "the" elbow joint; rather, we must specify whether we are referring to the humeroulnar and humeroradial joints or to the proximal radioulnar joint.

The strength of the **humeroulnar** and **humeroradial joints,** which act as a single joint in flexion and extension, depends primarily upon the muscles, especially the brachialis and triceps, that cross it, and upon

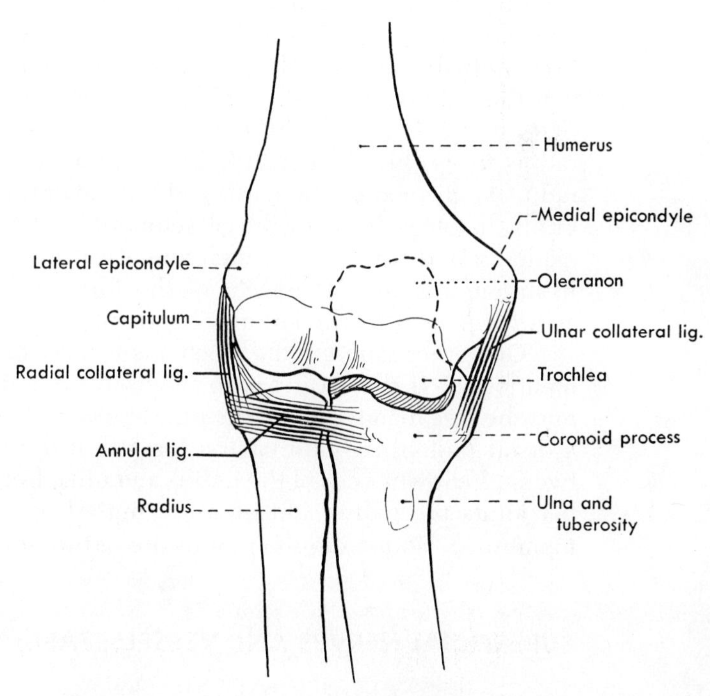

Figure 6–3. The right elbow joint viewed from in front. The thin capsule is not shown. Note that the ulna with its posteriorly projecting olecranon forms a hinge joint with the humerus, while the head of the radius is free to rotate within the annular ligament.

the shape of the articular surfaces of the humerus and ulna. The articulation is essentially a hinge joint although, due to the obliquity of the trochlea, the extended forearm is not brought into a straight line with the humerus. The articular capsule at the elbow joint is thin, lax and rather redundant, thus allowing free movement here. Anteriorly and posteriorly the capsule is protected by muscles rather than ligaments, but medially and laterally special ligaments occur. The medial or ulnar collateral ligament arises from the medial epicondyle and fans out to insert on the coronoid process and the olecranon (fig. 6-4*a*). Its edges are distinctly thicker than its middle. (A transverse band, of no importance, stretches across the lower part of this ligament from the olecranon to the coronoid process.) The lateral or radial collateral ligament arises from the lateral epicondyle (fig. 6-4*b*); it fans out less than does the medial ligament, and attaches mostly into the annular ligament (a strong attachment to the radius would interfere with pronation-supination).

The **proximal radioulnar joint** is one allowing rotation of the head of the radius to produce the movements of pronation and supination. The important ligament of this joint, the annular ligament of the radius, is attached at both ends to the coronoid process and forms about four fifths of a circle, the remaining fifth of the articular surface being provided by the radial notch of the ulna. Since the synovial membrane of the elbow joint extends downward around the neck of the radius deep to the annular ligament, the radius can rotate freely within this circle. While a purely ringlike form, such as is implied by the term "annular ligament," would suffice to allow for this movement it would offer no resistance against distal displacement of the head of the radius. Actually such displacement is provided against through the fact that the annular ligament is shaped not so much like a ring as like a portion of a cup with the bottom broken out of it; the head of the radius fits within the expanded lips of the cup while the neck of the radius is grasped by the narrowed bottom of the cup. The cup is held firmly in place through the attachment of the ligament to the ulna medially and through the attachment into it of the radial collateral ligament laterally. A muscle of the forearm, the supinator muscle, arises in part from the annular ligament.

One other aspect of the mechanism of the radioulnar joint must be mentioned. If the radius is to move in pronation and supination, it must be free to move about the ulna at its lower end also. Thus there is a distal radioulnar joint cavity, located at the wrist, intervening between the lower ends of the radius and ulna; between the two radioulnar joints the radius and ulna are united by a flexible interosseous membrane. Both radioulnar joints are of the trochoid or pivot type.

SUPERFICIAL NERVES AND VESSELS; FASCIA

The superficial fascia (subcutaneous tissue) of the arm contains a variable amount of fat, and in it run the superficial nerves and vessels. Deep to this, enclosing the muscles of the arm, is a tough membranous

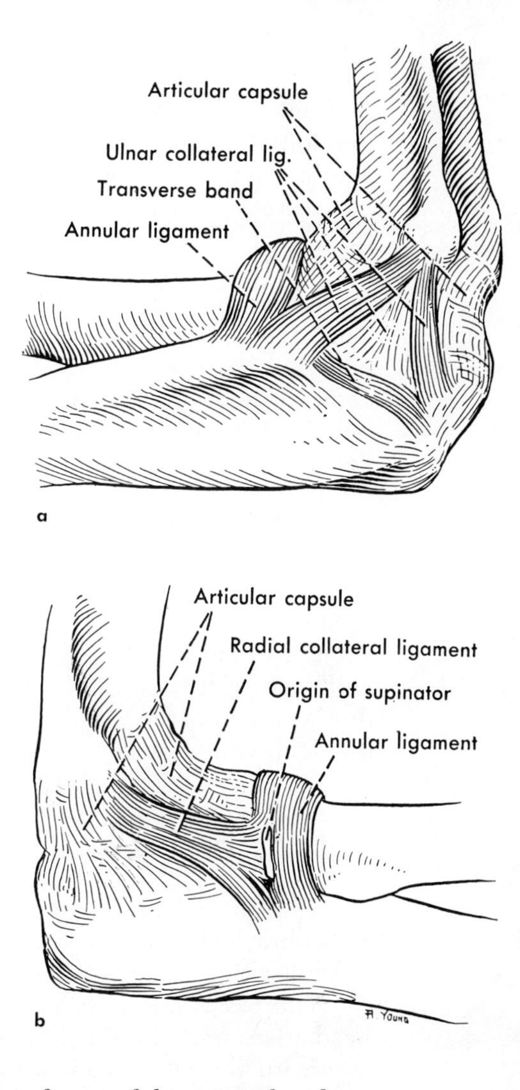

Figure 6–4. Medial, *a,* and lateral, *b,* views of the ligaments of the elbow joint.

layer of fascia, the brachial fascia or deep fascia of the arm. This deep fascia forms a sheath completely around the arm, but is loose-fitting anteriorly in order to allow for the bulging of the muscles during contraction; posteriorly it is fused to the flatter muscle here. On the medial and lateral sides of the arm, especially in its lower part, it is attached to the humerus between the anterior and posterior muscles, forming medial and lateral intermuscular septa. The brachial fascia is penetrated by many of the superficial nerves and vessels.

The cutaneous nerves of the arm emerge laterally, medially and posteriorly. They include upper and lower lateral brachial cutaneous branches from the axillary and radial nerves, respectively, a posterior brachial cutaneous branch of the radial, a medial brachial cutaneous nerve (from the medial cord), an intercostobrachial from the second, or second and third, intercostal nerves, and branches from the medial antebrachial cutaneous nerve (from the medial cord). The two large nerves perforating the brachial fascia are the medial antebrachial cutaneous, which emerges on the medial side of the arm at about the junction of the upper two thirds and lower third (accompanying the basilic vein), and the lateral antebrachial cutaneous, which emerges in front of the elbow just lateral to the strong superficial tendon (tendon of the biceps) that crosses the front of the joint. The posterior antebrachial cutaneous, smaller than the other two antebrachial nerves, emerges posterolaterally in the lower third of the arm. All three nerves continue into the forearm to supply skin here.

There are two main superficial veins in the arm. The **cephalic vein** lies in the superficial fascia of the forearm and arm, lying in the latter along the anterolateral surface of the biceps muscle, and frequently being visible here through the skin. In the upper part of the arm the cephalic vein passes between the deltoid and pectoralis major muscles to empty into the axillary vein. The **basilic vein** also lies superficially in the forearm and medial side of the lower part of the arm; in the hollow of the elbow (cubital fossa) there is usually a prominent communication, the vena mediana cubiti (median cubital vein), from the cephalic to the basilic vein. The prominence and accessibility of the superficial veins at the front of the elbow make them particularly convenient vessels from which to withdraw blood. At about the junction of the middle and lower thirds of the arm the basilic vein passes deep to the brachial fascia and courses upward deep to this; as it enters the axilla it becomes the axillary vein, which receives the deep veins and the cephalic vein.

MUSCLES OF THE ARM

The musculature of the arm consists of the biceps brachii, the coracobrachialis, and the brachialis anteriorly (fig. 6-5), and the triceps brachii with its associated anconeus (fig. 6-6) posteriorly.

The **biceps brachii** has two heads, as its name implies. The short head arises from the tip of the coracoid process of the scapula in common with the coracobrachialis muscle, while the long head arises from the supraglenoid tubercle of the scapula and traverses the cavity of the shoulder joint to run in the groove (intertubercular) between the greater and lesser tubercles. An intertubercular synovial sheath, a diverticulum of the shoulder joint, follows it downward in the intertubercular groove. The two heads unite in the lower part of the arm and form a strong tendon that passes across the front of the elbow joint to insert on the prominent radial tuberosity on the upper end of the radius. As the tendon passes downward it gives off a strong expansion, the bicipital aponeurosis, which blends with the fascia over the flexor muscles of the forearm and passes with this fascia to the ulna. Since the tuberosity on the radius is somewhat on the ulnar surface of this bone, the biceps, in producing flexion at the elbow, will also rotate the radius so as to produce supination. Both heads of the biceps are so situated as to be able to flex the arm at the shoulder, and when the humerus is externally rotated the long head is in a position to help in abduction. However, electromyographic evidence indicates that while the biceps does probably contract for flexion, it does not do so for abduction.

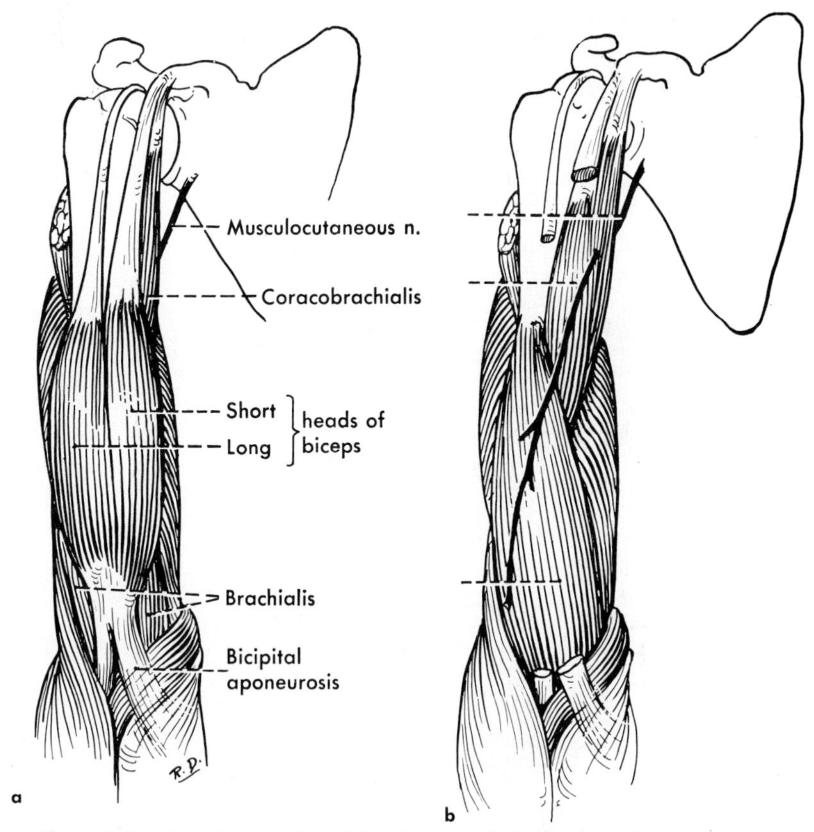

Figure 6–5. Anterior muscles of the right arm. In both views the insertion of the deltoid, parts of the medial and lateral heads of the triceps, and the origins of some of the forearm muscles are also seen. In *b* the biceps has been removed to show the brachialis better.

The brachial artery with its accompanying veins and the large median and ulnar nerves, accompanied by smaller cutaneous nerves, leave the axilla to lie on the medial side of the biceps, in the groove between it and the brachialis (fig. 6-7). The ulnar nerve gradually diverges posteriorly to pass behind the medial epicondyle, but the brachial artery and median nerve enter the forearm on the front of the brachialis muscle. The musculocutaneous nerve lies behind the biceps, between that and the brachialis, having attained this position by passing through the coracobrachialis. The fourth major nerve in the arm, the radial, starts posteriorly around the arm as it leaves the axilla, but reaches an anterolateral position, just behind the lateral part of the brachialis muscle, some distance above the elbow, and passes in front of the elbow joint to reach the forearm.

The **coracobrachialis** arises from the coracoid process with the short head of the biceps and inserts on the anteromedial surface of the humerus about its middle; since it does not cross the elbow joint it can have no action there, but is a flexor and adductor of the arm. The musculocutaneous nerve leaves the axilla by running through the muscle. The **brachialis** arises from much of the lower half of the anterior surface of the humerus and from both intermuscular septa between it and the

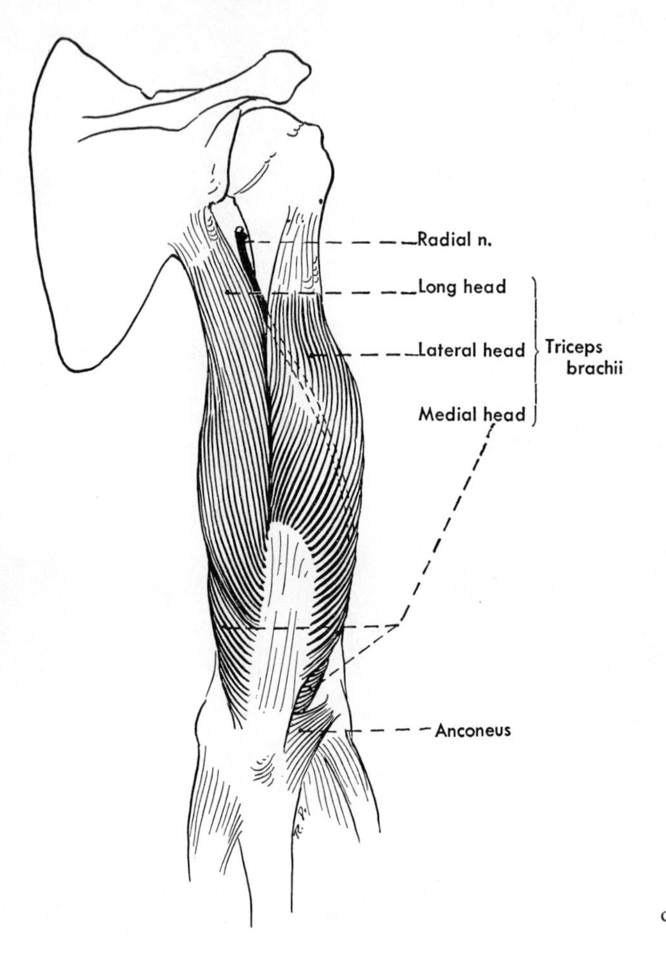

Figure 6–6. Posterior muscles of the right arm.

triceps; covering the front of the elbow joint, it inserts upon the ulnar tuberosity just distal to the coronoid process. Since it is rotation of the radius that produces supination and pronation, the brachialis can only flex the forearm. The brachialis is an equally effective flexor whether the forearm is pronated or supinated, since it inserts on the ulna. However, if the forearm is fixed in pronation, the effectiveness of the biceps is reduced by the vain attempt to supinate as it flexes. Thus in doing pull-ups, or "chinning" one's self, the most effective flexion is obtained when the forearm is supinated, the bar being grasped with the palms toward the body.

The three muscles just described, that is, all the anterior muscles in the arm, are innervated through the musculocutaneous nerve (fig. 6-1). Before it leaves the axilla by running through the coracobrachialis it supplies this muscle; and as it lies between the biceps and the brachialis it supplies both of these. It contains fibers from the fifth, sixth and seventh cervical nerves. The lateral part of the brachialis often receives a twig from the radial nerve, although the functional importance of this is not clear. It is in part a branch to the elbow joint, but there is disagreement on whether it supplies motor fibers to a small lateral part of the brachialis.

Of the three heads of the **triceps brachii** muscle (fig. 6-6), the long head arises from an infraglenoid tubercle on the lateral border of the scapula just below the glenoid cavity, and passes downward in front of the teres minor but behind the teres major (fig. 5-13). The lateral head of the triceps arises from the humerus above and lateral to the groove for the radial nerve, and from the lateral intermuscular septum, and unites with the long head to form the superficial tendinous part of the insertion of the muscle. The third or medial head arises also from the humerus, but medial to and below the spiralling groove for the radial nerve, so that it comes to cover the entire posterior surface of the lower part of the bone, where it arises also from both intermuscular septa; it attaches into the deep surface of the combined lateral and long heads. The radial nerve and the deep brachial artery pass between the long head and the humerus and then spiral around the back of the humerus between the origins of the lateral and medial heads, approximately in the so-called groove for the radial nerve but usually lying upon upper-

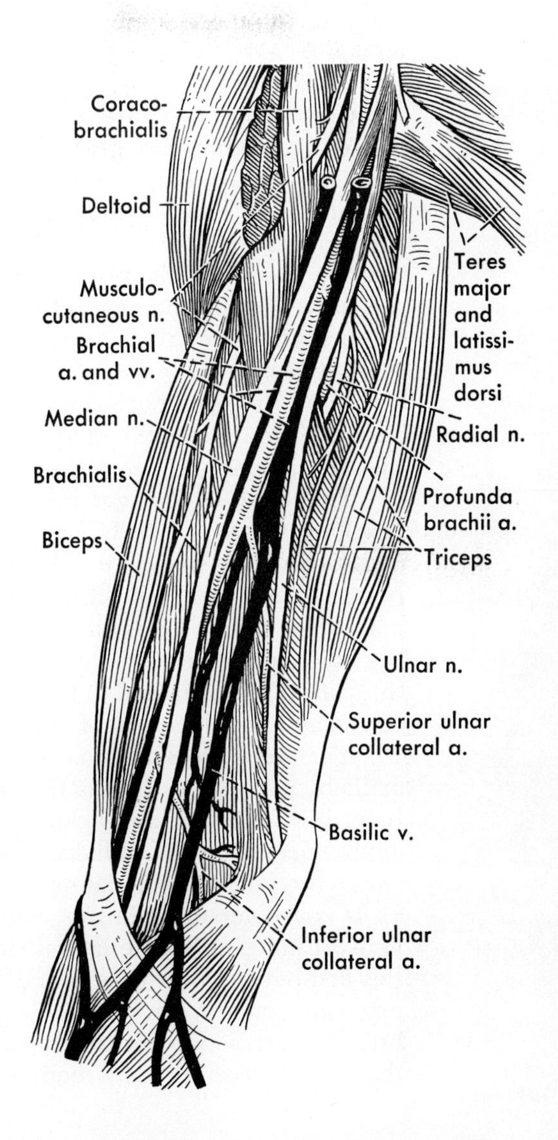

Figure 6–7. Nerves and vessels of the arm. In this figure, veins are black, arteries are transversely shaded.

most fibers of the medial head rather than directly against the bone. All three heads of the triceps insert together on the proximal end of the olecranon, and the triceps constitutes, therefore, the chief extensor of the elbow joint. Since the long head of the triceps, but not the other two heads, crosses the shoulder joint, the long head also aids in extension and adduction of the arm.

The **anconeus** is a small triangular muscle attached to the lateral epicondyle and to the lateral side of the olecranon and adjacent part of the ulna; while it is too small to supply much power, it apparently contracts not only for extension but also whenever the joint needs to be stabilized against flexion or pronation-supination. The triceps is innervated by the radial nerve which passes posteriorly under cover of this muscle to attain the lateral side of the arm; a branch of the nerve to the medial head of the triceps is continued downward to innervate the anconeus.

NERVES AND VESSELS

Of the four main nerves traversing the arm (fig. 6-7), namely, median, ulnar, radial and musculocutaneous, the first two give off no branches to the muscles of the arm. After their origins from the brachial plexus these two nerves run down the medial side of the arm, the median nerve at first anterolateral but later medial to, the ulnar nerve posterior to, the brachial artery. Just above the elbow the median nerve lies on the front of the brachialis muscle and passes with the brachial artery in front of the elbow joint. The ulnar nerve passes down the arm posterior to the median nerve and brachial artery; it gradually diverges posteriorly, penetrates the medial intermuscular septum, and runs on the medial head of the triceps to pass behind the medial epicondyle.

The **musculocutaneous nerve** arises from the lateral cord of the brachial plexus (C5 to C7) and passes through the substance of the coracobrachialis muscle to lie between the biceps and brachialis. Just before or as it penetrates the coracobrachialis it supplies this muscle, and subsequently gives off branches to both heads of the biceps and to the brachialis. Thereafter, as the lateral antebrachial cutaneous nerve, this nerve passes lateral to the biceps tendon, penetrates the brachial fascia, and supplies skin of the forearm.

The **radial nerve,** the continuation of the posterior cord, leaves the axilla by passing posteriorly in a wide spiral around the humerus, first lying between the long head of the triceps and the humerus, then approximately in the radial groove on the back of the humerus, between the origins of the lateral and medial heads of the triceps (fig. 6-8). It is accompanied in this course by the deep brachial artery and veins. Emerging on the lateral side of the humerus, the radial nerve lies first between the triceps and brachialis, then passes in front of the extensor forearm group to lie between the brachioradialis and brachialis, in which position it passes into the forearm (fig. 9-3). While the radial nerve is on the medial side of the arm it usually gives off a

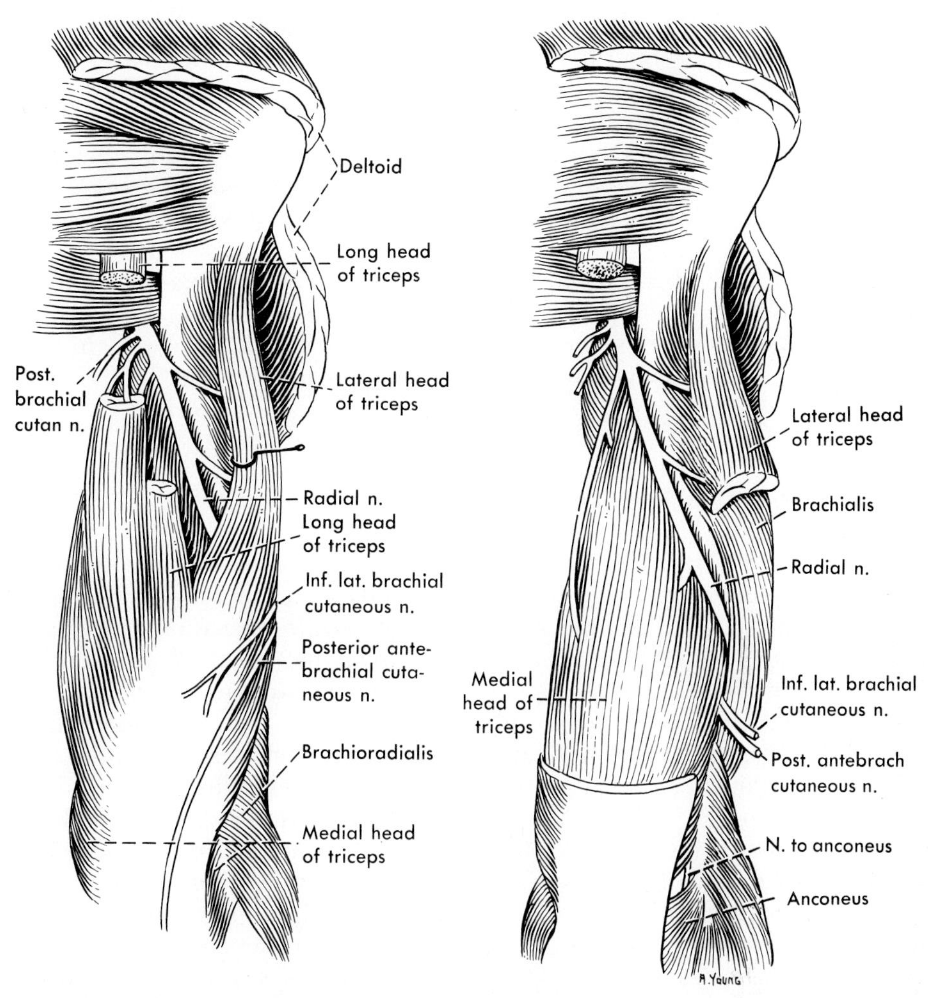

Figure 6–8. The radial nerve in the arm.

branch to the long head of the triceps and a second branch that descends parallel to the ulnar nerve to reach the medial head of this muscle. In its spiral course deep to the triceps it gives additional branches to all three heads of this muscle, and in the lower part of the arm may give a twig into the brachialis (p. 120).

The radial nerve supplies the triceps (and anconeus) with fibers derived primarily from C6, 7, and 8. While injury to the radial nerve may abolish all active extension at the elbow, it should be noted that some of the branches to the triceps usually arise before the nerve has left the axilla. Injury to it as it lies in the radial groove therefore affects primarily extension of the wrist and fingers.

The **brachial artery** (fig. 6-7), the continuation of the axillary artery, passes down the medial side of the arm and then runs with the median nerve in front of the elbow, lying on the brachialis muscle. It gives off branches to the brachial muscles, including one (the deep brachial) that accompanies the radial nerve in its posterior course around the humerus, and also a nutrient artery to the humerus and anastomotic branches about the elbow joint. (One of the latter passes behind, and one in front of, the medial epicondyle, and branches of the deep brachial form similar anastomoses behind and in front of the lateral epicondyle.) Occasionally the brachial artery is double during part or all of its course in the arm; when this occurs one of the vessels usually lies superficial to the median nerve and is known as a superficial brachial artery.

The brachial artery is accompanied by two brachial veins, which not infrequently blend into one for a part of their course. In addition to the deep veins, there are two important superficial veins, the cephalic and the basilic. These have already been described.

MOVEMENTS AT THE ELBOW JOINT

In considering the movements at the elbow joint it should be remembered that flexion and extension occur between the humerus and both the ulna and the radius while pronation and supination involve rotation of the radius about the ulna. Flexion (fig. 6-9) is brought about especially through the actions of the biceps and brachialis. (Since the biceps supinates as it flexes, flexion from the pronated position is carried out by the brachialis alone unless there is strong resistance). A paralysis of these muscles, as caused, for instance, by injury to the musculocutaneous nerve, does not, however, abolish the ability to flex the elbow, since forearm muscles that cross the front of the elbow are innervated by other nerves. The most superficial muscle of the lateral forearm group, the brachioradialis, arises from the lateral border of the humerus some distance above the elbow and crosses well in front of the elbow joint; it is a particularly good flexor of this joint when the hand is held so that the thumb is up, although it normally participates in flexion primarily when the movement is a fast one. Other muscles on the extensor side, especially the extensor carpi radialis longus, may assist the brachioradialis in its flexor action but arise too low on the humerus to be as important in flexion. Of the flexor muscle mass in the forearm the pronator teres has the highest origin from the humerus, although not as high as that of the brachioradialis. It is accordingly a much weaker flexor of the forearm upon the arm, and may or may not be able, working alone, to carry out this movement against gravity. Other muscles of the flexor forearm group arise from the medial epicondyle and have no significant action in flexing the elbow joint. Supination is carried out normally by the supinator muscle in the forearm, but loss of the biceps markedly weakens this action.

The extensor muscles of the forearm are the anconeus and the triceps (fig. 6-6); the anconeus apparently acts first to extend or stabilize the elbow, and the various heads of the triceps are recruited, as needed, for more strength—first the medial, then the lateral, then the long one.

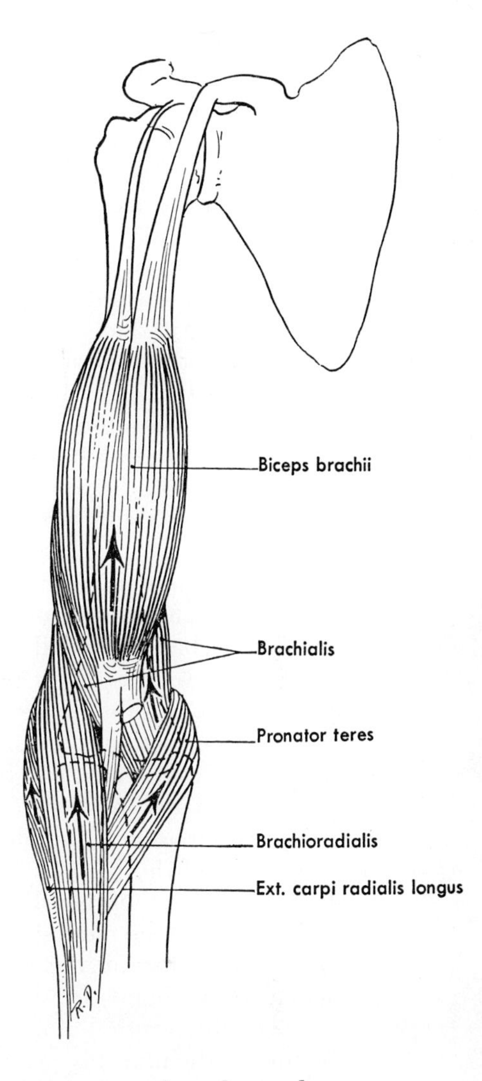

Figure 6–9. The flexors of the forearm.

The movements of pronation and supination can best be understood after the muscles of the forearm have been studied. They are therefore discussed on page 158.

REVIEW OF THE ARM

The anatomy of the living arm can be adequately reviewed rather quickly.

The coracoid process, already identified in the study of the shoulder, the body of the humerus, the medial and lateral epicondyles, and the olecranon, are the bony landmarks that can be palpated.

Of the anterior **muscles** of the arm, the most familiar is the biceps, which little boys use to "show their muscle"; its tendon of insertion can be readily felt in front of the flexed elbow, and the upper and medial edge of the bicipital aponeurosis is also easily palpable. The two heads of origin are not easily separable, but the short head and the coracobrachialis can be identified in the lower part of the axilla where they lie behind the insertion of the pectoralis major; they can be distinguished from each other because the more anterior rounded tendon of the biceps is prominent in forcible flexion of the elbow, while the broader and posteriorly lying coracobrachialis is particularly prominent when the arm is adducted. The fact that the biceps contracts strongly for combined flexion and supination of the forearm, but little or none at all for either movement alone, unless it is resisted, can easily be demonstrated. The brachialis is somewhat more difficult to palpate; it is perhaps most easily recognized by palpating on the medial side of the tendon of the biceps while flexion of the supinated forearm against opposition is being carried out. The triceps is easily identifiable when the forearm is extended against opposition, and its long head can be felt in the axilla when the arm is adducted. The anconeus is not identifiable.

Of the **vessels** and **nerves,** parts of the cephalic and basilic veins in the arm are usually identifiable, particularly in front of the elbow where the median cubital vein can be seen connecting them. They can be brought out still more plainly by interfering with their flow by wrapping a band—a handkerchief will do—tightly about the arm above the elbow. In the upper part of the arm, the brachial artery can be palpated against the humerus—it is here that it is occluded by a blood pressure cuff when a physician "takes" the blood pressure—and the accompanying nerves can be rolled against the bone. The artery can also be palpated on the front of the brachialis muscle, just medial to the biceps tendon at the elbow. It is here that the physician listens for the sound of the blood returning to the artery as the pressure of the blood pressure cuff is released. The ulnar nerve can be palpated as it passes behind the medial epicondyle. The sensation induced by pressing upon or hitting the nerve here has given rise to the colloquial name "funny bone" for the medial epicondyle.

The distribution of nerves to the muscles of the arm, and the chief actions of the muscles, are summarized in Table 6-1. Figure 6-1 shows diagrammatically the distribution of the musculocutaneous nerve. The distribution of the radial nerve, which is to many more forearm than arm muscles, is shown in figure 7-3.

TABLE 6–1. Nerves and Muscles of the Arm

NERVE AND ORIGIN*	MUSCLE		
	Name	*Segmental innervation**	*Chief action*
Musculocutaneous C5–C7	Biceps	C5, 6	Flexion-supination of forearm
	Coracobrachialis	C5–C7	Adduction-flexion of arm
	Brachialis	C5, 6	Flexion of forearm
Radial C5–C8	Triceps	C5–C8	Extension of forearm
	Anconeus	C6, 7	Extension of forearm

*A common segmental origin or segmental innervation. The composition of both the chief nerves and their muscular branches varies somewhat among persons.

7 Forearm and Hand: General Considerations

The muscles of the forearm act on the elbow, wrist, and digits (fingers and thumb), while the muscles in the hand act on the digits alone. In the upper part of the forearm the muscles form fleshy masses below the medial and lateral epicondyles, but their mass rapidly tapers off toward the wrist, where the muscle bellies are replaced by long, strong tendons that continue into the hand. The reduction in bulk obtained by the transformation of the muscles into tendons allows a far greater number of muscles to have access to the hand than would otherwise be possible. The muscles in the hand form two masses in the palm, one at the base of the thumb (thenar eminence), the other at the base of the little finger (hypothenar eminence). Other muscles of the palm are situated more deeply, behind the long tendons and in association with the long bones (metacarpals) of the hand.

The forearm muscles can be clearly divided into flexor and extensor groups. The former arises in large part from the medial epicondyle and occupies the medial border and anterior (flexor) surface of the forearm, from which many muscles continue into the palm of the hand. The extensor group is particularly prominent in the region of the lateral epicondyle and occupies the lateral border and the posterior (extensor) surface of the forearm. Many of the muscles of this group send tendons onto the back (dorsum) of the hand.

The terminology of the forearm muscles may seem at first unnecessarily complex and confusing, but can be comprehended readily if an attempt is made to understand it rather than simply memorize it. The muscles are named, for the most part, from their chief actions, often with qualifying words that indicate their relative locations or sizes. Thus we find that there are two pronators (muscles that pronate the hand), a teres (round) and a quadratus (quadrangular in shape); there is a supinator, or muscle that supinates; there is a flexor digitorum superficialis and a flexor digitorum profundus (superficial flexor and deep flexor of the digits). Similarly, "flexor carpi ulnaris" means "ulnar flexor of the wrist" and implies that there is a radial flexor of the wrist, as indeed there is. Since there are two radial extensors of the wrist, one is called the extensor carpi radialis longus (long extensor on the radial side of the wrist) while the other is the extensor carpi radialis brevis (short radial extensor of the wrist).

If the following terms are understood, the student should have no difficulty with the terminology of the muscles of the forearm and hand: pronator, supinator, flexor, extensor, abductor, and adductor all refer to movements which will be defined in the next paragraph; radialis and ulnaris (which are self-evident), superficialis (superficial) and profundus (deep) are adjectives of position; longus (long), brevis (short), and teres and quadratus (defined above) are adjectives descriptive of shape; pollicis (of the thumb or pollex), indicis (of the index finger), digiti minimi (of the little finger), carpi (of the carpus or wrist) and digitorum (of the digits or fingers) are qualifying nouns explaining upon what member or joint the muscle exerts its action.

Pronation and supination have already been defined as movements occurring at the elbow that result in turning the palm downward or upward, respectively, when the flexed forearm is held horizontal. Flexion at the wrist is the act of bending the palm of the hand toward the forearm. Extension is the movement of straightening the flexed wrist; when this movement of extension is continued to bend the wrist backward, it is called hyperextension or, more commonly, dorsiflexion (that is, bending toward the dorsum, or back, of the hand). Obviously, flexion is freer than is dorsiflexion. In addition to flexion and extension, movements at the wrist can take place in the plane of the extended hand. These movements may be described with reference to their relation either to the midline of the body or of the hand itself; the latter is preferable to avoid misunderstandings. Thus we can refer to such a movement toward the little finger side of the hand as being adduction or, better, ulnar abduction of the hand; similarly, movement at the wrist toward the thumb side is abduction, or, better, radial abduction. Ulnar abduction far exceeds radial abduction in its fullness of movement. As the joint between radius and wrist bones, although concavo-convex in shape, is ellipsoidal rather than of the ball-and-socket type, rotation at the wrist is barely possible. A limited amount may accompany pronation and supination.

Movements of the thumb are best described separately from those of the four fingers. The thumb is thought of as being in the normal position when its palmar surface is almost at a right angle to the palm of the hand, and movements of the thumb are described with reference to this position. Flexion of the thumb is bending it in a plane parallel to the plane of the palm (*not* in doing this and at the same time so rotating it that its pad comes in contact with the palmar surface of the fingers); extension is, of course, the opposite movement. Abduction is raising the thumb away from the other fingers, in a plane perpendicular to the palm, while adduction is bringing it back toward the palm.

As you observe these movements, note that the long bone at the base of the thumb (the first metacarpal) moves with the thumb and contributes more to these movements, except in flexion and extension of the distal joint of the thumb, than do the joints of the thumb itself. Note also that movements of the metacarpal in flexion and extension are not pure movements but involve also rotation and usually adduction or abduction. Thus as the thumb as a whole is flexed it is also rotated medially and adducted so that its pad can come in contact with the pads of the fingers; this combination of movements is known as opposition. The movement away from opposition, involving extension, external rotation, and usually abduction, is conveniently termed "reposition."

Movements of the four fingers include flexion, or closing the hand as in making a fist, and extension, or straightening the fingers. Some hyperextension or dorsiflexion is also possible. Also, the fingers may be spread apart in the plane of the palm, abducted, or they may be brought together, adducted, in this plane. Since the metacarpophalangeal joints allow flexion and extension, abduction and adduction, a finger as a whole may also be circumducted. In addition to the above-described movements, which occur at the joints of the fingers, the little and ring fingers have metacarpals that, although far less mobile than the metacarpal of the thumb, can flex to help cup the hand.

The muscles of the forearm are innervated by the median, ulnar, and radial nerves, and those of the hand are innervated by the median and ulnar nerves. Since both the median and the ulnar nerves are derived from the anterior divisions of the brachial plexus, they are distributed to the anterior musculature of the forearm and hand. The median nerve supplies most of the anterior (flexor) musculature of the forearm, but only a few muscles in the hand (fig. 7-1). The ulnar nerve supplies only about one and one half muscles in the forearm, but the majority of those in the hand (fig. 7-2). The radial nerve, the only derivative of the posterior division of the brachial plexus to reach either the arm or the forearm, supplies all the posterior (extensor) muscles of the forearm (fig. 7-3); there are normally no posterior muscles in the hand. Between them, these three nerves innervate most or all of the skin of the hand, but only the radial innervates also skin of the forearm (p. 135).

The arteries to the forearm and hand are the radial and the ulnar, the terminal branches of the brachial. Both run down the anterior side of the forearm and end in the palm; the posterior side of the forearm and the dorsum of the hand are supplied by branches of these two vessels.

BONES AND JOINTS

In preparation for a study of the soft tissues of the forearm and hand certain essentials of the skeletal anatomy need to be understood. Study of the bones of the forearm, begun in connection with the arm

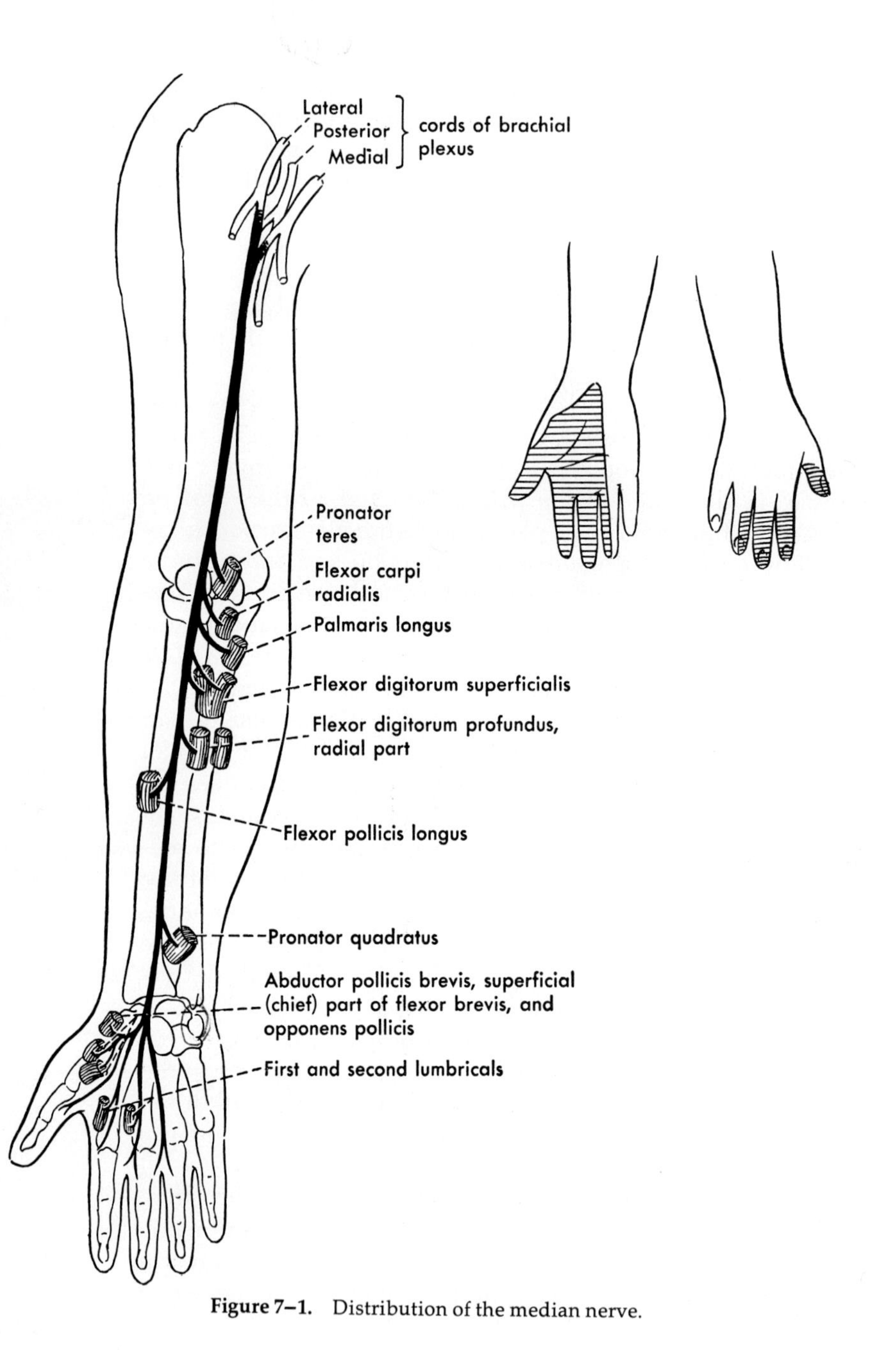

Figure 7–1. Distribution of the median nerve.

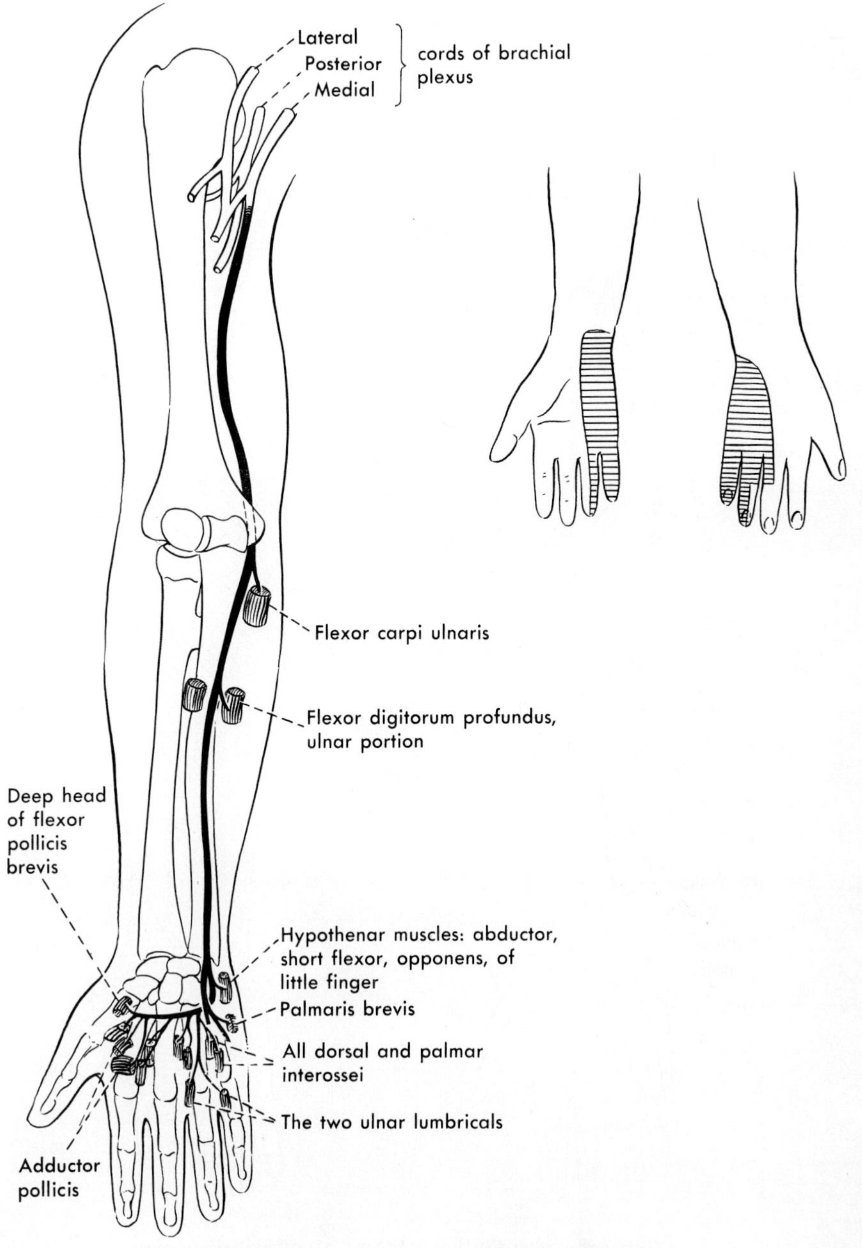

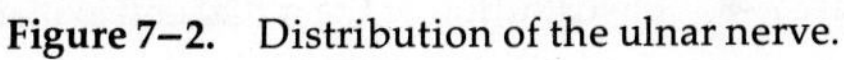

Figure 7–2. Distribution of the ulnar nerve.

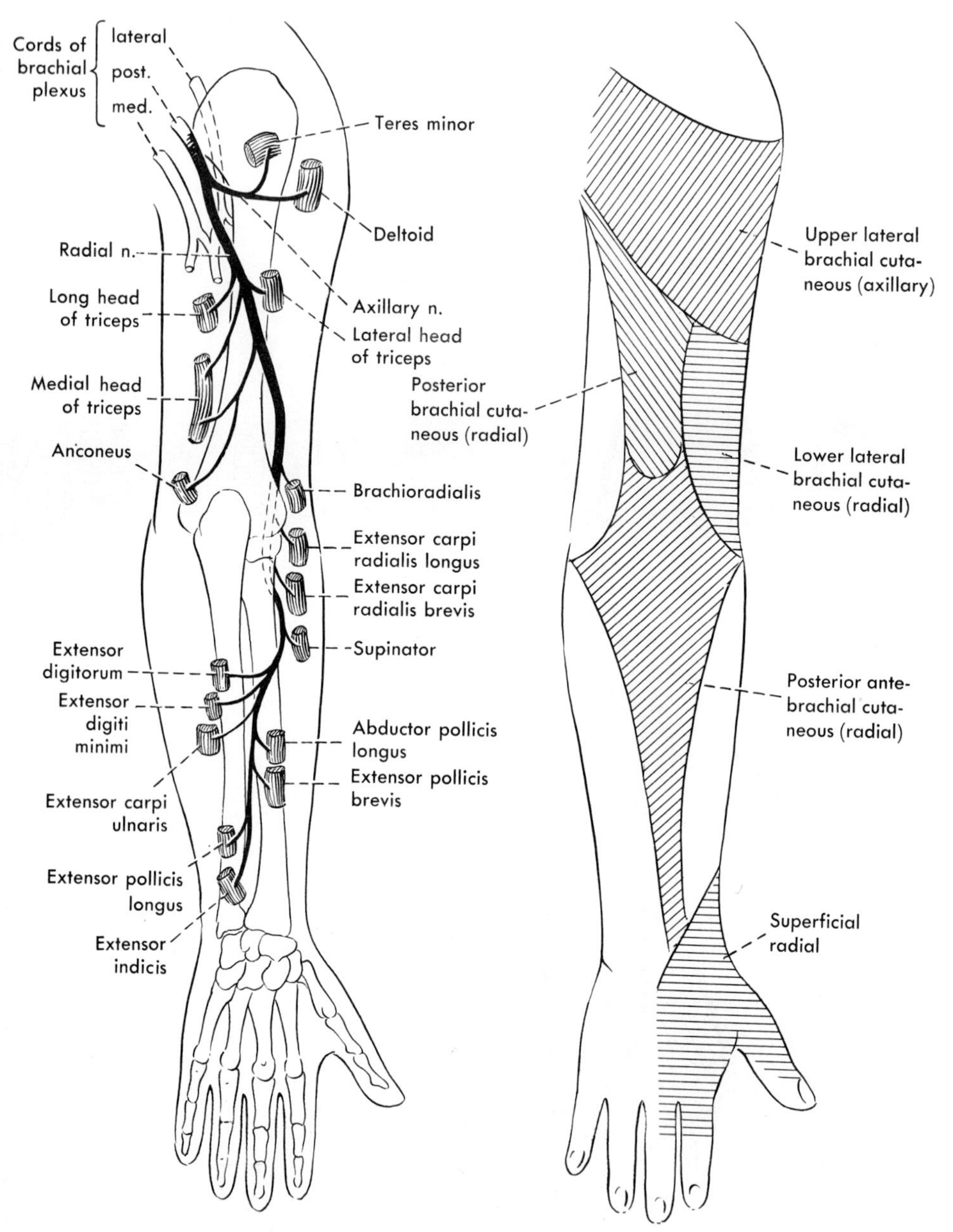

Figure 7–3. Distribution of the radial and axillary nerves.

(p. 114), should be completed; and since so many of the muscles of the forearm extend into the hand a preliminary study of the bones and joints of the wrist and hand should be undertaken now. The details can best be left for consideration when the hand itself is studied.

The upper ends of the bones of the forearm (fig. 7-4) the radius on the side of the thumb and the ulna on that of the little finger, should be reviewed. The articulation of the head of the radius with the ulna and the annular ligament is particularly important, since this allows the radius to rotate and cross the ulna in the movement of pronation. The bodies of these bones merit no particular comment. The chief marking on each is its interosseous border (medially on the radius, laterally on the ulna). Between these borders is stretched during life a strong interosseous membrane that largely fills the interval between the bones and thus completes the separation between anterior and posterior structures of the forearm. Above the upper end of the interosseous membrane there is a gap between the radius and ulna that transmits to the posterior side of the forearm the chief artery here (posterior interosseous). In the lower part of the membrane there is a small aperture through which a branch of another vessel passes from the anterior to the posterior sides of the forearm. The strongest fibers of the interosseous membrane run obliquely downward from the radius to the ulna. In a fall on the hand (where most of the weight bears upon the radius) the interosseous membrane transmits some of the force to the ulna, and thus prevents the head of the radius from being jammed so forcibly against the humerus.

The **radius** expands markedly at its lower end, being particularly broad from side to side. The cartilage-lined concavity at its end is the carpal (wrist) articular surface; on the medial (ulnar) side of its end is the small ulnar notch (incisure) for articulation with the distal end of the ulna, and the projection from the anterolateral part of its end is the styloid process.

The lower end of the **ulna** is its head (thus the head of the radius is proximal, that of the ulna distal). Much of its circumference and its distal surface is smooth for articulation with the radius and the fibrocartilaginous disk that separates the distal radioulnar joint from the wrist joint proper (the ulna does not participate in the latter joint). The nipple-like projection from the posteromedial side of the head is the styloid process of the ulna.

The **wrist** is formed by carpals, of which there are eight arranged in two rows. Each is named, but they can be conveniently studied later. The bones are so arranged that they form a concave anterior surface, the carpal groove. The bulk of the **hand** is formed by long bones, the metacarpals, which are designated by Roman numerals, that of the thumb being metacarpal I or the first metacarpal. The skeleton of the **fingers** is formed by phalanges; the three phalanges of each finger are called proximal, middle, and distal phalanges, respectively, the two of the thumb proximal and distal. The digits are numbered like the metacarpals and are named. Thus the first digit is the thumb (in Latin, the pollex); digits II to V are also called the index, middle, ring, and little digits or fingers; since, strictly speaking, there

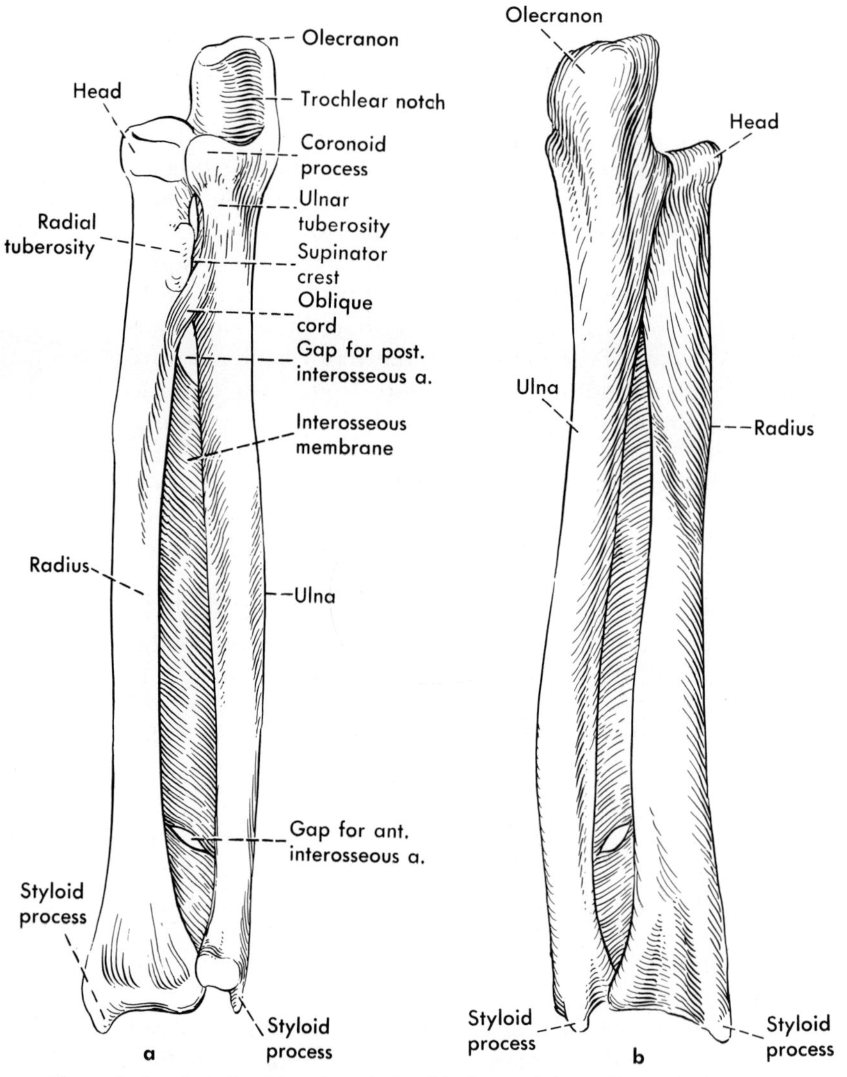

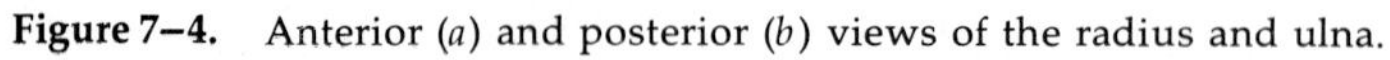

Figure 7–4. Anterior (*a*) and posterior (*b*) views of the radius and ulna.

is no middle finger, "long finger" more accurately describes the middle digit. The skeleton of the wrist and hand is shown in figures 11-1 and 11-2, and figure 11-3 illustrates the joints at the wrist.

The **joints** of the wrist and fingers can also be studied more profitably later, and will be found described in the chapter on the hand (p. 166). The large joint at the wrist is the radiocarpal joint (between the radius and the carpal bones), but there are also intercarpal joints (joints among the carpals) and these contribute to mobility at the wrist. The carpometacarpal joints (between the bones named) of the second and third digits are essentially immovable, but the fourth metacarpal can be moved slightly and the fifth still more; the carpometacarpal joint of the thumb is very movable and responsible for much of the movement of the thumb. The metacarpophalangeal joints of the fingers form the knuckles, and are much more movable than is that of the thumb. There are two interphalangeal joints, proximal and distal, for each finger, and a single one for the thumb. Movements at these joints have already been defined.

Superficial Nerves and Vessels; Fascia

The cutaneous innervation of the forearm is through three nerves. Running down the back of the forearm is the posterior antebrachial cutaneous nerve, a branch of the radial arising above the elbow. The lateral antebrachial cutaneous nerve, the continuation of the musculocutaneous, supplies skin on both anterior and posterior surfaces of the lateral aspect of the forearm. The medial antebrachial cutaneous nerve, a nerve arising directly from the medial cord of the brachial plexus, similarly supplies skin on the anterior and posterior surfaces of the medial side of the forearm. These nerves usually end close to the wrist, but may continue a variable distance onto the hand, either as independent branches or through anastomosis with radial or ulnar nerve branches to the hand.

The cutaneous innervation of the hand is primarily by the median, ulnar and radial nerves (figs. 7-1 to 7-3). The median nerve typically supplies much of the skin on the palmar surface of the hand, including that of the thumb, index, long, and the radial half of the ring finger. In addition to this palmar distribution, the median nerve sends branches toward the dorsum to supply the bases of the nails and most of the skin over the middle and distal phalanges of index, long, and half of the ring fingers. The ulnar nerve typically supplies the ulnar side of the palm of the hand, the ulnar half of the ring finger and all the palmar surface of the little finger. In addition, through its dorsal branch to the hand, it supplies at least the corresponding fingers on their dorsal surfaces and a similar region on the back of the hand, and it frequently supplies or helps to supply the adjacent dorsal surfaces of the proximal phalanges of ring and long fingers. Thus the ulnar nerve supplies skin on one and a half fingers on both palmar and dorsal sides and often the proximal parts of other fingers on the dorsum.

The radial nerve through its superficial branch supplies the remaining surface of the dorsum of the hand, thus including the dorsum of the thumb and the proximal portion of one and a half or two and a half adjacent fingers. It should be noted that the digital branches of the radial nerve do not supply the more distal portions of the fingers.

The superficial veins of the hand form a dense network on both surfaces of the fingers; the veins on the palmar surface drain primarily backward onto the dorsum of the hand, where the chief venous network of the hand occurs. Superficial veins are scarce in the palm of the hand. From the extensive venous network lying primarily on the dorsum two veins, both of which have already been seen in the arm, take origin and run upward in the superficial fascia. The cephalic vein arises largely from the radial side of this network and runs upward, winding around the radial side of the forearm to reach the lateral side of the front of the elbow; here it communicates with the basilic vein (through the median cubital vein). Its further course in the arm has already been described. The basilic vein arises more from the ulnar portion of the dorsal veins and runs upward on the medial border of the anterior surface of the forearm. A third vein, usually much smaller than the preceding two but of varying size, may run up the middle of the anterior surface of the forearm to communicate with the basilic or cephalic or both; this is the median forearm vein.

The deep fascia of the forearm (antebrachial fascia) resembles that of the arm in being a tough fibrous membrane that surrounds the underlying muscles. In its upper part it receives the bicipital aponeurosis and gives origin to the more superficial fibers of both flexor and extensor muscles of the forearm. Septa passing from it between the muscles give further attachment to these in the upper part of the forearm. In the lower part of the forearm the fascia more loosely surrounds the muscles, and on the anterior surface of the wrist it is partially split into two layers, between which pass the palmaris longus tendon and the ulnar vessels and nerve. The more superficial layer at the wrist, covering the above-mentioned structures, may be thin and poorly developed. The deeper layer of fascia at the wrist is markedly strengthened by transverse fibers that stretch across the carpal groove between its higher medial and lateral sides, and thus convert the groove into the carpal canal (also called carpal tunnel). These transverse fibers form the **flexor retinaculum** (see also page 172); most of the tendons going into the palm of the hand pass behind this retinaculum, therefore in the carpal canal. On the dorsal side of the wrist the fascia is similarly thickened to form the **extensor retinaculum,** but as this stretches from one side of the wrist to the other it is attached to the underlying bones by septa that divide the space deep to the ligament into a number of separate compartments for the tendons going onto the dorsum of the hand (fig. 11-13).

The deep fascia of the hand is continuous through the retinacula with the antebrachial fascia. That of the palm needs special description, which will be found on page 172. Until the hand is studied, it need only be understood that a central part of the palmar fascia is particularly thick and tendinous, and is called the **palmar aponeurosis.** The major tendons, nerves, and vessels of the palm lie mostly behind the palmar aponeurosis.

Flexor Forearm 8

The superficial nerves and vessels and the fascia of the forearm have been considered in the previous chapter.

MUSCULATURE

The flexor muscles of the forearm may be conveniently divided into a superficial and a deep group with one muscle, the flexor digitorum superficialis, forming an intermediate mass between these groups. There are four muscles in the superficial group (fig. 8-1): the pronator teres, the palmaris longus, the flexor carpi radialis, and the flexor carpi ulnaris. These muscles are in part fused where they arise from the medial epicondyle, and thus share a tendon of origin (called the common flexor tendon); they arise also from the antebrachial fascia covering them and from intermuscular septa between them.

The uppermost member of the superficial group, therefore contributing to the fleshy mass distal to the medial epicondyle, is the **pronator teres.** This muscle arises mostly from the humerus above and down to the medial epicondyle but has also a second head of origin from the coronoid process of the ulna. The median nerve passes between the two heads of the muscle, after which they unite to insert on the lateral side of the radius close to its middle. The brachial artery divides into radial and ulnar arteries on the brachialis muscle above the pronator teres; the radial artery runs downward behind the brachioradialis but across the insertion of the pronator (fig. 8-4); the ulnar artery passes behind both heads of the muscle. The muscle is supplied by branches of the median nerve arising just before and as the nerve passes through it. Because it is wrapped around the radius its chief action is to roll the radius medially, thus to pronate the forearm. By virtue of its relatively high origin on the humerus it is also a weak flexor at the elbow.

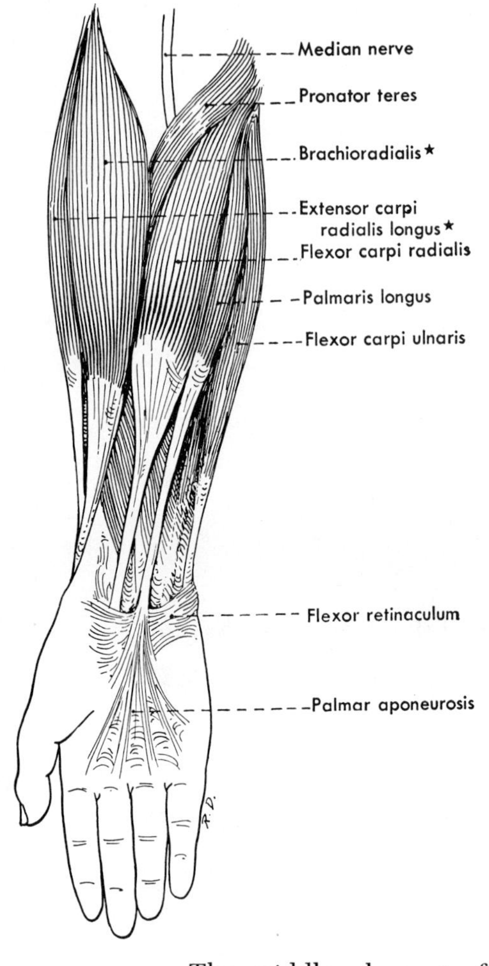

Figure 8–1. The more superficial flexor muscles of the right forearm. Those muscles that are starred are anterior muscles of the extensor group.

The middle element of the three remaining members of the superficial flexor group is the **palmaris longus.** This is a slender muscle arising from the medial epicondyle with other muscles of the flexor group (thus from the common flexor tendon) and ending in a long but narrow tendon that passes superficially across the wrist to end in the palmar aponeurosis. The muscle is lacking on one or both sides in about 12 per cent of persons and its absence seems to be determined by hereditary factors. When present, the palmaris longus aids in flexion at the wrist, and may also assist in pronation. It is innervated by a single branch of the median nerve.

Lying to the radial side of the palmaris longus and partly covered at its origin by the humeral head of the pronator teres is the **flexor carpi radialis.** This arises from the common flexor tendon and passes obliquely downward across the wrist and into the hand to insert upon the base (proximal end) of the second metacarpal. At the wrist the tendon of the flexor carpi radialis possesses a synovial sheath that extends almost to its insertion, protecting it as it passes through the radial attachment of the flexor retinaculum (not behind the retinaculum with the flexor tendons of the digits). The muscle is a flexor of the wrist and because of its radial insertion it probably may also act as a radial abductor; in addition, its obliquity allows it to assist in pronation. This muscle, like the two preceding, is innervated by the median nerve, sometimes through two branches.

On the ulnar side of the palmaris longus is the **flexor carpi ulnaris,** which arises both from the humerus through the common flexor tendon and from an aponeurosis that passes across the surface of the deeper-lying flexor digitorum profundus to attach to the ulna. Between the two heads of origin the ulnar nerve passes into the forearm. The tendon of insertion of the flexor carpi ulnaris attaches to the pisiform bone (a carpal bone of the proximal row—fig. 11-1) on the ulnar side of the hand. Through the pisohamate and pisometacarpal ligaments (that connect the pisiform to a carpal of the distal row, the hamate bone, and to the fifth metacarpal) the action of this muscle is continued across the entire wrist joint. It is, therefore, a better flexor of the wrist than would appear from an inspection of its insertion. The muscle is also an ulnar abductor. In contrast to most of the flexor muscles of the forearm, the flexor carpi ulnaris is innervated by the ulnar nerve through two to four branches.

The **flexor digitorum superficialis** (fig. 8-2) forms an intermediate layer between the superficial and deep groups. Its origin is by two

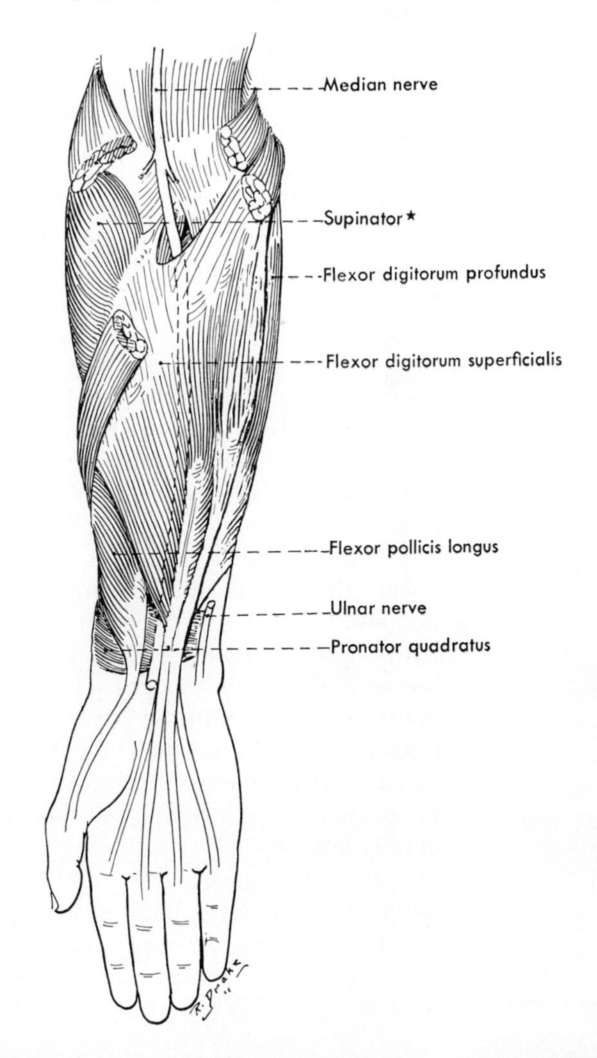

Figure 8–2. The deeper flexor muscles of the right forearm. Starred muscle is an extensor visible from the anterior side.

heads, a humeroulnar head with the common flexor tendon from the medial epicondyle and the coronoid process of the ulna, and a radial, broader but thinner, head from the upper half of the radius below the radial tuberosity. Between the two heads of the muscle the median nerve and the ulnar artery pass more deeply. Behind the muscle the median nerve clings to its posterior surface and runs almost straight distally, while the ulnar artery runs obliquely toward the ulnar side and passes deep to the flexor carpi ulnaris. From the combined muscular belly four tendons arise; as these tendons reach the wrist they are arranged in two layers, the two anterior tendons being destined for long and ring fingers while the posterior two go to index and little fingers. At the wrist the tendons of this muscle pass deep to the flexor retinaculum, where they are surrounded by a large tendon sheath (the common flexor tendon sheath, containing also the tendons of the deep flexor of the fingers) that facilitates their free movement in this position. The four tendons of the flexor digitorum superficialis diverge after passing behind the flexor retinaculum and run out along the digits to attach to their middle phalanges. In its course on the finger each tendon is enclosed in a digital tendon sheath (p. 172) with the tendon of the deep flexor to that finger. It splits to allow this tendon (flexor digitorum profundus) to pass through; the two bands interchange some fibers behind this deep tendon and then each inserts onto the sides of the palmar surface of the base of the middle phalanx. The flexor digitorum superficialis is primarily a flexor at the proximal interphalangeal joint. It is supplied by several branches from the median nerve.

The deep muscles of the flexor group are the flexor digitorum profundus, the flexor pollicis longus, and the pronator quadratus (fig. 8-3), the latter lying behind the two former. The **flexor digitorum profundus** has an extensive origin from the anterior and medial surface of the proximal two thirds or more of the ulna, and arises also from the aponeurosis through which the flexor carpi ulnaris attains attachment to the ulna. Like the superficialis, this muscle ends in four tendons, but unlike the superficialis the four tendons are arranged at the wrist in the same plane. These tendons pass into the common flexor tendon sheath at the wrist and thus lie behind both the flexor retinaculum and the superficial flexor tendons of the fingers. As the tendons diverge toward the fingers after passing beyond the flexor retinaculum they lie immediately behind the superficial flexor tendons, and within the tendon sheaths on the fingers they run through the divided portions of the superficial tendons (fig. 11-8). Passing across the interphalangeal joints, they insert on the bases of the distal phalanges of each of the four fingers. The portion of the muscle going to the index finger is usually separate from the rest of the muscle for some distance in the lower part of the forearm, but the parts going to the other fingers may or may not form a single tendon almost to the level of the wrist. In the former case, flexion of the distal phalanges of one of these fingers without flexing all is impossible.

The ulnar nerve lies on the front of (anterior to) the flexor digitorum profundus, between that muscle and the overlying flexor carpi ulnaris, as soon as it rounds the medial epicondyle and enters the forearm. The ulnar artery joins it deep to the flexor carpi ulnaris to run downward on the profundus lateral to the nerve. The muscle is primarily a flexor at the distal interphalangeal joints but is also secondarily a good flexor at the proximal interphalangeal joints.

The innervation of the flexor digitorum profundus is double, a radial portion of the muscle being supplied by the median nerve (through its anterior interosseous branch) while an ulnar portion is supplied by the ulnar nerve. The exact amount of the muscle supplied by each nerve varies from one individual to another. In one study about 50 per cent of the subjects were found to have the flexor profundus to the index and long fingers supplied exclusively by the median nerve and that to ring and little fingers exclusively by the ulnar nerve. Where the muscle was not equally divided between median and ulnar nerve innervation the median nerve more frequently supplied or helped supply some of the ulnar half of the muscle, while the ulnar nerve less commonly supplied the portion of the muscle connected with the long finger. The part of the profundus to the index finger is apparently always supplied by the median nerve.

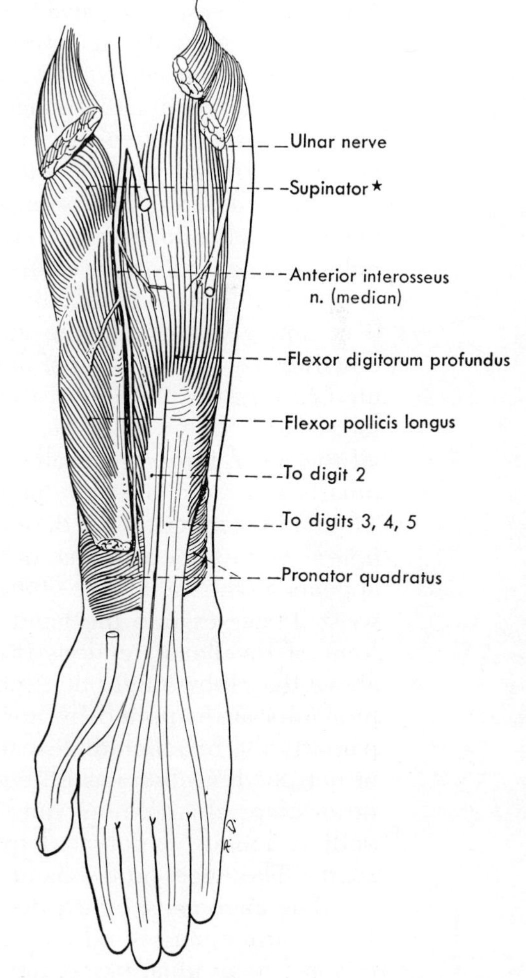

Figure 8–3. The deepest flexor muscles of the right forearm. Starred muscle is an extensor visible from the anterior side.

The **flexor pollicis longus** arises from about the middle half of the anterior surface of the radius and the adjacent interosseous membrane; its tendon has a separate tendon sheath behind the flexor retinaculum, on the radial side of the common flexor tendon sheath, and passes to an insertion on the base of the distal phalanx of the thumb. This muscle is innervated by the anterior interosseous branch of the median nerve, and is the flexor of the distal phalanx of the thumb.

The **pronator quadratus** is a flat quadrangular muscle passing almost transversely but with a slightly distal slant from approximately the distal fourth of the ulna to a corresponding part of the radius. It is supplied by the anterior interosseous branch of the median nerve and initiates pronation, being assisted by the pronator teres when more speed or power are required.

NERVES AND VESSELS

Aside from cutaneous branches, the nerves of the anterior aspect of the forearm (figs. 8-2 to 8-4) are only two in number, the median and the ulnar. The **median nerve** passes into the forearm with the brachial artery, lying medial to it on the surface of the brachialis muscle. While it supplies no muscles of the arm, branches to both the pronator teres and the flexor carpi radialis may arise slightly above the elbow. As and after it passes between the two heads of the pronator, it gives off branches to that muscle and to the palmaris longus, flexor carpi radialis, and flexor digitorum superficialis in no regular order; the nerve to the palmaris may arise with a branch to the flexor carpi radialis. The median nerve then passes between the two heads of the flexor digitorum superficialis, may give off additional branches to that, and gives rise to the anterior interosseous nerve. This runs downward along the front of the interosseous membrane to be distributed to a lateral portion of the profundus and to the flexor pollicis longus and the pronator quadratus. The median nerve thus supplies all the flexor forearm muscles with the exceptions of the flexor carpi ulnaris and a variable ulnar portion of the flexor digitorum profundus. The main stem of the median nerve continues down the forearm adherent to the deep surface of the flexor digitorum superficialis, and appears on the radial side of the tendons of this muscle just above the wrist. It passes into the hand behind the flexor retinaculum but in front of the flexor tendons (fig. 11-11). Severing the median nerve above the elbow might be expected to prevent pronation, since both pronators are supplied by it, but only weakens the movement: apparently the brachioradialis can pronate, whether it does so normally or not. Such section has little effect on flexion at the wrist, since the flexor carpi ulnaris, innervated by the ulnar nerve, and the abductor pollicis longus, innervated by the radial nerve, can produce that action. The effect on the hand is discussed later.

The **ulnar nerve** enters the forearm between the two heads of the flexor carpi ulnaris. Under cover of this muscle it gives off branches to this and to an ulnar part of the profundus and, continuing downward, crosses superficial to the flexor retinaculum to enter the hand. Above the wrist it gives rise to a small palmar branch that supplies skin of the hypothenar eminence, and a larger dorsal branch that gives rise to the ulnar nerve's dorsal digital branches. As it enters the palm the ulnar nerve divides into superficial and deep branches; the former gives rise to the palmar digital branches, while the latter disappears into the muscles of the hypothenar eminence. At the level of the medial epicondyle the ulnar nerve is subject to damage by being stretched across the epicondyle, from a roughness of the ulnar groove, or from compression by a fibrous band extending between the two heads of the flexor carpi ulnaris. The condition has been treated by transplanting the nerve to a shorter course in front of the epicondyle, resecting the epicondyle, or dividing the fibrous band and a little of the adjacent muscle. Almost all the effects of ulnar nerve injury are in the hand. Paralysis of the flexor carpi ulnaris is hard to detect, but there may be some weakness in ulnar deviation of the hand.

Since the median nerve arises from both medial and lateral cords of the brachial plexus, it can contain fibers from all of the spinal nerves contributing to the plexus. It does receive fibers from all, but has relatively few motor fibers from the fifth cervical. The ulnar nerve, arising from the medial cord, contains fibers from those segments contributing to the medial cord, the eighth cervical and first thoracic nerves, and sometimes gets some from the seventh cervical by a communication from the lateral cord. Of the flexor muscles in the forearm, the pronator teres and flexor carpi radialis are usually supplied mostly by sixth and seventh cervical nerve fibers (through the median nerve), the palmaris longus by seventh and eighth. The flexor digitorum superficialis usually receives fibers from C7, C8, and T1 and so does the flexor pollicis longus, while the other muscles—the flexor carpi ulnaris, the flexor digitorum profundus, and the pronator quadratus—are usually supplied by fibers from the eighth cervical and first thoracic nerves. Thus lesions of the lower part of the brachial plexus may badly cripple flexor movements of both wrist and fingers, while lesions of the upper part of the plexus will affect, but not abolish, movements of pronation, wrist flexion and radial abduction.

The **radial artery** is one of the two terminal divisions of the brachial artery (fig. 8-4). After giving off a recurrent branch that runs upward in front of the lateral aspect of the elbow, it runs downward, at first under cover of the brachioradialis but later covered only by skin and fascia, on the radial side of the anterior surface of the forearm. At the wrist it winds dorsally deep to the extensor tendons of the thumb to follow a course that is described in connection with the hand.

The **ulnar artery,** the other terminal branch of the brachial, is at first larger than the radial; after passing behind the pronator teres it continues between the two heads of the flexor digitorum superficialis, deep to which it gives off recurrent branches about the elbow that anastomose with branches of the brachial artery, and also a large common interosseous artery. Escaping from under cover of the flexor superficialis the ulnar artery then passes downward under cover of the flexor carpi ulnaris and in company with the ulnar nerve, and enters the hand superficial to the flexor retinaculum, on the radial side of the

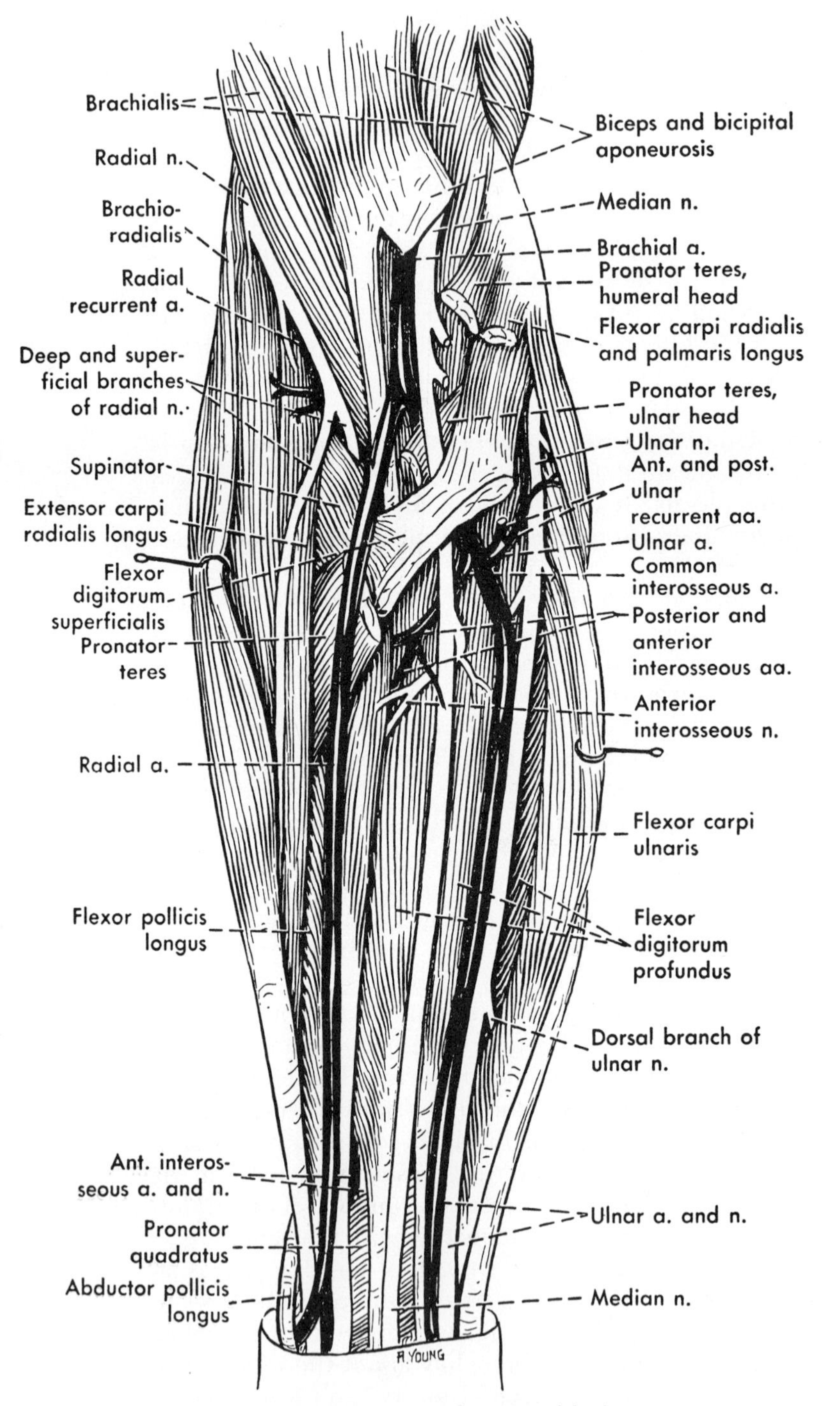

Figure 8–4. Anterior nerves and arteries of the forearm.

ulnar nerve. The common interosseous artery divides into anterior and posterior interosseous branches. The posterior artery passes between the radius and ulna, in the gap above the interosseous membrane, to supply extensor muscles of the forearm. The anterior interosseous artery passes down the forearm in company with the corresponding branch of the median nerve. At the lower end of the forearm it supplies branches to the palmar surface of the carpus and a larger perforating branch that passes through the interosseous membrane to supplement or unite with the lower end of the posterior interosseous artery.

REVIEW OF THE FLEXOR FOREARM

There is not much bony anatomy to be reviewed on the living forearm. The ulna is subcutaneous throughout its length, and can be felt to separate the anteromedial flexor muscles from the muscles on the posterior side of the forearm. The rounded head of the ulna produces a bulge on the posterior side of the forearm just above the wrist. The upper part of the radius is covered not only laterally and posteriorly but also anteriorly by the extensor muscles, but its head can be felt immediately below the lateral epicondyle and its distal half can be palpated and traced to its expanded lower end. Its rather broad styloid process can best be felt on the extreme radial border of the anterior surface of the wrist, just anteromedial to the prominent tendons extending to the base of the first metacarpal, and at about the level of the proximal crease at the wrist. Ridges on two of the carpal bones (scaphoid and trapezium) form the rounded projection, largely distal to the distal crease of the wrist, at the base of the thenar eminence. On the ulnar side the slightly movable pisiform bone (one of the carpals) can be felt at the level of the distal crease, while just distal to it, at the base of the hypothenar eminence, the unyielding projection of another carpal (the hamate bone) can be palpated. The flexor retinaculum, although it cannot be distinctly felt, stretches between the bony prominences on the radial and ulnar sides. Thus the flexor tendons palpable proximally cannot be felt at this level.

Muscles and Tendons

The hollow in front of the elbow between the lateral and medial muscle masses is the cubital fossa; the biceps tendon has already been traced into it, and the bicipital aponeurosis traced medially over the medial muscle mass. The fact that the lateral mass is composed of extensor muscles can be readily confirmed by palpating it when the wrist is dorsiflexed (hyperextended). Similarly, the fact that the medial mass is composed of flexor muscles and arises in part from the medial epicondyle can be confirmed by palpating it when the fingers are clenched or the wrist flexed or both, but few of the individual muscles can be identified at this level. Perhaps the easiest to identify is the pronator teres, which can be felt, when the forearm is slightly flexed and strongly pronated, as the medial border of the cubital fossa. The posteromedial border of the flexor carpi ulnaris can be identified, and the muscle traced to its tendon at the wrist, by palpating deeply in front of the ulna when the hand with extended fingers is sharply abducted ulnarward. Contraction of the flexor digitorum profundus can be recognized by palpating in this same place, between the ulna and the flexor carpi ulnaris, and strongly flexing the fingers.

Several tendons can be recognized without difficulty at the wrist. Unless the muscle is missing, the rather thin sharp tendon of the palmaris longus can usually be both palpated and seen in the midline of the wrist when the hand is slightly flexed. The tendon is particularly prominent because it passes in front of the flexor retinaculum. Very close to it on its radial side is the broader tendon of the flexor carpi radialis, best brought out by flexing the hand against resistance. It may seem to end at the bony prominence at the base of the thenar eminence, but actually runs just medial (ulnarward) to this. The most ulnar tendon on the flexor side of the wrist is that of the flexor carpi ulnaris; it is most easily identified when the wrist is simultaneously slightly flexed and slightly abducted in the ulnar direction, and can be traced to the pisiform bone. Between the palmaris longus and the flexor carpi ulnaris, lying more deeply, flexor digitorum superficialis tendons can be felt.

Vessels and Nerves

The superficial veins of the forearm can often be seen fairly clearly. Of the arteries, the radial can be palpated, as is usually done in feeling the pulse, by pressing it lightly against the radius where it is superficially located in the lower part of the forearm; it lies lateral (radial) to the tendon of the flexor carpi radialis. The ulnar artery is difficult to palpate, but at the wrist it lies just radial to the tendon of the flexor carpi ulnaris and the pisiform bone. Knowing the locations of these vessels at the wrist, and the position of the parent brachial artery just medial to the biceps tendon at the elbow, one should easily visualize the approximate courses of the vessels.

The ulnar nerve can be identified behind the medial epicondyle just before it enters the forearm. It is under cover of or too close to the flexor carpi ulnaris to be palpable elsewhere, but at the wrist emerges from behind the radial border of the tendon of this muscle and, with the ulnar artery radial to it, passes across the radial side of the pisiform bone. Its course is therefore practically straight down the forearm. The median nerve also runs practically straight down the forearm but in its middle. Although it is not palpable, its course can be visualized by remembering that it lies at the elbow just medial to the tendon of the biceps, and at the wrist lies behind or behind and slightly radial to the tendon of the palmaris longus.

The innervation of the muscles of the flexor group in the forearm,

TABLE 8–1. Nerves and Muscles of Flexor Forearm

NERVE AND ORIGIN*	MUSCLE *Name*	*Segmental innervation**	*Chief action*
Median C5–T1	Pronator teres	C6, 7	Pronation of forearm
	Pronator quadratus	C7–T1	Pronation of forearm
	Flexor carpi radialis	C6, 7	Flexion at wrist
	Palmaris longus	C7, 8	Flexion at wrist
	Flexor digitorum superficialis	C7–T1	Flexion of middle phalanges of fingers
	Flexor pollicis longus	C7–T1	Flexion of distal phalanx of thumb
	Flexor digitorum profundus, radial part	C8, T1	Flexion of distal phalanges of digits II and III
Ulnar C8, T1	Flexor digitorum profundus, ulnar part	C8, T1	Flexion of distal phalanges of digits IV and V
	Flexor carpi ulnaris	C8, T1	Flexion-adduction at wrist

*A common segmental origin or innervation. The composition of both the chief nerves and their muscular branches varies somewhat among persons–the median nerve may contain no fibers from C5 or none from T1; the ulnar nerve frequently contains fibers from C7, but their distribution is not known.

and the chief actions of these muscles, are summarized in Table 8-1. The actions of the muscles of the forearm as a whole are discussed in more detail in Chapter 10, and the innervation of the flexor group is shown diagrammatically in figures 7-1 and 7-2.

9 Extensor Forearm

The superficial nerves and vessels and the fascia of the posterior side of the forearm have already been described (p. 135). In introduction to the musculature, the student may recall that all the extensor muscles are not actually posteriorly placed in the forearm, for some arise from the front of the lower end of the humerus, pass across the front of the elbow joint, and are visible anteriorly (for instance, fig. 8-1). However, some are entirely posteriorly placed, and the insertions of the extensor muscles are almost entirely on the radial and extensor sides of the limb. It might also be recalled that the tendons of the extensor muscles that cross the wrist are housed in a number of separate compartments deep to a special thickening of the deep fascia, the extensor retinaculum (fig. 9-1). Each compartment has a single tendon sheath, so where two or more tendons share a compartment they also share the sheath.

MUSCULATURE

All the muscles of the extensor surface of the forearm are innervated through the radial nerve, and many of them have a common origin from the lateral epicondyle of the humerus. The extensor muscles are conveniently divided into two groups; a superficial and a deep. All the members of the superficial group have an origin from the humerus, but most of those of the deep group do not. In the lower part of the posterior side of the forearm some of the members of the deep group (muscles to the thumb) become superficial and cover some of the superficial muscles. The superficial extensor forearm muscles are the brachioradialis, the extensor carpi radialis longus, the extensor carpi radialis brevis, the extensor digitorum, the extensor digiti minimi, and the extensor carpi ulnaris.

The **brachioradialis** (barely visible in fig. 9-1, and seen better in fig. 8-1) is the most anterior member of the superficial group; it arises anteriorly from the lateral border of the humerus well above the epicondyle and from the lateral intermuscular septum of the arm, and inserts on the lateral side of the lower end of the radius. It is primarily a flexor at the elbow, being used particularly to add speed or power. It has been said to pronate from a position of supination, and to supinate from one of pronation; it is not efficient, and may not be normally used in either action, but apparently can pronate better than it can supinate.

Partly under cover of the brachioradialis and arising largely above the lateral epicondyle from the humerus and the lateral intermuscular septum is the **extensor carpi radialis longus** (fig. 9-1). Its lowest fibers arise from the lateral epicondyle by a common extensor tendon which it shares with other muscles of the superficial group. Closely associated with it is the **extensor carpi radialis brevis.** It and the remaining muscles arise not only from the epicondyle but also from intermuscu-

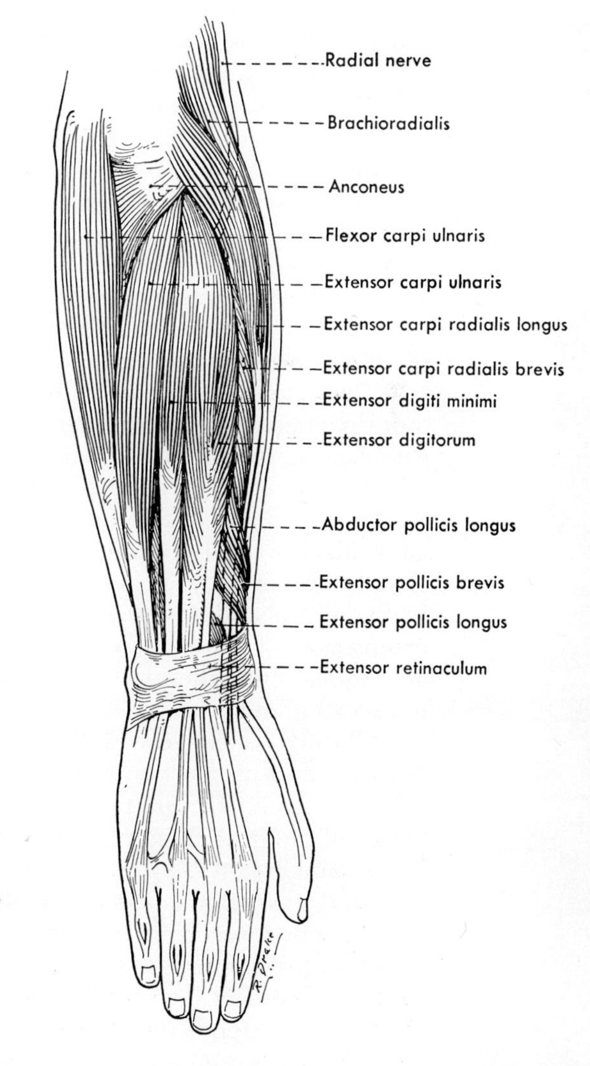

Figure 9–1. The superficial extensor muscles of the right forearm.

lar septa and the covering antebrachial fascia. The two muscles can be traced downward deep to the extensor muscles of the thumb, then deep to the extensor retinaculum, where they have a common compartment and tendon sheath, to their insertions on the bases of the second and third metacarpals (the longus on the second, the brevis on the third). Both muscles extend or dorsiflex the hand, and the long extensor, at least, helps to abduct it radially; in extending the hand, the short extensor apparently acts alone unless more speed or power is required. The long extensor probably, the short one perhaps, can contribute to flexion of the forearm.

The radial nerve runs under cover of these muscles on the front of the humerus, between them and the brachialis, and while it is still anteriorly situated supplies the brachioradialis and both extensors. In this position also it divides into superficial and deep branches. The superficial branch runs obliquely downward deep to the brachioradialis, to emerge on the dorsal side of the muscle's tendon a little above the wrist and continue onto the hand; the deep branch enters the supinator muscle (see below) and spirals around the radius in this muscle to the posterior side of the forearm. It emerges deep to the extensor digitorum and immediately breaks up into a number of branches for the remaining muscles of the extensor group.

The **extensor digitorum** occupies much of the posterior surface of the forearm; it arises from the lateral epicondyle, intermuscular septa, and the antebrachial fascia, and splits into three or four tendons as it reaches the wrist. These tendons pass deep to the extensor retinaculum in a tendon sheath common to them and another muscle (extensor indicis). On the hand they diverge to the four fingers, but the tendons are united by obliquely placed bands that limit the independent movement of any one tendon. The tendon of the little finger is typically small and may not be present.

The extensor tendons to the index and to the little finger, when there is one, unite with the tendons of the extensor indicis and the extensor digiti minimi, respectively (these are among the muscles yet to be examined). On the fingers, the tendons receive the insertions of the interossei and of the lumbricals (muscles in the hand) and thereafter attach to both middle and distal phalanges—see also page 191. Expansions from them form the posterior capsules of the metacarpophalangeal and interphalangeal joints. The extensor digitorum is an extensor of all joints of the fingers, but it can extend the interphalangeal joints only when the metacarpophalangeal joints are kept from hyperextending.

Closely associated with the extensor digitorum and appearing indeed as an ulnar portion of this muscle is the **extensor digiti minimi.** Arising in common with the extensor digitorum, the tendon of the extensor of the little finger diverges at the wrist and passes through its own compartment deep to the extensor retinaculum. On the dorsum of the hand its tendon is usually double; the radial tendon or the undivided tendon receives the extensor digitorum tendon to the little finger, or a slip from the tendon to the ring finger, and the combined tendons insert on the middle and distal phalanges of the fifth digit. The muscle extends the little finger alone, and also abducts it.

The **extensor carpi ulnaris** has an origin similar to the preceding muscles, and a more extensive origin from somewhat more than the upper half of the posterior border of the ulna. Its tendon of insertion passes in its own compartment deep to the extensor retinaculum, to an attachment upon the base of the fifth metacarpal. The extensor carpi ulnaris aids in extension and ulnar abduction of the hand.

The multiple branches of the deep radial nerve to these three muscles enter their deep surfaces, fanning out from the nerve just as it emerges from the lower border of the supinator. One branch of the deep radial, the posterior interosseous nerve, runs downward on and between the deep group of muscles in company with the posterior interosseous artery to supply these muscles, sometimes giving an additional supply to the extensor digitorum and extensor digiti minimi.

The deep extensor muscles are the supinator, three thumb muscles—the abductor pollicis longus, extensor pollicis brevis, and extensor pollicis longus—and the extensor indicis (fig. 9-2). Of these, the **supinator** is the most proximal; it arises from the posterolateral surface of the ulna just below the radial notch, from the lateral epicon-

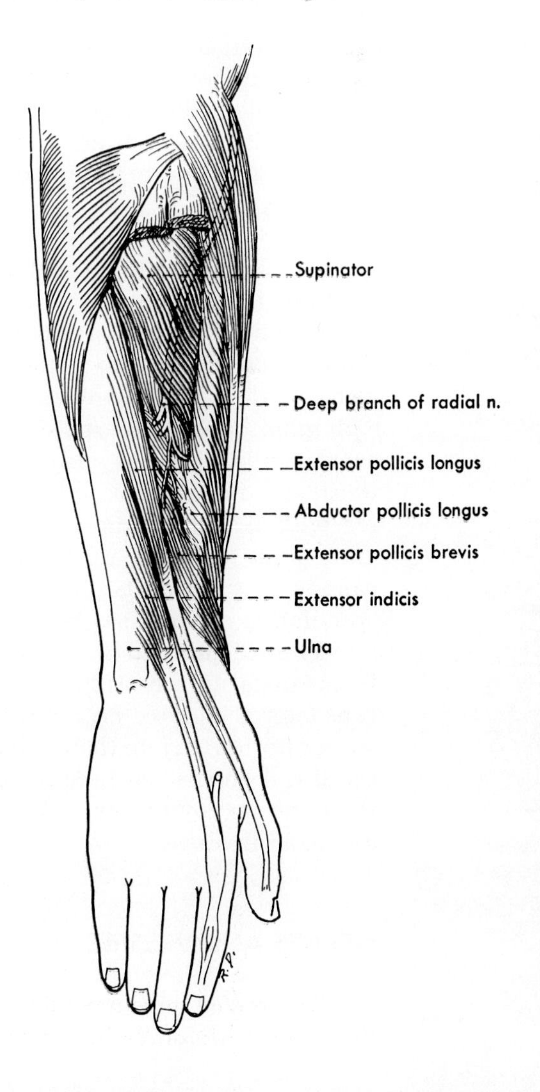

Figure 9–2. The deeper extensor muscles of the right forearm.

dyle, and from the radial collateral and annular ligaments; it passes obliquely downward and laterally across the arm to insert on the lateral and adjacent posterior and anterior aspects of the radius for a considerable distance below the radial head. The deep radial nerve separates its fibers into a superficial and a deep lamina. The muscle supinates, as its name indicates.

Of the three muscles of the thumb, the first two emerge between the extensor digitorum and the radial extensors and cross superficial to the latter. The **abductor pollicis longus** takes origin from the posterior surface of the ulna below the origin of the supinator, from the posterior surface of the radius below the insertion of the supinator, and from the intervening interosseous membrane; it passes to an insertion on the front (radial side) of the base of the first metacarpal. Often the tendon splits to attach also to the trapezium or to fascia or muscles of the thenar eminence. The **extensor pollicis brevis** is partly covered by the abductor longus; it arises from the radius and the interosseous membrane below the radial origin of the long abductor, and goes to an insertion on the proximal phalanx of the thumb. It and the long abductor usually share a compartment and tendon sheath deep to the extensor retinaculum. The **extensor pollicis longus** arises from about the middle third of the ulna and the adjacent interosseous membrane, largely below the origin of the long abductor; it crosses the wrist obliquely to proceed along the thumb to the distal phalanx. It also crosses superficial to the radial extensors, but at the wrist rather than in the forearm.

The abductor pollicis longus extends and externally rotates the first metacarpal, and thus restores this bone to its normal position (repositions it) after opposition of the thumb; by virtue of its location at the wrist it is also both a radial abductor and a flexor of the hand. The extensor pollicis brevis extends both the proximal phalanx and the metacarpal of the thumb, and the extensor pollicis longus extends both phalanges and extends and adducts the metacarpal. The extensor pollicis brevis also helps radially abduct the hand at the wrist, and the extensor pollicis longus helps extend it. Both the abductor pollicis longus and the extensor pollicis longus have been regarded as assisting in supination (the extensor brevis obviously cannot, in spite of its obliquity, since it arises from the radius), but stimulation of them has failed to produce supination.

The **extensor indicis** arises from the ulna and interosseous membrane distal to the origin of the extensor pollicis longus; its tendon runs laterally across the wrist, within the tendon sheath for the extensor digitorum, to join the medial side of this muscle's radial tendon at the distal end of the second metacarpal. This muscle therefore, like the extensor digitorum, extends all joints of the index finger; it also adducts this finger.

NERVES AND VESSELS

The **radial nerve** emerges from its position deep to the triceps on the lateral side of the lower part of the arm to lie between the brachialis muscle and, first, the brachioradialis and then the extensor carpi radialis longus mucles (fig. 9-3). In this position it gives off branches to both the brachioradialis and the extensor carpi radialis longus; the nerve supply to the brevis arises a little lower, from the radial nerve proper or from one of the two chief branches of this nerve. Soon after it enters the forearm the radial nerve splits into superficial and deep branches. The superficial branch lies under cover of the brachioradialis muscle for much of its course but emerges from deep to the tendon of this muscle in the lower part of the forearm to be distributed to skin on the dorsum of the hand, and to a variable number of joints. The deep branch of the radial nerve plunges into the supinator muscle, supplying this as it passes through, and following the muscle around the

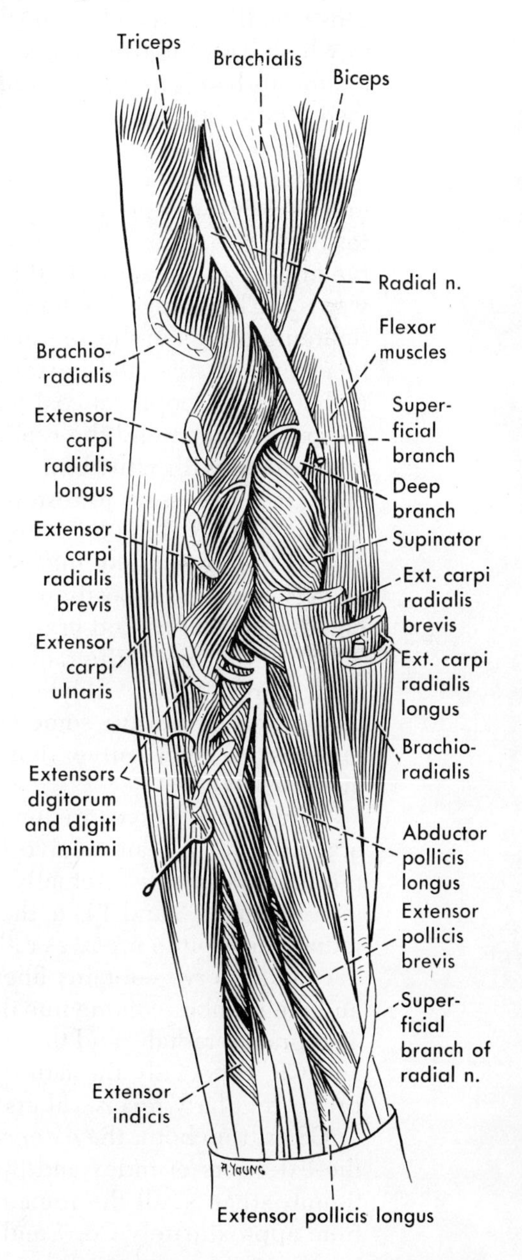

Figure 9–3. The radial nerve in the forearm.

radius to reach the posterior aspect of the forearm. Here, under cover of the superficial posterior extensor muscles it breaks up into a number of branches to the remaining extensor muscles just as it emerges at the lower border of the supinator. The continuation of the deep radial, called the posterior interosseous nerve, extends downward in company with the posterior interosseous artery across the superficial surface of the abductor pollicis longus and gives one or more branches into each of the long thumb muscles and the extensor of the index finger; a filament continues deep to the extensor pollicis longus, on the interosseous membrane, to the wrist joint.

The radial nerve is especially subject to injury in fracture of the body of the humerus, because of its close association with that bone. This usually occurs when the fracture is in the lower third of the bone, in which case the triceps is spared. The nerve may also be entrapped by fibrous bands associated with the lateral head of the triceps or with the entrance of the deep branch into the supinator. In the former case many of the nerves to the triceps are spared; in the latter, as in fracture of the upper third or half of the radius, the brachioradialis and the two radial extensors would also be spared. Paralysis of the radial extensors results in wristdrop, evident when the forearm is held horizontal and the fingers are relaxed—in this position the hand hangs loosely downwards. (Clenching the fingers pulls upon their extensor tendons, tightening them and extending the wrist.) The phalanges also cannot be extended, since their extensors also would be paralyzed; however, if the metacarpophalangeal joints are fixed in extension, muscles in the palm will extend the remaining phalanges.

A hand with wristdrop is useless, so if the nerve does not regenerate, it is necessary to restore extension at the wrist by other means. This is usually done by transferring flexor tendons to the dorsum to extend both wrist and fingers. However, one flexor of the wrist must always be left in position; otherwise the wrist will be so sharply dorsiflexed that a good grasp is impossible.

If the lesion to the radial nerve is above the origin of the superficial branch, there will be some loss of sensation on the dorsum of the hand. This is always somewhat limited, sometimes to an area not much larger than a silver dollar situated between the first and second metacarpals.

The radial nerve, arising as it does from the posterior cord of the brachial plexus, can receive fibers from all the nerves entering into the brachial plexus. Actually, however, the posterior division of the lower trunk (C8 and T1) to the posterior cord is usually small, and the radial nerve often receives only eighth cervical fibers through it. Thus the radial nerve contains fibers derived mostly from C5, 6, 7 and 8, and the number coming into the nerve from C5 appears to be variable. The brachioradialis and the supinator are supplied primarily from C5 and C6, especially the latter; the extensor carpi radialis longus and brevis regularly receive fibers from C6 and 7, and often from either C5 or C8 or from both; the extensor digitorum, the extensor carpi ulnaris, the extensors of index and little fingers, and the long muscles of the thumb—that is, all the remaining muscles—are supplied with fibers from approximately C6, 7 and 8, mainly C7. Obviously, then, lesions of the brachial plexus centering around C6 and C7 will markedly affect the extensor forearm muscles, while injuries above or below these levels may have little or no effect on these muscles.

The more anterior extensor muscles are supplied by a recurrent branch from the radial artery that runs up along them and the radial nerve in front of the lateral epicondyle, but the **posterior interosseous artery** (fig. 9-4) is the chief vessel supplying the extensor muscles on the posterior side. After leaving the common interosseous artery on the anterior side of the forearm it passes between the radius and ulna to reach the posterior aspect deep to the supinator muscle, and then courses downward on the interosseous membrane, giving off branches to the various muscles. At the wrist it may be reinforced by the perforating branch of the anterior interosseous artery, and the latter vessel or the common terminal stem formed by the two arteries supplies branches to the dorsal aspect of the wrist.

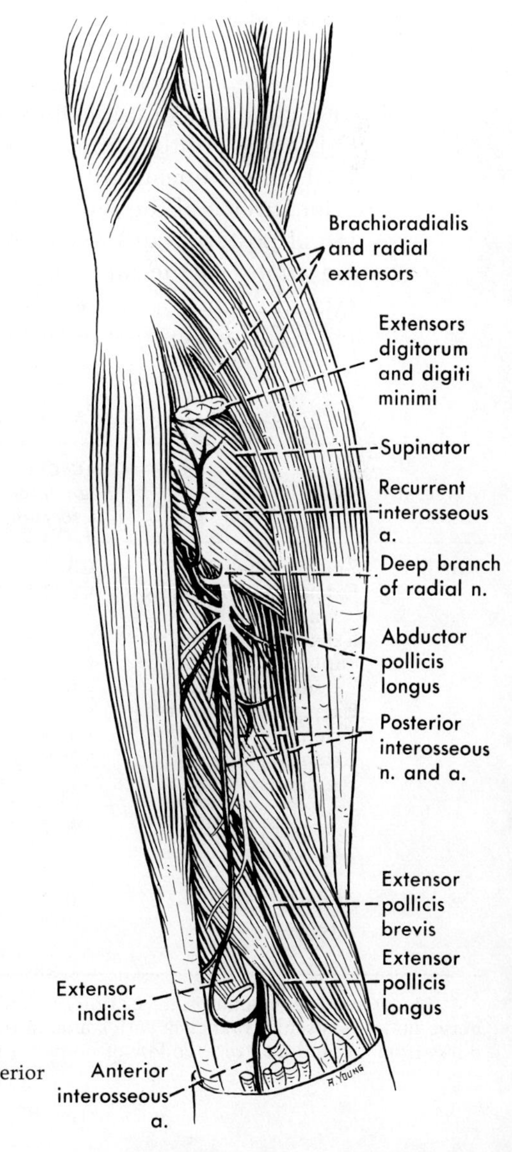

Figure 9–4. Nerves and arteries of the posterior aspect of the forearm.

REVIEW OF EXTENSOR FOREARM

The bony anatomy of the forearm has already been reviewed (p. 145). The only prominent feature of the posterior surface is the rounded head of the ulna; the styloid process can be felt with some difficulty as it projects distally at the junction of posterior and ulnar surfaces of the wrist.

The muscular bulge that forms the lateral border of the cubital fossa is produced by the brachioradialis and the two radial extensors. The brachioradialis can be differentiated by forceful flexion of the forearm with the thumb up; it then stands out at the elbow lateral to the biceps tendon. Contraction of the radial extensors can be felt when the wrist is extended. On the posterior side of the limb, contraction of the extensor digitorum can be felt when the fingers are extended, but otherwise the posterior muscles are difficult to identify in the forearm. The tendons of the extensor digitorum and of the extensors of the index and little fingers can, however, be palpated and frequently seen beneath the skin of the dorsum of the hand.

The tendons palpable or visible at the wrist are, beginning on the ulnar side, that of the extensor carpi ulnaris, largely covering the styloid process of the ulna; the extensor digitorum in the middle of the wrist; the extensor pollicis longus, running obliquely from the posterior surface of the wrist onto the dorsum of the thumb, at the junction of posterior and radial surfaces of the wrist; and, anterior to this and almost at the junction of the radial and anterior surfaces of the wrist, the tendons of the extensor pollicis brevis and abductor pollicis longus. The depression on the radial side of the wrist between the

TABLE 9–1. Nerves and Muscles of Extensor Forearm

NERVE AND ORIGIN*	MUSCLE *Name*	*Segmental innervation**	*Chief action(s)*
Radial C5–C8	Brachioradialis	C5, 6	Flexion at elbow
	Extensor carpi radialis longus and brevis	C6, 7	Extension-abduction at wrist
	Extensor carpi ulnaris	C6–C8	Extension-adduction at wrist
	Supinator	C5, 6	Supination of forearm
	Extensor digitorum	C6–C8	Extension all joints of digits II to V
	Extensor digiti minimi	C6–C8	Extension all joints of digit V
	Extensor indicis	C7, 8	Extension all joints of digit II
	Extensor pollicis longus	C7, 8	Extension of both phalanges, extension and adduction of metacarpal, of thumb
	Extensor pollicis brevis	C6, 7	Extension of proximal phalanx and metacarpal, of thumb, abduction at wrist
	Abductor pollicis longus	C6, 7	Extension and abduction (reposition) of thumb, flexion and abduction at wrist

*A common segmental origin or innervation. The composition of both the radial nerve and its muscular branches varies among persons; the radial nerve may receive fibers from T1, and may receive few or no motor fibers from C5.

tendon of the extensor pollicis longus and those of the extensor brevis and abductor longus is often called the "anatomic snuff box"; in it the radial artery can frequently be palpated as it runs onto the dorsal surface of the hand. The tendon of the extensor pollicis brevis overlies that of the long abductor at the wrist, but is thinner and when the thumb is extended can usually be traced some distance along the metacarpal. The broader tendon of the long abductor seems to end at the base of the metacarpal, and is best brought out by slight flexion and radial abduction of the hand.

Except for the radial artery in the "anatomic snuff box," neither vessels nor nerves associated with the extensor muscles can be palpated. Table 9-1 shows the innervation of the extensor musculature and the chief actions of the muscles. These are discussed in more detail in following chapters. The distribution of the radial nerve is shown diagrammatically in figure 7-3.

10 Radioulnar and Wrist Movements

From the descriptions already given, it should be obvious that the forearm muscles may act on the elbow joint, the wrist joint, or the fingers and thumb. The part played by the brachioradialis and other forearm muscles in flexion at the elbow has already been pointed out, and movements of the fingers can best be considered after the muscles of the hand have been studied. Here, however, we must consider the movements of pronation and supination, and movements at the wrist joint.

Movements at the Radioulnar Joints

These movements are those of pronation and those of supination, used in such common actions as turning a doorknob or a screwdriver. Pronation can be brought about (fig. 10-1) by the action of a number of muscles. The pronator quadratus pronates alone until further strength or speed is needed, at which time the pronator teres also contracts. The flexor carpi radialis will apparently pronate after it has first flexed the wrist, and the palmaris longus contracts during forceful pronation, although it may contribute little to that. The brachioradialis apparently contributes little normally to pronation, but is a better pronator than it is a supinator; it apparently participates in pronation only when the movement is resisted or when the pronator teres and quadratus are paralyzed. All these muscles except the brachioradialis are innervated by the median nerve, so that lesions of the median nerve above the elbow markedly weaken pronation, but do not abolish it.

Supination is a much stronger movement than pronation, for it is brought about (fig. 10-2) not only by the supinator but also by the biceps brachii. The supinator may act alone, but the biceps supplies most of the power; it is most effective when the forearm is flexed, and contracts for supination of the extended forearm only when the movement is resisted. (The greater strength supplied by the biceps is reflected in the design of screws; a righthanded person must supinate to drive the screw into wood.) The extensor carpi radialis longus and the brachioradialis have both been said to supinate in part from the pronated position when stimulated and thus usually are listed as supinators; however, electromyographic studies of the brachioradialis, which is in a slightly better position than the extensor longus to supinate, have indicated that its participation is probably incidental to its use in flexing the elbow. Frequently the long extensor and the long abductor of the thumb are also listed as supinators, but, as already mentioned, stimulation of them has not produced this movement.

The supinator muscles are supplied by both the musculocutaneous nerve (to the biceps) and the radial, and since all receive fibers from C5 and 6, much of the strength of this important movement of

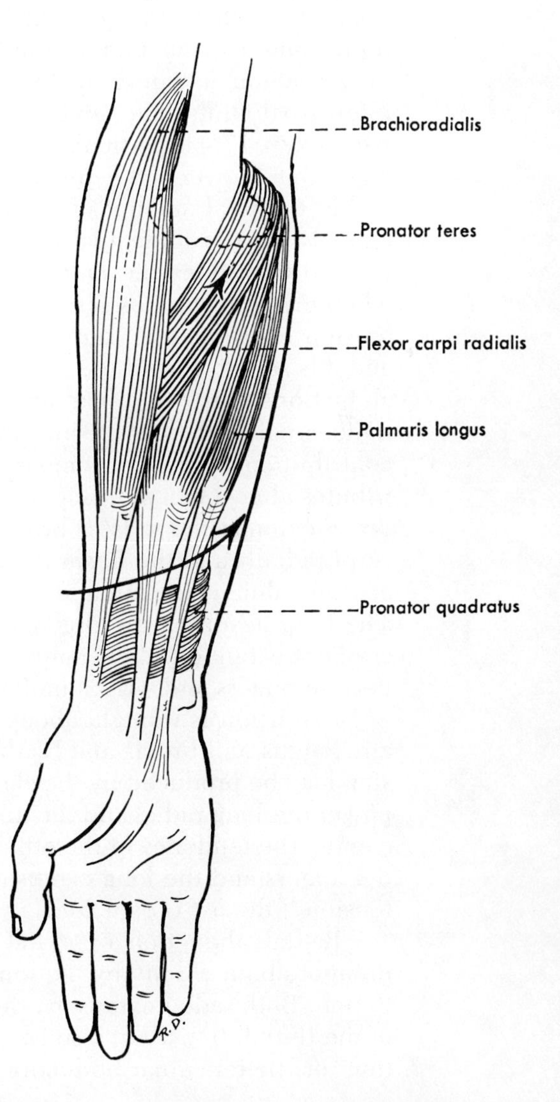

Figure 10–1. Pronators of the forearm. Note their general direction of pull, owing to the fact that they run obliquely from ulnar to radial side.

supination may be expected to be lost in damage to the upper trunk or lateral cord of the plexus. The supinator and the extensor carpi radialis longus usually receive also fibers from other cervical nerves, especially the seventh.

Movements at the Wrist Joint

These movements have already been defined as flexion, extension, and abduction (radial or ulnar). The amount of movement in any of these directions varies much from one person to another, varies appreciably according to whether the hand is pronated or supinated, and may even vary somewhat between the two hands of the same person. Because of the ellipsoidal nature of the radiocarpal joint, rotation here is obviously minimal.

With the single exception of the flexor carpi ulnaris all the muscles acting upon the wrist pass across the carpals to attach to the metacarpals, and the flexor ulnaris may be considered as inserting upon the fifth metacarpal through a pisometarcarpal ligament that represents a distal part of the tendon. Thus all these muscles exert an action not only on the radiocarpal joint but also across the intercarpal joints. The joint between the two rows of carpals (midcarpal joint, p. 170) is particularly important, for the additional movement occurring between the proximal and distal rows increases considerably the total amount of movement at the wrist. Thus it is usually agreed that the midcarpal joint contributes more to flexion (palmar flexion) than does the radiocarpal joint; there is disagreement as to whether it contributes less or more than the radiocarpal to hyperextension (dorsiflexion), but none that it does contribute appreciably; and it is said to contribute almost all of the limited movement of radial abduction. Only in ulnar abduction, which occurs mostly at the radiocarpal joint, does the midcarpal joint fail to make a very significant contribution to the movement, and according to one report it contributes almost half of this if the hand is supinated instead of pronated.

Flexion at the wrist is brought about primarily through the flexor carpi radialis and the flexor carpi ulnaris; also assisting in this action are the palmaris longus, and the long abductor of the thumb (fig. 10-3). The long flexors of the digits assist in wrist flexion only if the digits are kept extended; their range of action is too short to allow them to flex the fingers and wrist simultaneously.

Extension is brought about (fig. 10-4) by the extensor carpi radialis longus and brevis and by the extensor carpi ulnaris; pure extension may be produced by the short radial extensor alone, but for more power the long radial and the ulnar extensor both contract, each overcoming the tendency of the other to abduct the hand. The extensors of the fingers and the long extensor of the thumb can assist in wrist extension if the fist is clenched.

Radial abduction or radial deviation (or simply abduction) is brought about chiefly by the long abductor and short extensor of the thumb. Both radial extensors, the radial flexor, and the long extensor of the thumb (fig. 10-5) also contract during radial abduction. Adduction, or, better, ulnar abduction, is brought about by the combined

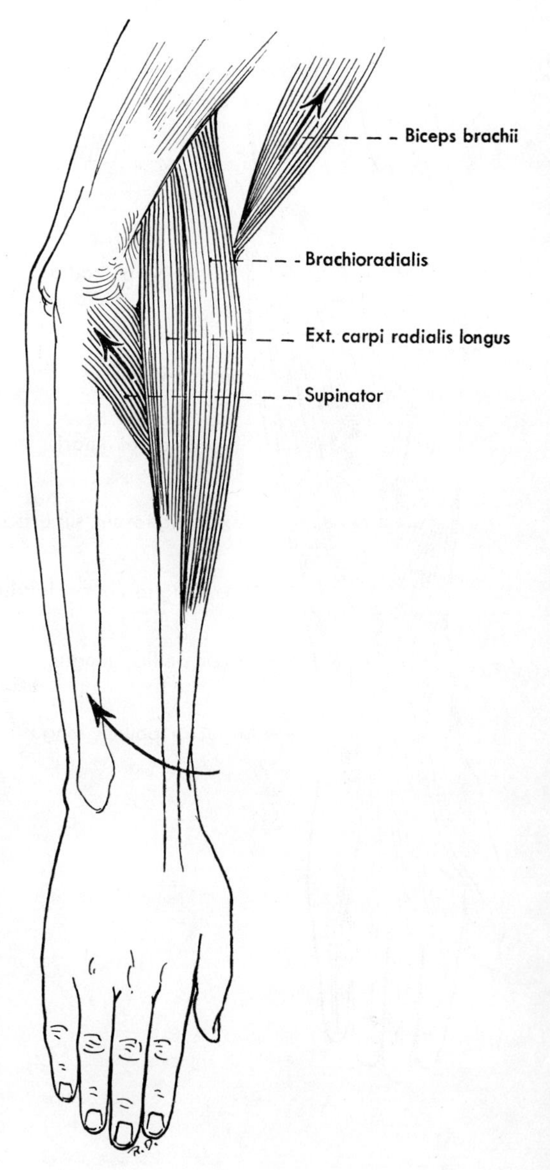

Figure 10–2. Supinators of the forearm. Note that the prevailing direction of these, like that of the pronators, is an oblique one from ulnar to radial side. The obliquity of the brachioradialis and of the extensor carpi radialis longus is considerably increased by pronation; therefore, these muscles can presumably supinate no farther than to about the neutral position between pronation and supination. They certainly contribute little strength, and whether they are normally used in supination is doubtful. The biceps is a strong supinator because of its insertion on the anteromedial surface of the radius.

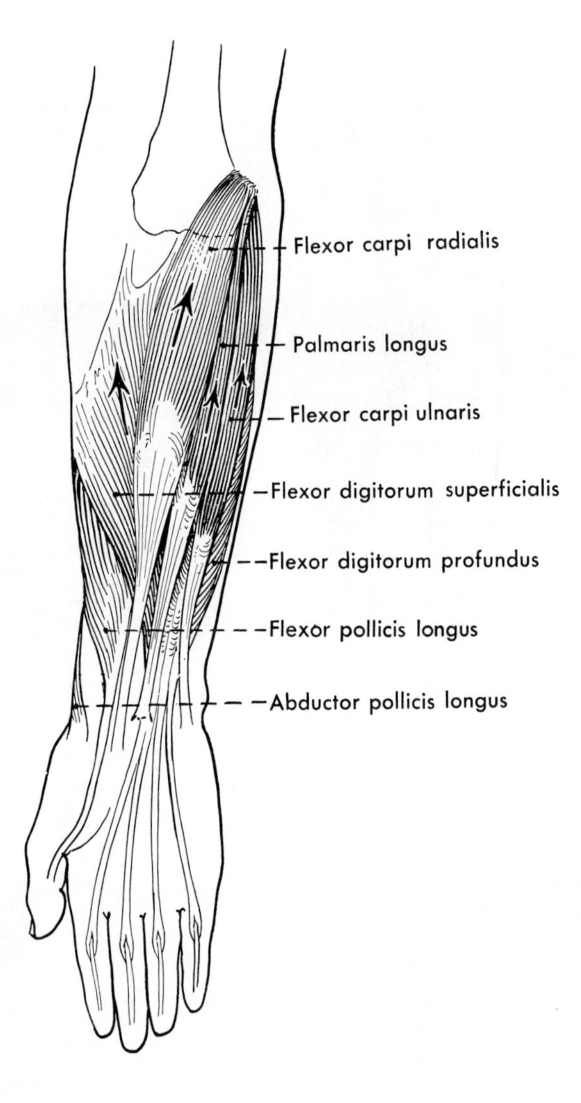

Figure 10–3. Flexors at the wrist.

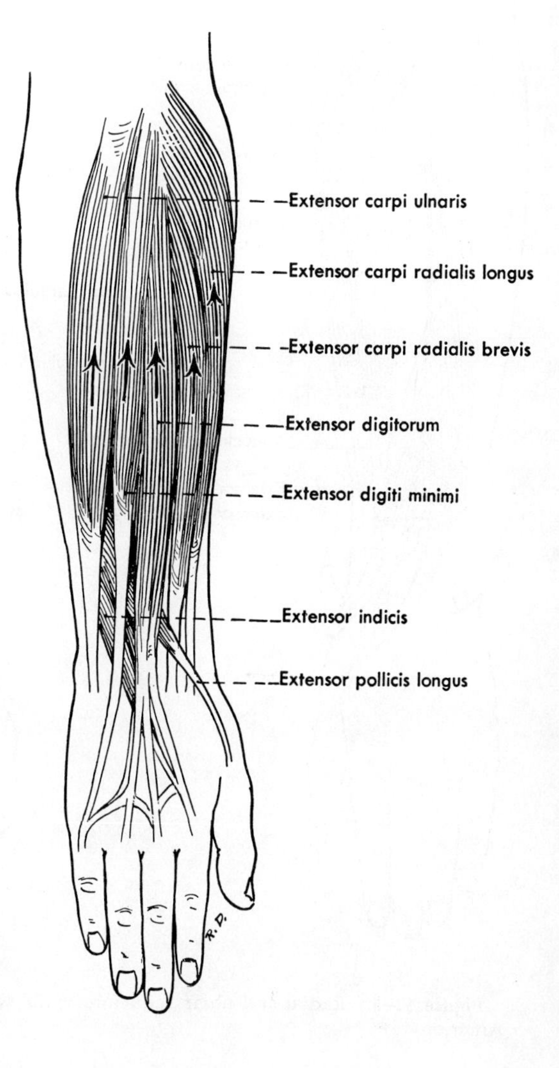

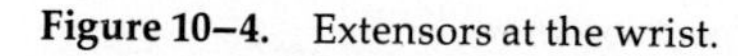

Figure 10–4. Extensors at the wrist.

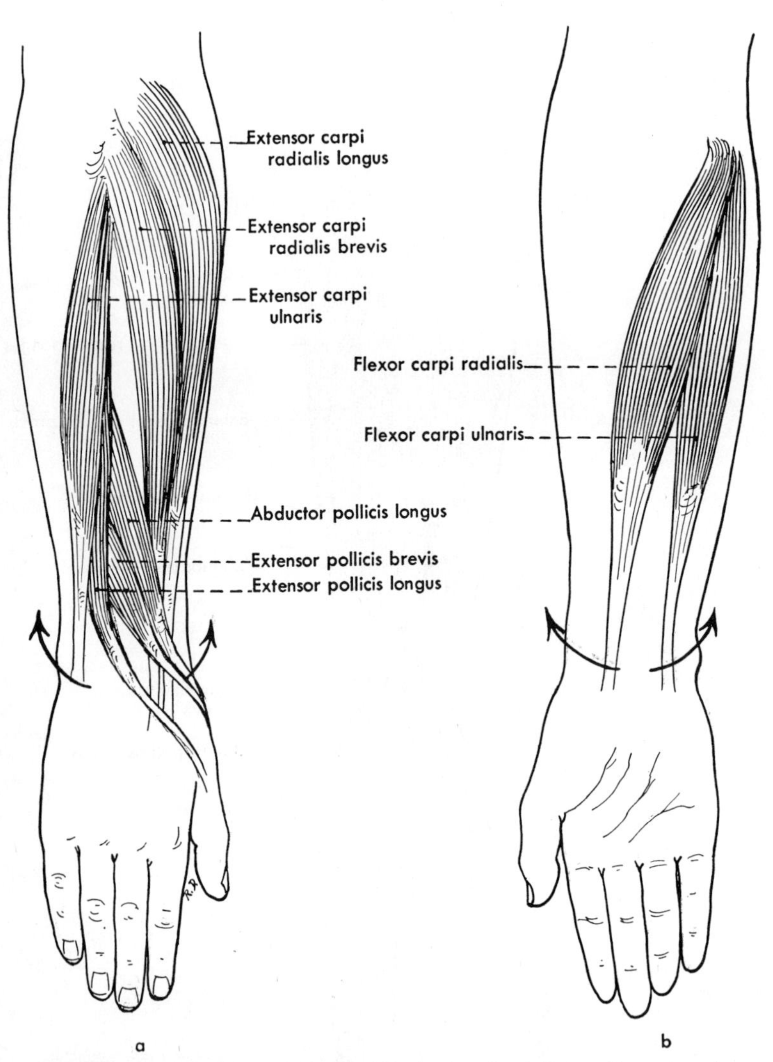

Figure 10–5. Radial and ulnar abductors at the wrist. *a*. Posterior view. *b*. Anterior view.

actions of the extensor carpi ulnaris and the flexor carpi ulnaris (fig. 10-5).

The flexors of the wrist joint are innervated by all three nerves of the forearm—median (most of them), ulnar (flexor carpi ulnaris and part of flexor digitorum profundus), and radial (abductor pollicis longus). The abductor pollicis longus will flex the wrist when it alone is acting, therefore even combined injury to both median and ulnar nerves does not abolish wrist flexion. The flexors of the wrist receive most of their innervation from C7, C8 and T1, however. Consequently, injuries to the lower part of the brachial plexus will more markedly affect wrist flexion than will injuries to the upper part of the plexus.

The extensors of the wrist are all supplied by the radial nerve with fibers from about C6 to C8. The extensors of the fingers are so short that when an injury to the radial nerve has produced wristdrop (p. 154) the wrist can still be straightened by making a fist, as already noted; the finger flexors then tighten the extensors as if they were ligaments.

Radial abduction involves both median and radial nerves, while ulnar abduction involves ulnar and radial nerves. The radial abductors as a group are supplied by most segments contributing to the brachial plexus; the two ulnar abductors may both receive fibers from C8, but otherwise have no innervation in common.

As a group, the forearm muscles acting across the wrist (and also, as will appear later, the intrinsic muscles of the hand) receive most of their innervation through the lower portion of the brachial plexus. We have already seen how Erb's paralysis, or injury to the upper portion of the plexus (C5 and C6), affects especially muscles of the shoulder and arm. Similarly, Klumpke's paralysis, involving injury to C7, C8 and T1, affects most of the muscles of the forearm and hand and is characterized, therefore, by severe crippling of the wrist and fingers.

11 The Hand

The hand is a particularly complicated organ, for in it are concentrated not only the tendons of the long muscles that we have already considered but also a large number of intrinsic muscles (ones confined to the hand) together with important nerves and vessels.

The terms "lateral" and "medial," or "radial" and "ulnar" are used to distinguish the thumb and little-finger sides of the hand just as they distinguish these sides of the forearm; instead of anterior and posterior surfaces, however, the hand is described as having palmar and dorsal ones. The thenar eminence, it will be remembered, is the prominence that the thumb muscles form on the lateral side of the hand, while the hypothenar eminence is that formed by the little-finger muscles on the medial side.

BONES AND JOINTS

As has already been noted, carpal bones compose the carpus or wrist; the metacarpals form the skeleton of the major part of the hand; and the phalanges are the bones of the digits.

The eight **carpal bones** are arranged in two rows (figs. 11-1 and 11-2). The three large bones of the proximal row, beginning at the radial side, are the scaphoid, lunate, and triquetral bones; the smaller pisiform bone sits on the palmar surface of the triquetral. The distal row, also named from the radial side, consists of the trapezium, trapezoid, capitate, and hamate bones. Some of these bones, such as the scaphoid, lunate, pisiform, and hamate can be easily recognized by their shape, while the identification of others involves more attention to details. The scaphoid, lunate, and triquetral bones articulate with the radius and the articular disk on the ulna and form a convex surface upon which movement at the wrist occurs; the carpals also articulate with each other, and as a whole the dorsal surface of the carpus is convex, the palmar surface concave; this concavity, accommodating the long flexor tendons that enter the hand, is called the carpal groove. Distally, the carpals articulate with the elongated metacarpals; distal to the metacarpals are the phalanges, two in number for the thumb and three for the other digits. The proximal end of each metacarpal and phalanx is its base, the distal end is its head, and the body intervenes between base and head. Two sesamoid bones, one on each side of the front of the metacarpophalangeal joint of the thumb, articulate with the head of the first metacarpal.

The joints at the wrist are actually multiple and include the distal radioulnar joint, the radiocarpal joint, often thought of as "the" wrist joint, the intercarpal joints, and the carpometacarpal joints (fig. 11-3). The **distal radioulnar joint** is L-shaped; the vertical portion of its synovial cavity is interposed between the lower ends of the radius and ulna, while the transverse portion lies between the lower end of the ulna and its articular disk. This is the lower of the two trochoid joints between radius and ulna that make possible pronation and supination; it is surrounded by a weak articular capsule.

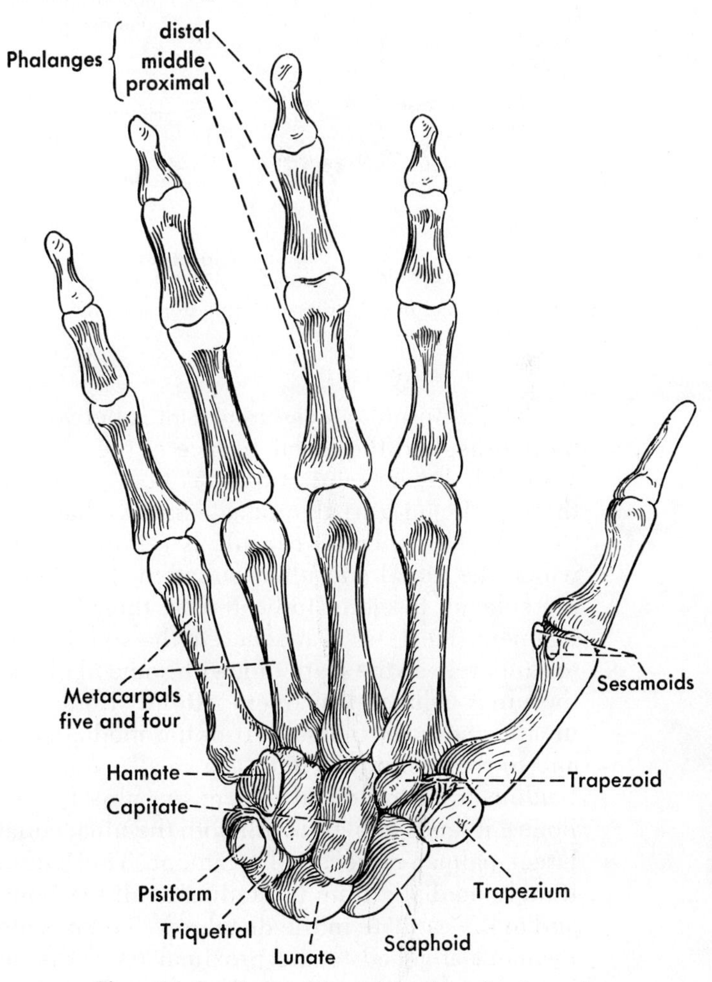

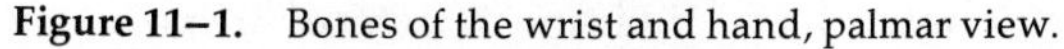
Figure 11–1. Bones of the wrist and hand, palmar view.

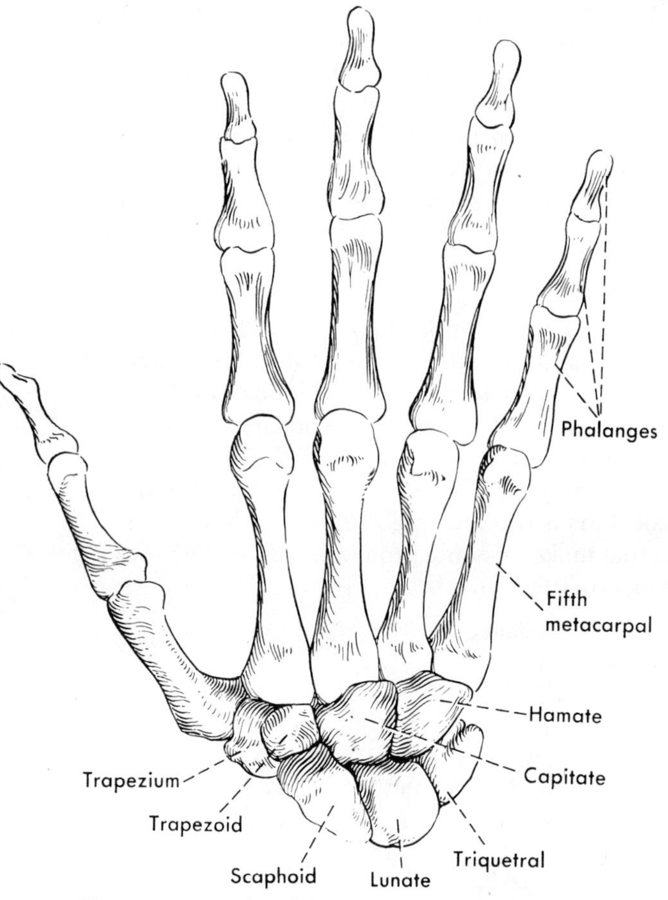

Figure 11–2. Bones of the wrist and hand, dorsal view.

The ellipsoidal **radiocarpal joint** is between the articular surface of the radius (and the distal surface of the ulnar articular disk) and the scaphoid, lunate, and triquetral bones. The movements here, and those at the midcarpal joint (see below), have already been discussed (p. 160). The articular capsule is reinforced by special ligaments, of which the radial and ulnar collateral ligaments are narrow bands on the sides of the joint indicated by their names. The radial collateral ligament (fig. 11-4) is attached to the styloid process of the radius and to a tubercle on the scaphoid, with some fibers reaching the trapezium; the ulnar collateral ligament extends from the styloid process of the ulna to the nonarticular part of the medial surface of the triquetrum, and to the pisiform. There is a small palmar ulnocarpal ligament extending from the distal end of the ulna to the lunate and triquetral bones; it tends to blend with both the ulnar collateral ligament and the larger palmar radiocarpal ligament. The latter ligament extends obliquely medially from the radius to all the bones of the proximal row and to the capitate in the distal row. There is also a dorsal radiocarpal ligament attached to the proximal row of bones. Since the fibers of both these ligaments extend ulnarward as they pass from the radius

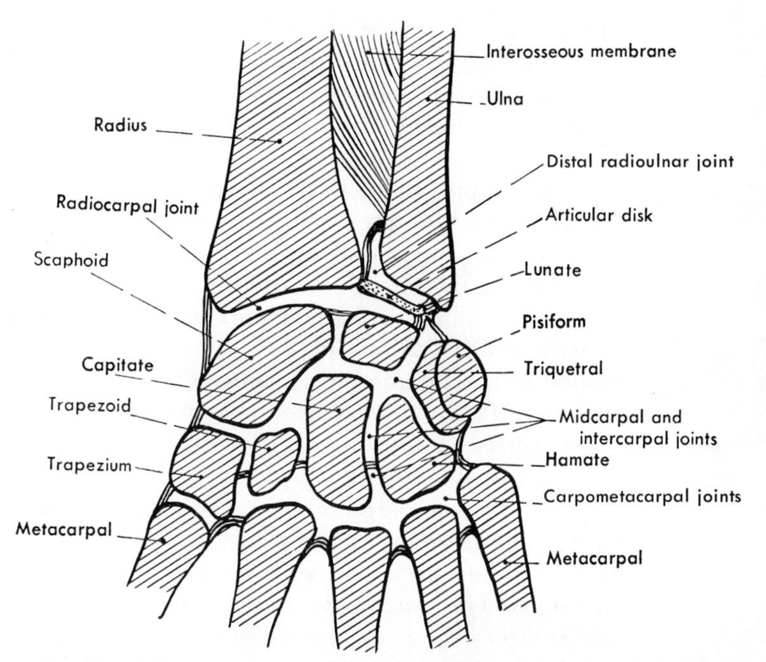

Figure 11–3. The bones and the joint cavities of the wrist. Note that the radiocarpal and inferior radioulnar joint cavities are separate and distinct; that the midcarpal joint is continuous with the intercarpal joints between both the proximal and distal rows of the carpals; and that the carpometacarpal joints, except for that of the thumb, are continuous with the intermetacarpal joints and with the distal parts of the intercarpal joints, but have no communication with the midcarpal joint.

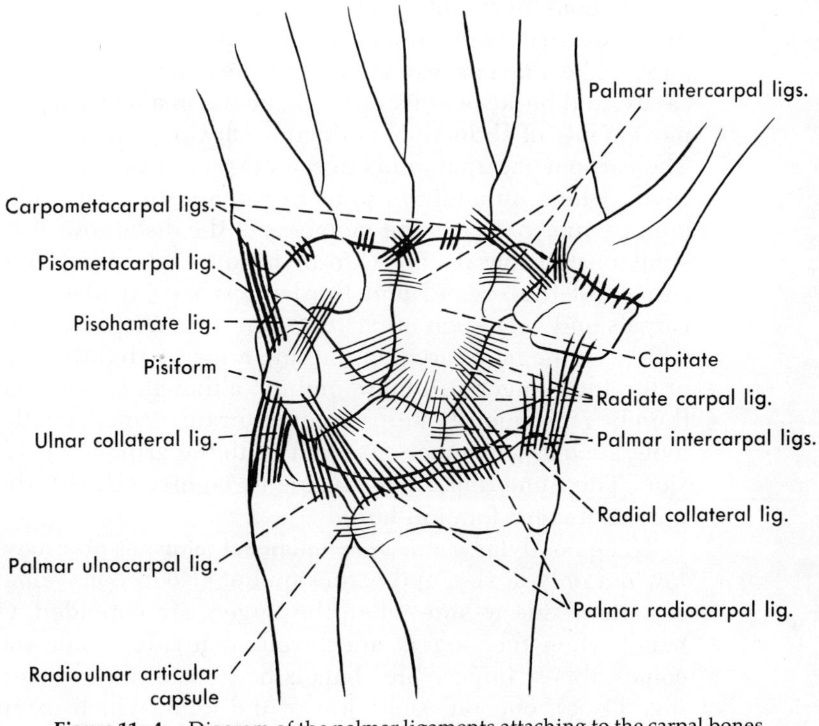

Figure 11–4. Diagram of the palmar ligaments attaching to the carpal bones.

to the proximal row of carpal bones, they ensure that the hand moves with the radius during pronation and supination: during pronation the fibers of the dorsal radiocarpal ligament carry the hand with the radius, and during supination those of the palmar radiocarpal ligament do so.

The carpals in each row are bound together by small intercarpal ligaments on both their palmar and their dorsal surfaces, and also by interosseous intercarpal ligaments (situated between the palmar and dorsal ligaments) that interrupt the continuity of the intercarpal joints. The pisiform has pisohamate and pisometacarpal ligaments, generally considered extensions of the tendon of the flexor carpi ulnaris, attached to it. The **intercarpal joints** between the members of the proximal row of carpals, and also those between the members of the distal row of carpals, allow only a little gliding movement between the bones in one row. Between the two rows, however, lies a larger intercarpal joint cavity, the **midcarpal joint;** through this the distal row of carpals moves rather freely on the proximal row. This joint is a single cavity that separates the two rows of bones. It sends expansions between the members of the proximal carpal row to form the intercarpal joints here, and shorter expansions between the members of the distal row to form proximal portions of the intercarpal joints of this row. Interosseous ligaments intervene between the radiocarpal joint and the intercarpal joints of the proximal row, and the intercarpal joints of the distal row are divided into proximal and distal parts by similar interosseous ligaments; the proximal portions of the distal intercarpal joints are, as just stated, continuous with the midcarpal joint, while the distal portions are proximal extensions from the carpometacarpal joints. Because of these interosseous ligaments there is usually no communication between the midcarpal joint and either the radiocarpal or carpometacarpal joints. The **carpometacarpal joint** of the thumb is a separate synovial cavity, and because of its sellar shape the first metacarpal can undergo movements of abduction, adduction, flexion, extension, and rotation. The carpometacarpal joints of the other four digits comprise a single cavity which, in addition to its proximal extensions to help form the joint cavities between the members of the distal row of carpals, sends similar extensions distally to form the intermetacarpal joints. The metacarpals of the second and third digits so articulate with the distal carpals and with each other that almost no movement of them is possible, but the metacarpal of the ring finger is slightly mobile, and that of the little finger still more mobile, although less so than that of the thumb. This mobility helps account for the firmness with which many tools, such as a hammer, are held, with the grip primarily on the ulnar side. The numerous small ligaments connected with these joints do not merit consideration here.

The condylar **metacarpophalangeal joints,** as you may observe, allow not only flexion and extension but also free movement of the fingers from side to side when the fingers are extended. On the other hand, when the fingers are flexed such side-to-side movement becomes almost impossible. This is because of the collateral ligaments (fig. 11-5*c*), one on each side of the joint, which extend obliquely distally and palmarward from the dorsum of the side of the metacarpal to the palmar aspect of the side of the proximal phalanx. These ligaments become so placed during flexion that they check the rocking movement of the digits at this joint. In addition to the collateral ligaments, each metacarpophalangeal joint is protected on its palmar surface by a dense fibrocartilaginous pad called the palmar ligament. Dorsally the joint is protected by an expansion of the long extensor tendon, usually called the "extensor hood" (fig. 11-5*b*). The heads of the metacarpals of the four fingers are also connected together by strong transverse bands, the deep transverse metacarpal ligaments, that attach also to the palmar ligaments. There is no such band between thumb and forefinger; its presence would much restrict the mobility of the first metacarpal.

The **interphalangeal joints** (fig. 11-5*a* and *c*) differ from the metacarpophalangeal ones in that they are hinge joints, and allow only flexion and extension. The ligaments in connection with these joints are identical with the ligaments of the metacarpophalangeal joints. The palmar ligaments of the interphalangeal joints are, however, of especial importance, as they prevent hyperextension of these joints; if one of them is ruptured, the associated phalanx may become locked in hyperextension.

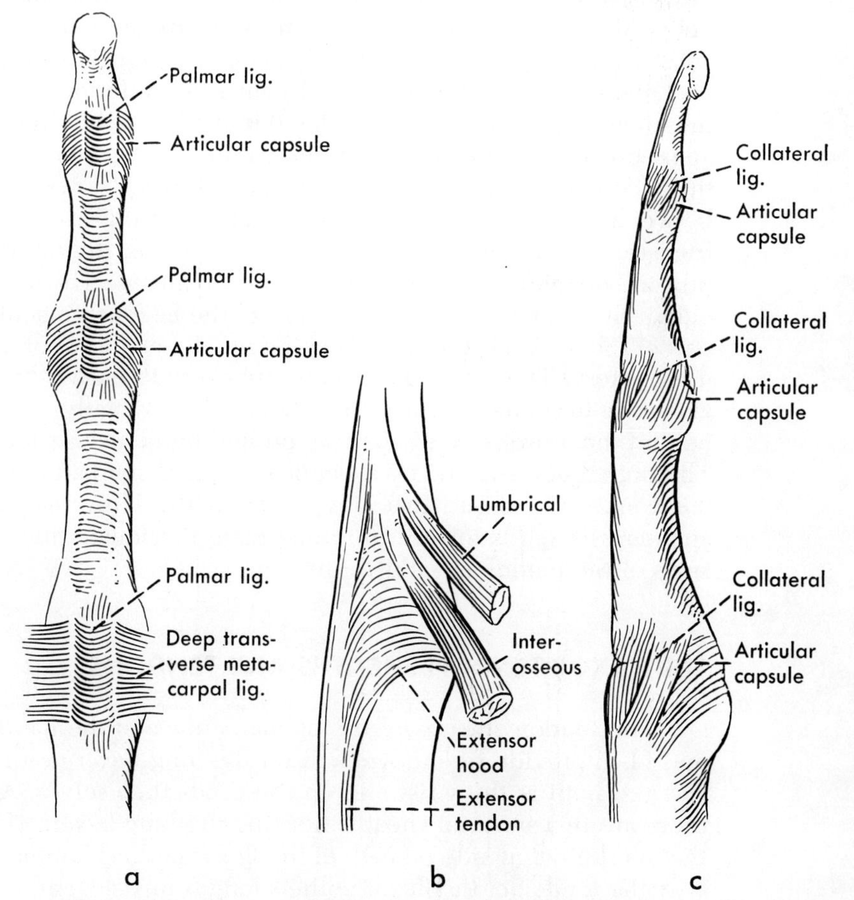

Figure 11-5. Metacarpophalangeal and interphalangeal joints. *a* is a palmar view; *b* a lateral view of the metacarpophalangeal joint with the extensor tendon in place; *c* a lateral view of a digit.

THE PALMAR FASCIA

The heavy fibrous **palmar aponeurosis** bridges the center of the palm of the hand; this aponeurosis is continuous above with the lower edge of the flexor retinaculum, the strong transverse thickening of the antebrachial fascia at the wrist that converts the carpal groove into a carpal canal (carpal tunnel), and it also receives the insertion of the palmaris longus. Distally it gives rise to slips to each finger which not only attach to the metacarpals and palmar ligaments, around the fibrous sheaths for the long flexor tendons, and to the front of the sheaths, but may also extend to the proximal phalanges. Thus it is that fibrosis and contracture of the palmar aponeurosis lead to marked flexion of the digits, especially at the metacarpophalangeal joint (Dupuytren's contracture). Medially and laterally the palmar aponeurosis sends septa to attach to the first and fifth metacarpals; these septa pass medial to the thenar muscles and lateral to the hypothenar ones, and form the walls of a central palmar compartment. A less dense fascia covers the muscles of the thenar and hypothenar eminences.

Within or adjacent to the central palmar compartment, bordered by the palmar aponeurosis and the intermuscular septa to the first and fifth metacarpals, lie the tendons of the long flexor muscles of the digits, associated short muscles connected with these tendons, arterial arches that supply the hand and fingers, and branches of the median and ulnar nerves. The superficial palmar arch, formed primarily by the ulnar artery, runs across the hand immediately behind the palmar aponeurosis and in front of the flexor tendons (fig. 11-11). It gives off three common digital branches that run forward toward the spaces between the fingers; each artery, after receiving a communication from the deep palmar arch, divides into proper digital arteries that proceed out along the adjacent sides of two fingers. The median nerve enters the central compartment behind the flexor retinaculum, and a part of the ulnar nerve enters by piercing the medial septum with the ulnar artery. The two nerves may connect with each other, but each gives rise to common digital branches that run distally, usually at first behind the arteries, to divide into proper digital nerves for the digits. The digital nerves and vessels appear in the distal part of the hand in the connective tissue that lies between the slips that the palmar aponeurosis sends toward the fingers; on the fingers they lie on the sides of the tendon sheaths there.

THE FLEXOR TENDON SHEATHS AND TENDONS

The tendon sheaths of the palmar surface of the hand (figs. 11-6 and 11-7) provide free movement for the long flexor tendons and are situated both at the wrist and on the digits themselves. At the wrist there are two synovial sheaths housing the long flexors of the digits; that on the radial side (sheath of the flexor pollicis longus) surrounds only the tendon of the flexor pollicis longus muscle, and is continued around this tendon almost to its insertion on the distal phalanx of the thumb. Thus the digital flexor sheath of the thumb is simply a continu-

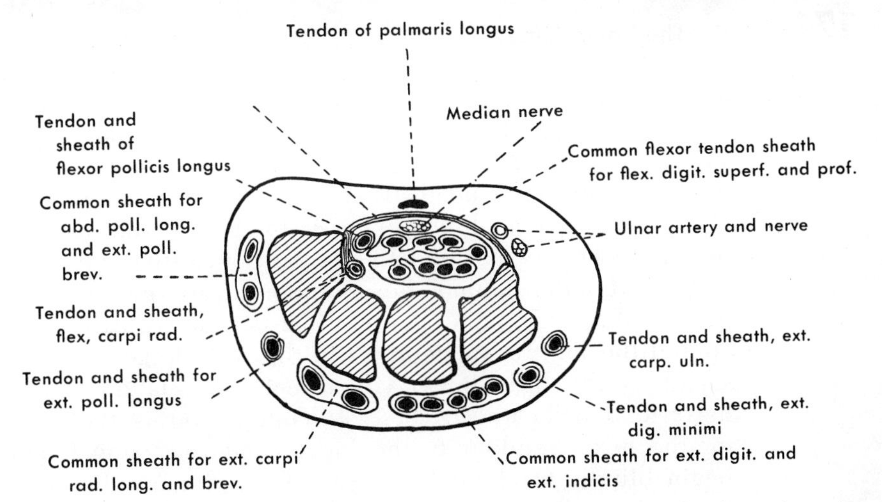

Figure 11–6. The carpal canal and the arrangement of tendon sheaths at the wrist, illustrated in a diagrammatic cross section of the right wrist seen from below. The extent and arrangements of mesotendons vary somewhat from one hand to another; some tendons have therefore been shown with, others without, mesotendons as they lie in their synovial sheaths.

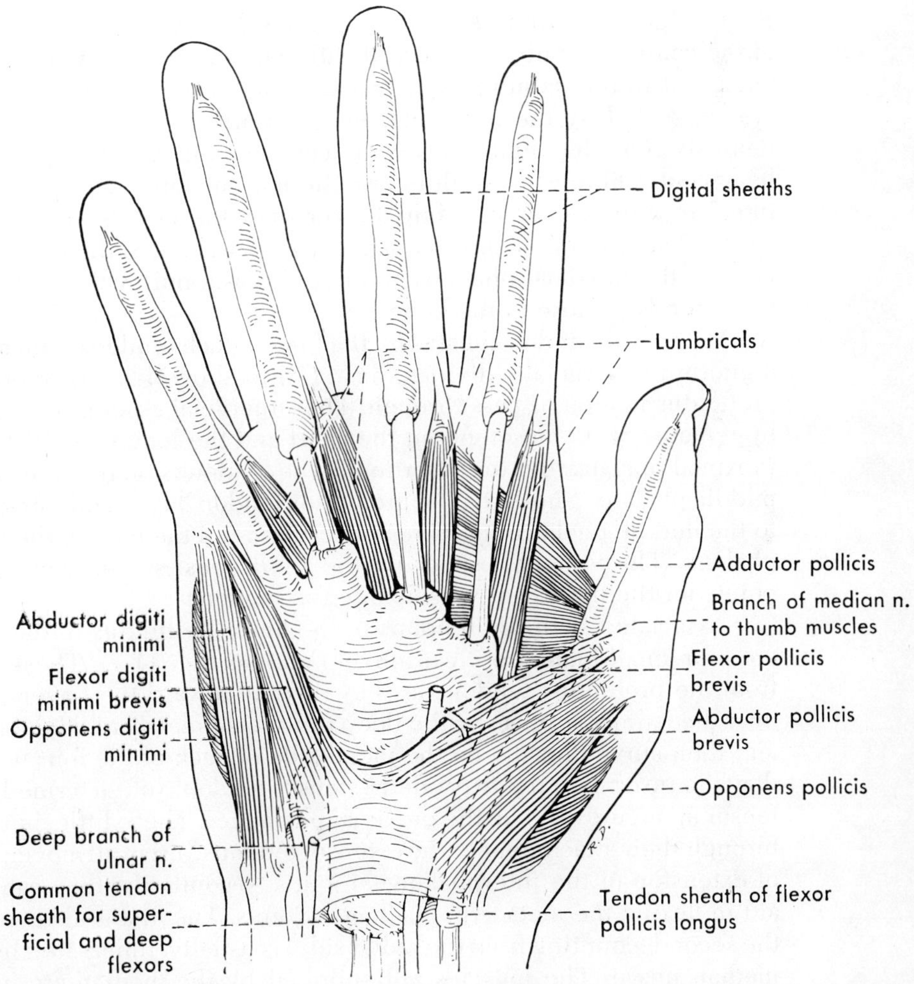

Figure 11–7. Short muscles of the thumb and little finger, and flexor tendon sheaths of the hand.

ation of the flexor sheath at the wrist. The much larger sheath on the ulnar side of the flexor pollicis surrounds the tendons of both flexor superficialis and flexor profundus, and therefore is the common flexor tendon sheath. This sheath also begins just proximal to the upper edge of the flexor retinaculum and in passing behind this occupies most of the space between it and the carpal bones (the carpal canal). The larger part of the common flexor tendon sheath stops about the middle of the palm of the hand, but an ulnar portion typically continues out around the long flexor tendons to the little finger. Thus the digital sheath of the little finger is a direct continuation of the common flexor tendon sheath (fig. 11-7). In contrast to this, the digital sheaths for the flexor tendons to the index, long, and ring fingers usually begin blindly at about the bases of the fingers, therefore distal to the common flexor tendon sheath, and normally have no connection with the tendon sheath at the wrist. Because of this discontinuity, infections within the tendon sheaths on the index, long, and ring fingers can extend upward toward the wrist only by rupture of the sheaths; infections within the tendon sheath of the little finger or the thumb, however, routinely present themselves at the wrist because of the continuity between these sheaths and those at the wrist. At the wrist and in the palm the synovial tendon sheaths are thin, for they are supported by the overlying flexor retinaculum and palmar aponeurosis. On the digits, however, each tendon sheath acquires a heavy outer fibrous layer, thinner in front of the joints where it might interfere with movement, but thicker over the bodies of the phalanges. The tendon sheath on a finger is particularly thick over the body of the proximal phalanx, and here forms a pulley that holds the flexor tendons close to the bone.

Within the tendon sheaths on the fingers each tendon of the flexor digitorum superficialis divides (fig. 11-8) to allow the corresponding profundus tendon to pass through. The superficial tendon is attached by remains of its mesotendon, the short and the long vincula, to the proximal phalanx, but the insertion of the tendon is on the base of the middle phalanx. Similarly, the profundus tendon has vincula attaching to the middle phalanx but its real insertion is on the base of the distal phalanx. The vincula, especially the short ones, serve primarily as points for the ingress of vessels into the tendons.

Associated with the tendons of the flexor profundus in the hand are four small **lumbrical** ("wormlike") **muscles** (fig. 11-7). These arise from the profundus tendons as they diverge toward the fingers, pass over the palmar surfaces of the deep transverse metacarpal ligaments, and then curve dorsally on the radial side of each of the four medial digits to insert into the expanded extensor tendons (often termed "extensor aponeuroses") on the proximal phalanges. These little muscles, through their attachment to the extensor tendons, aid in all movements of extension of the interphalangeal joints; secondarily, they can also aid in flexing the metacarpophalangeal joints. The first is always, and the second (counting from the radial side) is usually, innervated by the median nerve. The muscles not supplied by the median are innervated by the deep branch of the ulnar, or both nerves occasionally supply one muscle.

FASCIAL SPACES OF THE PALM

Deep to the flexor tendons and their associated lumbrical muscles is an area of loose connective tissue; the potential space here is bounded on the radial side by the septum passing from the palmar aponeurosis to the first metacarpal and on the ulnar side by the similar septum passing to the fifth metacarpal. Distally it ends about at the point at which the flexor tendons assume their digital sheaths. Most accounts describe a more or less marked septum that tends to separate this subtendinous area into two compartments, of which the more ulnar is known as the **midpalmar space,** the more radial as the **thenar space.** The associated flexor tendons and palmar fascia bound these spaces superficially, while deeply they are bounded by the fascia on the interosseous muscles and the adductor muscle of the thumb. (A different description is that there is only one palmar fascial space, although it is subdivided distally into compartments for the flexor tendons, the lumbrical muscles, and the digital nerves and vessels.) The palmar fascial spaces are of importance in infections of the hand, as a considerable amount of pus can collect in the very loose connec-

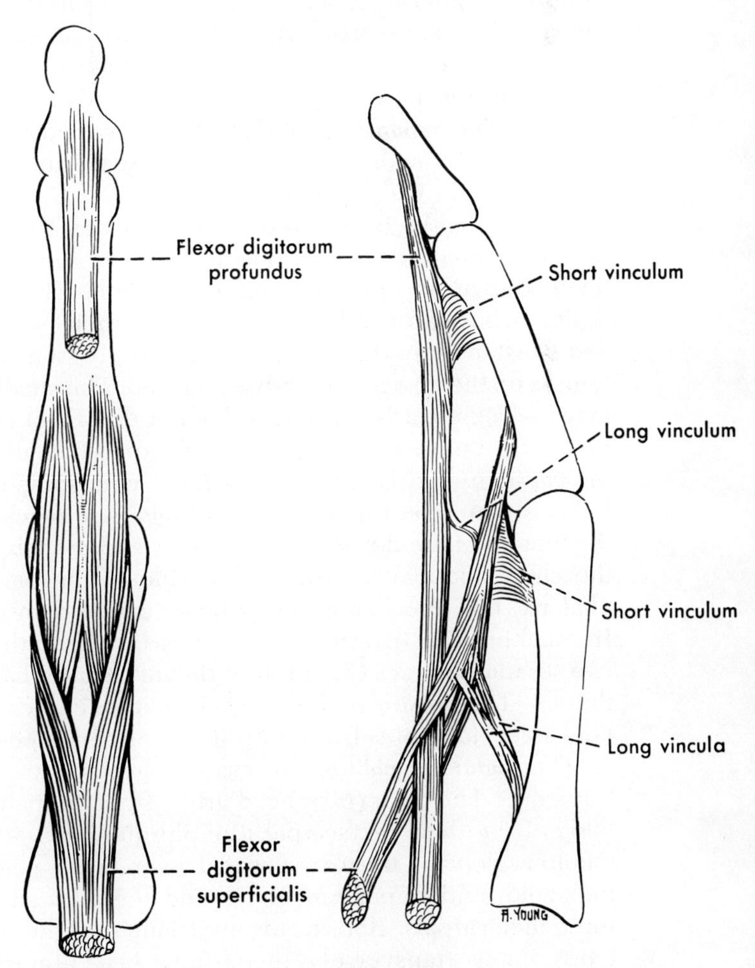

Figure 11–8. The flexor tendons on a finger, anterior and lateral views.

tive tissue that they contain. They may be infected directly through penetrating wounds of the hand, or indirectly through rupture of the flexor tendon sheaths into them.

THE MUSCLES OF THE THUMB

The four short muscles of the thumb (pollex) interact with the long thumb muscles and greatly increase the usefulness of this important organ. Three of them, the short abductor, the opponens, and the major part of the short flexor, form the thenar eminence (fig. 11-7); the fourth, the adductor, lies deeply in the palm behind the long flexor tendons, where it forms the posterior wall of the thenar fascial space. Associated with it is the deep part of the short flexor.

The **abductor pollicis brevis** is a flat muscle arising from the flexor retinaculum and the scaphoid and trapezium bones and inserting on the radial side of the base of the proximal phalanx; usually, also, a portion of the muscle inserts on the tendon of the extensor pollicis longus. It is a true abductor of the thumb in the sense that it moves the thumb almost perpendicularly away from the plane of the palm. Since it lies more palmarly than dorsally on the side of the metacarpophalangeal joint it is also a flexor at this joint; since it inserts in part on the long extensor tendon, it aids in extending the distal phalanx. The **opponens pollicis** is largely covered by the short abductor; it, like the abductor, arises from the flexor retinaculum and the trapezium bone but inserts along most of the length of the first metacarpal on its radial side. It thus draws the first metacarpal across the palm of the hand, rotating this bone as it contracts and producing the movement known as opposition of the thumb. The **flexor pollicis brevis** typically has two heads of origin, a superficial and a deep. The large and constant superficial head arises mostly from the flexor retinaculum, as do the muscles already described. The small deep head, which may be lacking, arises from the floor or dorsal wall of the carpal canal, from one or more of the distal row of carpals (usually the trapezoid and the capitate) and is closely associated with some of the origin of the following muscle, the adductor pollicis. The two heads unite deep to the tendon of the flexor pollicis longus and the muscle inserts close to the short abductor on the radial side of the base of the proximal phalanx, but more upon the palmar surface than does the abductor. In reaching this insertion, it attaches in part to the radial one of the two sesamoid bones (fig. 11-1) of the metacarpophalangeal joint of the thumb. The flexor pollicis brevis not only flexes the metacarpophalangeal joint but also aids in adduction and opposition of the thumb.

The **adductor pollicis** also has two heads, a transverse and an oblique one. The transverse head arises from the palmar surface of the body of the third metacarpal; the oblique head arises from the ligamentous floor of the carpal canal over distal parts of the capitate, trapezoid, and trapezium bones and the adjacent bases of the first three metacarpals. Both heads are triangular; the transverse head extends almost transversely, the oblique head almost parallels the first metacarpal, and the two heads come together to insert into the ulnar side of the palmar surface of the base of the proximal phalanx of the thumb. Their tendon of insertion attaches in part to the ulnar sesamoid of the metacarpophalangeal joint of the thumb, and a smaller part continues to the long extensor tendon. The adductor pollicis both adducts and flexes the thumb at the carpometacarpal joint, and flexes the metacarpophalangeal joint; it should be able to help extend the interphalangeal joint but apparently is not usually active in this movement.

As the median nerve emerges from behind the flexor retinaculum it gives off an important motor branch into the muscles of the thumb. The exact distribution of this branch varies from one individual to another; through it, however, the median nerve commonly supplies the short abductor, the opponens, and the large superficial head of the short flexor. The adductor pollicis and the deep head of the short flexor are usually supplied by the deep branch of the ulnar nerve which runs transversely across the hand to end in these muscles.

THE MUSCLES OF THE LITTLE FINGER

Lying in the fascia over the hypothenar eminence is the small **palmaris brevis,** a transversely arranged bit of muscle that arises from the medial border of the palmar aponeurosis and inserts into the skin of the ulnar border of the hand. The muscles of the little finger (fig. 11-7) are only three in number; movements of this finger are less complex than those of the thumb. The most superficial muscle on the ulnar border of the palm is the **abductor digiti minimi**, arising largely from the pisiform bone and inserting on the ulnar aspect of the base of the proximal phalanx. The **flexor digiti minimi brevis** arises from the flexor retinaculum and the projecting hamulus or hook of the hamate bone and joins the abductor to insert with it on the proximal phalanx, but more onto the palmar surface. The **opponens** lies deep to these muscles; it, also, arises from the flexor retinaculum and the hook of the hamate, but it inserts into the ulnar border of almost the entire length of the body of the fifth metacarpal. The abductor and short flexor aid in both abduction and flexion of the little finger while the opponens aids in opposition of this finger to the thumb, and therefore also in cupping the hand and in grasping tools firmly. All three of these muscles are supplied by the deep branch of the ulnar nerve as it passes among them to reach a deep position in the palm of the hand.

THE INTEROSSEI

These are deep-lying muscles that are largely situated, as their name implies, between the bones (the metacarpals) of the hand. The adductor pollicis lies in front of the interossei on the radial side of the third metacarpal, but the interossei on the ulnar side of this bone form most of the posterior wall of the midpalmar fascial space. At their proximal ends the interossei are crossed by the deep branch of the ulnar nerve, extending laterally from the hypothenar muscles, and by the deep palmar arterial arch which enters the palm through the most

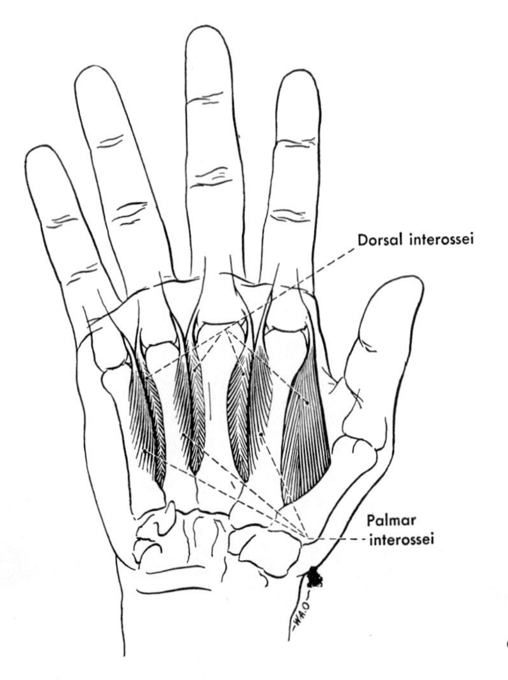

Figure 11–9. The interossei, the deep muscles of the palm of the hand.

radial interosseous and runs medially across the hand (fig. 11-12). Branches from both the arch and the nerve run distally on the surfaces of the interossei.

The interossei (fig. 11-9) are divided into two groups, palmar and dorsal. In contrast to the lumbricals, with which they have certain actions in common, all the interossei pass dorsal to the deep transverse metacarpal ligaments as they run distally to their insertions. Further, they are arranged about the midline of the hand, which runs through the middle of the middle digit or long finger, in such a fashion that they abduct and adduct the fingers about this midline. The **palmar interossei** are the adductors; since there are three fingers and a thumb to be adducted toward the middle digit it takes four muscles to carry out this movement. However, the thumb has an adductor of its own, therefore there are only three palmar interossei; these, inserting on the index, ring and little fingers, are sufficient to carry out the movement of adduction, working with the adductor of the thumb. The three palmar interossei arise from the second, fourth, and fifth metacarpals respectively, that is, the metacarpals of the three fingers on which they insert. They pass across the metacarpophalangeal joints on the side nearest the middle digit and their tendons then pass dorsally to join the extensor tendons on the proximal phalanges. These muscles therefore adduct the index, ring, and little fingers; they also flex the metacarpophalangeal joints, and thereafter can help extend the interphalangeal joints.

The **dorsal interossei** are arranged so as to abduct the fingers from the midline of the hand. There are thus two of them arranged about the middle digit, so that this finger may be abducted in either a radial or ulnar direction. It takes four more abductors to move the remaining four digits, but both little finger and thumb have abductors of their own. Thus there are only four dorsal interossei in all, one for the index and one for the ring finger in addition to the two attaching to the middle digit. The four dorsal interossei arise from the adjacent surfaces of two metacarpals; the first dorsal interosseous arises from both first and second metacarpals, the second from second and third metacarpals, and so forth. The first dorsal interosseous has usually a strong attachment on the radial side of the base of the proximal phalanx of the index finger, and little or no extension to the extensor tendon. The second and third interossei are attached in part to the base of the proximal phalanx of the long finger on its radial and ulnar sides, respectively, but also send strong expansions to the extensor tendon. The fourth dorsal interosseous arises from the fourth and fifth metacarpals, and, passing on the ulnar side of the metacarpophalangeal joint of the ring finger, inserts like the preceding muscles into both the proximal phalanx and the extensor tendon of this finger. The four dorsal interossei in conjunction with the abductors of the thumb and little finger abduct all the digits. With the usual exception of the first dorsal interosseous they are also, like the palmar interossei, extensors of the interphalangeal joints when the metacarpophalangeal joints are flexed; and with the palmar interossei they are the primary flexors at the metacarpophalangeal joints.

The deep branch of the ulnar nerve usually supplies all the interossei; the only common exception is the first dorsal interosseous, which in some 3 to 10 per cent of cases is supplied partially or completely by the median nerve.

VESSELS AND NERVES OF THE PALM

The vessels and nerves of the palm (shown diagrammatically in figure 11-10) are somewhat complicated, for the ulnar and radial arteries, each accompanied by smaller paired veins, are still large vessels when they reach the hand, and both the ulnar and median nerves are distributed to the hand. The branches of both the arteries and the nerves are in general divisible, however, into ones that lie in front of the flexor tendons and those that lie behind them on the front of the interossei. The ulnar artery, the median nerve, and the superficial branch of the ulnar nerve are the chief contributors to the more superficial vessels and nerves; the radial artery and the deep branch of the ulnar nerve are the chief contributors to the deeper-lying structures.

The **ulnar artery** passes into the hand superficial to the flexor retinaculum and forms an arch, the **superficial palmar arch,** across the hand (fig. 11-11). This arch lies between the palmar aponeurosis and the long flexor tendons. It ends in the muscles of the thumb where it usually is completed by a branch from the radial artery. From the arch are given off a proper digital artery to the ulnar side of the little finger and three common digital arteries which, after being joined near the heads of the metacarpals by branches from the deep arch, divide between the fingers to supply arteries to the adjacent sides of the little and ring, ring and long, and long and index fingers. The radial side of the

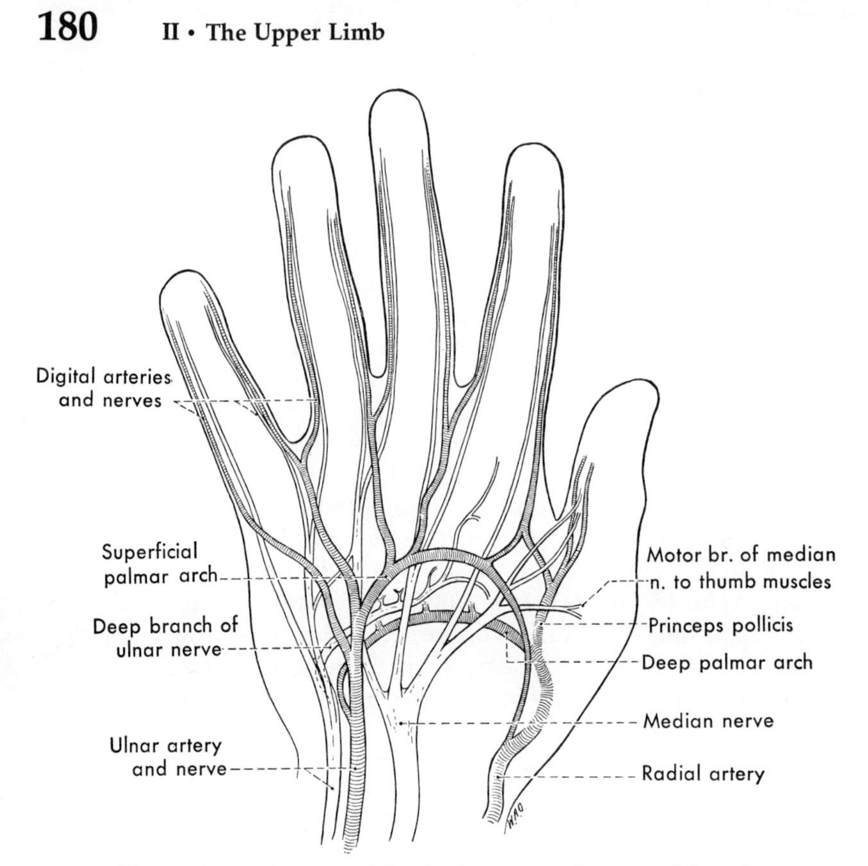

Figure 11–10. Diagram of the chief arteries and nerves of the palm.

index finger, and both sides of the thumb, are regularly supplied by branches from the radial artery. These branches may or may not be joined by one or more branches from the superficial palmar arch.

The **radial artery** enters the palm by a more devious course. Just below the area in which it is usually palpated in taking the pulse, the radial artery gives off a superficial palmar branch into the thumb muscles (this is the branch that may complete the superficial arch) and then turns sharply dorsally deep to the long abductor and the extensor tendons of the thumb to reach the dorsum of the hand. Here the artery gives off a small dorsal carpal branch (that helps to give rise to dorsal metacarpal and digital arteries) and then passes into the palm of the hand (fig. 11-12) by traversing the space between the two heads of the first dorsal interosseous muscle. As it passes through the muscle it gives off the princeps pollicis artery, which supplies the thumb and may help to supply the radial side of the index finger. After reaching the palm of the hand it passes as the **deep palmar arch** toward the ulnar side; here it lies against the palmar surfaces of the interossei, and therefore at first behind the adductor pollicis, then on the floor of the midpalmar fascial space. This arch is often completed by the small deep branch of the ulnar artery which supplies muscles of the little finger and may emerge to join it. The metacarpal branches of the deep arch join the common digital branches of the superficial arch to help supply the digits.

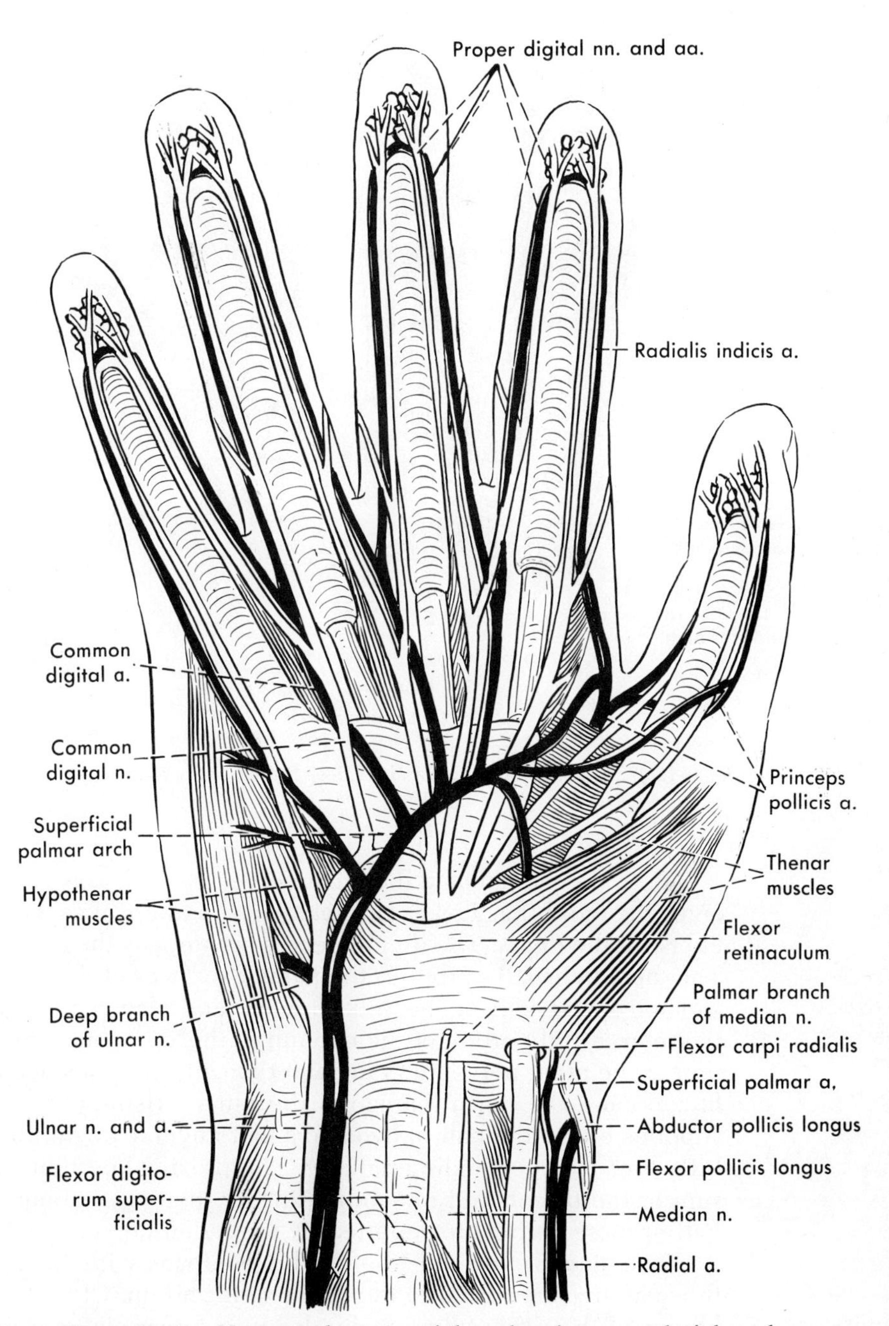

Figure 11–11. Nerves and arteries of the palm after removal of the palmar aponeurosis.

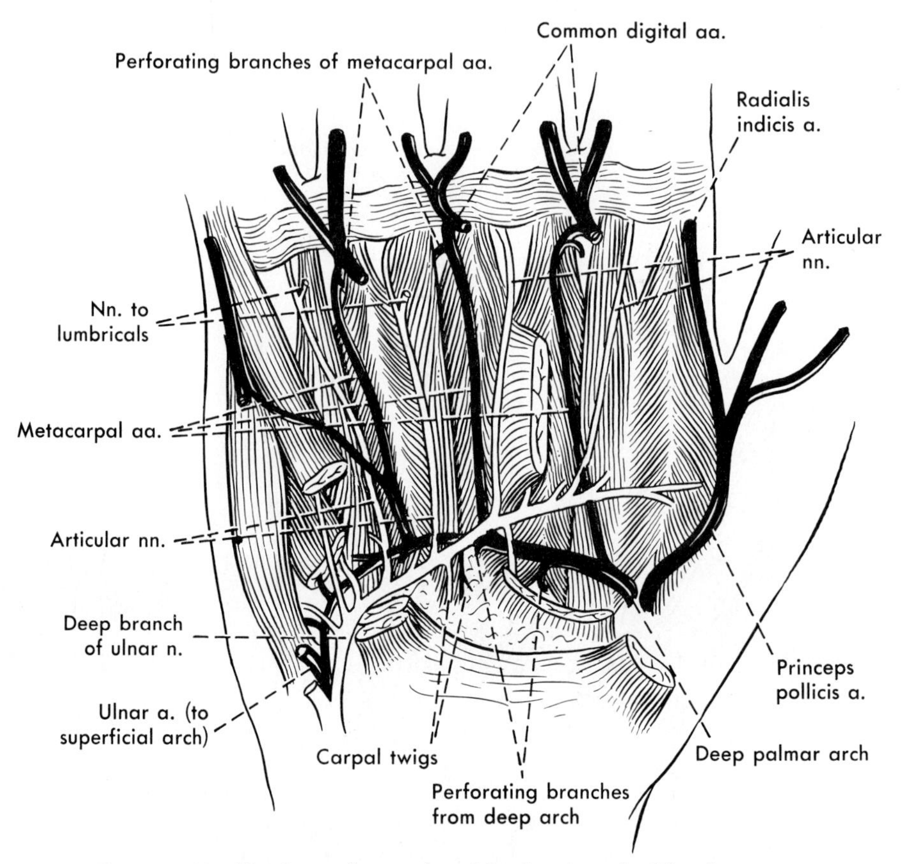

Figure 11–12. The deep palmar arch and the deep branch of the ulnar nerve.

Both the median and the ulnar nerves are still large trunks as they enter the hand (fig. 11-11), but many of the fibers of each are sensory fibers for the digits. The digital nerves accompany the digital arteries. The median usually supplies the palmar surfaces of the thumb, index and long, and half of the ring, fingers, and also sends branches to the more distal portion of the dorsum of these fingers. The continuation of the ulnar nerve into the palm typically supplies one and a half fingers (and the dorsal branch of the ulnar, arising above the wrist, supplies one and a half to two and a half fingers), beginning with the little finger and extending radially. The proximal portions of the dorsum of the thumb and remaining fingers (that is, of about two and a half or more digits) are supplied by the radial nerve.

The **ulnar nerve** enters the hand in company with the ulnar artery, thus passing superficial to the flexor retinaculum. Close to the lower border of the retinaculum it divides into superficial and deep branches. The superficial branch supplies the palmaris brevis muscle and divides into the palmar digital branches already mentioned; the deep branch passes deeply among the hypothenar muscles, which it supplies in its course, and then runs across the palm of the hand in company with the deep palmar arterial arch.

Here it supplies the third and fourth lumbricals and all the interossei, and ends in muscles of the thumb, usually only the adductor and the deep portion of the short flexor. In its deep palmar course it also sends branches to a variable number of the metacarpophalangeal joints.

The ulnar nerve, or its deep branch, is sometimes injured as it crosses the flexor retinaculum. There is little difference between the results of lesions here and of those at the elbow. In the latter case paralysis of the flexor carpi ulnaris is usually not evident; aside from a variable disability to flex the distal phalanges of the fourth and fifth digits (because the profundus to these digits may or may not be partly supplied by the median nerve), the findings in the hand are the same. If the lesion is severe, there should be some wasting of the hypothenar muscles, and weakness in abduction-adduction of the fingers because of the effect on the interossei. These movements are not impossible even with complete paralysis of the interossei, since the extensor digitorum abducts as it extends and the long flexors of the fingers adduct as they flex. The most important loss is that of the flexing action of the interossei at the metacarpophalangeal joints. Because of this, the extensors will draw the proximal phalanges into as much hyperextension as the ligaments of the joints will allow; with their pull concentrated on the proximal phalanges, the extensors lose their effect on the other two phalanges, and the long flexors, which become stretched, flex them. This condition is referred to as a claw hand; the greater the ligamentous laxity at the metacarpophalangeal joints, the greater the clawing. (It is usually less in the index and long fingers, since the lumbricals can flex the proximal metacarpals and extend the distal ones, and those of the index and long fingers are commonly innervated by the median nerve.)

The **median nerve** enters the palm behind the flexor retinaculum, between this and the common flexor tendon sheath. In this position it is subject to compression between the flexor tendons and the unyielding retinaculum by anything that decreases the space in the carpal canal—sometimes a carpal dislocation, but more commonly rheumatoid thickening of the synovial membrane of the tendon sheaths. The resulting condition (sensory changes, perhaps atrophy of the thenar eminence) is called "carpal tunnel syndrome," and if an operation is called for, this consists of slitting the retinaculum. (The flexor tendons remain in place when the digits are flexed because the wrist must be dorsiflexed for digital flexion to occur.)

Close to the distal edge of the retinaculum the median nerve gives off its digital branches; motor branches to the first two lumbricals usually arise from the branches to the index and long fingers. The most important motor branch of the median is that turning into the muscles of the thumb and supplying the abductor brevis, the opponens and most of the short flexor.

The effects of injuries to the median nerve on movements at the elbow and wrist have already been recounted. The effect on the hand varies somewhat with the level of the lesion. A lesion above the elbow affects digital movements less than might be expected, since the flexor digitorum superficialis, part of the flexor digitorum profundus, the flexor pollicis longus, and some of the thenar muscles should all be affected. However, loss of the flexor superficialis is unimportant, since the profundus routinely flexes the middle phalanges as it flexes the distal ones. Moreover, the ulnar side of the profundus is supplied by the ulnar nerve, and in most cases this can flex the three medial digits — either because the ulnar nerve helps to supply the parts of the muscle going to these digits, or because the tendons separate so low that contraction of the part going to the little finger, or to the little and ring fingers, pulls also on the tendon to the long finger. The result therefore is that the loss of the long flexors typically results only in inability to flex the middle and distal phalanges of the index finger and the distal phalanx of the thumb.

Loss of the thenar muscles innervated by the median nerve would be expected to abolish opposition, since the short abductor and the opponens are the principal muscles involved, and it may do so; however, in one careful study of 13 cases in which the median nerve was completely severed above the elbow, opposition was found to be normal in 4, almost normal in 4 others, with the metacarpal being abducted less than usual, thus leaving only 5 with loss of an opposable thumb.

A lesion of the median nerve below the origin of the anterior interosseous branch could affect only the thenar muscles, as could also one at the wrist. A lesion of the anterior interosseous branch could affect only flexion of the distal phalanges of the thumb and one or two adjacent fingers.

There are at least two reasons why opposition and abduction of the thumb are not always lost with a complete lesion of the median nerve. One is that the "rule" that this nerve innervates the short abductor, opponens, and superficial head of the short flexor is only generally true; any or all of these muscles may be innervated by the ulnar nerve, or receive through the ulnar nerve median nerve fibers that have joined the ulnar, as they are known to do sometimes, in the forearm. (Here the level of the lesion, whether above or below the communication, would obviously make a difference.)

The second reason is that the long abductor (radial nerve) and the deep head of the short flexor (ulnar nerve), or the long abductor and the adductor (ulnar nerve) can often substitute satisfactorily for the opponens and short abductor. While tendon transfers to produce flexion of the distal phalanges, and perhaps opposition, are usually necessary with high lesions of the median nerve they are less often necessary with lesions at the wrist. In any case, if the nerve is completely interrupted and regeneration does not occur, the loss of sensation is very crippling.

In lesions of both the median and the ulnar nerves all the short muscles of the hand are paralyzed, and the long abductor and short and long extensors, innervated by the radial nerve, will gradually rotate the first metacarpal into the plane of the palm, a condition referred to as "ape hand."

THE DORSUM OF THE HAND

The extensor tendons to the dorsum of the hand are retained at the wrist by the extensor retinaculum, and are provided with tendon sheaths as they pass between this and the underlying bones of the wrist. With the exception of the tendons of the extensor digitorum and extensor indicis, which run together in a single sheath, of the long abductor and short extensor of the thumb, which usually share a single sheath, and of the two radial extensors of the wrist, which regularly do, the various tendons each have individual sheaths as they pass deep to the retinaculum (fig. 11-13). In contrast to the palm, however, these sheaths are not continued much beyond the distal edge of the retinaculum, so the long tendons lie, for the most part, in direct contact with the loose connective tissue of the dorsum. As there are no intrinsic muscles of the dorsum of the hand, the tendons to the digits lie almost directly upon the metacarpal bones and the dorsal surfaces of the dorsal interossei, which appear here between the metacarpals. On both the dorsum of the hand and the digits the tendons are subcutaneous. There are no extensor digital tendon sheaths, hence, except for a limited length at the wrist, blood vessels

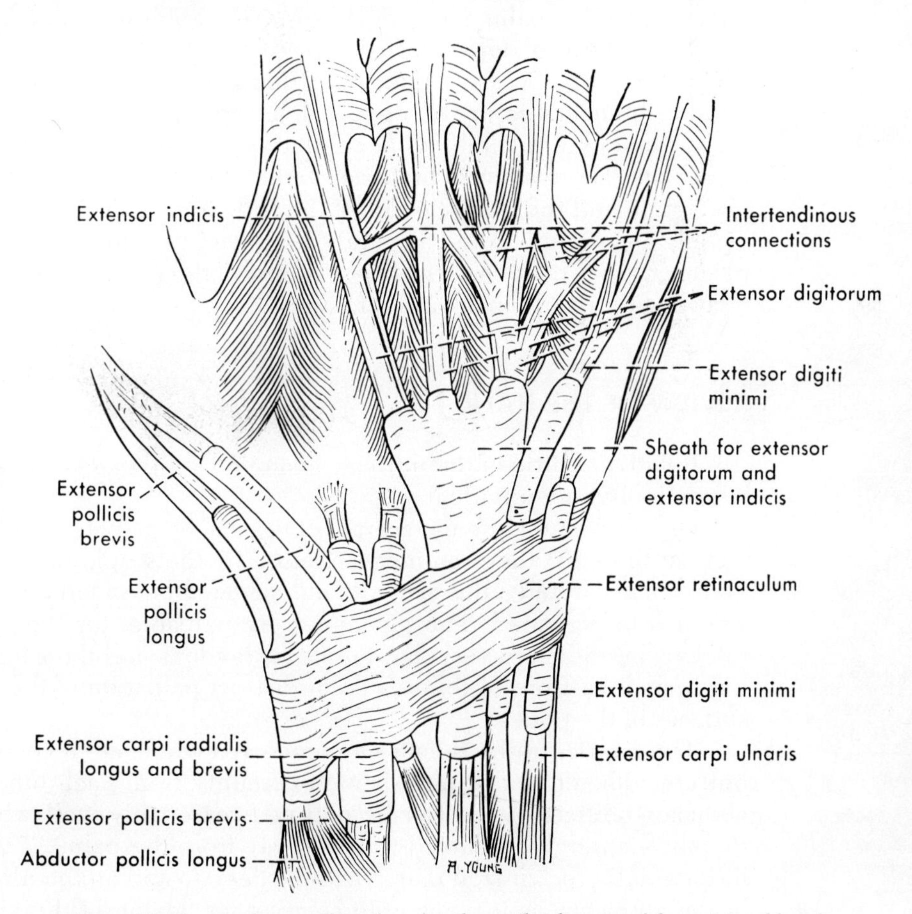

Figure 11–13. Tendons and tendon sheaths on the dorsum of the wrist and hand.

can enter the extensor tendons over much of their surfaces, instead of being limited to the narrow ingress afforded by a mesotendon or a vinculum. Also, the extensor tendons are much more intimately associated with the metacarpophalangeal and interphalangeal joint cavities than are the flexor tendons. The extensor tendons and expansions from them form the chief dorsal protection of these joints, and if the extensor tendons are torn off the joint cavities are usually laid open.

The cutaneous nerves to the dorsum of the hand (fig. 11-14) include the dorsal branch of the ulnar supplying one and a half to two and a half fingers, the radial nerve supplying the proximal portions of the dorsum of the thumb, index, and long fingers, and to some extent antebrachial cutaneous nerves which may pass for variable distances onto the hand. The sensitive tissue deep to the nails is in every case supplied by branches from the palmar digital nerves. The palmar branches of the median nerve also supply most of the skin over the two distal phalanges of index and long fingers and the radial portion of the ring finger.

As there is relatively little tissue to be supplied with blood on the dorsum of the hand, the blood vessels here are small and insignificant. The dorsal carpal branch given off by the radial artery before it passes into the palm joins twigs from the interosseous and ulnar arteries to form a dorsal carpal rete (network). This gives off four small metacarpal vessels that, after being reinforced by small perforating branches from the deep palmar arch and the palmar metacarpal vessels, divide into proper digital vessels to the adjacent sides of two digits. There are also branches to the radial side of the dorsum of the thumb and the ulnar border of the little finger. The dorsal digital vessels are minute, and can rarely be traced beyond the first interphalangeal joint; much of the blood supply of the dorsum of the fingers is received from the palmar digital vessels.

REVIEW OF THE HAND

Relatively little of the intricate anatomy of the hand can be made out in the living individual.

On the dorsum, the metacarpal bones can be palpated, and on the palm at the wrist the prominences raised by the scaphoid and trapezium bones laterally, the pisiform and hamate bones medially (p. 145) can also be palpated. The phalanges are palpable, for the extensor tendons on them are very flat; the flexor tendons and the connective tissue pads in front of them prevent distinct palpation of the palmar surfaces of the bones.

The **muscles** of the thenar eminence as a whole are recognizable, but are difficult to distinguish with certainty from each other. The abductor pollicis brevis can be outlined reasonably well when the thumb is strongly abducted (raised away from the palm). The short flexor can be recognized more vaguely, deep to and on the ulnar side of the abductor, as it is made to contract for flexion of the proximal phalanx; the contraction is much stronger if the thumb is opposed, but it is then impossible to know how much is due to the flexor and how

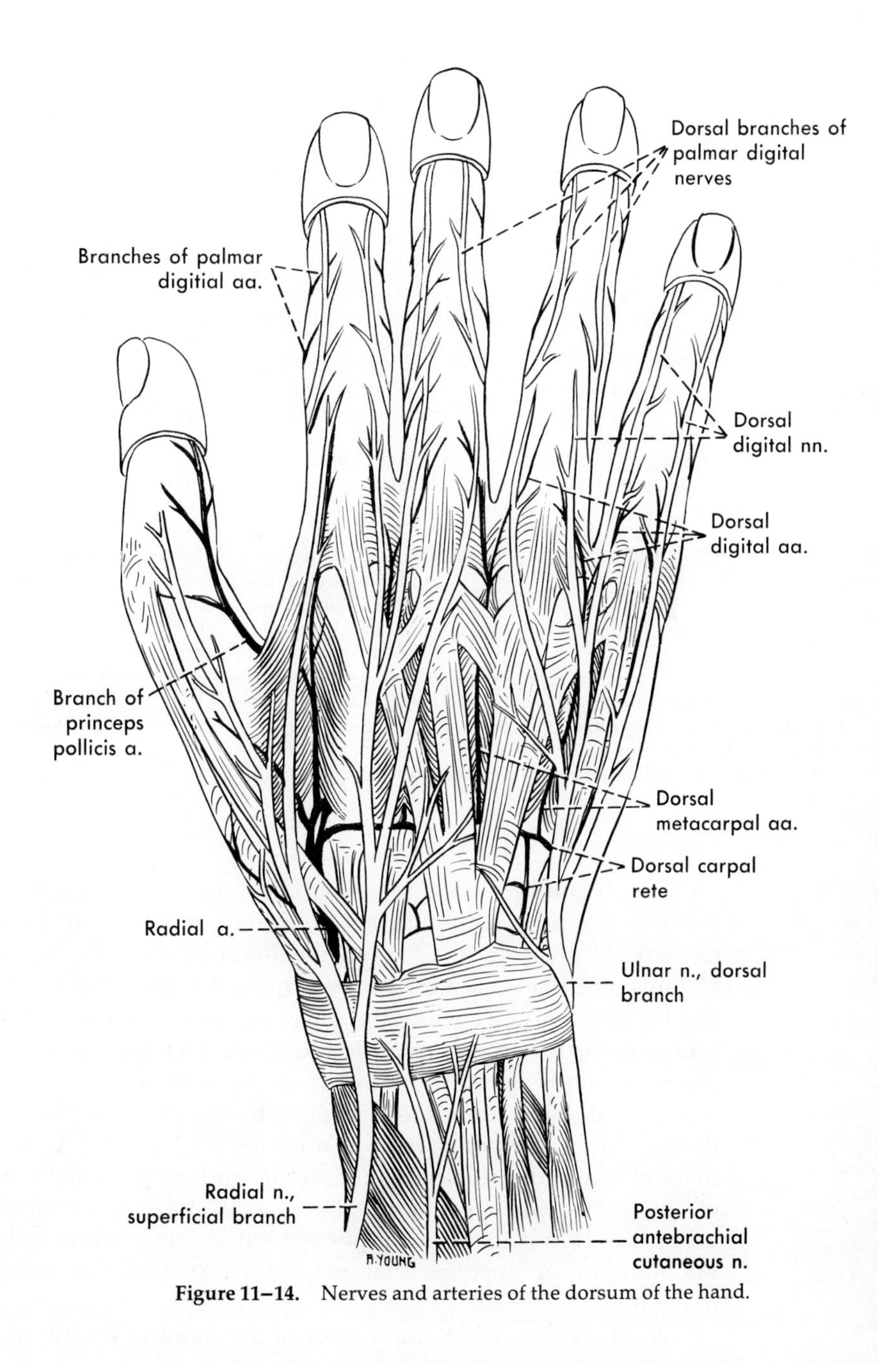

Figure 11–14. Nerves and arteries of the dorsum of the hand.

TABLE 11–1 Nerves and Muscles of the Hand

NERVE AND ORIGIN*	MUSCLE Name	Chief action(s)
Median C5–T1	Abductor pollicis brevis	Abduction of thumb
	Flexor pollicis brevis, superficial head	Flexion at metacarpophalangeal joint of thumb
	Opponens pollicis	Opposition of thumb
	Lumbricals I and II	Extension at interphalangeal joints, flexion at metacarpophalangeal joints, of digits II and III
Ulnar C8, T1	Flexor pollicis brevis, deep head	Flexion at metacarpophalangeal joint of thumb
	Adductor pollicis	Adduction of metacarpal, flexion at metacarpophalangeal joint, of thumb
	Palmaris brevis	Wrinkling skin of hypothenar eminence
	Abductor digiti minimi	Abduction of digit V
	Flexor digiti minimi brevis	Flexion at metacarpophalangeal joint of digit V
	Opponens digiti minimi	Cupping hand
	Lumbricals III and IV	Extension at interphalangeal joints, flexion at metacarpophalangeal joints, of digits IV and V
	Palmar interossei	Adduction of digits II, IV, and V; flexion at their metacarpophalangeal joints; extension at their interphalangeal ones
	Dorsal interossei	Abduction of digits II, III, IV; other actions similar to palmar interossei

*The common segmental origins—see footnote of Table 8-1. The segmental innervations of the individual muscles are not listed because there is strong evidence that all of them are supplied by C8 and T1; however, some authors list the muscles supplied by the median nerve as receiving fibers from C6 and C7.

much to the underlying opponens. The opponens cannot be identified, nor can the adductor with any certainty, although it forms part of the muscle between the first and second metacarpals. On the dorsum of the hand between these metacarpals the first dorsal interosseous can be felt distinctly when the index finger is abducted; it is the only interosseous that can be plainly recognized. Of the muscles of the hypothenar eminence, the abductor of the little finger can usually be recognized along the ulnar border of the hand when the finger is abducted, but the other muscles cannot be recognized. As will become evident when the following chapter is studied, weakness or paralysis of individual muscles of the hand, especially those of the thenar group, is often difficult or impossible to assess accurately because of the number of muscles that may assist in carrying out a specific movement.

None of the **vessels** and **nerves** of the hand is visible or palpable, except for the superficial venous network on the dorsum. However, it is not difficult to locate the major vessels and nerves in relation to the surface. Remembering that the ulnar artery passes into the palm on the radial side of the pisiform bone, its course can be approximated by drawing a line distally to about the proximal palmar crease; the superficial arch follows this crease across the palm. The radial artery, after passing through the "anatomic snuff box," dives palmarward between the bases of the first and second metacarpals; the deep arch lies an inch or more proximal to the superficial arch, about at the level of the palm where thenar and hypothenar eminences come together. The course of the median nerve can be approximated by recalling its branches, and the fact that it gives rise to them at about the distal edge of the flexor retinaculum; similarly, that of the ulnar nerve may be approximated by projecting its digital branches from just medial to the pisiform bone to the sides of the fingers it supplies, while the deep branch, of course, tends to parallel the deep palmar arch, already located.

The usual nerve supply and the chief actions of the muscles in the hand are shown in Table 11-1. They are also discussed in more detail in the following chapter. The nerve supply is also shown diagrammatically in figures 7-1 and 7-2.

12 Movements of the Digits

Movements of the fingers and thumb involve complicated integrations of the long and short muscles attaching to them. The exact share that any one of these muscles may play in a particular movement is frequently not well understood. As a rule, several muscles participate in a given movement, and this movement may or may not be partly distorted by the absence of some member of the group. The following analysis represents some generally accepted ideas as to how these muscles cooperate.

Flexion of the fingers is brought about through the flexor digitorum profundus, the superficialis, and palmar muscles (fig. 12-1). The flexor digitorum profundus, through its attachment on the distal phalanges, is primarily a flexor at the distal interphalangeal joints. By continued action it will also flex the proximal interphalangeal joints and finally the metacarpophalangeal joints. The action of the profundus on the proximal interphalangeal joints depends, however, upon these joints being already in slight flexion. If, by rupture of the palmar ligament of the joint, the proximal interphalangeal joint is in slight extension, contraction of the profundus increases this extension and the joint is therefore locked in a hyperextended position. Normally, the integrity of the palmar ligaments allows the flexor profundus to act not only on the distal interphalangeal joints but on the other more proximal joints, and while other muscles can assist, it alone is commonly used in making a fist. It is so efficient in flexing the proximal interphalangeal joints that it was once routine for it to be left as the sole flexor of these joints, the superficialis being removed, when the flexor tendons were injured on the fingers.

The flexor digitorum superficialis, through its attachment to the middle phalanges, acts upon the proximal interphalangeal joints and, by continued action, will also aid in flexing the metacarpophalangeal joints. Flexion at the metacarpophalangeal joints has traditionally been described as being due to the short muscles of the hand, the lumbricals and interossei. Recent studies have indicated that the lumbricals will flex the metacarpophalangeal joints only after they have extended the interphalangeal ones; their primary function seems to be the latter action, for they assist in this regardless of the direction of movement at the metacarpophalangeal joints. The interossei, however, are regularly active during flexion of the metacarpophalangeal joints, whether the joints are held in flexion or being flexed. They are, therefore, the primary flexors of these joints. The long flexors, so essential to the power of grip of the fingers and hand, are at a mechanical advantage only when the wrist is extended; flexion at the wrist therefore markedly interferes with flexion of the fingers. For this reason, when extensive paralysis of flexor muscles necessitates transferring extensor tendons to the palm and joining them to the distal ends of flexor tendons, it is necessary that at least one good wrist extensor be left intact. Similarly, if the wrist is to be arthrodesed (fused to make it immobile), it is always fixed in slight hyperextension.

Extension of the digits usually involves the cooperation of two sets of muscles. The tendons of the first set, the extensor digitorum with the associated extensors of the little and index fingers, expand over the metacarpophalangeal joints of each of the four fingers, covering the joint capsules here, and are so attached to the palmar ligaments that, once they have moved proximally a certain distance, they hyperextend the metacarpophalangeal joints even though their actual insertions are on the middle and distal phalanges. The extensor digitorum

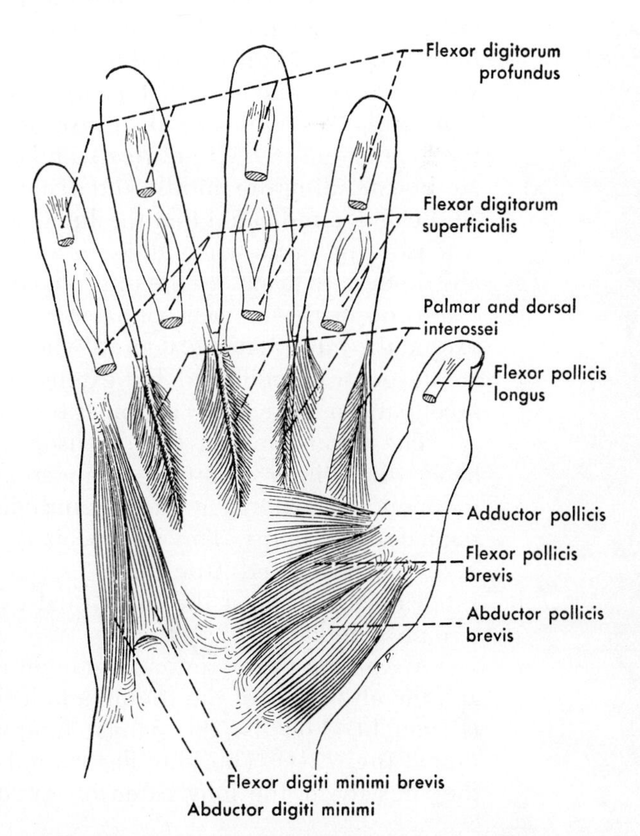

Figure 12–1. The flexors of the interphalangeal and metacarpophalangeal joints. (The lumbricals, secondary flexors at the metacarpophalangeal joints, are omitted.)

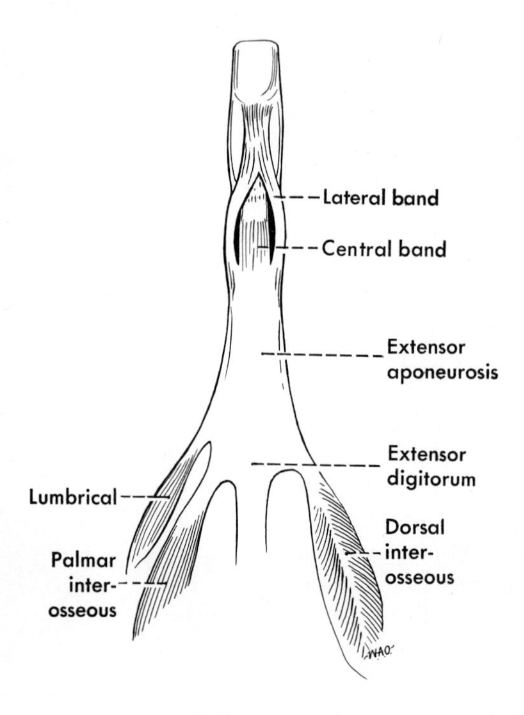

Figure 12–2. The extensors of a typical digit. The one shown here is the fourth; others have slightly different arrangements of the interossei, and some, of course, have proper extensors, but the principle is the same for the four medial digits.

and the special extensors of the index and little fingers that join it are therefore the sole extensors of the metacarpophalangeal joints. Distal to these joints, however, the extensor tendons are joined by a second set of tendons, those of the lumbricals and the interossei (fig. 12-2). Over the proximal phalanx the tendons blend together to form a single tendon that is often referred to as the extensor aponeurosis. This divides into a central band that inserts on the middle phalanx and two lateral bands that converge to an insertion on the distal phalanx; hence the long extensor, the lumbricals, and the interossei should all cooperate in extending the middle and distal phalanges. Electromyographic studies have indicated that the lumbricals typically contract with the extensor digitorum when all joints are extended at once, and presumably both help to extend the interphalangeal joints and prevent hyperextension at the metacarpophalangeal ones; they and the interossei extend the interphalangeal joints when the metacarpophalangeal ones are flexed or being flexed, but extension by the interossei seems to be secondary to their flexor action at the metacarpophalangeal joints.

The bands uniting the extensor tendons on the dorsum of the hand (the "intertendinous connections" in figure 11-13) hinder independent extension of the individual fingers at the metacarpophalangeal joints. This is especially true of the long and ring fingers; the index and little fingers are less hampered in independent extension because of the individual extensors with which they are also provided.

Abduction of the digits is brought about by the dorsal interossei and the abductors of the thumb and little finger (fig. 12-3); adduction is brought about by the palmar interossei and the adductor of the thumb (fig. 12-4). The long flexor tendons also adduct the fingers as they flex them, the long extensor tendons abduct the fingers as they extend them. These long tendons, however, act upon all four fingers at once and therefore do not allow for individual abduction and adduction of a given finger. The extensor indicis can independently adduct the index finger and the extensor digiti minimi can abduct the little finger.

Movements of the thumb, and to a lesser extent those of the little finger, are more complicated than are movements of the fingers as a whole. Further, electromyographic studies have indicated that most of the short muscles of the thumb contract during any movement of that part, either to assist directly or to steady the movement; and the diagnosis of injury to the musculature or nerves of the thumb is made more complicated in clinical practice because the innervation of the muscles of the thumb may vary, a given muscle being supplied in some cases by the median nerve and in others by the ulnar nerve. Flexion of the distal phalanx of the thumb can be brought about only by the long flexor of the thumb. Flexion at the metacarpophalangeal joint of the thumb is more limited and usually is accompanied by marked movement of the metacarpal. The flexor pollicis brevis and the adductor produce flexion at the metacarpophalangeal joint, and if they move the metacarpal produce also adduction and opposition of the thumb. The short abductor also aids in flexion at this joint.

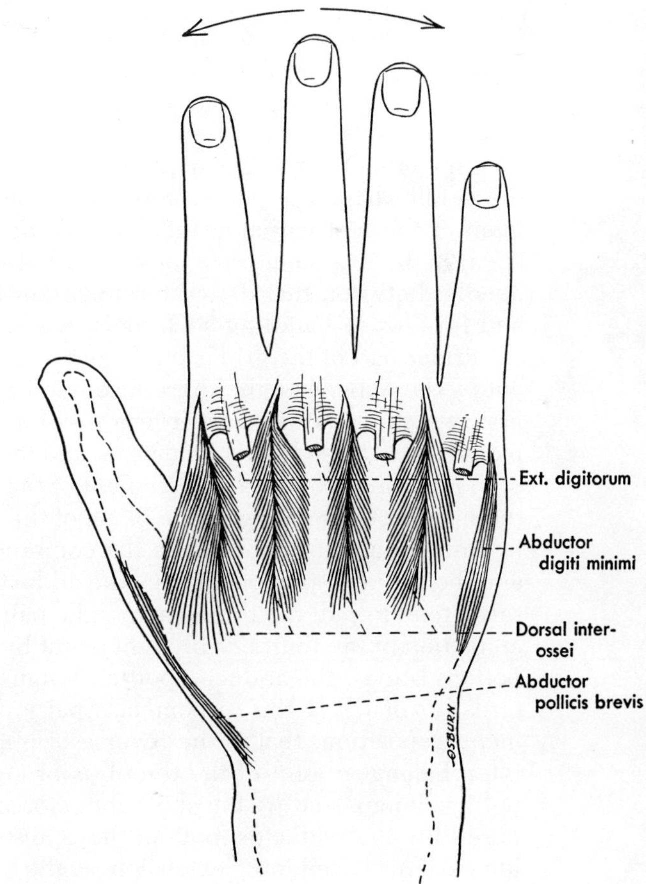

Figure 12–3. Dorsal view of the chief abductors of the digits.

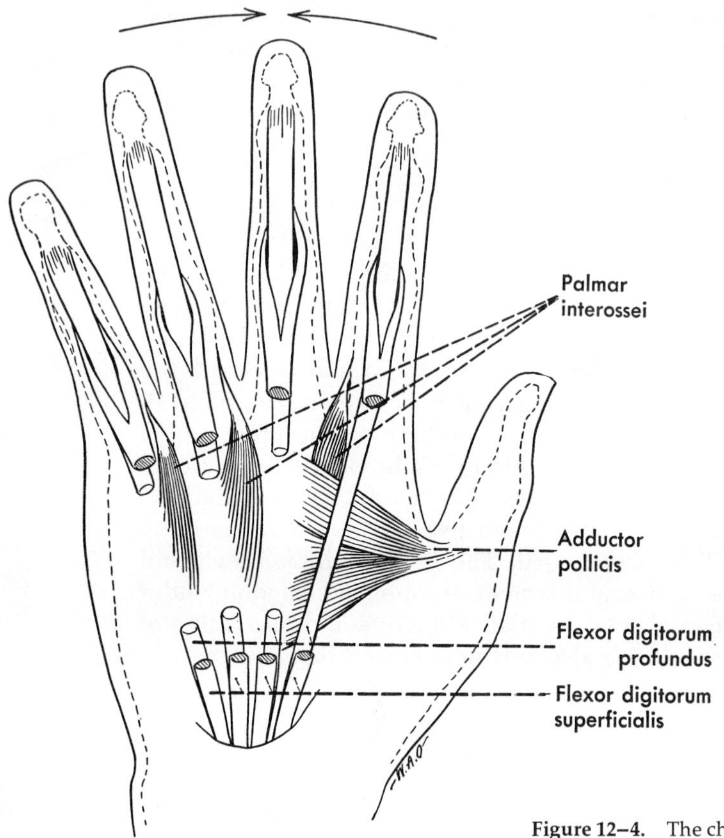

Figure 12–4. The chief adductors of the digits.

In contrast to the fingers, the most movable joint of the thumb is the saddle-shaped carpometacarpal joint. Opposition, involving movement of the metacarpal at this joint, is the most useful movement of the thumb; it is necessary for one to hold a small object, such as a needle, between thumb and forefinger, or to grasp a baseball firmly, and it helps in handling such tools as a hoe or hammer (in this case the firmer part of the grip is on the ulnar side of the hand). It is carried out by the opponens, the short flexor, and the short abductor, with the first two contributing most if firm pressure is demanded; this movement is aided also by the long flexor and the adductor. Pure adduction of the thumb involves the movement of the thumb toward the palm of the hand at right angles to the plane of the palm. This movement can be brought about only through the combined actions of the adductor and the extensor pollicis longus; the adductor acting alone will necessarily tend to pull the thumb across the palm as well as adduct it. Pure abduction of the thumb is brought about by the action of the abductor pollicis brevis; the abductor pollicis longus abducts, extends, and externally rotates at the carpometacarpal joint—brings about a movement (reposition) that is the reverse of opposition. Extension at the interphalangeal joint of the thumb is brought about by the extensor pollicis longus and by the short abductor and, if the movement is opposed, by the adductor (both of these muscles insert in part into the long extensor tendon); extension at the metacarpophalangeal joint is carried out by the extensor pollicis brevis, which may also join the long extensor and thus help to extend the distal phalanx. Flexion and extension at the carpometacarpal joint of the thumb are regularly accompanied by rotation of the metacarpal.

Both the abductor and the short flexor of the little finger flex the metacarpophalangeal joint, and also abduct the finger; the opponens produces a certain amount of rotation of the fifth metacarpal.

The multiplicity of muscles acting upon the digits, the various and sometimes varying nerve supplies of these muscles, plus the considerable substitution by undamaged muscles, all make especially difficult the assessment of the role of any one nerve supply in a particular movement of the fingers, and complicate the clinical problem of judging the extent of injury to nerves and muscles of the hand.

The effects of injury to the radial, median, and ulnar nerves have already been recounted. In summary of the effects on the hand: a radial nerve injury may weaken or abolish extension at the metacarpophalangeal joints, but extension of the interphalangeal joints of the fingers by the lumbricals and interossei and of the interphalangeal joint of the thumb by virtue of the attachment of the short abductor and the adductor to the long extensor tendon is still possible. If the lesion is high enough to affect the extensors of the wrist, an effective grip is prevented by the flexion of the wrist accompanying digital flexion.

An injury to the median nerve above the elbow may produce no more disability in the hand than loss of flexion of the middle and distal phalanges of the index finger and the distal phalanx of the thumb. However, there may also be loss of flexion of the long finger, and there may be poor or no opposition of the thumb. A lesion at the wrist may or may not produce loss of opposition of the thumb, depending on which muscles receive another innervation, and how effectively they can compensate for any paralyzed ones.

Except for its effect on the ulnar part of the profundus, resulting usually in no more than inability to flex the distal phalanx of the little finger, the sequelae of a high lesion of the ulnar nerve and those of a lesion at the wrist are similar. Because of paralysis of the interossei, abduction and adduction of the fingers is much weakened; also, in the absence of the interossei's important action at the metacarpophalangeal joints the extensor digitorum draws the proximal phalanges into as much hyperextension as the joints will allow. This both abolishes the pull of the extensor on the middle and distal phalanges and stretches the long flexors, so that a claw hand is produced. The effect on the thumb may be difficulty in holding a sheet of paper between the extended thumb and forefinger or, depending on the innervation and whether the ulnar nerve at the wrist contains median nerve fibers, grave deficit or no deficit at all.

13 The Back

The skeleton and musculature of the back are, of course, largely responsible for the support and movements of the trunk as a whole. In addition to this, the back must be prepared to support and stabilize the upper limb and head so that these members can move smoothly and evenly or support strains put upon them. Since the weight borne by the back increases as one descends to the lumbar region, and greater leverage is also exerted in this region, it is obvious that the lower part of the back is subjected to very great stress. The enormous strains on this part account for the extreme commonness of low back pain. In view of the importance of the musculature of the back, and of the bony vertebral column, in stabilizing the body as a whole, it should be easy to realize that crippling of the back results not in simple crippling of one portion of the body but rather in crippling of the body as a whole—it may make any posture or locomotion painful or difficult.

The movements of the vertebral column are flexion, or forward bending; extension, or dorsiflexion; lateral flexion; and rotation. These various movements do not go on with equal freedom in all portions of the vertebral column, and certain movements are quite regularly a combination of two of these primary movements.

SURFACE ANATOMY

Relatively few bony structures can be felt in the cervical region, as most of the spinous processes (fig. 13-1) are buried deeply in the musculature on the back of the neck. The spinous process of the second cervical vertebra (axis) can sometimes be felt somewhat vaguely in the posterior midline of the neck about two fingerbreadths beneath the occipital protuberance (the bump on the back of the head), but little more can be felt posteriorly until the sixth and seventh cervical spinous processes project at the base of the neck. Laterally in the neck the transverse process of the first cervical vertebra can be felt just behind and below the mastoid process (the bony bump behind the ear), but the transverse processes of the other cervical vertebrae are difficult to distinguish. In the thoracic region the rather sharp spinous processes of the thoracic vertebrae can be seen or palpated, and thrown into greater prominence by flexion of the trunk. The broader lumbar spinous processes can also usually be identified; these and the lower thoracic ones lie in an increasingly deep groove formed by the increasingly heavy mass of the back muscles. The back muscles of the cervical and lumbar regions are easily identified as a whole; those of the thoracic region are thinner and therefore less prominent.

THE VERTEBRAL COLUMN

The vertebral column, often called the "spinal column" or "the spine," consists of a series of bones termed vertebrae, numbering about thirty-three at birth. Of these there are in the adult typically seven separate vertebrae in the neck or cervical region, twelve connected with the ribs and classed as thoracic vertebrae, and five, known as the lumbar vertebrae, in the lower part of the back. Usually the next five are fused together to form the sacrum, and the remaining three or four, rudimentary in character, form the coccyx. The vertebral

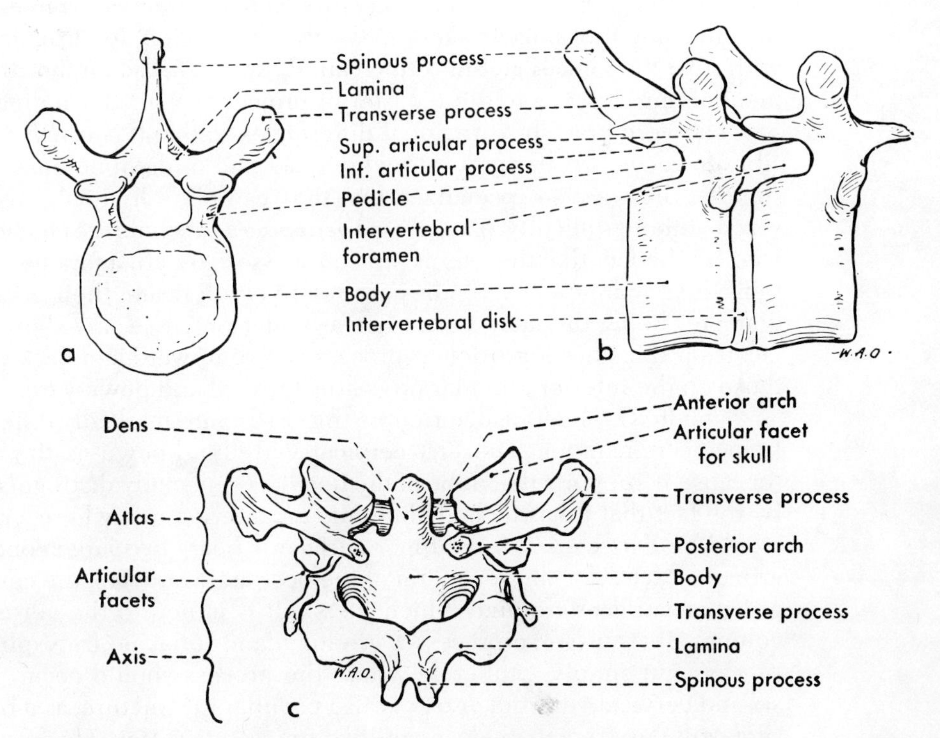

Figure 13–1. Structure of the vertebrae, *a.* Thoracic vertebra viewed from above. *b.* Two adjacent vertebrae, lateral view. *c.* The first two cervical vertebrae, posterior view; the posterior arch of the atlas has been partially cut away, and the ligaments that hold the dens in place are not shown.

column serves two chief purposes: that of supporting the trunk and that of protecting the spinal cord.

A typical vertebra (fig. 13-1*a*) consists of several named parts fused together to form a single bone. The heavy, roughly cylindrical base of the vertebra is called the **body.** It is largely spongy bone covered generally by a thin layer of cortical or dense bone, but the ends of the bodies are mostly covered by a thin layer of hyaline cartilage (which persists throughout life, but is responsible for the growth in length of the vertebrae during the period of growth). The body bears upon its posterior surface paired pillars or **pedicles** (roots of the arch) which in turn support the **laminae,** the pedicles and laminae together forming the **arch** of the vertebra; between the vertebral arch and body is the **vertebral foramen.** Successive vertebral foramina form the **vertebral canal,** in which lie the spinal cord and its adnexa. In most vertebrae the pedicle presents a deep inferior vertebral notch on its lower border; with the much slighter depression on the upper surface of the pedicle next below, this notch forms the **intervertebral foramen** (fig. 13-1*b*) through which a spinal nerve makes its exit from the vertebral canal. At about the junction of the pedicle and lamina there are **articular processes** or **zygapophyses** which bear smooth facets for articulation with the vertebrae above and below; there are thus paired superior and paired inferior articular processes. The exact position and the planes of articulation of these processes vary with the region in which the vertebra occurs. Projecting laterally from about the point of union of the pedicle and lamina on either side is a **transverse process** for the attachment of muscles and, in the thoracic region, for articulation with ribs; the **spinous process** (often called "spine"), also for the attachment of muscles, is a midline posterior projection from the laminae.

The vertebrae show regional differences so that it is usually possible to recognize the group to which any one of them belongs, and some of them are so specialized that they can be individually recognized without difficulty. The **cervical vertebrae** as a whole are characterized by the fact that their transverse processes contain a foramen, the transverse foramen. The bodies are relatively delicate, their greatest diameter being the lateral one. The articular processes are short; the facets on the superior articular processes face upward and backward, those on the inferior articular processes forward and downward.

The first two cervical vertebrae (fig. 13-1*c*) are markedly different from the remainder. The first cervical vertebra, known as the **atlas** (because it supports the globe of the head), is especially distinguished by the fact that it has no body but only an anterior arch where a body would be expected. On its upper surface it bears two large concave articular facets for the reception of the occipital condyles, the convex articular surfaces through which the skull is joined to the vertebral column. Its transverse processes are long, and it has no true spinous process but simply a tubercle where the process should occur. The second cervical vertebra or **axis** is also peculiar in structure, as it bears projecting up from its body a toothlike process, the **dens.** This process articulates with the anterior arch of the atlas, to which it is firmly held by ligaments in such a fashion (fig. 3-2*f*) that it acts as a pivot around which the atlas rotates. The dens represents the body of the atlas which has separated from the rest of this bone and fused to the axis.

The remaining cervical vertebrae show a gradual increase in size as they are followed downward, but otherwise no particularly individual characteristics. The spinous processes of the middle cervical vertebrae are usually bifid. The seventh cervical vertebra (vertebra prominens), the last of the series, may be somewhat transitional between a typical cervical and a typical thoracic vertebra, and regularly has a rather long spinous process with a somewhat bulbous extremity.

As the vertebral column is followed downward, the bodies of the **thoracic vertebrae** (figs. 13-1*b* and 13-2) continue the gradual increase in size that was observed in the cervical region. The superior and inferior articular facets are almost in the frontal plane, especially in the midthoracic region; the superior facets face posteriorly, the inferior ones anteriorly. In contrast to the transverse processes of the cervical vertebrae, those of the thoracic vertebrae are solid. On the anterolateral surfaces of the ends of the transverse processes of the upper ten thoracic vertebrae there are smooth articular surfaces or costal pits for articulation with the ribs. Similar costal pits (costal foveae) occur on the sides of the vertebrae at about the junction of the pedicle and body. The first vertebra and the tenth, eleventh, and twelfth usually have a complete costal pit on their upper edges for articulation with the head of the rib. On the others, the costal pits are shared by two adjacent vertebrae, so that most of the thoracic vertebrae have halfpits

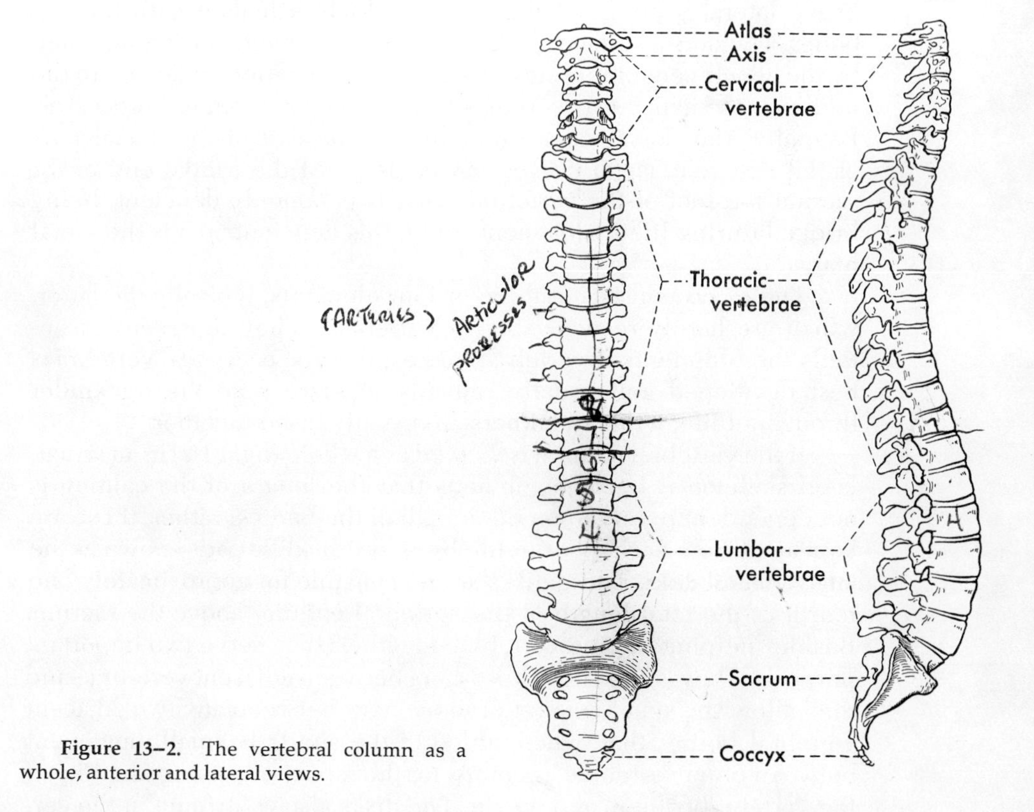

Figure 13–2. The vertebral column as a whole, anterior and lateral views.

(superior and inferior foveae) situated at both the upper and lower borders of the bone. The spinous processes of thoracic vertebrae tend to be long and slender and directed markedly downward so that they overlap each other. Those of the lower thoracic vertebrae are, however, broader and directed more posteriorly, in this respect being transitional between typical thoracic and typical lumbar vertebrae.

The bodies of the **lumbar vertebrae** are again more massive than are those of the thoracic region, and the spinous processes are much heavier, and broad when viewed from the side (fig. 13-2). In contrast to the cervical and thoracic vertebrae, the height of each lamina is less than that of the body, so that a considerable space exists between adjacent laminae. The transverse processes tend to be long and slender. The facets of the superior articular processes are directed upward and posteriorly but largely medially, and those of the inferior processes downward and anteriorly but largely laterally; therefore these joints approach the sagittal plane. On the posterior surfaces of the superior articular processes appear large irregular protuberances, the mammillary bodies, for additional attachment of muscles.

The five sacral vertebrae are, in adult life, fused to form a single bone, the **sacrum;** on its pelvic (anteroinferior, concave) surface are four slight ridges that mark the lines of fusion between the five elements entering into the bone. Laterally, in line with these ridges, are the pelvic sacral foramina through which the ventral rami (p. 224) of the first four sacral nerves make their exit. The modified transverse processes separating these foramina fuse again laterally to form the heavy lateral portions of the sacrum which articulate with the hip bones. The dorsal surface of the sacrum is convex and much roughened by the attachment of muscles; the irregular posterior projection in the midline (median crest) represents rudimentary spinous processes. Laterally, the dorsal sacral foramina for the exit of the dorsal rami of the first four sacral nerves can be seen. At the caudal end of the sacrum the roof of the vertebral canal is commonly deficient, being bridged during life by ligaments only; this deficient area is the sacral hiatus.

The **coccyx** consists of three or four elements, typically the latter, which are hardly recognizable as vertebrae. They represent essentially the rudiments of vertebral bodies; the first coccygeal vertebra is best developed and usually remains separate from the remainder throughout life, while the others are usually fused together.

If the vertebral column is viewed as a whole (fig. 13-2) in an articulated skeleton it becomes obvious that the length of the column is not dependent purely upon the length of the bodies; rather, these are kept apart from each other in life by heavy disklike pads known as the **intervertebral disks.** These disks are responsible for approximately one fourth of the total length of the vertebral column above the sacrum. Besides helping to bind vertebrae together, they serve two important purposes: they act as shock absorbers between adjacent vertebrae and they allow the small amount of movement between any two adjacent vertebral bodies that, when added to the similarly small movement between other vertebrae, accounts for the rather surprising mobility of the vertebral column as a whole. The disks also contribute in the cervical and lower lumbar regions to the curvatures typical of those parts of the column. They are discussed further on page 203.

In early fetal life the vertebral column as a whole is somewhat C-shaped, and the thoracic and sacral curvatures of the adult, with their forward-looking concavities, represent the remains of this original curve. On the other hand, the cervical and lumbar segments of the column in the adult are so curved that their concavities look posteriorly. These two reverse curves develop as a means of better balancing the weight of the body on the vertebral column; the cervical curvature develops as the infant tries to hold up his head, the lumbar curvature develops during the stages of learning to sit, stand, and walk. The lumbar curvature is due to unequal growth of the anterior and posterior borders of both the bodies and disks, the cervical curvature to the latter alone. The part that the disks play in the curvatures is strikingly shown by comparison of an articulated vertebral column with one strung on a string. The general forward concavity of the latter is reminiscent of the stoop of old age, which is also due in part to changes in the intervertebral disks.

The joints connecting most of the vertebrae are of three kinds: synovial, of the plane type, formed by the apposition of the articular processes; fibrous, connecting other parts of the arches; and cartilaginous, between the bodies, formed by the union of the intervertebral disks with the bodies.

The synovial joints formed by the articular processes allow only simple gliding movements. Although the capsules of these joints are lax, and therefore allow more movement than might be expected from examination of the bony elements only, the directions in which the articular surfaces face have much to do with determining what type of movement (flexion-extension, lateral bending, or rotation) is allowed between any two adjacent vertebrae. These joints are supplied by branches of the dorsal rami of spinal nerves (p. 224) and arthritic changes in them or undue strain put upon them as a result of abnormal postures or movements are common causes of back pain.

The ligaments that form the fibrous junctions between the arches are the **supraspinous** (or supraspinal), stretched across the tips of the spinous processes, the **interspinous** (or interspinal) between one spinous process and the next, and the **ligamenta flava,** paired ligaments that connect adjacent laminae and almost completely fill the spaces between them (fig. 13-3).

The supraspinous and interspinous ligaments blend together where they are adjacent; in the neck, between the seventh cervical vertebra and the skull, the supraspinous ligament is represented by the **ligamentum nuchae,** strong superficially where it gives rise to muscles, but extending deeply as a thin midline septum between other muscles of the two sides.

Each of the paired ligamenta flava extends from the anterior surface of one lamina to the posterior surface of the next lamina below. The pairs are separated from each other by a midline gap through which pass veins that connect a venous plexus inside the vertebral canal with one around the spinous processes. Laterally, the ligamenta flava blend more or less with the capsules of the synovial joints. The

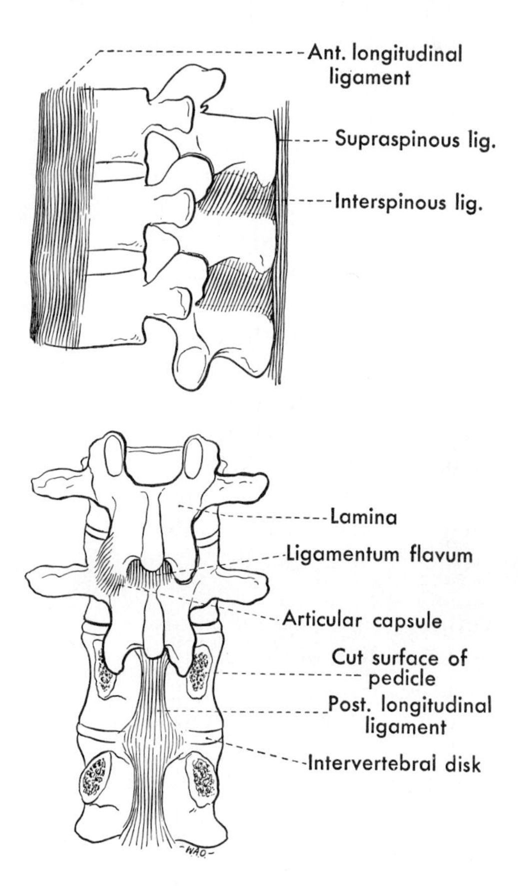

Figure 13–3. The chief ligaments of the vertebral column.

ligamenta flava thus form a part of the posterior wall of the vertebral canal. In the cervical and thoracic regions, where the laminae overlap or almost do, they are largely hidden in posterior view by the laminae, but in the lumbar region they fill in an appreciable space between laminae. Flexion widens still more the spaces between laminae, so when a needle is to be introduced between two of them in the lumbar region (the procedure of lumbar puncture) this can be done most easily when the back is flexed; the needle is then pushed through a ligamentum flavum or between the two.

The supraspinous and interspinous ligaments are largely collagenous but the ligamenta flava are composed primarily of yellow elastic tissue (flavum means yellow); thus they stretch during flexion of the back and yet do not become folded, but rather remain taut, in extension. Violent hyperextension of the neck, such as may result from an automobile accident, may, however, produce a movement that exceeds the elasticity of the ligamenta flava so that they do become folded, and in such cases may apparently injure the spinal cord.

The bodies also are firmly united to each other by strong collagenous ligaments, which partly cover the intervertebral disks. Anteriorly, there is a broad band, the **anterior longitudinal ligament,** which stretches from the sacrum to the occipital bone of the skull. The anterior longitudinal ligament is especially strong, for it is thick and consists of fibers of varying length. The deepest fibers extend only from one vertebra to the next; other fibers extend across a single vertebra, or across two or three, while the most superficial fibers extend over four or five vertebrae. This ligament gives important support to the vertebral column and is of great clinical importance in fractures of this column. Such fractures usually involve crushing of the anterior portion of one or more bodies, but without injury to the anterior longitudinal ligament. In these cases flexion allows further crushing of the vertebra or vertebrae involved, and posterior displacement at the level of injury with consequent danger to the spinal cord. In hyperextension, however, the pull of the taut anterior longitudinal ligament will realign the fragments and hold them in position. Use is therefore made of the potential splinting action of this important ligament whenever it is suspected that a fracture of the vertebral column may have occurred. In all potential cases of fracture of the back (except in the occasional cervical fractures that result from hyperextension) it is necessary to keep the patient in hyperextension at all times; "first aid" that allows flexion may produce irreparable injury. (Fractures or fracture-dislocations of the cervical vertebrae are particularly dangerous, since a displacement here can easily cause death. "First aid" when such fractures are suspected consists, therefore, of being sure that there is no movement of the head and neck.) The extensions of the anterior longitudinal ligament on the sides of the bodies are composed of short fibers only, and are much thinner than is the main part of the ligament.

On the posterior aspect of the bodies, and thus within the vertebral canal, is a second longitudinal band, the **posterior longitudinal ligament.** This commences as a broad band attached to the occipital bone, and remains fairly broad throughout most of the cervical region. In the thoracic and lumbar regions it becomes narrowed over the centers of the bodies and expanded over the intervertebral disks so as to resemble a series of hourglasses. It is firmly attached to the intervertebral disks and the adjacent portions of the bodies but is separated from the middle of each body by veins and small arteries that leave and enter the vertebra. The posterior longitudinal ligament also consists of long and short fibers, the longest fibers being placed most superficially and extending over a number of vertebrae while the shortest fibers extend only from one vertebra to the next.

The most important connections between the bodies are the **intervertebral disks;** these have been mentioned already several times and must be described in more detail here. Each intervertebral disk consists of two portions — an anulus fibrosus, or outer fibrous layer, and a nucleus pulposus, or soft center. Some workers regard the thin cartilaginous plate that separates the spongy bone of the vertebral body from the nucleus pulposus as a third part of the disk, although developmentally it is the epiphyseal plate of the vertebral body.

The **anulus fibrosus** (once annulus) consists of a number of layers of dense fibrous tissue and fibrocartilage which are firmly attached to the ends of the bodies adjoining the disk. The fibers of each of these layers run obliquely, but those of any two adjacent layers run at an angle to each other so that their fibers cross like the limbs of an X. When the anulus is compressed, the Xs become shorter and broader; when the compression is relieved, the Xs become taller and narrower.

The **nucleus pulposus** is a semigelatinous mass situated somewhat eccentrically, slightly closer to the posterior edge of the disk. When the bodies and disks are cut the nucleus pulposus bulges outward, giving evidence of the disk's elasticity and of the compression to which it has been subjected. Even in the supine position, when the vertebral column is not supporting the weight of the body, the disk is maintained under pressure by the ligaments connecting the arches. It has been shown that it takes about 30 pounds of pressure to restore a disk to its former size after it has been freed and allowed to expand.

The nucleus pulposus contains a very high percentage of water—from 70 to more than 80 per cent—and is therefore itself essentially incompressible. However, its softness allows it to change shape easily, and it is the change in shape that accounts for the compressibility of the intervertebral disk as a whole. Thus when the vertebral column is bent in any direction the nucleus pulposus becomes somewhat wedge-shaped, with its thin edge in the direction of bending, while the part of the anulus fibrosus toward this side bulges out and that on the opposite side is put upon a stretch.

Because the central parts of the disks, the nuclei pulposi, have such a high water content, the disks are subject to dehydration as a result of the pressure placed upon them. In fresh disks placed in a mechanical press, water droplets can be squeezed out of the disk, and standing and moving have the same effect save, of course, that as the water is squeezed out in minute quantities it is absorbed into the blood stream. As the disks lose water they of course become thinner, and sufficient dehydration can occur during a day's activity to result in a loss of height of 3/4 inch in an adult man. During rest in bed and sleep, when the pressure upon the disks is least, water is reabsorbed from the blood stream by the disks, and the original height is regained. However, over the years a little less water is reabsorbed than is lost, so the water content of the disks becomes somewhat less with age, and the disks become very slightly thinner. This is one of the reasons (the increased stoop with age is the other) why old persons are typically less tall than they were in young adulthood.

Dehydration results in an imperceptible change in the thickness of any one disk, and marked narrowing of a disk (visible in a roentgenogram—x-ray—because the intervertebral disks do not interfere with the rays as do the adjacent vertebral bodies) is therefore a sign of massive loss of the substance of the disk. This loss may occur as a result of rupture of the nucleus pulposus through the cartilaginous plate into the spongy bone of the vertebral body, or may occur through a break in the anulus fibrosus that forms the periphery of the disk. Degenerative changes begin to appear in the disks in early adulthood, and a weakened anulus fibrosus subjected to excessive strain will bulge or break. While massive protrusions of the disk substance may cause narrowing of the intervertebral space, relatively small protrusions or ruptures that produce no apparent narrowing may cause distressing symptoms and markedly incapacitate the individual.

Because the anulus is thinner posteriorly, protrusion or rupture of the disk (sometimes called "slipped disk") is more likely to occur posteriorly, and it is posterior or posterolateral protrusions and ruptures that cause symptoms and are therefore recognized clinically. Large posterior protrusions may exert pressure upon the spinal cord and cause paralyses and anesthesias similar to those produced by tumors in the vertebral canal. However, the most common protrusions and ruptures are posterolateral ones, around or through the thin lateral edge of the posterior longitudinal ligament, which reinforces the anulus more strongly close to the midline than elsewhere. Such posterolateral protrusions and ruptures, more common in the lower lumbar and lower cervical regions than elsewhere, are very likely to press upon the roots of a spinal nerve just before it leaves the vertebral canal. Pain is the most common result of such pressure, and is interpreted as coming from the area to which that nerve is distributed. There is little loss of sensation because of the overlap between spinal nerves, and little motor deficit because of the multiple segmental innervation of muscles.

The interpretation of the pain as coming from the area of distribution of the nerve is spoken of as "reference" of pain to a part, or "radiation" of pain down a nerve. Thus protruded lower cervical disks usually produce pain that seems to come from the hand, which is supplied especially by the lower cervical nerves, and protruded lower lumbar disks produce pain referred to the lower part of the leg and to the foot and seeming to extend along the sciatic nerve, the large nerve passing down the back of the thigh to the leg and foot. Thus sciatic pain is suggestive of a protruded lower lumbar disk. Flexing the thigh at the hip while the leg is held straight at the knee draws the nerve roots contributing to the sciatic nerve forward, and therefore presses them more firmly against a protruded disk if one is present. In consequence, this "straight-leg-raising test" can be expected to reproduce or accentuate sciatic pain due to a protruded or ruptured disk.

The pain produced by a protruded disk also produces reflex spasm of the muscles of the back, usually more on one side than on the other, so that the back is somewhat laterally flexed. Most commonly the flexion is toward the side of the protrusion or rupture, but sometimes it is toward the opposite side or even alternates, being at one time toward, at another time away from, the painful side.

In addition to the pain produced by pressure upon a nerve root, protrusion of a disk may produce local pain, apparently as a result of stretching of the posterior fibers of the anulus and the fibers of the posterior longitudinal ligament; spasm of the back muscles is also a source of pain; and massive loss of the substance of an intervertebral disk may place undue strain upon the synovial joints between the affected vertebrae and this may be still another source of pain.

In addition to the joints between typical vertebrae, such as have already been described, there are special joints at the upper end of the cervical column to facilitate movements of the head. The articular pits on the upper surface of the atlas receive the rounded occipital condyles of the skull to form the atlantooccipital joints. The **atlantooccipital joints** are surrounded by articular capsules that are rather weak and do not contribute much to the stability of the joint. The atlas and skull are also united by anterior and posterior atlantooccipital membranes which bridge the gap between the first cervical vertebra and the skull and thus complete the vertebral canal. The **anterior atlantooccipital membrane** is essentially the continuation upward to the skull of a portion of the anterior longitudinal ligament; the **posterior atlantooccipital membrane** is broader and thinner and runs only from the posterior arch of the atlas to the occipital bone. The suboccipital and other muscles attaching to the skull, and the shape of the joint itself, provide strength to the atlantooccipital joint. This joint allows for a flexion-extension movement of the head such as is involved in nodding.

The atlas is connected to the axis by laterally placed synovial joints essentially similar to those found elsewhere in the vertebral column, and also by the special articulation of the dens (p. 198). The dens projects upward from the body of the axis to lie just posterior to the anterior arch of the atlas. It is held in place against the atlas by a heavy transverse ligament, and there are also special ligaments extending from the dens to the skull. Between the anterior surface of the dens and the arch of the atlas, and between the posterior surface and the transverse ligament of the atlas, are synovial joints that allow the atlas to rotate on the dens as a pivot. The atlantoaxial articulation thus allows for considerably more rotation than can be found between other cervical vertebrae, and is responsible for much of the freedom of the movement of shaking the head, as in signifying "no." The dens can also slip downward and upward between the atlas and the transverse ligament, so the atlantoaxial articulation also contributes to flexion and extension of the neck.

The lateral atlantoaxial joints are not strong enough to prevent gradual anterior dislocation of the atlas and skull upon the axis; therefore when the dens is fractured, congenitally absent, or destroyed by disease, displacement does occur. The vertebral foramen of the atlas is so much larger than is the spinal cord at this level that considerable displacement may occur before the cord is damaged, but eventually pressure upon the cord will produce increasing paralysis and loss of sensation. Interruption of the spinal cord at this high level is incompatible with life; the usual cause of death by hanging is fracture of the axis with dislocation between the axis and the third cervical vertebra and severance of the cord.

MOVEMENTS AND STABILITY

The movements of the skull upon the atlas and of the atlas upon the axis have already been briefly discussed. It remains, however, to consider the movements of the remainder of the vertebral column. These movements are limited not only by the various ligaments already described as attaching the vertebrae to one another, but also by the positions of the articular facets, the shape and slant of the spinous processes, the relative sizes of the intervertebral disks, and various other factors. Movement between any two adjacent vertebrae is exceedingly limited in all portions of the vertebral column; however, the total amount of movement in a given region may be considerable.

In the cervical region the intervertebral disks are relatively large in comparison with the sizes of the bodies; moreover, the lower surfaces of the bodies are concave anteroposteriorly, and can thus overlap the convex anterior border of the upper surface of the succeeding body during flexion, while lateral flexion is facilitated by the lateral convexity of the lower surfaces of the bodies and the corresponding concavity of their upper surfaces. In this region all movements are especially free; the positions of the articular facets allow flexion and extension, lateral flexion when accompanied by rotation, and fairly free rotation.

In the thoracic region the intervertebral disks are relatively small, and the flatness or slight concavity of both upper and lower ends of the bodies further limit the usefulness of the disks in allowing free movements. Flexion and extension are limited also by the fact that the articular facets are almost in the frontal plane. Further, movement in the thoracic region is hindered by the attachment of the ribs, and extension is interfered with by the overlapping of the spinous processes. Flexion and extension, therefore, are especially limited in the thoracic region; lateral flexion, with the ribs being brought closely together on one side and spread somewhat on the other, is the freest movement of all; rotation is more limited.

In the lumbar region the large intervertebral disks, the posterior direction of the spinous processes, and the almost sagittal direction of the facets of the articular processes allow especially good flexion and extension; lateral movement here is also rather free, with the inferior facets of a vertebra sliding down on one side, up on the other; thus the upper portion of the trunk can be circumducted by these movements in the lumbar region. Rotation in the lumbar region varies among persons, but is always very limited, for the articular processes soon lock together. The movements of the lumbar vertebral column are especially free in the lower lumbar segments.

The **stability** of the vertebral column depends upon a number of factors. Of these, the relationship to a vertical line representing the center of gravity is among the most important, since when weight is properly balanced on the vertebral column minimal muscular activity is necessary. A constantly maintained position in which the weight is not reasonably well balanced results in structural changes and a permanent deformity in the growing child, and constant muscular strain in the adult.

In the first place, the line of gravity should lie in the midsagittal plane of the body. If it does not do so as a result, for instance, of unequal length of the lower limbs (which would result in a lateral tilt of a straight vertebral column) the only way in which balance can be restored is by a lateral bending (scoliosis) toward the side of the long limb. Similarly, if a structural scoliosis develops as a result of unequal growth of the two sides of the vertebral bodies, it so throws the center of gravity to one side that it must be compensated for by purposely curving some other part of the vertebral column to the opposite side.

(The primary scoliotic curve is now thought to be often due to weakness of muscles of the back on the convex side, particularly as a result of poliomyelitis. If uncorrected, it becomes progressively worse until growth ceases.) The lateral curvature is regularly accompanied by a rotation of the vertebrae forming the curve, with their bodies turning toward its convexity, their spinous processes toward its concavity. If it is in the lumbar region, pain usually results from the rotation—for little rotation is allowable here, as already noted—while if it is in the thoracic region the entire chest becomes markedly distorted.

In the second place, the weight of the head and trunk is more nearly balanced in the frontal plane when the line of gravity passes behind the bodies of most of the cervical vertebrae but through the bodies at the junction of the cervical and thoracic parts of the column (then the posterior curve of the cervical column buttresses the weight of the head); when it passes again through the bodies at the thoracolumbar junction (necessarily passing in front of the bodies of most of the thoracic vertebrae); and when it passes not too far from the center of the body of the fourth lumbar vertebra. This distribution of weight approaches the ideal except that the greater weight in front of most of the thoracic vertebral bodies tends to increase the thoracic curvature. Because of the limited movements of flexion and extension of the thoracic column this cannot be corrected by posture, but fortunately it is largely the vertebrae and their ligaments rather than the back muscles that prevent further flexion here (although with paralysis of the back muscles in this region, flexion does become increasingly severe). Kyphosis (in its severest state, humpback) is an increase in the anterior curvature of the thoracic region. It commonly results from a collapse of one or more vertebral bodies (usually from tuberculosis) or, if congenital, from absence of one or more bodies, but a mild degree accounts for the stoop of elderly persons.

Since the lumbar column supports all the weight above it, it is this part of the column that commonly adjusts to forward and backward shifts in the line of gravity. Kyphosis, for instance, shifts the line forward, and the resulting tendency to fall forward is most easily overcome by an increase in the lumbar curvature (lordosis) so that it better buttresses the weight thrown upon it. The lordotic curve is then a compensatory one for the kyphotic curve. In the same way, a temporary lordosis develops in the later stages of pregnancy, to counteract the forward movement of the line of gravity produced by the weight of the fetus and its adnexa. If on the other hand the center of gravity is moved backward, the lumbar curve becomes somewhat flattened.

The lumbar curve also compensates for the obliquity of the upper end of the sacrum (which always slants downward and forward). The sacrum is slightly more oblique in women than in men, and consequently the lumbar curve is slightly greater in women than in men. Similarly, downward rotation of the pelvis, which increases the sacral obliquity, is accompanied by an increased lumbar curve, and upward rotation by a decreased one; and a permanent abnormal increase in sacral obliquity is accompanied by a permanent lordosis.

The compensatory changes of the lumbar curve are brought about by the back muscles, but once brought about they tend to stabilize the column. Movements of the trunk from the most stabile position necessarily shift the relations of the line of gravity, and as soon as they occur the muscles acting upon the column (usually the back muscles, since the common shifts of the line of gravity are forward or to one side) must spring into action. In the ultimate analysis, therefore, it is the back muscles that are primarily responsible for the stability of the vertebral column. When these are paralyzed it is impossible to maintain a balance, for although the ligaments and articular facets will help to check extreme movements they necessarily must allow an appreciable amount of movement, far more than enough to upset the line of gravity.

There are apparently a number of causes of **low back pain,** but a factor common to much such pain is the strain that may be put upon the lumbar portion of the vertebral column and its muscles. The farther the line of gravity is shifted forward, the more active the muscles of the back necessarily become; thus bad posture is a common, and correctable, cause of low back pain. Since the muscles use the vertebral column as a lever, the forces involved are very much greater than the actual weight borne by the lumbar part of the column. Measurements of the pressure on the disks below the third and fourth lumbar vertebrae have shown that in the erect standing position the weight borne by them is in the range of 200 to 300 pounds; and the leverage is such that weights of 22 pounds held in each hand may produce an added load of 230 pounds on the disk when one leans forward approximately 20 degrees. (The tension exerted by the ligamenta flava, approximately thirty pounds, and that due to the minimal muscular activity in quiet standing, must be added to the weight of the body above the level of measurement; when one leans forward, the musculature necessarily becomes more active, thus increasing the pressure on the disk, and with further leaning the pressure increases still more.) When these figures are considered, it is easy to see how the muscles of the lower part of the back may be strained, and why degenerative changes in the disks, particularly common in the lumbosacral one, can lead to protrusions that cause crippling low back pain. Using the back as a lever in picking up objects can obviously put enormous strains on the vertebral column and its musculature. These strains can be minimized if one will crouch, holding the back as straight as possible, and use the strong muscles of the buttock, thigh, and calf to do the lifting.

Asymmetry of articular facets (most common at the lumbosacral junction) is another cause of low back pain, presumably because the strains of movement are unequally distributed between the two sides. Still other causes referable to the vertebral column are variations and abnormalities; these most frequently affect the last lumbar vertebra.

Abnormalities of the fifth lumbar vertebra occur in between 5 and 6 per cent of individuals. The important types of congenital abnormalities here are nonfusion of the two sides of the laminae, resulting in spina bifida, and separation of the inferior articular processes from the rest of the vertebra, a condition known as spondylolysis. Spina bifida occurs in slightly more than 1 per cent of persons; in so far as its presence weakens the back (and in one series of patients with low back pain the incidence of spina bifida was 10 per cent), such weakness is due primarily to the lessened area provided for the attachments of muscles. Spondylolysis was found in more than 4 per cent of a series of approximately 750 vertebral columns, and in three fourths of these it was bilateral. Since the inferior articular processes of the fifth lumbar vertebra form the chief anchorage of this to the first sacral segment, bilateral spondylolysis produces pronounced weakness in the lower part of the back. With the loss of the anchoring effect of its inferior articular processes the fifth lumbar vertebra (and therefore all the vertebral column above it) may slide forward on the sacrum (spondylolisthesis), thus throwing the entire column out of line.

MUSCULATURE OF THE BACK

With the minor exception of a few very small anterolateral muscles, the posterior muscles of the back are innervated by the dorsal rami of the spinal nerves. These muscles were originally segmental, extending from one vertebra to the next; in the course of ontogenetic and phylogenetic development the more superficial fibers have united to form longer bundles extending over a number of segments. Thus the musculature of the back is formed by a number of incompletely separated layers of muscles, distinguishable in part by the direction of their fibers and in part by their length. The longer muscles are placed superficially, the intermediate muscles more deeply, and the shortest muscles lie immediately against the vertebrae. The muscles of the back are rather completely covered by the musculature of the upper limb which has spread over the back to attain attachment to the spinous processes (fig. 5–11).

The true back muscles (fig. 13-4) present a very complicated picture, as they are composed of numerous converging and diverging fascicles that are arbitrarily grouped together in certain fashions and described as individual muscles. They are, in fact, not muscles in the sense that one speaks of a muscle in the limbs; they could be very easily subdivided into many more component groups than are usually named, or could be classified in only a few great groups. From the functional standpoint there is little reason to subdivide the back musculature in any detail, since the muscles usually work together in large groups. The subdivision and naming of the back muscles are largely, therefore, efforts to systematize them for descriptive purposes.

In the cervical region, deep to the trapezius and the rhomboids, there are two back muscles that can easily be seen to form a special group because of the direction of their fibers. These are the **splenius capitis** and the **splenius cervicis**. In contrast to the other back muscles, which run either roughly parallel to the midline of the vertebral column or run toward it as they are traced upward, the splenius muscles arise medially and pass laterally as they are traced upward. The splenius capitis arises from the lower half of the ligamentum nuchae and the spinous processes of the seventh cervical and upper three or four thoracic vertebrae and inserts laterally on the mastoid process and occipital bone of the skull (figs. 21-1 and 21-2). The splenius cervicis arises from the spinous processes of about the third to sixth thoracic vertebrae and inserts laterally on the upper two to four cervical transverse processes. The muscles of one side, acting together, will

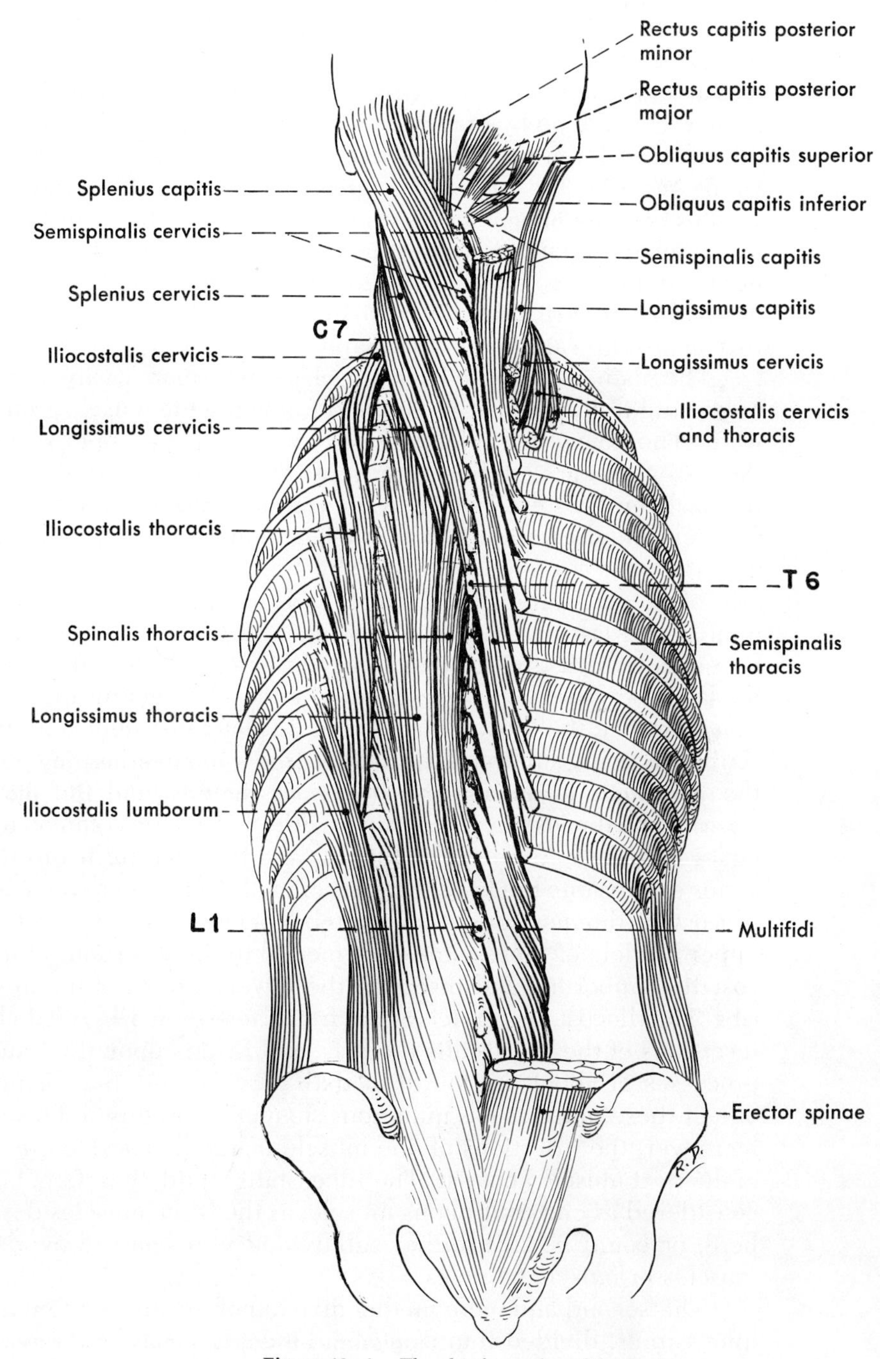

Figure 13–4. The chief muscles of the back.

rotate the head and cervical vertebral column toward the same side; when they act bilaterally they aid in extension of the head and neck.

In the upper and lower thoracic regions there are two muscles that cover the true back muscles, lying between these and the muscles of the upper limb, but themselves belonging to neither group. These are the superior and inferior serratus posterior muscles, connected with movements of the ribs. The **serratus posterior superior** arises from the lower part of the ligamentum nuchae and the spinous processes of the seventh cervical and upper two or three thoracic vertebrae and inserts on upper ribs, usually the second to the fifth. This muscle is supplied by branches from the first three or four intercostal nerves, therefore from ventral rami of spinal nerves, and assists in elevating the ribs and therefore increasing the size of the thorax. The **serratus posterior inferior** arises from the lower two thoracic and upper two lumbar spinous processes and inserts on the lower four ribs. Its nerve supply is from the eighth to eleventh intercostal nerves, and it draws the lower ribs downward to enlarge the thorax and steady these ribs against upward pull of the diaphragm.

The main mass of the back muscles can be more easily studied by beginning in the lumbar region and following the muscle groups upward. The heavy musculotendinous mass over the upper sacral and the lower lumbar vertebrae represents the origin of a large segment of the back musculature, known as the **erector spinae** because of its action in extending the vertebral column. This muscle group, indivisible at its origin, has a tendinous and fleshy attachment to the posterior surface of the sacrum, the iliac crest, and the spinous processes of the lumbar and last two thoracic vertebrae. As the erector spinae is followed upward it divides into three series of muscles, of which only the lateral two are well developed. The most lateral upward continuation is termed the iliocostalis system; this lateral system is in turn described as being subdivided into three linear but overlapping portions, the **iliocostalis lumborum,** the **iliocostalis thoracis,** and the **iliocostalis cervicis.** The iliocostalis lumborum, while sharing the common tendon, arises especially from the iliac crest and the sacrum; it inserts by a series of slips into the lower borders of the lower six or seven ribs. The iliocostalis thoracis consists of a number of fascicles that arise from the upper borders of these same ribs, medial to the insertions of the iliocostalis lumborum, and insert on the lower borders of the upper six ribs. The iliocostalis cervicis arises from these same ribs, medial to the insertions of the iliocostalis thoracis, and inserts upon the transverse processes of about the fourth to sixth cervical vertebrae. Thus each one of these muscles has numerous segmental origins and insertions; moreover, the insertions of one muscle markedly overlap the origins of the next muscle above it. The iliocostalis could, therefore, be fairly considered as one continuous muscle, as the three muscles described here, or could be still further subdivided into some 18 overlapping muscles or muscle fascicles.

The second and more medial division of the erector spinae is the longissimus, divided into **longissimus thoracis, longissimus cervicis,** and **longissimus capitis**—in other words, it is described as the longest muscle and divided into thoracic, cervical and head portions. The fascicles composing the lowest of the three divisions of this muscle arise from the common tendon of the erector spinae and insert into the lower nine or ten ribs and the adjacent transverse processes of the vertebrae. The longissimus cervicis arises from the transverse processes of the upper four to six thoracic vertebrae and inserts into the transverse processes of the second to sixth cervical vertebrae. The longissimus capitis arises medial to the upper end of the cervicis, partly with its tendons and partly from the articular processes of the lower four cervical vertebrae. It runs slightly laterally to insert into the mastoid process of the skull. Like the iliocostalis, therefore, each division of the longissimus is composed of a number of fascicles, and these fascicles so overlap each other, and the divisions so overlap each other, that the muscle has attachments upon almost every segment.

The most medial and last division of the erector spinae is the **spinalis** muscle. This is always poorly developed and is usually represented mostly by a slender muscle in the midthoracic region, arising from the common tendon and lower thoracic spinous processes and inserting upon upper thoracic spinous processes. This portion is called the **spinalis thoracis.** A slip of muscle representing a spinalis cervicis may occur in the cervical region, attached to the ligamentum nuchae or to this and lower cervical and upper thoracic spinous processes. The spinalis capitis is not a separate muscle, but is a smaller medial part of the semispinalis capitis, one of the following muscles.

When the erector spinae is removed a deeper more continuous set of fibers is seen. This is the **semispinalis muscle,** divided into **thoracis, cervicis,** and **capitis** according to the insertion of the muscle bundles. Much of the semispinalis consists of muscle fibers that arise from transverse processes and run upward and medially to attach to spinous processes; in contrast to the overlying erector spinae these fibers are directed not so much upward as upward and inward, taking an oblique course toward the midline. The fascicles of which this muscle is composed are of varying lengths, for the fibers that arise from any one transverse process attach to about four spinous processes, some of them running for as much as eight segments, others no more than five. The fibers of the semispinalis cervicis have their heaviest insertion upon the second cervical spinous process and do not reach the atlas; the semispinalis capitis covers the semispinalis cervicis, and runs almost straight upward to the occipital bone of the skull. The medial part of this muscle, usually having some origin from spinous processes, is called the spinalis capitis.

Deep to the semispinalis are the **multifidi,** once called multifidus, which much resemble the semispinalis except that the muscle fascicles are shorter (fig. 13-5), being only from two to four segments in length, and thus more obliquely placed in relation to the vertebral column. The fascicles run upward and medially from transverse processes to spinous processes; these muscles form a continuous mass from the upper end of the sacrum to the second cervical vertebra. The multifidi are particularly heavy in the lumbar region, thinner in the

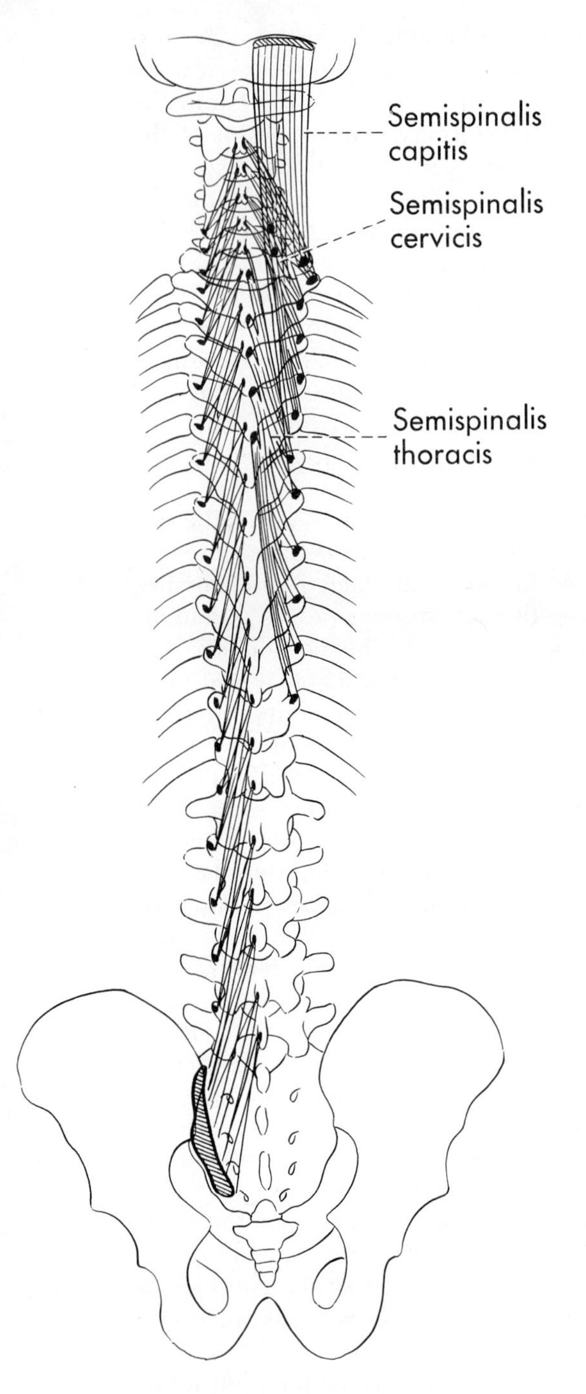

Figure 13–5. The semispinalis muscles (right) and the multifidi (left). Note the pronounced difference in length and obliquity of the longest fiber bundles.

thoracic, and thicken again in the upper cervical region, but do not extend to the skull.

The remaining muscles of the back are small isolated ones lying immediately against the vertebrae and passing either from one vertebra to the next above or, skipping one vertebra, to the second above. These small muscles are the **interspinales,** which are between spinous processes and are not present throughout most of the thoracic region;

the **intertransversarii,** which are between adjacent transverse processes but are poorly developed or absent in the thoracic region and are double in the lumbar region where there are medial and lateral intertransversarii, and in the cervical region where there are anterior and posterior ones; and the **rotators,** extending from the level of the sacrum to that of the second cervical vertebra. The rotators are obliquely set, and shorter than the shortest fibers of the multifidi. They are small muscles that can be divided into two groups, the long and the short rotators; each has a single origin, from a transverse process, and a single insertion, into the base of a spinous process. The short rotators, like the intertransversarii and the interspinales, pass only from one vertebra to the next above; the long rotators pass from one vertebra to the second above. Because of their origins and insertions, the semispinalis, the multifidi, and the rotators are grouped together as the **transversospinalis muscle.** It should be noted that all the elements of this muscle tend to slant inward, and that the deeper any component lies the greater is the slant and the shorter the muscle bundles.

In the uppermost part of the neck just below the skull there are several short muscles that extend between the axis and the occipital bone, the atlas and the occipital bone, or the axis and atlas. The posterior ones of these, visible in figure 13-4, lie deep to the other back muscles and are innervated by the dorsal branch of the first cervical nerve; lateral and anterior ones are innervated by the ventral branch of this nerve. These muscles (obliquus capitis inferior and superior; rectus capitis posterior major and minor, rectus capitis lateralis, and rectus capitis anterior) are classified as muscles of the head and neck, and assist the larger muscles in moving the head. More important than these small muscles, which are sometimes grouped together as the suboccipital muscles, are two long ones, the longus colli (collum, like cervix, means neck) and the longus capitis, that lie on the anterior surfaces of the vertebral bodies and help flex the neck and head. The longus colli both arises and inserts on the vertebrae, while the longus capitis arises from vertebral bodies but inserts on the occipital bone anterior to the foramen magnum. Also situated in the neck, and acting as flexors, lateral flexors, and rotators of it are the scalene muscles, which are described with other muscles of the neck in Section V.

There are no muscles attached anteriorly to the thoracic region of the vertebral column. In the lumbar region the psoas major, although a limb muscle (p. 254), has a direct action upon the vertebral column, for it takes its origin from bodies and transverse processes. Both muscles become active, taking their fixed points from below when one leans back from a sitting position, and the muscle on the opposite side becomes active when one leans to one side. Similarly, the two muscles can help flex a partially flexed vertebral column. It should be noted, however, that when the individual is supine and attempts to flex the lumbar column as in sitting up without the use of his hands, the first action of the psoas is to pull forward the lumbar column and thus increase the lumbar curvature; only secondarily does it flex the lumbar column on the pelvis and the pelvis upon the femurs. Thus, if one is attempting to exercise the anterior abdominal muscles by sitting-up exercise, one should start by curling the head, shoulders, and trunk upward — otherwise the stronger psoas will do all the work. A posterior abdominal muscle, the quadratus lumborum (p. 375) is also attached to the lumbar portion of the vertebral column (to the transverse processes); it is a lateral flexor of the column.

The true back muscles are primarily extensors and rotators of the vertebral column. The various muscles act specifically on the parts indicated by their names – muscles called "capitis" moving the head, those called "cervicis" moving the neck, and so forth – but many movements of the vertebral column involve the simultaneous contraction of many muscles.

Almost all the muscles of the back, acting bilaterally, extend (dorsiflex) the vertebral column or the head – this is true of the three major groups, the splenius, erector spinae, and transversospinalis systems. The splenius muscles, acting unilaterally, rotate the head and neck toward the same side (that is, turn the face toward that side). The erector spinae is a strong extensor; it has also been regarded as a lateral flexor and as a rotator to the same side, but lateral flexion is carried out primarily by the quadratus lumborum and the anterolateral abdominal muscles, and rotation by the latter and by the transversospinalis. The slight activity of the erector spinae as these movements are started has been interpreted as helping to prevent concomitant flexion rather than assisting in the movement. The transversospinalis system rotates to the opposite side if acting unilaterally; the multifidi, at least, are bilaterally active in extension. The interspinales are obviously extensors only, the intertransversarii lateral flexors only. The activity in the deeper muscles has been interpreted as being concerned primarily with fine adjustments between vertebrae rather than with movement of the vertebral columns as a whole.

The chief function of the muscles of the back when one is erect is to resist gravity. Regardless of what muscles start the movement, once the vertebral column is bent far enough to allow gravity to become an important factor the muscles of the back that resist this movement, rather than muscles that promote it, must actively contract in order to prevent falling and to make the movement smooth and controlled. When flexion is complete, these muscles relax completely, leaving support to the ligaments; they must spring into action again to start extension.

It should again be mentioned that muscles other than those of the back, even muscles that have no attachment to the vertebral column, also play a very important part in movements of the back. Thus the anterolateral abdominal muscles (p. 372) are the chief flexors and lateral flexors of the trunk, and are also important in rotation; and the sternocleidomastoid, a muscle of the neck, flexes, laterally flexes, and rotates the head.

THE SPINAL CORD AND MENINGES

The spinal cord, continuous above with the brain, lies within the vertebral canal, and the spinal nerves arising from it make their exit between adjacent vertebrae, usually through the intervertebral foramina. The spinal cord is separated from contact with the bony vertebral canal and its connecting ligaments by a layer of fatty connective tissue and by special coverings collectively termed **meninges.** The fatty connective tissue contains vertebral venous plexuses that help to drain the vertebral column and the spinal cord and connect with veins about the outer surface of the column, and small arterial branches that supply the vertebrae, the connective tissue and meninges, and to a variable extent the spinal cord. The space occupied by this tissue is called the epidural space; injections of anesthetic agents are sometimes made into it, as in the technic of "caudal analgesia" for reducing the pain of childbirth.

The outer covering of the spinal cord is the **dura mater** (figs. 13-6 and 13-7), a tube of tough connective tissue that is continuous with a similar layer within the skull and tapers to a point at about the level of the second sacral vertebra. Beyond this it is drawn out into a slender

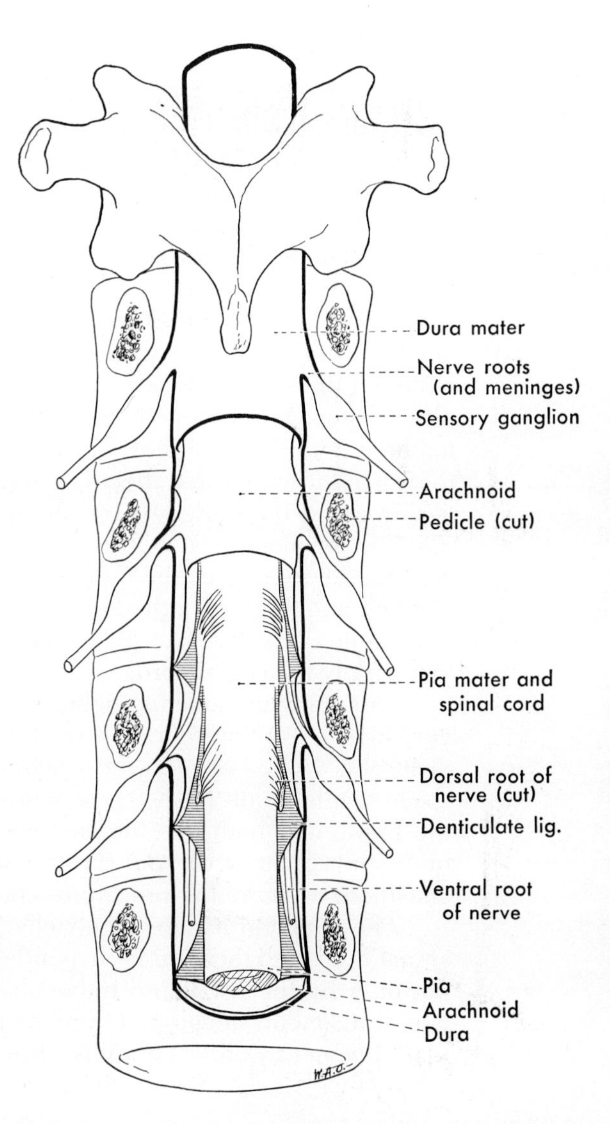

Figure 13–6. . The spinal cord within its coverings. From above downward, the cord is shown within an intact vertebral canal; the posterior portions of the vertebrae have been removed; the dura has been removed to show the arachnoid membrane; and this in turn has been removed to show nerve roots, the denticulate ligaments, and the pia mater.

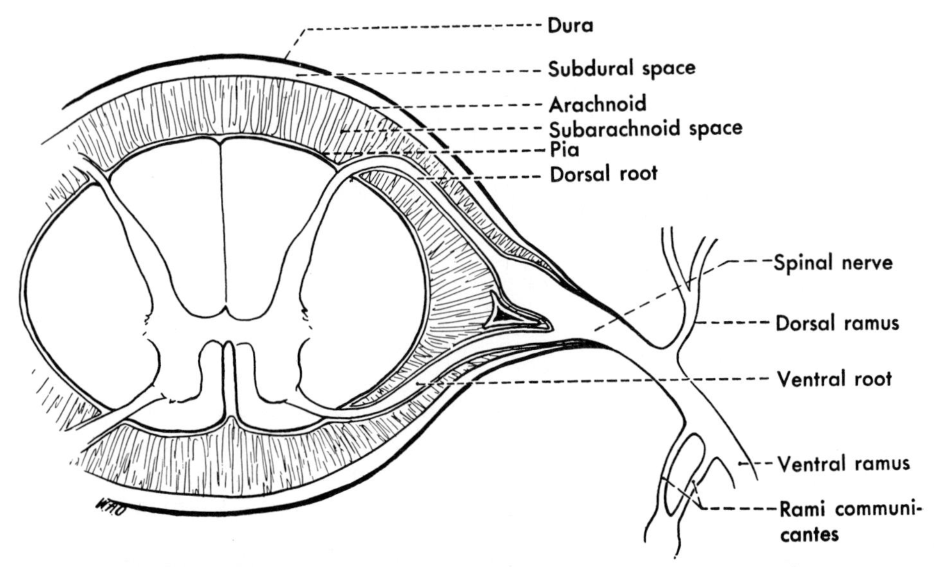

Figure 13–7. Relationships of the spinal cord and meninges in cross section.

thread, the filum of the dura, that anchors the lower end of the dura to the posterior surface of the coccyx. The dura also sends tubular sheaths around the roots of the spinal nerves as these leave the dural sac proper.

Immediately inside the dura, and separated from it only by a slitlike subdural space containing just enough fluid to keep the adjacent surfaces moist, is the **arachnoid.** This tubular membrane is much thinner than the dura; its outer surface is smooth but its inner surface, although lined by flattened cells, gives off numerous strands (trabeculae), also covered by flattened cells, that cross the underlying space to blend with the innermost layer of the meninges, the pia mater. It is these spider-web-like trabeculae that give the arachnoid its name. It terminates at the same level as the dura.

The **pia mater** is also thin, but is so tightly attached to the spinal cord that it appears as the outer surface of this; there is no space between it and the cord, although pieces of it can be stripped off the cord with some difficulty. The pia contains numerous small blood vessels that supply the spinal cord.

The dural and arachnoidal tubes or sacs surround the spinal cord very loosely, so that there is a relatively large space, the **subarachnoid cavity** (fig. 13-7), between the arachnoid and the pia. During life this space is filled with fluid, the **cerebrospinal fluid** (it is largely formed in the brain, and both fills the cavities of the brain and surrounds the brain and spinal cord). Anesthetic agents are introduced into the subarachnoid cavity in the procedure known as spinal anesthesia.

The cord and its closely attached pia are thus bathed in cerebrospinal fluid, and the cord is suspended within this protective medium not only by the arachnoid trabeculae but also by particularly tough lateral ligaments developed from the pia. These ligaments, the denticulate ligaments, project shelflike from each side of the spinal cord; at the intervals between each two adjacent spinal nerves, for the length of the spinal cord, pointed continuations from the lateral free edge of each denticulate ligament penetrate the arachnoid and attach into the dura (fig. 13-6). Viewed from behind or in front, therefore, a denticulate ligament looks somewhat like a saw.

The dura and arachnoid send sleeves outward around each dorsal and each ventral root of each spinal nerve as the nerve turns laterally to leave the dural-arachnoidal tube, and the pia also follows out on the surfaces of the nerve roots. These layers become continuous with the ordinary connective tissue of the nerve at about the level of the spinal ganglia, but the subarachnoid space surrounds the roots up to this point.

The **spinal cord** varies from oval to almost round in cross section. It tends in general to be larger at its upper end than at its lower, because its upper end contains more fibers that are either going to or coming from the brain. However, the number of cells at a given level also affects its size, so the levels of the cord that must supply much muscle and much skin – namely the levels that supply the limbs – show localized enlargements. These are the cervical intumescentia (swelling) for the upper limb, the lumbar one for the lower limb.

Just as the dural and arachnoidal sacs, once about as long as the vertebral column, have been drawn into a thread at their lower ends, so has the spinal cord. Even more than the meninges, it fails to grow as fast as the vertebral column, and in the adult its lower end lies at about the lower border of the first lumbar vertebra. From the tapered lower end of the cord, the conus medullaris, a threadlike strand of tissue that represents the drawn-out original end of the cord runs down to attach to the ends of the dural-arachnoidal sacs and the dural filum that anchors them to the coccyx. This thread of original spinal cord is the filum terminale.

Since the spinal cord has been pulled up more than the dura-arachnoid during growth, the lower part of the arachnoidal sac, between about the second lumbar and second sacral vertebral levels, contains no spinal cord but only the filum terminale and the roots of spinal nerves (which have to run caudally from the spinal cord to their levels of exit between vertebrae). This collection of nerve roots around the filum terminale is the cauda equina (horse's tail). The pia-covered roots here float in the cerebrospinal fluid, and it is in this location, between lower lumbar vertebrae, that spinal punctures (for the purpose of drawing off cerebrospinal fluid for examination, or introducing medicine or anesthetic agents into the subarachnoid cavity) are made. Punctures at higher levels risk damage to the spinal cord, but the nerves slip away from the needle.

The **tracts** of the spinal cord (fig. 13-8) consist of groupings of nerve fibers of similar function. Few of them can be distinguished in a normal cord, and all representations of them are diagrammatic; some of them shift position, others are found in only one part of the cord, each varies in size according to the level of the cord being considered, and there is much overlap between tracts. There is even disagreement as to how many tracts there are and uncertainty as to exactly where some of them begin or end. Although we have a useful working knowledge of the more important ones, there are many details yet to be resolved.

As already noted (p. 49), the tracts of the spinal cord connect different parts of the cord to each other, ascend to the brain, or descend from the brain. The chief interconnecting tract of the cord itself is believed to be the fasciculus proprius, surrounding the gray matter of the cord and consisting of both ascending and descending fibers. Collaterals given off from the long tracts arising in the cord also connect the various levels.

The chief long ascending tracts (right side of fig. 13-8) are divisible into those that conduct impulses to consciousness and those that do not. The two spinocerebellar tracts (dorsal and ventral, or posterior and anterior) are in the latter category. They arise from cells in the gray matter that receive afferent impulses mostly from muscles, and they transmit information concerning muscular activity to the cerebellum, where it is used in helping to control that activity at a subconscious level.

The major tracts that conduct to consciousness are the posterior funiculus, divided in the cervical region into the fasciculus cuneatus from the upper part of the body and the fasciculus gracilis from the lower part, and the spinothalamic tract or tracts. The posterior funiculus is composed of fibers that are the central processes of cells of the dorsal root ganglia of the same side, and transmit impulses concerned with touch, pressure, and appreciation of position and movement (fibers concerned with the latter activity are known as proprioceptive fibers); the information they carry is also essential to recognition of vibratory stimulation, as from a tuning fork placed on the shin, and judgment of the texture and shape of an object placed in the hand. They are thus concerned with all sensations except those of pain and temperature. Whereas the posterior funiculus conducts impulses from the same side of the body, these are relayed by cells whose fibers cross to the opposite side in the brain.

The spinothalamic tracts are sometimes regarded as a single structure, since they blend with each other rather than being widely separated as shown in the figure, but are often described as lateral and ventral (anterior), in which case the ventral spinothalamic tract is described as a pathway for touch, the lateral as the one for pain and temperature. The fibers of both tracts originate in the gray matter of the opposite side, cross anterior to the central canal of the cord, and then turn upward.

If only one side of the cord is cut in two, perception of pain and temperature over the opposite side of the body is lost almost up to the segmental level involved in the cut, because the lateral spinothalamic tract is interrupted (and surgeons sometimes purposely interrupt this tract to alleviate pain on the opposite side); appreciation of movement and of vibratory sensibility on the same side is lost because of the involvement of the posterior funiculus; but since impulses concerned with touch travel in part up the posterior funiculus of the same side, and in part up the ventral spinothalamic tract of the other side, appreciation of touch is not abolished.

The major descending or motor tracts (left side of fig. 13-8) are the pyramidal or corticospinal tracts, the vestibulospinal tract, and fibers, not shown in the figure, that lie in both the lateral and anterior funiculi and are called the reticulospinal tracts. The rubrospinal tract is an important tract in many lower animals, but if present at all in man is very small. The corticospinal or pyramidal tracts are two, lateral, and ventral or anterior. Both originate in the cerebral cortex or pallium; the lateral corticospinal tract is much larger and more important, and most of its fibers cross in a lower part of the brain as they run toward the cord. Thus one cerebral hemisphere (p. 343) controls primarily the other side of the body, just as it receives afferent impulses primarily from the other side of the body.

The specific functions of the vestibulospinal, reticulospinal, and other tracts are not yet clear, and therefore it is often convenient to group them as extrapyramidal fibers, that is, fibers that do not traverse the part of the brain known as the pyramids, as do the corticospinal or pyramidal fibers. The latter were once regarded as being the sole pathway by which the brain could initiate voluntary movements, and the spastic paralysis typical of severe upper motor neuron lesions was therefore attributed to interference with pyramidal function alone. However, there are few places where pyramidal fibers can be injured without affecting extrapyramidal ones, and therefore both sets are usually injured together. We know now, from observations on man and experiments on monkeys, that either system can carry out most movements, and that the deficit resulting from a pure pyramidal lesion is not a spastic paralysis, but only a loss of delicate movements such as those of the fingers and thumb. The extrapyramidal system is apparently primarily responsible for postural adjustments of the trunk and limbs, and both systems usually cooperate in carrying out most other movements of the limbs.

The general composition of the **spinal nerves** has already been discussed (p. 54). After the arachnoidal sac has been opened the dorsal

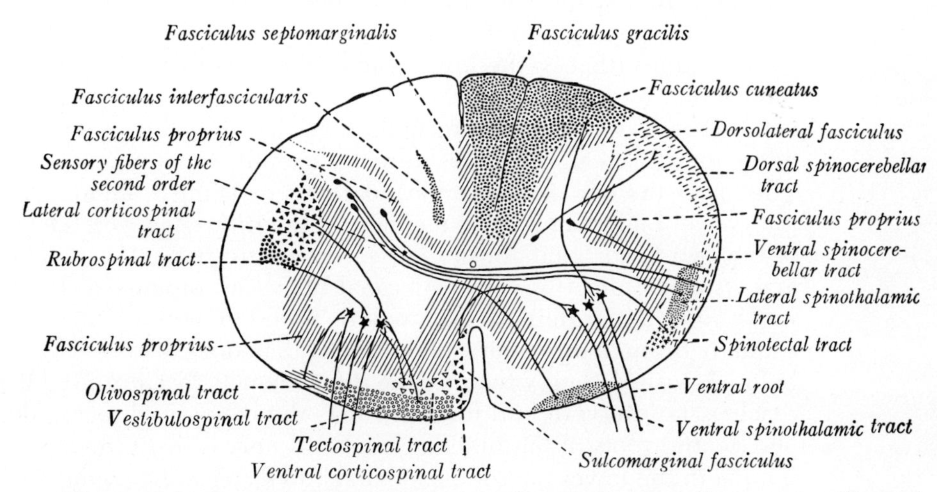

Figure 13–8. Some chief tracts of the spinal cord. The fasciculus proprius (diagonal lines) surrounds the gray matter of the cord (unshaded); otherwise, ascending tracts are shown on the right, descending ones on the left. (From Ranson, S. W., and Clark, S. L.: *The Anatomy of the Nervous System* [ed. 10]. Philadelphia, W. B. Saunders, 1959.)

(sensory) roots of the spinal nerves can be seen arising from the posterolateral aspect of the spinal cord. The ventral (motor) roots arise anterolaterally, and their origins cannot be seen unless the cord is removed or is twisted enough after the denticulate ligaments have been cut. However, lateral to the denticulate ligament the dorsal and ventral roots of each spinal nerve parallel each other, and go into their dural-arachnoidal sleeves very close together. As already noted, the sleeves terminate at about the level of the dorsal root ganglia and the two roots join at the distal end of the ganglion to form the typical mixed spinal nerve (fig. 13-7).

The 31 pairs of spinal nerves are divided into eight cervical, 12 thoracic, five lumbar, five sacral, and a single coccygeal, named according to their relations to the vertebrae. The first cervical nerve makes its exit between the skull and the first cervical vertebra and the eighth cervical nerve between the seventh cervical and the first thoracic vertebrae. However, below the cervical region each nerve leaves the vertebral canal below the vertebra of the same name and number—the eighth thoracic nerve leaves below the eighth thoracic vertebra, the fifth lumbar nerve below the fifth lumbar vertebra, and so forth. Except between the skull and the atlas, and the atlas and axis, the nerves leave through intervertebral foramina (fig. 13-1) which are bordered above and below by the pedicles of the vertebrae, anteriorly by the vertebral bodies and intervertebral disks, and posteriorly by the articular processes. The ganglia of the nerves lie regularly at the level of the intervertebral foramina, and the dorsal and ventral roots of a nerve therefore unite just as the nerve is making its exit from the vertebral canal.

The length of the spinal cord that gives rise to a spinal nerve is known as a spinal cord segment. If the spinal cord were as long as the vertebral column, spinal cord segments and vertebrae would correspond, and each nerve would run almost directly laterally to its exit. However, since the cord ends at about the lower border of the first lumbar vertebra, the spinal cord segments are shorter than the vertebral segments (fig. 13-9). This is especially true of the lumbar and sacral regions (that is, the lower end of the cord that gives rise to the lumbar and sacral nerves), where the cord segments are very short. In consequence of the shortening of the cord, it is only the upper cervical nerve roots that can run almost directly laterally to their points of exit. The lower the nerve, the more caudally its roots must run to reach its intervertebral foramen, for the greater is the discrepancy in the levels of the two fixed points of the nerve—its origin from the spinal cord and its exit from the vertebral canal. Thus as progressively lower nerve roots are examined, each will be found to have a longer course caudally through the subarachnoid cavity than does the preceding one before it turns laterally to enter the dural sleeves that surround it as it reaches its intervertebral foramen. It is the caudally directed dorsal and ventral roots of the lumbar and sacral nerves that form the cauda equina in the lower part of the subarachnoid space. Diagrams such as figure 13-7, which show very short dorsal and ventral roots, are therefore approximately accurate only for upper cervical nerves; the roots of the sacral nerves are inches long.

Because of the discrepancy in length between spinal cord and vertebral column (fig. 13-9), the clinician must bear in mind that injury at a given vertebral level will affect a different segmental level of the cord, and may injure several spinal nerves simultaneously. Thus, for instance, injury at the lower border of the tenth thoracic vertebra will affect the spinal cord segment that lies at this level—usually the twelfth. Further, it may also injure any or all the nerves arising at or above this level but leaving the vertebral column below it—in this case the tenth, eleventh, and twelfth thoracic nerves. The usual protruded or ruptured intervertebral disk (p. 205), however, lies too low to affect the spinal cord itself; it is also usually laterally situated, and typically involves the roots of a single spinal nerve inside the vertebral canal.

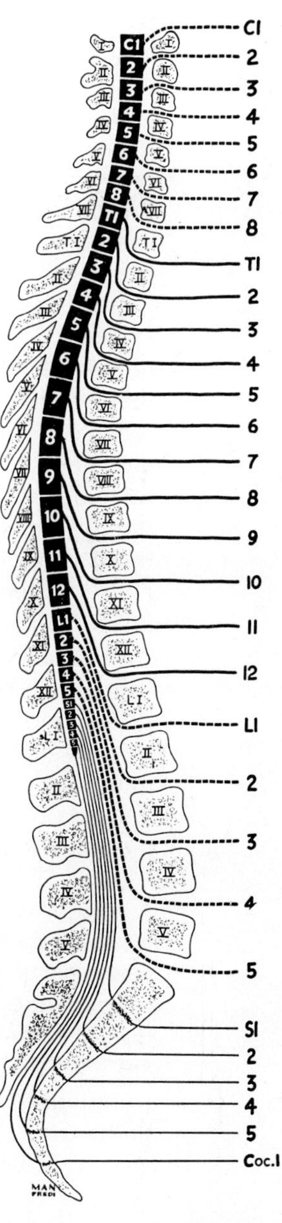

Figure 13–9. Relation of the spinal cord and nerve roots to the vertebral column. Vertebral bodies and spinous processes bear Roman numerals; spinal nerves and the segments of the cord from which each arises bear Arabic ones. Note that the cord segments are shorter than the vertebral ones, so that even in the lower cervical and upper thoracic region the nerve roots run downward to their exits. The lumbar and sacral segments of the cord are especially short, and the nerves arising here run markedly downward; below the end of the cord, these nerves constitute the cauda equina. (From Haymaker, W., and Woodhall, B.: *Peripheral Nerve Injuries* [ed. 2]. Philadelphia, W. B. Saunders, 1953.)

Shortly after the roots of the spinal nerve have united, the nerve divides into its two main branches, a dorsal ramus (ramus means branch) and a ventral ramus. This occurs outside the vertebral foramen. The dorsal ramus then simply turns posteriorly into the back muscles to supply them and continue to the skin of the back, while the ventral ramus continues laterally and anteriorly. (In the sacral region, where the transverse processes of the sacral vertebrae have fused together, the dorsal rami make their exits through the dorsal foramina of the sacrum and the ventral rami make their exits through the pelvic foramina.) A few of the dorsal rami exchange branches with each other, but for the most part they remain segmental. In contrast, many of the ventral rami enter into plexuses, in which the identity of the contributing nerves is lost. Thus, the upper cervical nerves form the cervical plexus (p. 360), the lower cervical nerves and the first thoracic form the brachial plexus (p. 76), and most of the lumbar and sacral nerves form the lumbosacral plexus (p. 229). The ventral rami of the thoracic nerves are however kept apart from each other by the ribs; because they lie between ribs, these ventral rami are known as **intercostal nerves.** Whether they enter a plexus or not, the ventral rami of the spinal nerves supply most of the muscle and skin of the body—essentially everything except the back muscles and the skin covering them.

It is also the ventral rami that are connected to the sympathetic system (p. 56). Those nerves that contain preganglionic sympathetic fibers (all 12 thoracic and the first two lumbar) give off to the sympathetic trunk a branch containing these fibers, and all the spinal nerves receive from the sympathetic trunk a branch containing postganglionic fibers. The postganglionic fibers are distributed through both dorsal and ventral rami to blood vessels, certain other smooth muscle, and sweat glands. The connections between a spinal nerve and the sympathetic trunk are known as **rami communicantes.** All the thoracic and the first two lumbar nerves should have two rami communicantes, as in figure 13-7, one composed of preganglionic fibers and one of postganglionic ones (that is, one leaving the nerve to go to the sympathetic trunk, the other leaving the trunk to join the nerve), and all the other spinal nerves should have only one, composed of postganglionic fibers. There actually is a good deal of variation in number of rami communicantes, for the preganglionic and postganglionic fibers may run together in a single branch or, more frequently, there is more than one postganglionic ramus connecting a spinal nerve to the sympathetic trunk.

Approximately the second, third, and fourth sacral nerves also contain preganglionic fibers, but they send these to the pelvic viscera—particularly urinary bladder and rectum—directly rather than through the sympathetic trunk (for these fibers are parasympathetic ones), so the branches that the sacral nerves give to the pelvic viscera are not called rami communicantes. The sacral nerves do receive postganglionic rami communicantes from the sacral part of the sympathetic trunk, and conduct these fibers to the blood vessels and sweat glands of the lower limb.

General Survey of the Lower Limb 14

The parts and regions of the lower limb ("inferior member") are the buttock (nates, clunes, or gluteal region), the hip (coxa), the thigh (the noun "femur" is used only as applying to the bone, although the adjective "femoral" is used in the sense of the thigh as a whole), the knee (genu) and popliteal region (region behind the knee), the leg (crus), the ankle (tarsus), and the foot (pes).

The lower limb develops in very much the same manner as the upper limb (p. 67); projecting first as a mesenchymal bud covered by a thin layer of epithelium, it develops flexures that indicate the positions of the knee and ankle, and digits on the foot. As is the case of the upper limb, the cartilaginous skeleton and the muscles develop from the mesenchyme of the bud, while the vessels and nerves grow into the bud. Because the base of the bud is broader than that of the upper limb, more spinal nerves grow into the lower limb—typically all the lumbar nerves and the first three or four sacrals, or eight to nine in all.

The big-toe side of the limb (lined in fig. 14-1) is at first directed cranially, like the thumb side of the upper limb, and the little-toe side (black in fig. 14-1) therefore directed caudally. If the limb could be abducted at this stage, the dorsal or extensor surface would face dorsally, the ventral or flexor surface ventrally. However, these relations are not long maintained, for the limb soon begins to rotate medially, gradually bringing the extensor side of the limb into an anterior instead of a posterior position (fig. 14-1*d*). This rotation is obvious from the fact that the muscles on the anterior or ventral surface of the thigh (extending also onto the lateral surface) extend the leg at the knee, and those on the posterior or dorsal side of the leg plantarflex the foot and flex the toes. Limb rotation is not complete at birth, allowing an infant

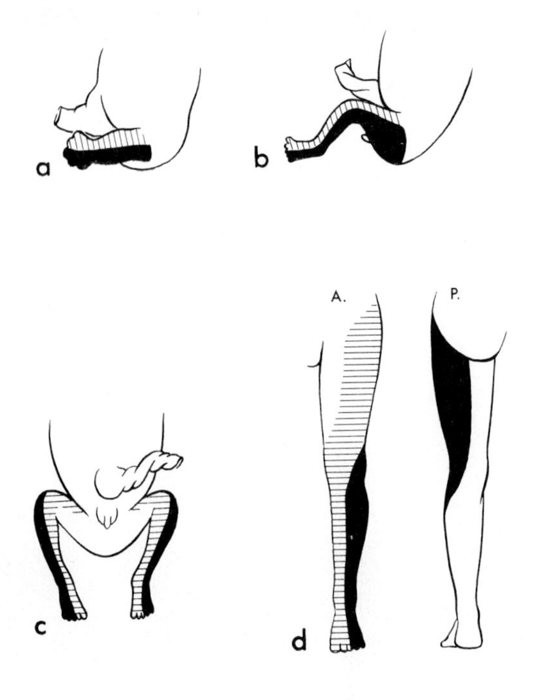

Figure 14–1. Rotation of the lower limb. *a.* The limb of a 7-week embryo. *b.* That of an 8-week one, seen from the side. *c.* The limbs at about the end of the third month. *d.* Anterior (A) and posterior (P) views of the adult limb. The original extensor or dorsal surface, directed laterally in *a,* is shaded to differentiate the cranial side (lined) from the caudal side (solid black); the original ventral part of the limb, visible in *c* and *d,* is unshaded.

to clap its feet together, but is completed as the infant learns to stand and walk.

The skeleton of the lower limb (fig. 14-2) is divided into the girdle and the skeleton of the free limb. The girdle, often called pelvic girdle, consists of two coxal (hip) bones, formerly known as the innominate bones. In contrast to the girdle of the upper limb, that of the lower limb is firmly attached to the axial skeleton (the sacral part of the vertebral column) at the sacroiliac joint. In addition, the two coxal bones articulate anteriorly through the pubic symphysis. The two os coxae thus form a strong arch completed dorsally by the vertebral column; this arch, only slightly movable at the vertebral column, transmits the weight of the body from the vertebral column to the femora, or thigh bones, and constitutes the bony pelvis ("basin"). In addition to the weight they receive from the vertebral column, the flared hip bones directly support some of the weight of the viscera. Muscles swinging across the pelvic floor, from one coxal bone to the other or to the sacrum, bridge the pelvic outlet (lower end of the bony pelvis) and thus help support the weight of the viscera.

Each coxal bone bears a deep cup-shaped fossa, the acetabulum, which receives the rounded head of the femur. The hip joint is thus a ball-and-socket joint. The femur is the bone of the thigh, corresponding to the humerus in the arm. The tibia and fibula are the two bones of the leg corresponding to radius and ulna in the forearm. The tibia is the larger bone situated on the medial side of the leg, and a portion of it is subcutaneous throughout its entire length. The fibula is the smaller bone on the lateral side of the leg; its upper and lower ends are subcutaneous but most of its body is buried in the leg muscles. In contrast to the upper limb, in which the ulna forms the chief articulation at the elbow while the radius forms the chief articulation at the wrist, the tibia articulates at both knee and ankle and transmits most of the weight. The fibula serves for the attachment of many muscles, and enters into the articulation of the ankle but not that of the knee. The knee joint, formed between the articular surfaces of the lower end of the femur and the upper end of the tibia, is largely a hinge joint. This is even truer of the ankle joint, formed by the lower ends of the tibia and fibula and the uppermost tarsal or ankle bone.

The tarsals are seven in number and are so arranged as to transmit weight both to the heel and to the ball of the foot; between these two weight-bearing points is the longitudinal arch of the foot. While the ankle joint proper allows little movement except flexion and extension, movement between the various tarsals allows some inversion and eversion of the foot and additional flexion and extension. Inversion of the foot is the movement which, if successful, would allow

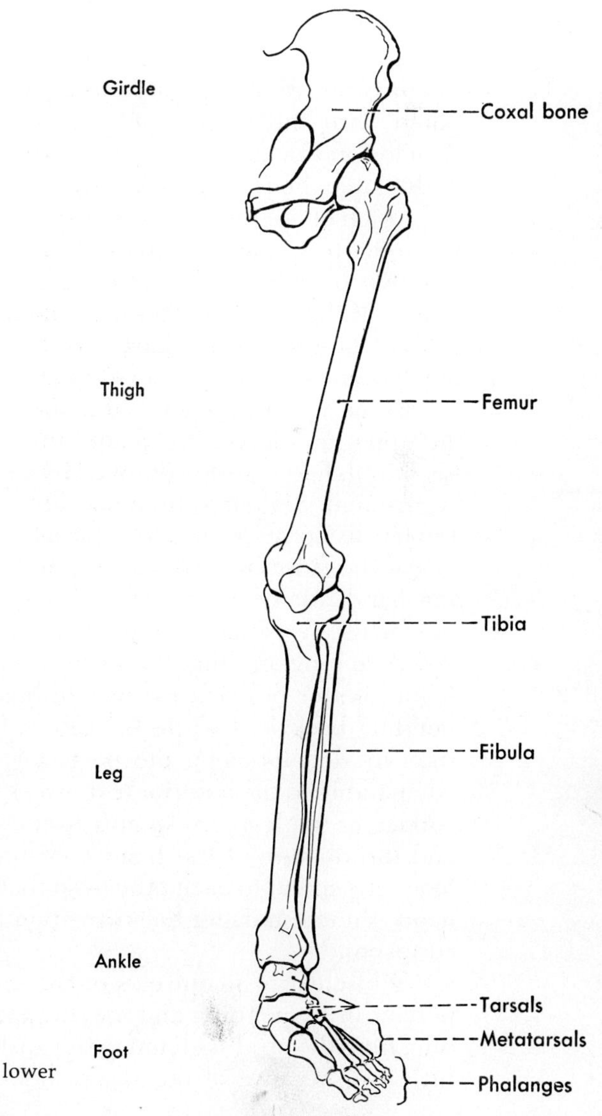

Figure 14–2. The skeleton of the lower limb.

turning the sole of the foot in so that the two soles could be placed together; it thus corresponds to supination of the hand. Eversion of the foot is turning the foot outward so that the weight falls on the inner rather than the outer border of the foot; it therefore corresponds to pronation of the hand. It should be noted that these movements in the lower limb are not carried out through movements of long bones, as in the upper limb, but are at the ankle and in the foot. Inversion and eversion of the foot are typically combined with other movements, but a foot in which eversion predominates is usually referred to by clinicians as a pronated foot, while one in which inversion predominates is similarly referred to as a supinated one. (Adjectives that carry similar meanings when applied to the foot, but that are also applicable to other parts of the lower limb, are *valgus* and *varus*. Both mean "bent," but valgus denotes an outer or lateral bending, varus an internal or medial bending. Thus "pes valgus" [pes = foot] denotes a foot bent outward, hence an everted or pronated foot, and since the arch then usually flattens, is also applied to flatfoot; "coxa vara" [coxa = hip] denotes a femur in which the angle between the neck and body of the femur is lessened, this internal bending shortening the limb as compared with the normal side; and "genu valgum" [genu = knee] denotes an exaggeration of the normal outward divergence of the leg at the knee, hence knock-knee.)

Much of the instep of the foot is formed by long metatarsals. With these the proximal phalanges of the digits are articulated; the big toe has two phalanges, the remaining toes three each. The distal phalanges of the toes, with the exception of the big toe, are rather insignificant bones; the phalanges of the toes as a whole are obviously nowhere nearly so well developed as those in the hand.

From what has been said it should be clear that the skeletons of the upper and lower limbs are built essentially upon the same plan. Each limb has a girdle, followed by a single bone: femur and humerus correspond. Similarly, tibia and fibula correspond to radius and ulna, the tarsals correspond to the carpals, and the metatarsals and phalanges of the foot obviously correspond to metacarpals and phalanges of the hand. Many other comparisons between the two limbs can be drawn also, especially in regard to the muscles. However, the torsion about its long axis that the lower limb has undergone in order to get it into better position for weight-bearing results in the knee joint's bending backward while the elbow bends forward. Thus, as is clear from its development, the posterior (extensor) surface of the arm is comparable to the anterior (extensor) surface of the thigh, the anterior surface of the leg corresponds to the extensor surface of the forearm, and the dorsum of the hand and the dorsum (upper surface) of the foot correspond to each other—so that, in the pronated position of the hand, thumb and big toe correspond, and little finger and little toe correspond.

While extrinsic muscles of the shoulder girdle are extremely important in suspending and moving this girdle, extrinsic muscles running from the axial skeleton to the girdle of the lower limb would be of little use because of the slight mobility of the sacroiliac joint. The girdle of the lower limb is so well stabilized that it gives origin to various muscles acting on the trunk; while it is acted upon in turn by a few of these muscles, none of these are, properly speaking, limb muscles. Only one muscle of the lower limb arises from the axial skeleton; it passes across the girdle to attach to the skeleton of the free limb. The other muscles of the hip region arise from the bony pelvis. Those in closest association with the girdle cover the posterior and lateral surfaces of the hip to form the musculature of the buttock, and act primarily across the hip joint.

Some of the muscles of the thigh act primarily at the knee joint, but those attached to the girdle have an action at the hip; these may or may not extend across the knee and have also an action at this joint. The muscles in the thigh are divisible into anterior, anteromedial, and posterior groups. The tendons of some of the posterior thigh muscles are easily felt behind the knee as they border the popliteal fossa, the depression behind the knee.

Muscles of the leg act primarily at the ankle and on the toes. They are divisible into muscles of the calf and those of the anterolateral part of the leg. Some of the muscles of the calf extend also across the knee joint and therefore have an action there. In a fashion similar to that of forearm muscles, many of the muscles of the leg continue into the foot by means of long tendons; there some of them are associated with the short muscles of the foot in movements of the toes, while others act upon the foot as a whole rather than the toes.

Nerves of the lower limb (fig. 14-3) are derived from two plexuses, the lumbar and the sacral; these adjoin, and together are referred to as the **lumbosacral plexus.** The lumbar plexus arises primarily from the ventral rami of the first four lumbar nerves; its two chief branches to the lower limb are the femoral and obturator nerves which pass to the front and anteromedial sides of the thigh, respectively, to supply the muscles there (figs. 16-9 and 16-10). The sacral plexus is formed by the union of ventral rami from the fifth lumbar and first three sacral nerves, usually joined by branches from the fourth lumbar and the fourth sacral. Most branches of the sacral plexus pass posteriorly between the sacrum and the coxal bone. The smaller branches supply muscles of the buttocks, but the largest portion of the sacral plexus is continued down the posterior aspect of the thigh as the sciatic nerve, the largest nerve in the body. As the sciatic nerve runs down the thigh it supplies the posterior muscles there and then divides a little above the knee into common peroneal and tibial nerves. The common peroneal nerve winds around the lateral surface of the leg to supply anterolateral leg muscles and be continued onto the dorsum of the foot; the tibial nerve runs down the posterior aspect of the leg and continues into the plantar aspect (sole) of the foot.

The great arterial stem of the lower limb is the femoral artery (fig. 14-3), situated anteriorly in the thigh. The femoral artery is the continuation of the external iliac, which is in turn the continuation and larger branch of the common iliac; the two common iliacs are formed by the bifurcation of the lower end of the aorta. The femoral artery runs down the anteromedial aspect of the thigh but above the knee

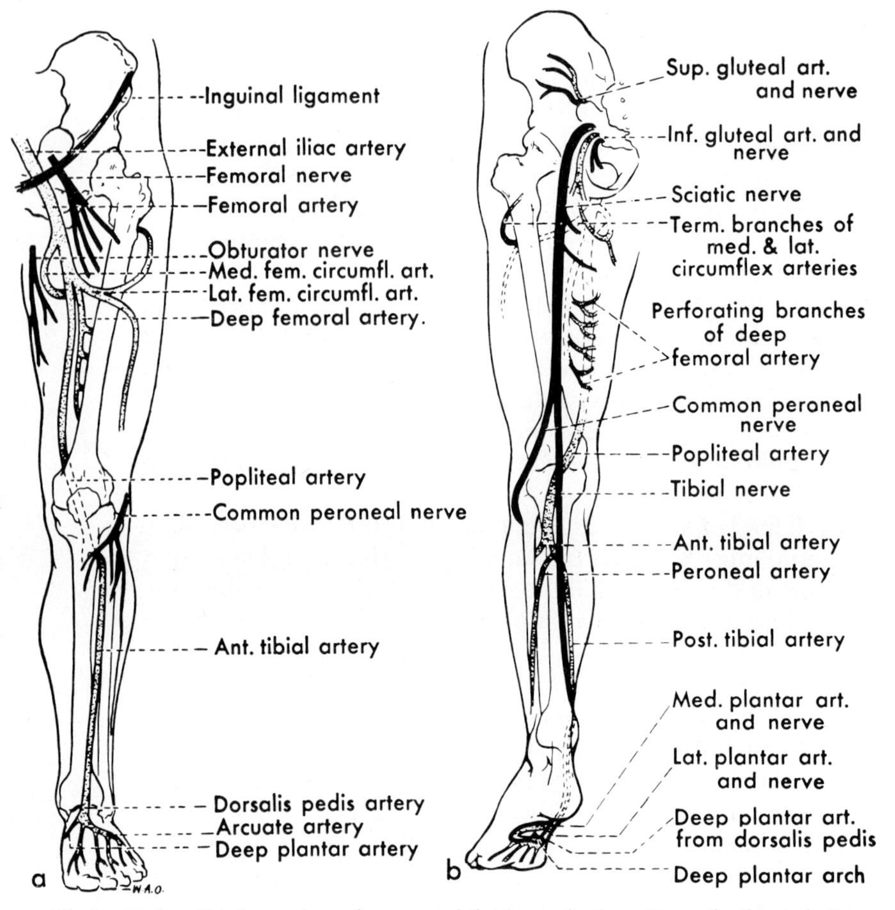

Figure 14–3. Chief vessels and nerves of the lower limb. *a*. From the front. *b*. From behind.

passes posteriorly around the medial surface of the femur to attain the popliteal fossa and become the popliteal artery. In the upper part of the leg the popliteal artery divides into anterior and posterior tibial arteries. The posterior tibial continues posteriorly down the leg to the plantar aspect of the foot, while the anterior tibial passes between the tibia and fibula to reach the anterolateral portion of the leg and continue onto the dorsum of the foot. The muscles of the buttock are supplied by arteries that are associated with the sacral plexus; these arteries arise not from the external but from the internal iliac artery, which is the chief artery to the pelvis.

The Bony Pelvis, Femur, and Hip Joint 15

THE BONY PELVIS

The bony pelvis is formed by the two paired coxal bones and the sacrum and coccyx. Each coxal bone consists of three separate bones in the fetus and infant; while these three bones are so fused in the adult that it is difficult to see any signs of their junction, their names are still retained in describing the coxal bone. Thus, even in the adult, ilium, ischium, and pubis are referred to as if they were separate bones.

The **ilium** is the superior element of the coxal bone (figs. 15-1 to 15-3); its great wing (ala) forms the projection of the hip, and the smooth inner and outer surfaces of this wing give attachment to muscles of the limb. The upper free edge of the wing, which gives attachment to abdominal muscles, is the iliac crest. The crest ends anteriorly in the anterior superior iliac spine and posteriorly in the posterior superior iliac spine. On the anterior border of the wing of the ilium below the anterior superior iliac spine is the anterior inferior iliac spine; on the posterior border, below the posterior superior spine and just above the smooth concavity (greater sciatic notch) on this border is the posterior inferior iliac spine.

On a level with the posterior inferior iliac spine the ilium bears on its medial aspect a smooth articular surface shaped somewhat like an ear and therefore called the auricular surface. Above and also behind the auricular surface is a larger rough area that serves for the attachment of heavy ligaments forming an important part of the sacroiliac joint. Below the posterior inferior iliac spine and the auricular surface of the ilium is a deep notch, the greater sciatic (greater ischia-

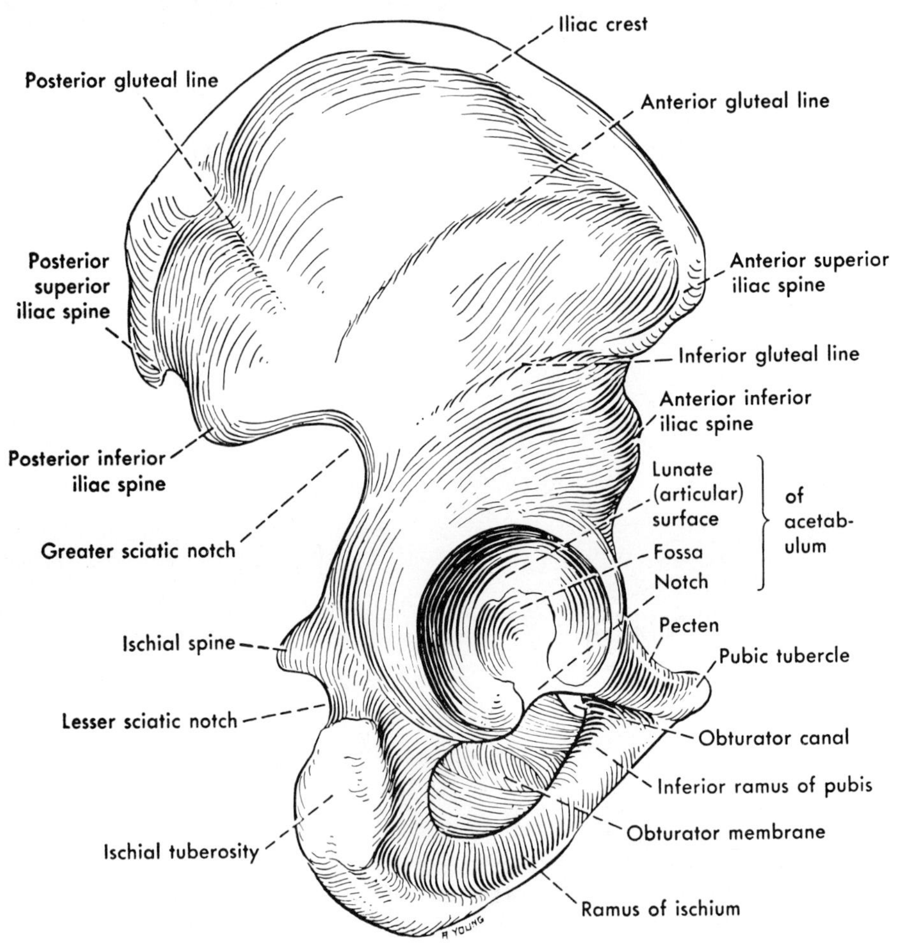

Figure 15–1. Lateral view of the coxal bone.

dic) notch. The lower and narrower part of the ilium, the body, extends down to about the level of a line drawn through the junction of the upper one third and the lower two thirds of the acetabulum, or hip socket.

Of the two inferior elements of the coxal bone, the **ischium** is the more posterior. Its body forms the posterior half of the lower two thirds of the acetabulum and its ramus forms the posterior wall and a part of the inferior wall of the large hole (obturator foramen) in the coxal bone, extending forward at the lower border of this foramen to join the pubis. On its posterior edge the ischium bears a pointed spine; above this spine the posterior border of the ischium forms a part of the greater sciatic notch which was noted on the ilium; below the ischial spine is a smaller notch, the lesser sciatic or lesser ischiadic notch. The heavy flattened posterior expansion of the ischium is the ischial tuberosity.

The **pubis** is the more anterior of the two lower parts of the coxal or hip bone; its superior ramus forms approximately the anterior half of the lower two thirds of the acetabulum, and its inferior ramus curves posteriorly and downward to join the ischial ramus and complete the wall of the obturator foramen. Its body, at the junction of the superior and inferior rami, has on its anterior superior surface a thickening, the pubic crest, which ends laterally in a more marked pubic tubercle; its medial surface is an articular one that enters into the pubic symphysis.

All three elements of the coxal bone help form the deep receptacle for the head of the femur, the **acetabulum.** As the acetabulum is examined it will be noted that only a part of its surface is smooth and obviously adapted for articulation; a deeper, rougher, portion is occupied by fat and a ligament. The lower edge of the acetabulum is deficient, thus making this fossa resemble a cup with a portion of the lip broken out.

The **obturator foramen** lies between the acetabulum and the conjoined ischial and inferior pubic rami. In the dried condition this is a large foramen; in life, however, the foramen is almost completely closed by a membrane that gives attachments to muscles on both of its surfaces; thus in life this foramen deserves its name of obturator (closed) foramen.

The word "pelvis" literally means "basin"; hence the pelvis consists not only of the two coxal bones, the girdle, but also of the inter-

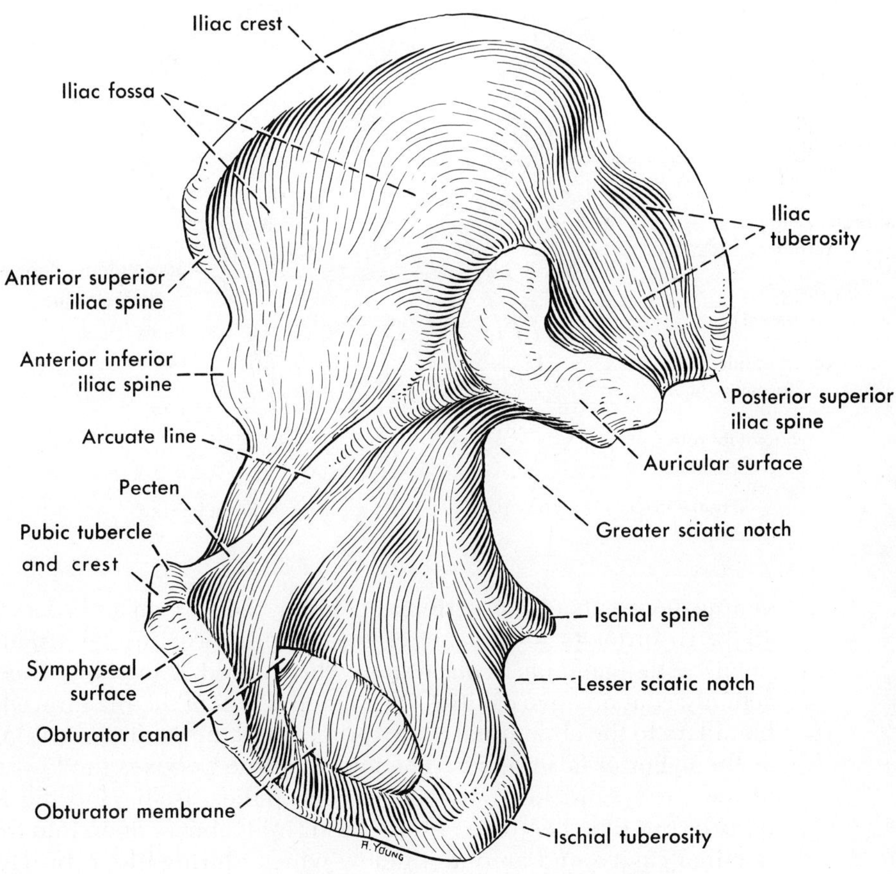

Figure 15–2. Medial view of the coxal bone.

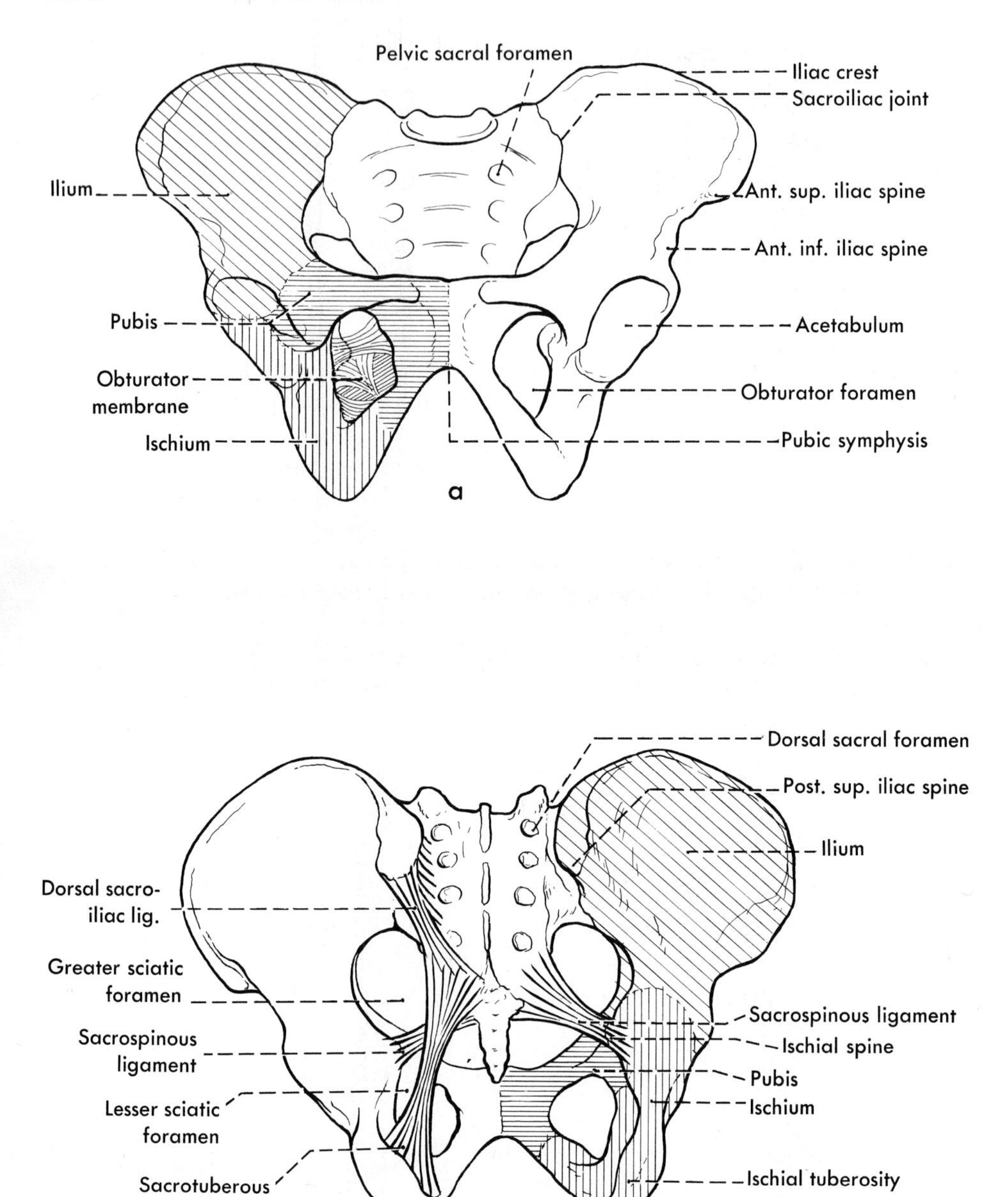

Figure 15–3. The bony pelvis. *a*. From the front. *b*. From behind.

vening segments of the vertebral column, the sacrum and coccyx (fig. 15-3). In order to form a better idea of this structure an articulated pelvis with ligaments in place should be studied. The pelvis thus seen actually contains two cavities. The flared wings of the ilia form a lower boundary to the abdominal cavity proper, and the cavity here is known as the major or false pelvis. The minor or true pelvis is the lower part of the cavity and lies between sacrum, pubes, ischia and the lower parts of the ilia. The true pelvis has an inlet above, open into the abdominal cavity, and an outlet below which, during life, is bridged by muscles.

The two coxal bones are firmly attached posteriorly to the sacrum through the sacroiliac articulations. In addition, the wide interval between the lower part of each hip bone and the coccyx and lower part of the sacrum is bridged by two strong fibrous bands named the sacrotuberous (sacrotuberal) and sacrospinous (sacrospinal) ligaments (fig. 15-3*b*). The **sacrotuberous ligament** is a broad band stretching from the sacrum and coccyx to the ischial tuberosity; the **sacrospinous ligament** is a shorter band, largely covered posteriorly by the sacrotuberous ligament, which extends from the sacrum and coccyx to the ischial spine. The sacrospinous ligament converts the greater sciatic notch of the dried bone into a greater sciatic foramen; the sacrotuberous ligament forms a lower boundary for the lesser sciatic notch, thus converting this notch into a lesser sciatic foramen.

The two pubes are united in front at the **pubic symphysis.** Firmly attached to the adjacent ends of the two pubes at the symphysis is a heavy fibrocartilaginous pad, the interpubic disk. The attachment of this disk to the pubis is strengthened by ligaments surrounding it. The pubic symphysis is barely movable during most of life, but in women it becomes much more movable during pregnancy.

The **sacroiliac joint** (fig. 15-4) consists of a relatively small joint cavity between sacrum and ilium and very powerful ligaments connecting these elements. The joint cavity is said to become partially or completely obliterated with age, especially in the male. However, as obliteration of the cavity is one of the most frequent findings in rheumatoid arthritis of the vertebral column, it is not clear to what extent this should be regarded as a normal accompaniment of age, and to what extent it is due to a more strictly pathologic process. The ventral sacroiliac ligament is relatively thin and lies across the front of the joint. The dorsal sacroiliac ligament is strong and blends deeply with the still stronger interosseous sacroiliac ligament which is attached to the roughened areas behind and above the joint cavity. In addition to these ligaments the pelvis is braced through its attachment to the sacrum by the sacrotuberous and sacrospinous ligaments (fig. 15-3), and to the last lumbar vertebra by paired iliolumbar ligaments.

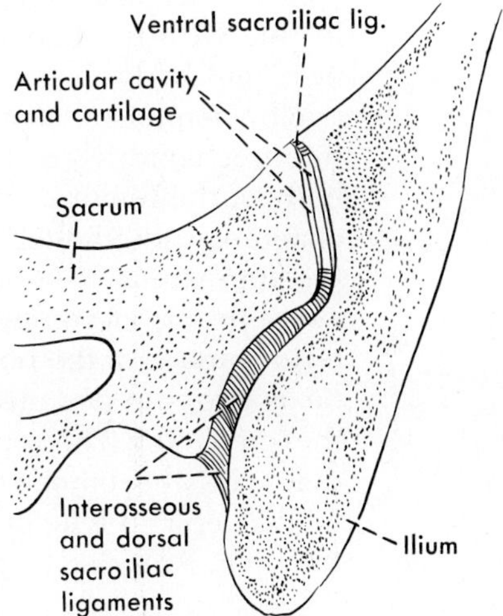

Figure 15–4. Schematic section through the sacroiliac joint.

The weight of the body bearing upon the sacrum tends to force its base, or upper end, downward and forward between the ilia, with its lower end (apex) tending to swing backward; the wedge-shaped sacrum would then tend to spread the two ilia farther apart or to escape posteriorly and inferiorly from between the ilia. However, these movements of the sacrum exert a greater pull upon the posterior ligaments, especially the interosseous ones, and this results therefore in more firm apposition at the sacroiliac joint. The pubic symphysis serves as the tie beam to resist flattening of the pelvis and consequent movement at the sacroiliac joint. Movement at the sacroiliac joint is also minimized by the fact that the opposed surfaces of the sacrum and ilium are wavy rather than plane. This joint is therefore an exceedingly strong one and allows little movement. As is the case at the pubic symphysis, however, the ligaments of the sacroiliac joint are loosened during pregnancy and the movement there is therefore much increased at that time.

Movements of the pelvis involve simultaneous movement of the lumbar portion of the vertebral column and movement at the hip joint. Upward rotation is the movement in which the anterior part of the pelvis is raised, and it therefore involves a decrease in the lumbar curvature. Downward rotation is an increased tilting of the pelvis, accompanied by an increase in the lumbar curvature (lordosis). Lateral rotation, to the same or the opposite side, involves swinging the pelvis, and therefore the body as a whole, upon one femoral head; this movement is of particular importance in walking. Lateral tilting of the pelvis is raising one side higher than the other, and involves a lateral bending of the lumbar part of the vertebral column.

FEMUR AND HIP JOINT

The rounded upper end or head of the **femur,** the bone of the thigh (figs. 14-2 and 15-5), is attached to the body by a neck that does not continue the direction of the body but rather projects medially at an angle. Where the neck joins the body a large protuberance, the greater trochanter, projects upward; the greater trochanter is therefore in line with the body. On the posteromedial side at the junction of neck and body is a smaller protuberance, the lesser trochanter. The two trochanters are connected on the posterior side of the femur by an intertrochanteric crest. The anterior surface of the upper part of the body of the femur is smooth but the posterior surface below the trochanters is roughened, presenting two ridges that run downward from the approximate regions of the lesser and greater trochanters to converge to form a roughened raised line, the linea aspera, on the posterior surface of the body. The lower end of the femur is enlarged and bears two rounded articular surfaces, the condyles, for articulation at the knee joint. The roughened medial and lateral surfaces of the condyles are the epicondyles for the attachment of muscles, and the medial epicondyle presents an additional projection, the adductor tubercle. Anteriorly, the articular surfaces of the two condyles come together to form a surface for articulation with the bone of the knee-cap (patella); posteriorly they are separated by a deep intercondylar fossa. The flat surface of the femur above the intercondylar fossa and between the two diverging lower ends of the linea aspera is the popliteal surface.

The deep acetabulum receives the globular head of the femur to form the **hip joint,** which is the best example in the body of a ball-and-socket joint. As compared with the shoulder, the hip joint has gained stability at the expense of some freedom of movement; quite obviously the deep ball-and-socket joint at the hip cannot allow as free movement as can occur between the very shallow glenoid cavity and

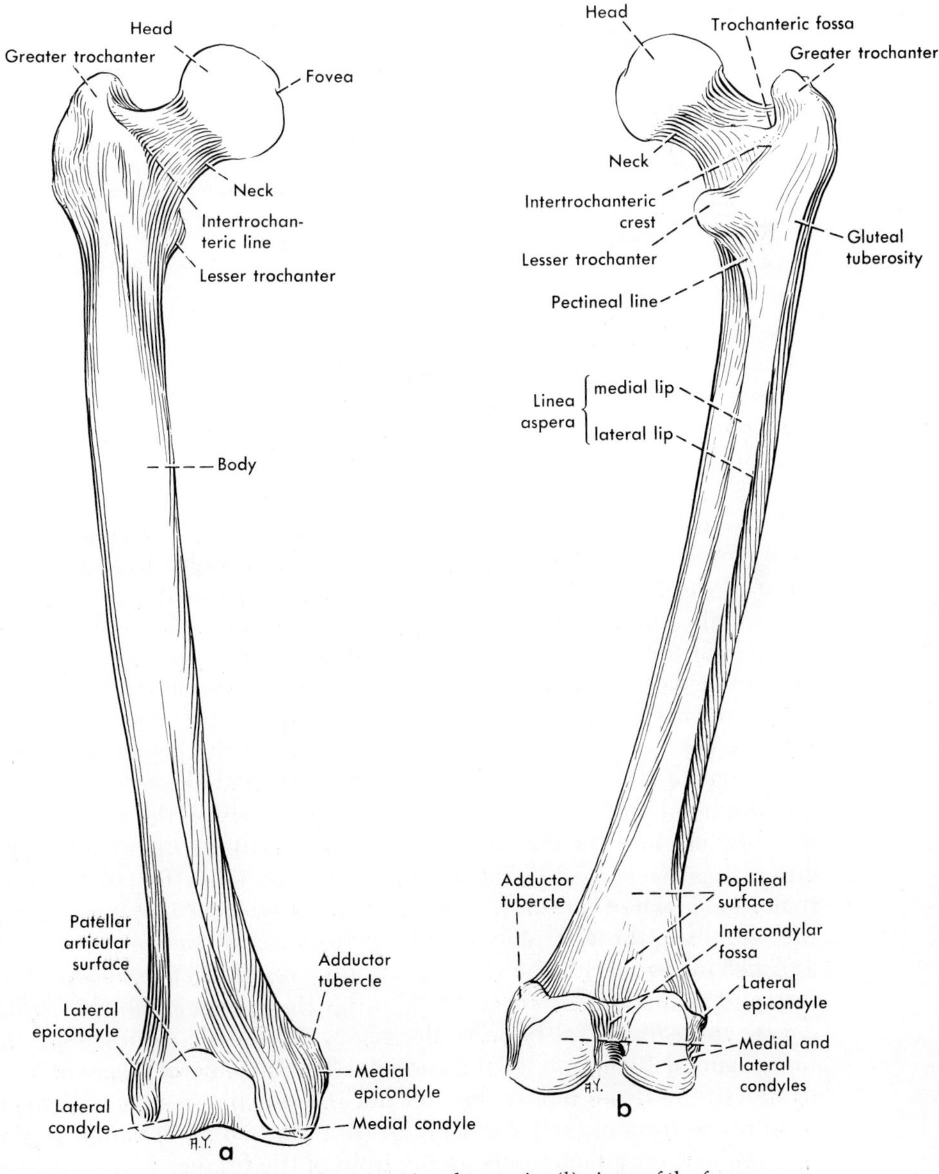

Figure 15–5. Anterior (*a*) and posterior (*b*) views of the femur.

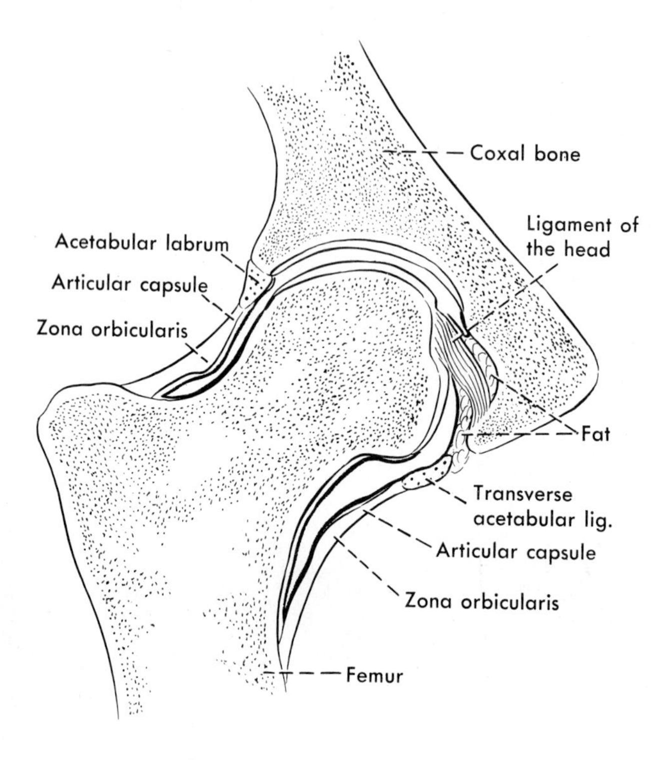

Figure 15–6. Frontal section through the hip joint. Synovial membrane is shown lining the articular capsule, and reflected around the neck of the femur and the ligament of the head to the articular cartilage. The zona orbicularis consists of circular fibers in the fibrous capsule.

the head of the humerus. At the same time this deep joint allows for a great deal of stability in the various positions in which the lower limb is placed; the lower limb must exert its weight-bearing function during many phases of its movement.

The smooth articular surface of the head of the femur occupies considerably more than a hemisphere and ends at the neck, but is interrupted at one point by a pit or fovea into which a ligament (ligament of the head) attaches. The similarly smooth articular surface of the acetabulum forms an inverted U (fig. 15-1), the open end of the U being continuous with the acetabular notch, and its cavity being occupied by a pad of fat. The acetabulum is made deeper by a fibrocartilaginous mass, the labrum (fig. 15-6), attached to its edge; the labrum bridges the notch as the transverse ligament of the acetabulum. The free edge of this acetabular rim extends beyond the equator of the femoral head and thus holds the head tightly within the acetabulum.

The strength of the hip joint lies primarily in the shape of the articular surfaces and in the ligaments of this joint (fig. 15-7), rather than in associated muscles. In contrast to the weak articular capsule of the shoulder, the articular capsule of the hip joint is very strong. It attaches lower on the anterior aspect of the femur than it does posteriorly and is marked by three thickenings that run a somewhat spiral course from the coxal bone to the femur. These thickenings are the three named ligaments of the capsule. The **iliofemoral ligament** is attached to the ilium below the anterior inferior iliac spine and covers most of the front of the joint. From this origin two main bands tend to diverge to their attachments on the front of the femur, thus giving the iliofemoral ligament somewhat the shape of an inverted Y; hence, it is known also as the "Y ligament of Bigelow." The fibers of the iliofemoral ligament spiral somewhat medially as they run downward; this same twist is maintained by the other two ligaments of this joint. The iliofemoral ligament, in consequence of its position in front of the hip joint, is particularly fitted to prevent undue extension at this joint. Since the weight of the body upon the femur tends to keep the extended hip joint extended, the iliofemoral ligament has as its special function the maintenance of the erect posture without constant muscular action.

The **pubofemoral ligament** arises from the pubic portion of the acetabular brim, and therefore on the anteroinferior aspect of the joint. Its fibers help to prevent excess abduction of the femur, and also assist the iliofemoral ligament in checking extension at the hip. Between the upper edge of the pubofemoral ligament and the medial edge of the upper part of the iliofemoral ligament the capsule of the hip joint presents a weak triangular area; this area during life is protected by the tendon of the iliopsoas muscle. In later adult life it is sometimes perforated to allow a communication between the hip joint and a bursa underlying the iliopsoas tendon.

The **ischiofemoral ligament** is less well developed than are the preceding two. It springs from the ischial rim of the acetabulum and therefore covers the lower posterior aspect of the joint. The upper fibers pass almost transversely toward the neck of the femur while the lower ones pass slightly upward to their attachment there. The spiral of this ligament is decreased by flexion at the hip, increased by extension. The tightening of the ligament during extension helps to make the extended position of the joint the most stabile one.

All three of these ligaments also have a common action in tending to limit internal rotation of the femur, as this movement would increase their spiral; on the other hand, external rotation tends to unwind their spiral and is therefore checked entirely by muscles. It should be noted that all the parts of the capsule are relaxed during flexion and external rotation of the thigh, and therefore dislocation of the hip can take place more easily in this position.

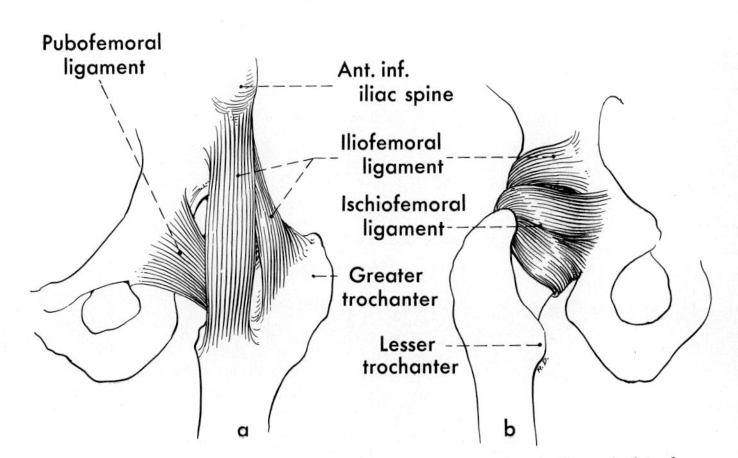

Figure 15–7. The left hip joint. *a*. From the front. *b*. From behind.

Within the hip joint there is a flattened band, the **ligament of the head** (once known somewhat incongruously as the ligamentum teres or round ligament), which is attached to the nonarticular surface in the acetabulum and to the pit on the head of the femur (fig. 15-6). This ligament should help check abduction but never becomes tense enough to do so. It is said that in the newborn child and in early infancy a normal ligament does check posterosuperior displacement of the head of the femur, a condition sometimes found in neonatal life.

The blood supply to the upper end of the femur is derived mostly from the two femoral circumflex arteries (fig. 14-3) that encircle it. The distribution to the head is of particular interest, for while a small part enters through the ligament of the head, most of the vessels to the neck and head pierce the capsule of the joint at its attachment to the femur and run upward along the neck. Thus fractures of the neck often tear the vessels, and make healing of the fracture difficult.

Movements at the hip joint consist of flexion, extension, abduction, adduction, circumduction, internal rotation, and external rotation. Flexion at the hip joint is the movement of bringing the thigh forward and upward as if to touch the abdomen; extension is a backward movement of the thigh. Abduction is drawing one limb laterally and therefore away from the other; adduction is bringing the limbs together. Circumduction is, of course, a combination of all four of these movements. Internal (medial) rotation is a rotation of the limb so that the kneecap points inward; external (lateral) rotation is the movement in the opposite direction, so that the knee is turned outward. (Because the head and neck of the femur are not in the long axis of the limb, the "rotation" is actually a swinging on the head of the femur, as a gate swings to and fro on a hinge.)

The Thigh and Knee 16

Many of the muscles of the thigh extend across both hip and knee joints, and therefore have actions upon both of these. The movements at the hip have been briefly discussed in the preceding chapter and will be considered again more fully in a following one. The movements at the knee are largely hinge ones, consisting of flexion and extension. Flexion is bending the knee, thus bringing the calf of the leg up toward contact with the posterior surface of the thigh, while extension is straightening it. A slight hyperextension at this joint is normal in some people. In the partially flexed condition a small amount of rotation between the tibia and femur may occur; this is said to be limited to about 40 degrees and is not very obvious upon inspection of the movements of the joint.

GENERAL ANATOMY

The few important landmarks of the thigh are, anteriorly, the anterior superior iliac spine, the pubic tubercle, the inguinal ligament extending between these two points, and the knee cap and adjacent enlarged ends of the femur. Except for the inguinal ligament, these are described in Chapter 15. The inguinal ligament is the lower edge of the aponeurosis of the most external of the flat muscles of the abdomen, and thus marks the boundary between abdomen and thigh. It lies deep to the crease between the two parts, and as it runs between the ilium and pubis is slightly convex downward. Laterally there is the greater trochanter and posteriorly the ischial tuberosity, also already identified. The hollow behind the knee is the popliteal fossa.

The muscles of the thigh are conveniently divided into three groups: an anterior, originally dorsal, group, concerned especially with flexion at the hip and extension at the knee; an anteromedial or adductor group, originally ventral, concerned especially with adduction and flexion of the thigh; and a posterior group, also originally largely ventral, concerned with extension at the hip and flexion at the knee. While there is some overlap in the functions and innervations of these groups it is usually convenient to think of each group as having certain chief actions and having its own particular nerve supply. The anterior muscles are innervated by the femoral nerve, the anteromedial ones by the obturator, and the posterior ones by the sciatic. The posterior muscles, nerves, and vessels are described in the following chapter.

Although the large nerves of the thigh enter it anteriorly, anteromedially, and posteriorly, there is only one important set of vessels to the thigh. These are the femoral vessels, which lie anteriorly at the groin with the femoral nerve. The femoral artery is a continuation of the external iliac artery, the change in name occurring at the inguinal ligament; the femoral vein, similarly, continues above the inguinal ligament as the external iliac vein. A little above the knee the femoral vessels pass through a gap in one of the muscles, close to the bone, to attain a position in the popliteal fossa, and have their names changed to popliteal vessels.

The femur, the bone of the thigh, has already been discussed (p. 236), but its lower end should be studied in more detail as the knee joint is studied. Since some of the muscles of the thigh cross the knee joint to attach to bones of the leg, the upper ends of these bones must also be studied now, although a more complete description of them can be reserved for a later chapter (p. 292).

In brief, the expanded lower end of the **femur** presents rounded medial and lateral condyles for articulation with the tibia, and the anterior confluence of the articular surfaces of the two condyles is the articular surface for the patella or knee cap. The intercondylar fossa of the femur gives attachment to ligaments that lie within the knee joint, and the medial and lateral epicondyles give attachment not only to muscles but also to the two important external ligaments of the knee joint.

The expanded upper end of the **tibia** (figs. 14-2 and 19-1) is formed by two tibial condyles that have almost flat upper articular surfaces which receive the weight transmitted from the femoral condyles; the articular surfaces are separated by a nonarticular area into which internal ligaments of the knee joint attach. The sides of the tibial condyles receive the attachment of the lower part of the capsule of the knee joint and certain muscles and ligaments. On the lower surface of the lateral condyle there is an articular facet for the joint between the tibia and the head of the fibula. On the anterior border of the body of the tibia below the condyles is a roughened raised area, the tibial tuberosity, that receives the patellar ligament (extending downward from the patella, and really the lower end of the tendon of the muscle that extends the leg at the knee—the quadriceps muscle).

The upper end of the slender **fibula** is the head, which rises to a pointed apex. It articulates with the lower surface of the lateral tibial condyle, and therefore does not enter into the knee joint.

The **patella,** the bone that is colloquially known as the kneecap, is triangular with its apex directed downward. Posteriorly it presents on its upper part a smooth articular surface for articulation with the femur, elsewhere it is rough for the attachment of tendon fibers and the entrance of blood vessels. The patella is the largest sesamoid bone in the body; it so interrupts the quadriceps tendon, in which it lies, that the part of the tendon between it and the tibia is known as the patellar ligament. By holding the tendon of the quadriceps farther forward, the patella adds a great deal to the effectiveness of the quadriceps in extending the leg. It is said that in the absence of the patella about 30 per cent more force is required to extend the leg completely.

THE KNEE JOINT

The knee joint (figs. 16-1 and 16-2) is complicated not so much because of the fibrous capsule, which is relatively simple, as by the structures lying within the joint. Anteriorly, the insertion of the quadriceps muscle on the patella, the patella itself, and the patellar ligament provide adequate protection for the knee joint, and the lining synovial membrane rests against parts of these. On either side of the quadriceps tendon, between the patella and the femoral and tibial condyles, the articular capsule is formed by fibers from the fascia lata (the deep fascia of the thigh) and expansions from the quadriceps

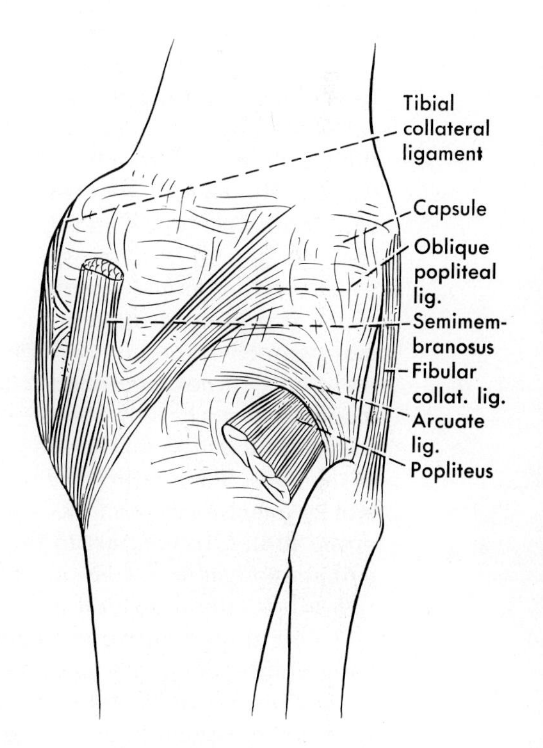

Figure 16–1. Posterior view of the knee joint.

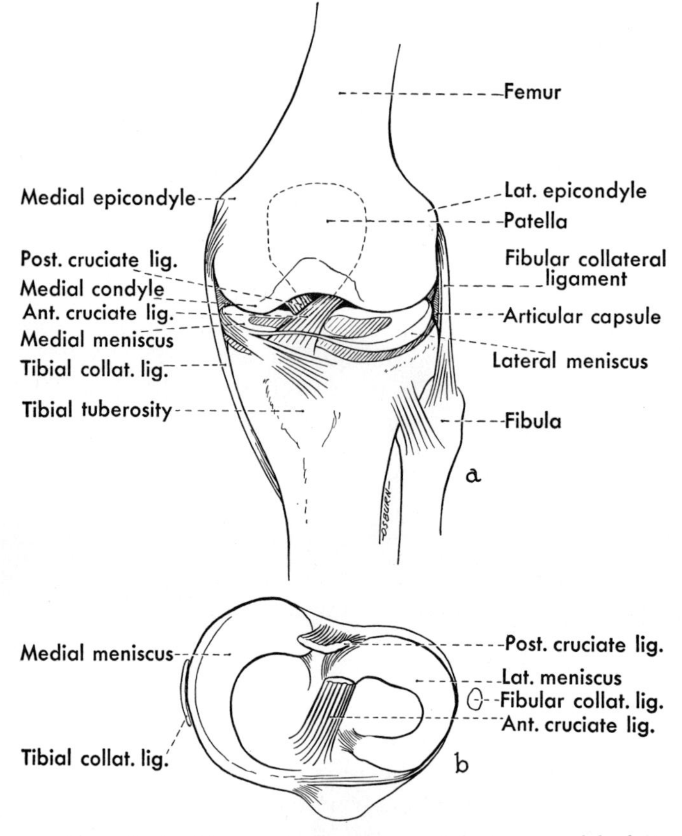

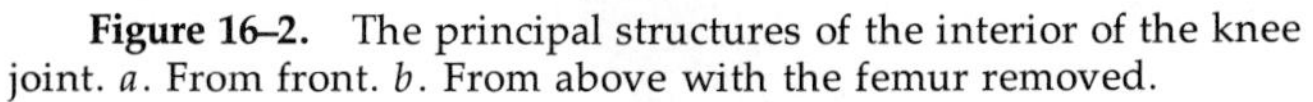

Figure 16–2. The principal structures of the interior of the knee joint. *a*. From front. *b*. From above with the femur removed.

tendon, these parts being called the medial and lateral patellar retinacula. Posteriorly the capsule consists of interlacing fibers that are reinforced by a strong attachment from the semimembranosus tendon; the band derived from this tendon runs obliquely upward and lateralward as the **oblique popliteal ligament.** In the posterior part of the capsule there is an opening to allow the exit of the popliteus muscle; the upper edge of this opening is strengthened by some arching fibers that constitute the arcuate popliteal ligament.

It is typical of hinge joints that they have special medial and lateral ligaments; the knee joint is no exception to this. The medial ligament of the knee, the **tibial collateral ligament,** is a broad band that is fused posteriorly with the capsule of the knee joint, passing from the medial epicondyle of the femur to the medial surface of the upper end of the tibia. The cordlike lateral ligament of the knee, the **fibular collateral ligament,** runs from the lateral epicondyle of the femur to the head of the fibula. In contrast to the medial ligament, however, the fibular collateral ligament does not blend at all with the joint capsule, but rather lies lateral to and free of this.

The fibular collateral ligament and the posterior part of the tibial collateral ligament are made taut by extension of the knee and relaxed by flexion; however, the anterior part of the tibial collateral ligament remains tense in all positions of the knee. The two ligaments together restrain rotation and lateral movement at the knee, especially in the extended position, and the tibial collateral ligament is an important stabilizer of the knee in all positions. Tear of this ligament as a result of a forcible blow to the outside of the knee is a common injury of football players.

The cavity of the knee joint is extensive; it passes some distance upward in front of the femur between this and the overlying quadriceps muscle and is usually continuous with a bursa, the suprapatellar bursa, which may, however, lie above its upper blind end. Below the patella and between the femur and tibia the cavity is much subdivided by structures lying within the capsule of the joint. A fold of synovial membrane sweeps downward and backward from the posterior surface of the patella, and, becoming wider as it does so, attaches to the inner border of the articular surfaces of both the femoral and tibial condyles. In consequence, the cavity of the knee joint between the femur and tibia is divided into medial and lateral parts that communicate only anteriorly, and are separated from each other by the structures extending between the intercondylar fossa of the femur and the intercondylar areas of the tibia—primarily the cruciate ligaments.

The medial and lateral parts of the cavity are in turn partly subdivided by semilunar cartilages, the **medial** and **lateral menisci** (fig. 16-2). These crescentic cartilages are wedge-shaped in cross section with the thin inner edge at the concave border of the cartilage. This border is a free one, and around it the part of the synovial cavity between a femoral condyle and a meniscus is continuous with that between the meniscus and the corresponding tibial condyle. On their convex or outer border both cartilages are attached to the synovial and fibrous capsules. However, while the medial meniscus is also anchored firmly to the strong tibial collateral ligament, the lateral meniscus has only slight attachments to the weak lateral portion of the capsule, from which it is partially separated by the tendon of the popliteus muscle, and it has no attachment to the fibular collateral ligament. This marked difference in strength of attachment may be one of the reasons why the medial cartilage is more often torn, in conjunction with tearing of ligaments at the knee, than is the lateral cartilage, the concept being that the medial cartilage is less mobile and therefore more likely to be caught and torn as the femur moves it back and forth, or rotates it, on the tibia. Both cartilages are anchored to the tibia by strong fibrous bands continuous with the ends of the cartilages, and these also limit their movement.

The exact function of the menisci is not known. They serve in small part to deepen the articular surfaces on the upper end of the tibia and thus adapt them better to the femoral condyles, and they apparently facilitate rotation at the knee. They are most likely to be torn by rotation of the femur on the supporting tibia when the knee is flexed—hence when a runner suddenly changes direction, rotating his body and therefore femur on the tibia while that limb is supporting weight. The torn part of a meniscus usually rolls up and locks the joint, and there is pain and swelling at the knee. Menisci have also been thought to be particularly important in maintaining an even film of synovial fluid and thus aiding in lubrication of the joint. If the cartilages are torn by violence they can be removed with no marked subsequent disability to the knee. However, the weight-bearing areas on the femur and tibia have been shown to be decreased by almost 50 per cent; this concentration of the weight on a smaller area, and perhaps a poorer lubrication, may account for the report that in time the articular cartilages of both the femur and tibia may show degenerative changes.

The **anterior** and **posterior cruciate** ("crossed") **ligaments** are especially important ligaments of the knee joint, lying within the joint capsule but covered anteriorly and on both sides by reflections of the synovial membrane. The anterior cruciate ligament ascends from the anterior area between the tibial condyles and runs upward, backward and somewhat laterally to attach toward the back of the medial surface of the lateral femoral condyle. The posterior cruciate ligament arises from the posterior intercondylar area and extends upward and somewhat forward and medially to attach to the lateral side of the medial femoral condyle. There has been considerable argument concerning the exact functions of these ligaments; both seem to be fairly tense in all positions, but particularly so in extreme extension and extreme flexion. The cruciate ligaments thus apparently contribute much to the stability of the knee joint in all positions, preventing anteroposterior displacement of the tibia and limiting rotation of the femur upon the tibia. They are, however, apparently not as important in limiting rotation as are the collateral ligaments; in one study it was found that sectioning both collateral ligaments increased the amount of rotation allowed almost two times as much as did section of both cruciate ligaments.

Rupture of the anterior cruciate ligament alone has been reported as a result from a hard block on the anterolateral side of the limb while the foot bearing all the weight was in slight medial rotation, thus differing little from the mechanism of rupture of the tibial collateral ligament. Indeed, the most common athletic injury to the knee, if it is severe, results in tear of the tibial collateral and anterior cruciate ligaments (and often of the medial meniscus). The resulting instability makes it difficult or impossible for a runner to suddenly change course, and surgical repair or replacement of the torn ligament or ligaments is necessary. The large muscle on the front of the thigh that extends the knee, the quadriceps femoris, must always be strengthened in such cases, since it rapidly loses strength in any disability of the knee. In the nonathlete with less severe injury, development of the quadriceps may be all that is necessary to restore adequate stability.

Rupture of the posterior cruciate ligament alone most often results from an automobile accident in which the tibia of an occupant comes in violent contact with the dashboard, thus forcing the tibia posteriorly. The usual test for a ruptured cruciate ligament consists of trying to displace the tibia anteriorly or posteriorly on the femur with the leg flexed. Rupture of the anterior cruciate ligament allows abnormal anterior displacement, while rupture of the posterior ligament permits abnormal posterior displacement.

Closely related to the posterolateral aspect of the joint is the tendon of the popliteus muscle, a muscle at the knee (p. 299). This tendon lies deep to the lateral part of the fibrous capsule, separated from the joint cavity only by a covering fold of synovial membrane, but attached to both the capsule and the lateral meniscus. As already noted, the muscle emerges through the posterior part of the capsule.

The **tibiofibular synovial joint** is of the plane type. The small synovial cavity is surrounded by a capsule with no particular distinguishing features. Occasionally it communicates with the knee joint.

SUPERFICIAL NERVES AND VESSELS; FASCIA

The superficial fascia of the thigh contains superficial nerves and vessels, the important superficial inguinal lymph nodes, and a greater or lesser quantity of fat, but otherwise presents no particular distinguishing features. The deep fascia of the thigh is the **fascia lata** (broad fascia) and resembles the fascia of the arm and forearm in that it is a tough layer loosely but completely surrounding the musculature of the thigh. In its lateral aspect it is thickened and strengthened by additional longitudinal fibers to form the **iliotibial tract.** This important band has a three-pronged origin: anteriorly from the attachment of the tensor fasciae latae muscle (p. 251); posteriorly from much of the insertion of the gluteus maximus (p. 264); and in between from the crest of the ilium through the fascia covering the gluteus medius. It is attached to the femur for much of its length by the lateral intermuscular septum, while below it re-enforces the capsule of the knee joint and attaches anterolaterally to the lateral tibial condyle. Posterolaterally, the fascia lata extends upward over the gluteus medius and on both sides of the tensor fasciae latae and the gluteus maximus, and thus attains attachment to the iliac crest, the sacrotuberous ligament and the ischial tuberosity. Anteromedially, it is attached to the inguinal ligament and the pubis. It sends medial and lateral intermuscular septa to the femur in the lower part of the thigh, and about the knee is attached to various bony prominences to help form the medial and lateral patellar retinacula.

The skin of the anterolateral aspect of the thigh is supplied by the **lateral femoral cutaneous nerve,** a branch of the lumbar plexus which penetrates the fascia lata a little below the anterior superior iliac spine. Most of the anterior and anteromedial surfaces of the thigh are supplied by branches of the **femoral nerve**; these pierce the fascia lata at various levels to ramify in the superficial fascia. A small area of skin on the medial surface of the thigh is usually supplied by a branch of the **obturator nerve,** while the posterior aspect of the thigh is supplied by the **posterior femoral cutaneous nerve.**

Many small veins form a network in the superficial fascia of the thigh, but the prominent vein here is the **great saphenous** (saphenous means "obvious," probably a reference to the usual involvement of this vein in varicose veins of the lower limb). The great saphenous vein ascends from the medial side of the leg to the medial side of the thigh and runs upward and slightly forward to attain the anterior surface of the thigh where, a little below the inguinal ligament, it penetrates the fascia lata and ends in the femoral vein. The gap in the fascia

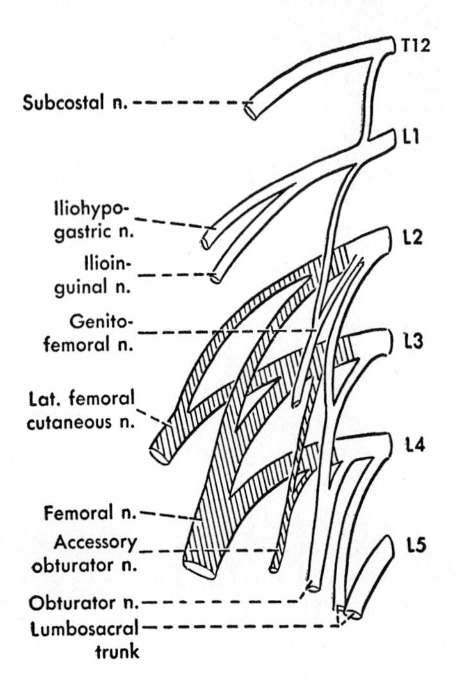

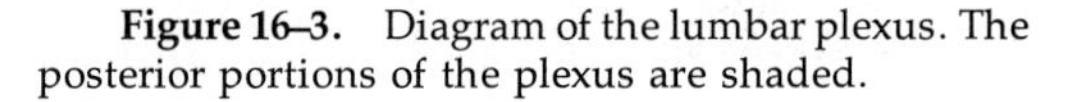

Figure 16–3. Diagram of the lumbar plexus. The posterior portions of the plexus are shaded.

lata is the saphenous hiatus; just before the great saphenous vein passes through the hiatus it usually receives veins from the lower abdominal and pudendal regions.

THE LUMBAR PLEXUS

The lumbar plexus (fig. 16-3) supplies the muscles and the skin on the anterior and medial sides of the thigh, and skin on the medial side of the leg and foot. Although it arises deep within the abdomen, its method of formation and branching must be studied at this time if the nerves in the thigh are to be understood. Similarly, the sacral plexus supplies posterior skin and muscles of the thigh as well as muscles of the buttock. Since the lumbar and sacral plexuses are connected, and both go primarily to the lower limb, they are frequently described together as the lumbosacral plexus (fig. 16-4). The sacral part of the lumbosacral plexus is described in the following chapter.

The origin of the lumbar plexus is from the ventral branches of the first, second, third, and a variable part of the fourth lumbar nerves, with usually a small communication from the twelfth thoracic nerve. Since it is formed at the lumbar level of the vertebral column, the plexus itself lies upon the inner surface of the posterior abdominal wall, and must be approached from the abdominal cavity; at its origin, it is imbedded in the psoas major muscle. The definitive peripheral nerves arising from the plexus make their exit from the psoas muscle and pass into the lower part of the anterior abdominal wall or into the thigh.

The plexus is usually arranged as follows: the first lumbar nerve, having received a communication from the last thoracic, divides into two branches. The upper one of these gives rise to the iliohypogastric and ilioinguinal nerves, to the lowermost part of the abdominal wall; the other joins a small branch from the second lumbar nerve to form the genitofemoral nerve to the scrotum (or labia majora) and some of the skin of the upper anterior surface of the thigh. Small anterior branches from the second, third and fourth lumbar nerves join to form the obturator nerve; branches from the third and fourth lumbars form also an accessory obturator nerve in about 8 to 10 per cent of plexuses.

From the larger, posterior, portions of the ventral rami of the second and third lumbar nerves go branches to form the lateral femoral cutaneous nerve. The remainder of the posterior portions of these two nerves join a part of the fourth lumbar nerve to form the largest branch of the plexus, the femoral nerve. The fourth lumbar nerve typically divides into two parts, one part going into the lumbar plexus to help form the obturator and femoral nerves, and the remainder passing downward to join the sacral plexus. There is considerable variation in the relative sizes of the contributions of the fourth lumbar nerve to the two plexuses; if the fourth lumbar nerve fails to participate in the lumbar plexus, but goes entirely to the sacral plexus, the plexus is known as a prefixed one. If all of the fourth lumbar nerve, and perhaps even

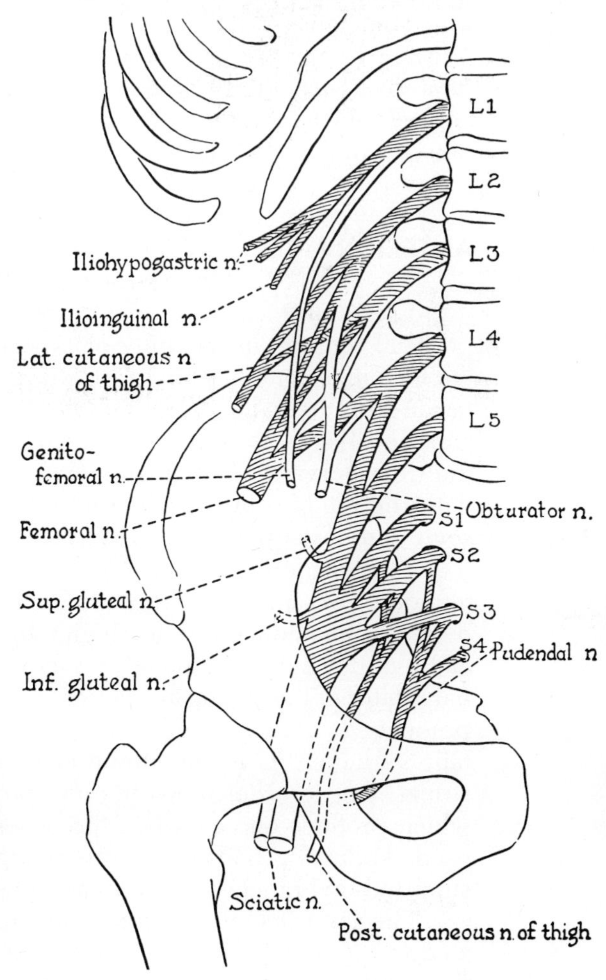

Figure 16-4. Diagram of the lumbosacral plexus. (From the Section of Neurology and the Section of Physiology and Biophysics, Mayo Clinic and Mayo Foundation: *Clinical Examinations in Neurology* [ed. 3]. Philadelphia, W. B. Saunders, 1971, p. 156.)

some of the fifth lumbar, goes into the lumbar plexus, this plexus is known as a postfixed one.

In the brachial plexus a division between anterior and posterior parts of the plexus is obvious, and we have seen how the anterior part, as exemplified by medial and lateral cords and their branches, supplies the anterior or flexor muscles of the limb while the posterior part, consisting of the posterior cord and its branches, supplies the posterior or extensor muscles. A similar division and distribution also exist in both lumbar and sacral plexuses, although it is somewhat less obvious than in the upper limb because of the extensive rotation that the lower limb has undergone during development. Thus the anterior portion of the lumbar plexus is distributed to the anterior abdominal wall or its derivatives (through the iliohypogastric, ilioinguinal and genitofemoral nerves) and to the adductor or anteromedial surface of the thigh (through the obturator nerve). This part of the thigh represents the cephalic part of the original flexor surface of the limb. Similarly, the lateral femoral cutaneous and femoral nerves, the posterior elements of the lumbar plexus, are distributed to the original posterior or extensor surface of the thigh. In the sacral plexus the gluteal nerves and the common peroneal, the larger parts of the posterior division of this plexus, are distributed also to original posterior or extensor muscles of the buttock and of the leg; the tibial nerve and its derivatives supply the more caudal muscles of the original flexor surface of the thigh and the flexor (original ventral or anterior) muscles in the leg and foot.

ANTERIOR MUSCLES OF THE THIGH

These muscles are shown in figures 16-5 and 16-6. The most superficial muscle on the anterior aspect of the thigh is the **sartorius** or "tailor" muscle. This long ribbon-like muscle winds across the anterior and medial surfaces of the thigh, arising from the anterior superior iliac spine and inserting on the medial surface of the body of the tibia below the tuberosity. Here its tendon is closely associated with the tendons of two other muscles, the gracilis medially and the semitendinosus posteriorly. Since it crosses anterior to the hip joint it is a flexor there, and since it usually crosses posterior to the axis of motion of the knee joint, it participates in flexion of the knee. Moreover, it is an abductor of the thigh (although a very weak one) and, because of its lateral origin and its medial position at the knee, an external rotator of the thigh. These four actions of the muscle together will produce the crosslegged sitting position commonly employed by tailors—hence its name. (None of these actions of the sartorius is a strong one; in order to assume the position of a tailor other muscles which produce one or two of the necessary movements must also be used. Abduction by the sartorius is especially weak.) The sartorius is supplied by branches of the femoral nerve.

The sartorius muscle forms the lateral boundary of a triangle situated in the upper part of the thigh, the **femoral triangle** (fig. 16-7). The floor and the medial wall are composed of other muscles of the thigh, and the femoral nerve and femoral vessels enter the thigh behind the inguinal ligament, the upper border of the triangle. The nerve breaks up into its branches in the triangle, and the femoral artery gives off its chief branch here (the profunda femoris or deep femoral artery) before continuing down the thigh deep to the sartorius muscle, separated from it by a heavy layer of fascia.

Arising from the iliac crest just behind the anterior superior iliac spine is a short straplike muscle enclosed between two layers of the fascia lata as they attach to the iliac crest. This muscle, the **tensor fasciae latae,** runs downward and slightly backward to insert into the iliotibial tract, and although anteriorly placed is actually a muscle of the buttock (p. 266). It receives its nerve supply from the superior gluteal nerve. It assists in flexion at the hip joint; because of its oblique pos-

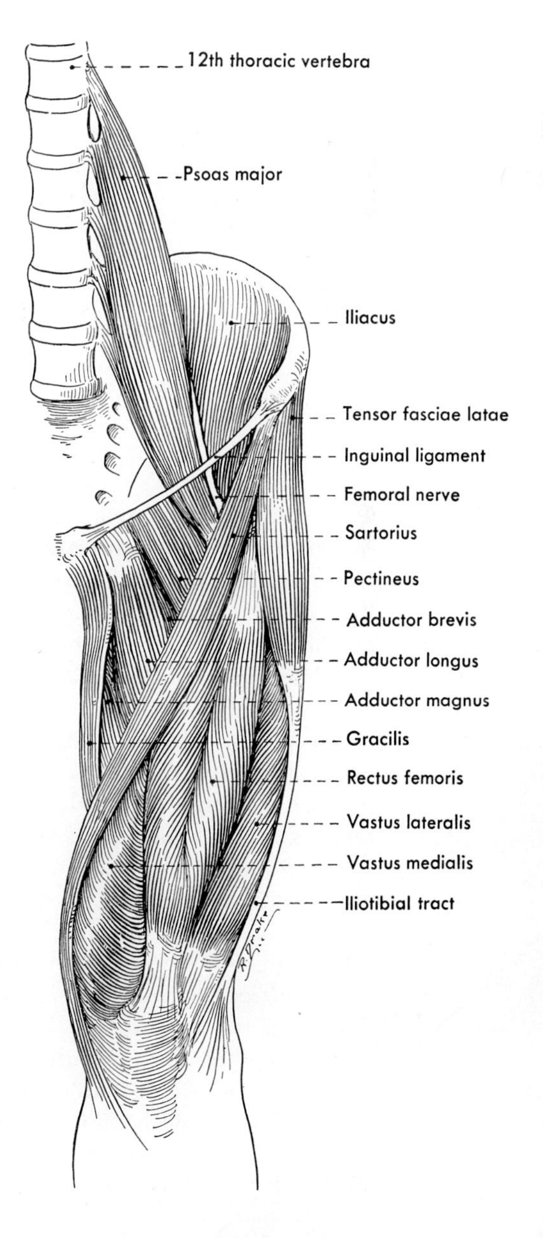

Figure 16–5. The more superficial muscles of the anterior aspect of the thigh. The space between the pectineus and the adductor longus is exaggerated so that the position of the adductor brevis can be shown.

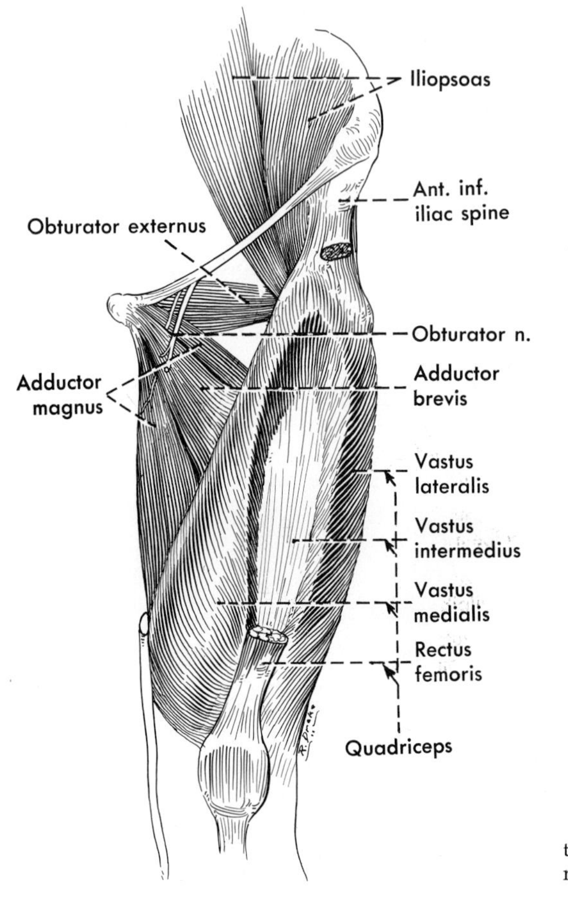

Figure 16–6. Deeper muscles of the anterior and medial aspects of the thigh after removal of the more superficial muscles.

terior direction it inwardly rotates as it flexes; and it also works with the gluteus medius and minimus in abduction of the thigh, or at least in preventing undue sagging of the opposite side of the pelvis when the weight is supported on one limb. It can contribute very little to the latter action, however.

The **quadriceps femoris** muscle is the large mass covering anterior, medial and lateral aspects of the femur, and divisible into four parts as its name indicates. The **rectus femoris** is the rounded, more anterior, head of the quadriceps, appearing as a separate muscle except at its insertion. It arises from the anterior inferior iliac spine and, by a posteriorly arching part of its tendon of origin, from the ilium just above the acetabulum. It combines with the other members of the quadriceps group in inserting upon the patella and hence through the patellar ligament upon the tibial tuberosity; it therefore shares with these others the function of extension at the knee. It is the only member of the quadriceps group that passes across the hip joint, and is a flexor at this joint.

The other three heads of the quadriceps muscle are the vastus medialis, the vastus lateralis, and the vastus intermedius; they are separable only with difficulty throughout much of their course, since both the medialis and the lateralis arise in part from septa that they share with the intermedius. The **vastus medialis** seems to cover much of the medial surface of the femur but is actually kept away from contact with the femur here by the vastus intermedius. The vastus medialis arises chiefly from the medial lip of the linea aspera—that is, from the posterior aspect of the femur. The **vastus lateralis** has a slight attachment on the anterior surface of the femur above the origin of the vastus intermedius but again arises primarily from the posterior aspect of the femur, along the lateral lip of the linea aspera. The **vastus intermedius** covers most of the anterior, medial and lateral surfaces of the body of the femur and has an extensive origin from this bone.

All three vasti muscles unite with the rectus femoris in inserting upon the patella, through which their pull is transferred to the patellar ligament and then to the tibia. The patella not only provides an enduring surface to withstand the friction that would otherwise be put upon the quadriceps tendon at the knee joint, but it also provides additional leverage for the quadriceps by holding the tendon away from the axis of motion. It has already been noted that the quadriceps must develop as much as 30 per cent more power in order to extend the knee after patellectomy; while exercise of the quadriceps is important after any injury to the knee, it is known that many persons cannot increase the strength of their quadriceps by that much. Thus methods have been developed to so repair the tendon after patellectomy that the additional strength needed is minimal.

All four heads of the quadriceps are extensors of the knee, and the

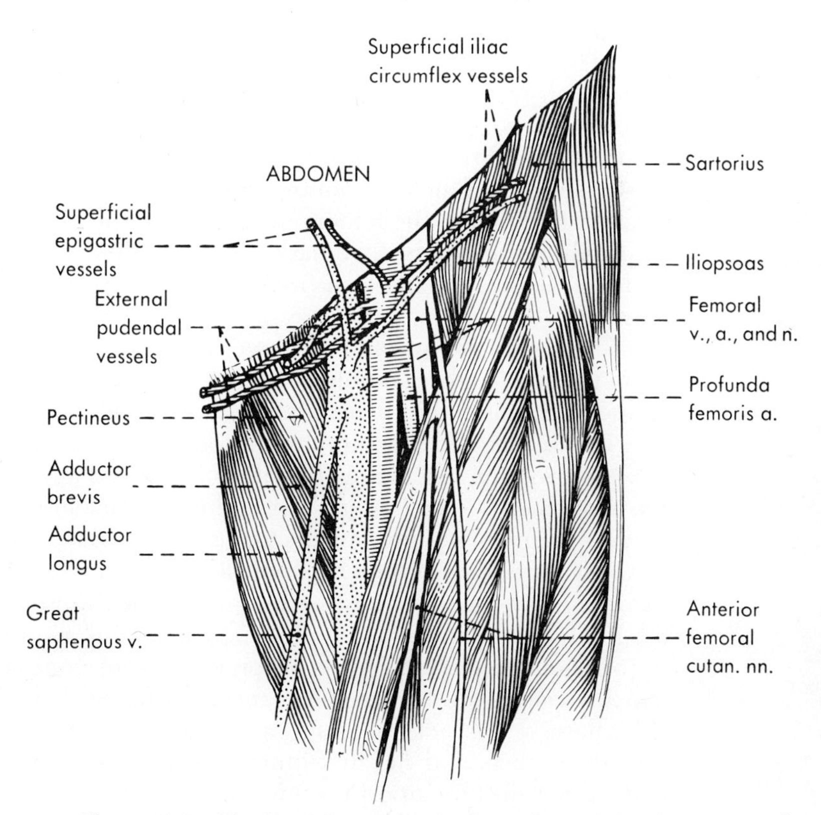

Figure 16–7. The femoral triangle; the femoral sheath has been removed, thereby destroying the femoral canal.

muscle as a whole forms the chief extensor at the knee. The last 15 degrees of extension are brought about by the three vasti. Each head of the muscle receives one or more branches from the femoral nerve.

Under cover of the lower part of the vastus intermedius a group of muscle fibers arises from the front of the femur and inserts into the upper part of the synovial membrane of the knee joint. These fibers form the **articularis genus,** which is innervated by the nerve to the vastus intermedius, and draws the synovial membrane upward as the leg is extended.

The **iliopsoas** muscle consists actually of two muscles, the psoas major and the iliacus, but they blend as they go to a common insertion and they have a common action. The **psoas major** arises within the abdomen from the anterolateral aspect of the lumbar vertebral bodies and from their transverse processes. At its origin it has imbedded in it the roots and branches of the lumbar plexus. The muscle descends, passing simultaneously somewhat laterally, and leaves the abdomen behind the inguinal ligament to attain the anterior aspect of the thigh. As it does so it is joined by the iliacus, and the conjoined muscle runs posteriorly around the medial aspect of the thigh to an insertion on and below the lesser trochanter. The iliopsoas passes across the front of the hip joint, and in this position a bursa (iliopectineal) usually intervenes between the muscle and the capsule of the hip; it may communicate with the hip joint.

The **iliacus** muscle, like the psoas, arises within the abdomen, but from the inner surface of the ilium rather than from the vertebral column. The muscle makes its exit behind the inguinal ligament in close association with the psoas major and runs with this to an attachment on the lesser trochanter. Both psoas major and iliacus muscles are innervated by branches from the femoral nerve; the psoas major also receives direct branches from the lumbar plexus. These muscles form a part of the floor of the femoral triangle, and the femoral nerve enters the thigh in the groove that marks their junction. The femoral vessels lie more medially, separated from the nerve by fascia that covers the muscles.

The iliopsoas muscle is a powerful flexor at the hip and, acting upon the free limb, can help rotate the femur externally, and slightly adduct it. In the infant it is apparently a very strong external rotator.

Taking their fixed points from below, the two iliopsoas muscles flex the trunk on the hip, as in sitting up in bed, and are essential to this movement. In so doing, their pull upon the anterior portion of the lumbar vertebral column results first in an increase in the normal lumbar curvature, thus producing lordosis or hyperextension in the lumbar region. In the erect posture the pull of the iliopsoas muscles upon the lumbar column can aid in flexion of the trunk against resistance.

Associated with the psoas major, lying on its ventral surface, there may be a small muscle known as the **psoas minor.** This arises from the anterolateral surfaces of only two or three vertebrae, usually the twelfth thoracic and the first lumbar, and ends in a long flat tendon that inserts on the superior ramus of the pubis. It assists in upward rotation of the pelvis, but is, of course, not really a muscle of the thigh.

Medial to the iliopsoas is the **pectineus** muscle, sometimes regarded as a member of the anterior group and sometimes as one of the adductor group. It arises from the superior ramus of the pubis and inserts on the femur just below the lesser trochanter. It is a flexor of the thigh, an adductor, and according to a recent report, an internal rotator. Although it is usually innervated by the femoral nerve, as are the other anterior muscles, it may be innervated by the obturator nerve, and is rather regularly innervated in part by the accessory obturator nerve when that is present.

THE ADDUCTOR GROUP OF MUSCLES

The adductor muscles (figs. 16-5 and 16-6) form an anteromedial group and are primarily adductors, flexors, and rotators at the hip joint. As just noted, the pectineus is sometimes included with this group, since it lies somewhat between the anterior and anteromedial groups, and is sometimes supplied by the obturator nerve, the nerve of the adductor group. The more superficial adductor muscles are the adductor longus and the gracilis, the deeper ones the adductor brevis, the adductor magnus, and the obturator externus.

The **adductor longus** muscle arises from the pubic tubercle and inserts on the linea aspera between the attachments of the vastus medialis and the adductor magnus to this line. It is both an adductor and a flexor of the thigh, and probably tends also to rotate internally or medially as it acts. It is supplied by the anterior branch of the obturator nerve.

Heavy fascia between the adductor muscles and the vastus medialis forms a canal here deep to the sartorius muscle and in front of all the adductor muscles—this is called the **adductor canal.** The femoral vessels plus some branches of the femoral nerve pass downward in this canal, therefore across the adductor longus close to its insertion, continuing downward from a similar position on the pectineus and adductor brevis, and thereafter lying on the adductor magnus (fig. 16-8). The deep femoral (profunda femoris) branch of the femoral artery, accompanied by a corresponding vein, passes downward behind the adductor longus and thus is separated from the femoral artery by this muscle.

The **gracilis** is a thin, straplike muscle on the medial surface of the thigh, extending from the inferior ramus of the pubis and the ramus of the ischium to an insertion on the medial surface of the upper end of the tibia close to the insertion of the sartorius and that of a posterior muscle of the thigh, the semitendinosus. It is a good adductor of the thigh, and a flexor of the leg; and it helps to rotate both the thigh and the flexed leg medially. If the leg is kept extended, it will help flex the thigh at the hip. Like the preceding muscle, it is supplied by the anterior branch of the obturator nerve.

The **adductor brevis** muscle lies under cover of the pectineus and the adductor longus. It arises from the body and the inferior ramus of the pubis and is inserted on the lower part of the line between the lesser trochanter and the linea aspera, and into the upper portion of the linea aspera. It is a flexor, adductor, and internal rotator at the hip

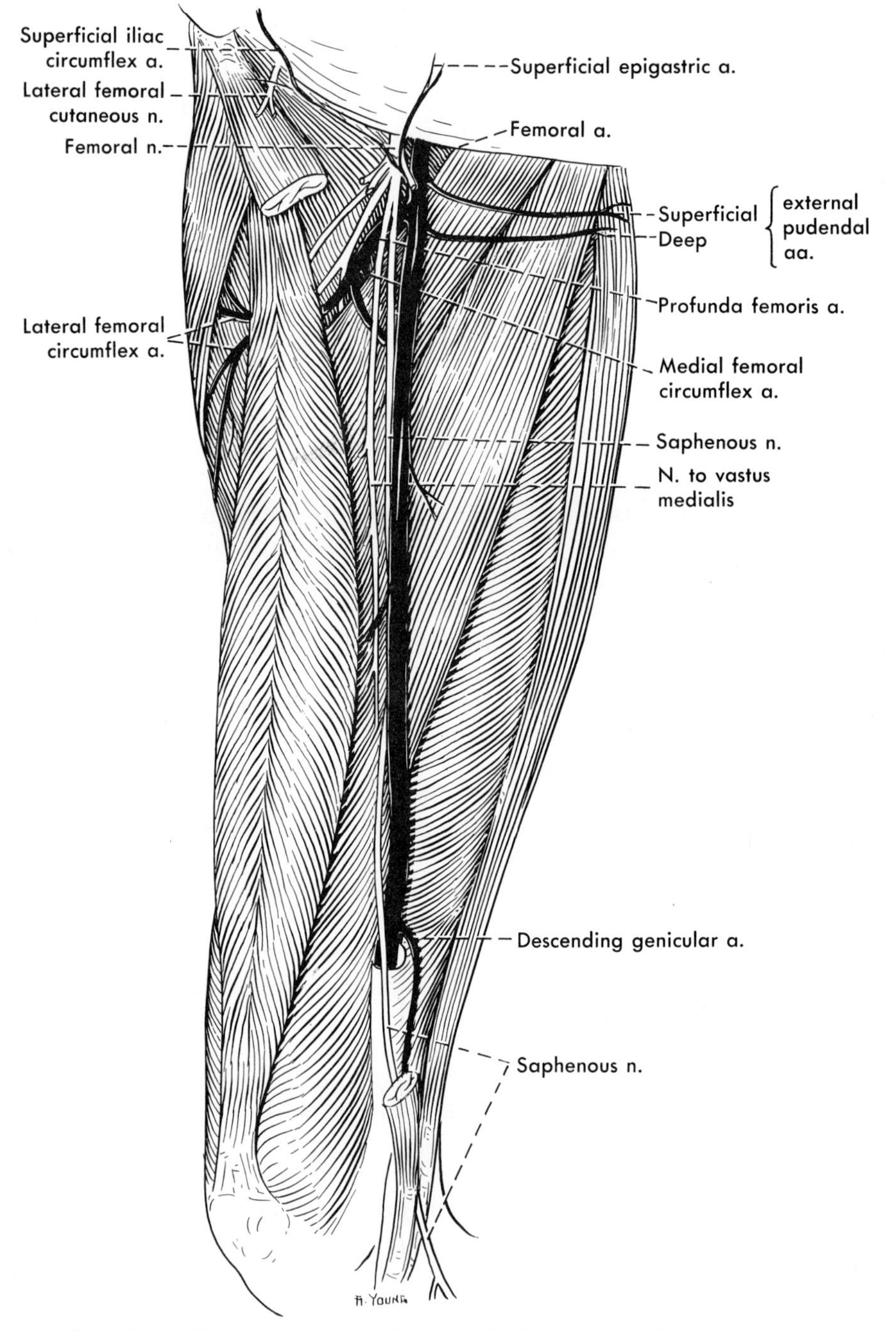

Figure 16–8. The femoral artery and nerve in the thigh. Many of the branches of the nerve have been removed.

joint. The anterior branch of the obturator nerve runs in front of the muscle, the posterior branch behind it, and either branch may supply it. Both the femoral and the deep femoral vessels run in front of it.

The **adductor magnus** muscle is by far the largest muscle of the adductor group. It arises from the inferior ramus of the pubis, from the ramus of the ischium, and from the ischial tuberosity. Its upper, anterior, fibers run almost horizontally to insert upon the linea aspera while its lower, most posterior, fibers run almost straight downward from the ischial tuberosity to the adductor tubercle at the lower medial end of the linea aspera; the intervening fibers spread out in a fan-shaped manner to insert between the upper and lower fibers along almost the whole length of the linea aspera. The adductor magnus has a double innervation, the anterior and more oblique fibers being innervated by the posterior branch of the obturator nerve while the straighter, more posterior, fibers are innervated by the sciatic nerve. The entire muscle adducts and, according to recent electromyographic evidence, medially or internally rotates the thigh. However, the fibers innervated by the obturator nerve assist the other adductors in flexion of the thigh while the fibers running from the ischial tuberosity to the adductor tubercle, and innervated by the sciatic nerve, work with the hamstrings in extension of the thigh. In action and in innervation, therefore, the adductor magnus is composed of two elements, an adductor, flexor, or obturator portion and a hamstring, extensor, or sciatic portion.

A little above the adductor tubercle the adductor magnus presents in its tendon a gap, the adductor or tendinous hiatus. Through this hiatus the femoral vessels, lying on the muscle's anterior surface, pass to the back of the thigh and leg where they are then called popliteal vessels.

The **obturator externus** lies deeply in the thigh, behind the pectineus and the upper ends of the three muscles named adductor. It arises from the outer surface of the pelvis, around the obturator foramen, and from the obturator membrane that almost fills that foramen. From this circular origin the muscle tapers and runs laterally and posteriorly, so that it has somewhat the form of a misshapen ice cream cone. It passes just below and then upward behind the hip joint to be inserted into a small pit (the trochanteric fossa) on the medial side of the greater trochanter, where it can be seen in a posterior dissection of the limb (p. 268). The muscle is an external rotator of the thigh. It is innervated by a branch of the obturator nerve given off before this nerve enters the thigh. The anterior branch of the obturator nerve usually runs anterior to the muscle as the nerve emerges from the obturator canal by which it leaves the pelvis, while the posterior branch runs through the muscle. The obturator artery largely ends in the muscle.

ANTEROMEDIAL NERVES AND VESSELS

The **femoral nerve, artery,** and **vein,** in that order from lateral to medial sides, pass behind the inguinal ligament to lie in the femoral triangle (fig. 16-7). The femoral nerve enters the thigh on the anterior surface of the iliopsoas muscle, in the same fascial compartment with that muscle, but the femoral artery and vein enter in a separate, more medial, compartment, and as they do so bring down with them a funnel-shaped continuation of the fascia lining the abdomen. This fascia surrounding the vessels is the femoral sheath. It contains three compartments, a lateral one for the artery, a middle one for the vein, and a medial one, empty except for a little loose connective tissue and a lymph node or two, which is the **femoral canal**. Since the femoral canal represents a part of the diverticulum from the abdominal fascia it is open above, and peritoneum and viscera may descend into it and enter the thigh as a femoral hernia.

A short distance below the inguinal ligament the **femoral nerve** breaks up (fig. 16-8) into muscular and cutaneous branches that are distributed to the quadriceps, sartorius, and pectineus muscles, and to the skin of the anterior surface of the thigh. One long cutaneous branch, the saphenous nerve, runs deeply in the thigh (with the femoral vessels in the adductor canal) but becomes subcutaneous just above the knee and is distributed to skin of the medial surface of the knee and leg, and the medial border of the foot as far as the base of the big toe. Interruption of the femoral nerve abolishes active extension of the knee, since the quadriceps is the sole muscle that can do this. However, the extended knee will support the body as long as the center of gravity is kept anterior to the axis of the knee joint.

Just below the inguinal ligament the **femoral artery** (also fig. 16-8) gives off small branches to the lower part of the abdomen, and soon thereafter at least one large branch, the profunda femoris (deep femoral) artery. Close to its origin the profunda femoris usually gives off two branches, the medial and lateral femoral circumflex arteries, which encircle the limb to anastomose with each other and with other vessels in the upper posterior part of the thigh. (Either or both of the circumflex arteries may arise from the femoral artery above the profunda instead of from the latter vessel.) The profunda femoris runs downward on the front of the pectineus, adductor brevis, and adductor magnus muscles, in that order, and behind the adductor longus, to give off a series of branches (usually four) that perforate the tendons of the adductor brevis and magnus close to the femur and supply the musculature on the back of the thigh.

After giving off its deep femoral branch the femoral artery continues down the anteromedial aspect of the thigh between the quadriceps and adductor group of muscles, lying in the adductor canal formed by these muscles and fascia between them. At the lower end of the adductor canal the femoral artery passes through the large gap, adductor hiatus, in the insertion of the adductor magnus, and thus attains the posterior aspect of the lower part of the thigh. Here it is known as the popliteal artery. This is described on p. 301.

The tributaries of the femoral vein and their names correspond to the branches of the femoral artery. The one exception is the great saphenous vein, to which there is no corresponding artery.

The **obturator nerve,** a branch of the lumbar plexus, passes through the obturator foramen and thus appears deeply within the adductor group of muscles (fig. 16-6). As it enters the thigh it divides into anterior and posterior branches; between them these branches supply the muscles of the adductor group, with the exception of the posterior portion of the adductor magnus. The nerve also supplies a limited amount of skin on the medial aspect of the thigh, gives one or more branches to the hip joint, and continues along the femoral artery to give twigs to the knee joint.

A lesion of the anterior branch of the obturator nerve would not seriously affect adduction, since the adductor magnus is innervated both by the posterior branch of the obturator nerve and by the sciatic nerve. Indeed this branch has been intentionally sectioned as part of an operation to relieve adduction contraction resulting from spastic cerebral palsy. However, a lesion of the entire obturator nerve, which would have to be above the inguinal ligament, would leave the limb so abducted and laterally rotated that the person could not walk.

The **obturator artery,** with its accompanying vein, emerges through the foramen with the obturator nerve but supplies chiefly structures immediately about the obturator membrane, therefore primarily the obturator externus muscle and the bone adjacent to the foramen. It also gives off a branch that enters the acetabulum to supply tissue here and continue through the ligament of the head of the femur to help supply that head.

REVIEW OF THE THIGH

The bony landmarks of the anterior part of the thigh are the anterior superior iliac spine, the pubis and the pubic tubercle, the greater trochanter, the patella, and the epicondyles of the femur, all of which are easily located and most of which have been studied previously. A little below the knee, on the sharp anterior border of the tibia, the tibial tuberosity is easily felt. Above it are the expanded tibial condyles. Just below the lateral tibial condyle the head of the fibula is subcutaneous.

Of the **muscles,** the quadriceps as a whole is easily identified on

TABLE 16–1. Anteromedial Nerves and Muscles of the Thigh

NERVE AND ORIGIN*	MUSCLE *Name*	*Segmental innervation**	*Chief action(s)*
L2–L4	Iliopsoas	L2–L4	Flexion at hip
Femoral L2–L4	Sartorius	L2, 3	Flexion and rotation at hip and knee
	Quadriceps	L2–L4	Extension at knee
	Articularis genus	L3, 4	Pulls synovial membrane of knee joint upward
	Pectineus	L2, 3	Flexion and adduction at hip
Obturator L2–L4	Pectineus (sometimes)	L2, 3	Flexion and adduction at hip
	Adductor longus	L2, 3	Adduction and flexion at hip
	Adductor brevis	L3, 4	Adduction and flexion at hip
	Gracilis	L3, 4	Adduction at hip, flexion at knee
	Adductor magnus (ant. part)	L3, 4	Adduction and flexion at hip
	Obturator externus	L3, 4	External rotation at hip

*A common segmental origin or innervation. The composition of both the chief nerves and their muscular branches varies somewhat among persons.

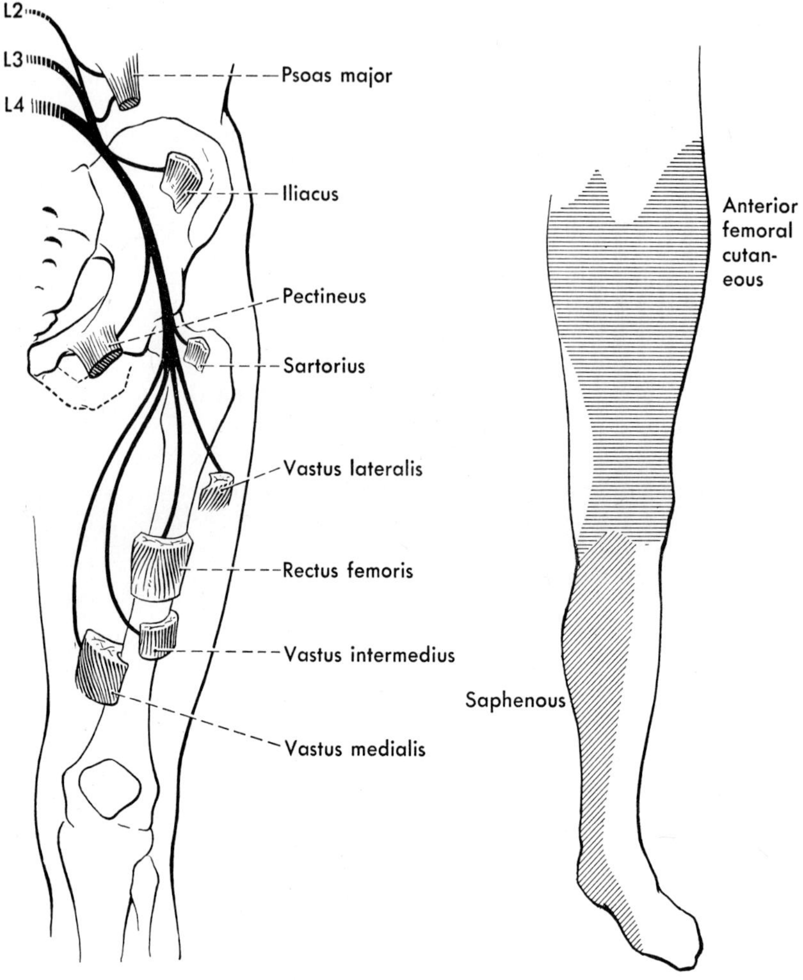

Figure 16–9. Distribution of the femoral nerve.

the front of the thigh and traced to the patella. The patellar ligament can also be traced between the patella and the tibial tuberosity. The lower ends of the vastus lateralis and the more bulky vastus medialis can easily be identified close to their insertions on the patella, and the upper end of the rectus femoris can be traced upward if thigh and leg are raised, with the leg extended, while in a sitting position. The sharp upper border of the sartorius can also be identified, and traced toward the anterior superior spine; this is easiest if, from a sitting position, the thigh is sharply flexed so as to lift the foot from the floor, and at the same time laterally rotated.

The adductor muscles can be felt to contract when the thighs are forcefully adducted. The most easily identified member of this group is the adductor longus, whose strong tendon of origin can be traced to the pubis. The gracilis is difficult to recognize close to its insertion because its flat tendon is closely applied on the medial side of the knee to the tendon of a posterior muscle of the thigh (semimembra-

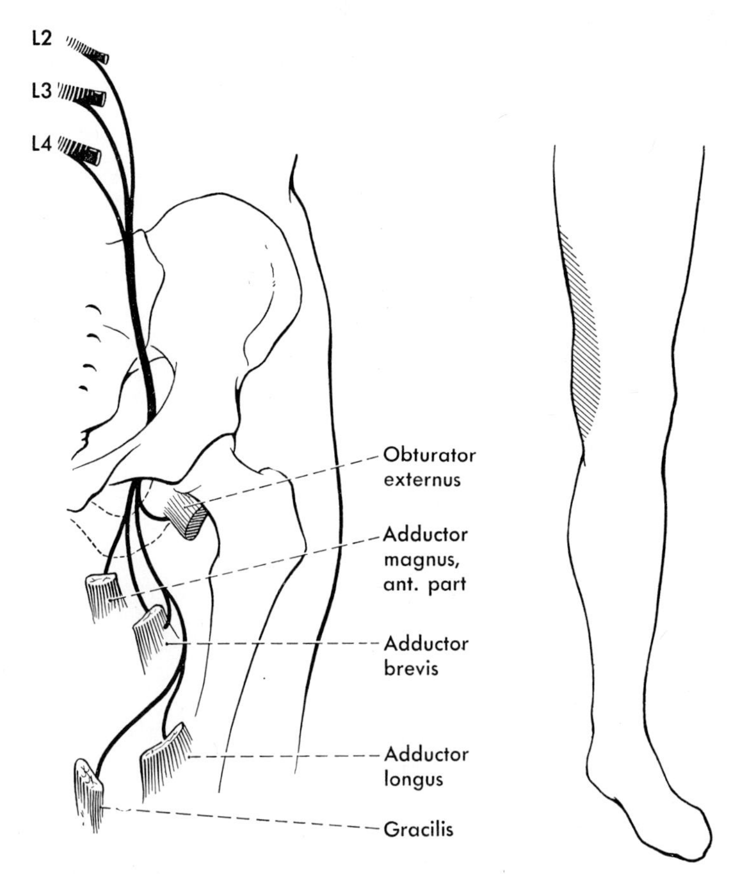

Figure 16–10. Distribution of the obturator nerve.

nosus). However, when the gracilis is contracted its sharp posterior border can be palpated in about the middle of the thigh with little difficulty. The anteromedial muscles of the thigh and their innervations are summarized in Table 16-1.

Of the **vessels,** the pulse of the femoral artery can be felt in the femoral triangle a little below the inguinal ligament. Parts of the greater saphenous vein may be visible through the skin. The deep femoral artery and most of the femoral artery lie too deep to be recognizable. The fact that it lies in the adductor canal should, however, make it easy to visualize the course of the femoral artery.

Neither of the two large anterior **nerves** of the thigh is distinctly palpable—the femoral because it breaks up into branches in the upper part of the femoral triangle, and the obturator because it lies too deep. The distributions of the femoral and obturator nerves are shown diagrammatically in figures 16-9 and 16-10.

17 Buttock and Posterior Thigh

THE SACRAL PLEXUS

The sacral plexus supplies the musculature of the buttock, and gives rise to the large sciatic (ischiadic) nerve that runs through the buttock to supply the posterior muscles of the thigh and all the muscles below the knee. The plexus is difficult to study in the laboratory, for it lies deeply, partly within and partly outside the pelvis, and passes through the greater sciatic foramen as it is taking form. Its method of formation can be seen from within the pelvis, where the nerve trunks contributing to it lie on the anterior surface of the sacrum. An even better concept of it and its branches can be obtained by approaching it from the posterior aspect, removing a lateral portion of the sacrum and the adjacent portion of the ilium after the dissection of the buttock is completed. A diagram of the plexus is shown in figure 17-1; a view of it and the lumbar plexus together is shown in figure 16-4.

The sacral plexus is typically formed by the union of a part of the fourth lumbar nerve with all of the ventral ramus of the fifth lumbar, to form a lumbosacral trunk, and by the union of this trunk with the ventral rami of the first three or four sacral nerves. Anterior branches from almost all these ventral rami, that is, L4, L5, S1, S2 and S3, usually unite to form the tibial nerve, the anterior component of the sciatic nerve. Posterior branches of the lumbosacral trunk and of the first two sacral nerves unite to form the posterior component of the sciatic, the common peroneal nerve. These two parts of the sciatic nerve regularly lie in a common connective tissue sheath in the thigh, the tibial nerve lying more medially and the peroneal more laterally, and form a single large nerve until they separate from each other a little above the knee. Occasionally they are separate from their origin at the plexus, but in any case it is usually not difficult in the dissecting room to separate the two parts of the sciatic nerve at their origin.

Other branches from the posterior part of the sacral plexus include the superior gluteal nerve, derived from the lumbosacral trunk and the first sacral nerve, the inferior gluteal nerve, derived from the lumbosacral trunk and the first and second sacral nerves, one or more twigs to the piriformis muscle, and a part of the posterior femoral cutaneous nerve. Branches from the anterior part of the plexus also contribute to the posterior femoral cutaneous nerve, and this nerve, derived from S1, S2 and S3, lies at the boundary between original dorsal and ventral surfaces of the thigh (now, because of the twisting that the limb has undergone, the posterior midline of the thigh—the original dorsal or posterior surface is lateral to this, the original ventral or anterior one is medial). The anterior parts of the lumbosacral trunk and of the first two sacral nerves also give rise to nerves to the small exter-

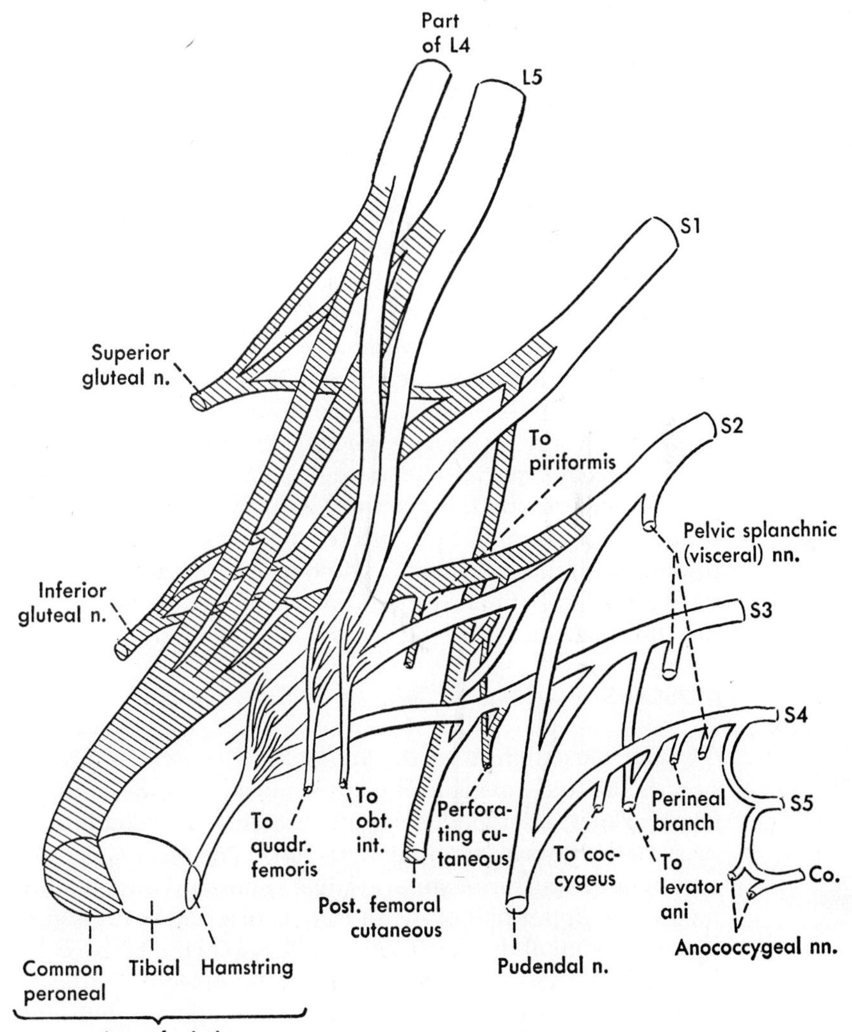

Figure 17–1. Diagram of the sacral plexus. The posterior parts of the plexus are shaded.

nal rotator muscles of the buttock; these two nerves supply four muscles, that to the quadratus femoris supplying also the inferior gemellus, that to the obturator internus supplying also the superior gemellus. Finally, anterior portions of the second and third sacral nerves unite with a part of the fourth sacral nerve to form the pudendal nerve, the lowest branch of the sacral plexus. This is distributed to the area (pudendal region) between the thighs—that is, to the anal and genital regions. If a very large continuation from the fourth lumbar nerve enters the sacral plexus the contribution of the fourth sacral to the pudendal nerve is often reduced or lacking; in other cases, a part of the fourth sacral nerve may contribute to the tibial nerve.

SUPERFICIAL NERVES AND VESSELS OF BUTTOCK; FASCIA

The subcutaneous tissue of the buttock is thick, for it is a favored area for the deposition of fat. The cutaneous nerves may therefore be difficult to find.

Dorsal branches of the first three lumbar nerves pierce the fascia of the back close together, just above the posterior part of the iliac crest, to run downward over the buttock and help supply skin of this area. Smaller dorsal branches from the first three sacral nerves appear closer to the midline over the sacrum and spread laterally. The posterior femoral cutaneous nerve as it runs down the posterior aspect of the thigh gives off recurrent branches that turn upward to supply skin of the buttock. There are no superficial vessels of any importance.

The gluteus maximus, the most superficial muscle of the buttock, is enclosed in a strong fascia that is especially well developed on the deep surface of the muscle. It is continuous with the heavy deep fascia of the thigh, the fascia lata, and much of the gluteus maximus is inserted into this fascia. The anterior part of the gluteus medius, where it is not covered by the maximus, is provided with a tendinous fascia from which some of the muscle fibers take origin; this also is continuous with the fascia lata. (The tensor fascia latae is also enclosed in the fascia lata, and inserts entirely into a special part of that fascia, the iliotibial tract.)

MUSCLES

The musculature of the buttocks is shown in figure 17-2. Most superficial, and covering the other muscles of this region, is the **gluteus maximus,** a large, coarse, quadrangular muscle. It arises from the sacrum, the dorsal sacroiliac ligaments, a small area of the ilium in the region of the posterior superior iliac spine, and the sacrotuberous ligament. The upper half of the muscle is inserted entirely into the strong lateral portion of the fascia lata, the iliotibial tract (also called iliotibial band). The fibers of the lower half of the muscle divide at their insertion, approximately half of them inserting into the iliotibial tract while the remaining, deeper, ones insert upon the upper lateral extension from the linea aspera, the gluteal tuberosity. Where the tendon of the muscle passes over the greater trochanter there is a bursa between the two. The gluteus maximus receives an independent nerve, the inferior gluteal nerve, from the sacral plexus; this leaves the pelvis below the piriformis muscle, and close to the lateral edge of the sacrotuberous ligament, where it is accompanied by the inferior gluteal artery and vein (fig. 17-3). Nerve and vessels penetrate the heavy fascia on the deep surface of the muscle before spreading out between the two, so that it is impossible to reflect the muscle to its origin without servering them. The muscle is an important extensor of the hip, being employed especially in straightening up from a bending position, in walking up stairs, and in other movements that require powerful extension at the hip joint. It is also an external rotator and, in the extended position of the limb, an adductor of the thigh. When the thigh is flexed, however, its lower portion aids in abduction. Because of its extensive insertion into the iliotibial tract and the attachment of the latter to the femur by an intermuscular septum, the gluteus maximus attains much greater leverage than can be obtained through its insertion on the gluteal tuberosity.

After the gluteus maximus is reflected, most of the muscles of the buttock, and the nerves and vessels to or passing through the buttock, can be seen. The posterior part of the **gluteus medius** lies under cover of the gluteus maximus, but the anterior part projects in front of this muscle. The gluteus medius arises from a major part of the upper lateral surface of the wing of the ilium, and the anterior part arises also from its covering fascia. Because of the convexity of the ilium, parts of the muscle lie in front of and behind the hip joint as well as lateral to it. The muscle inserts on the greater trochanter and is supplied by the superior gluteal nerve and vessels, which leave the pelvis above the upper border of the piriformis muscle and turn upward and laterally to run deep to the gluteus medius.

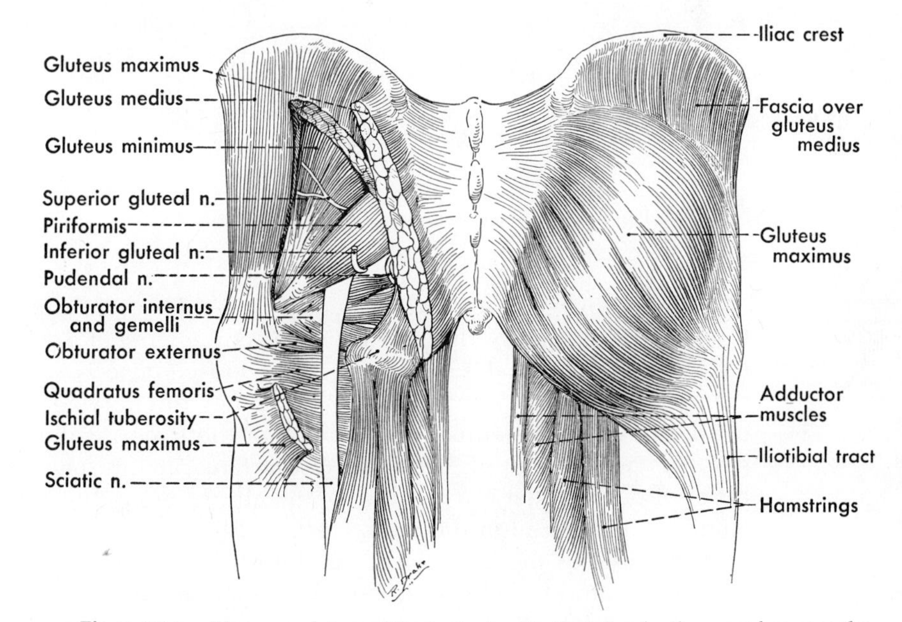

Figure 17–2. The musculature of the buttocks. On the left side, the space between the inferior gemellus and quadratus femoris is exaggerated so that the insertion of the obturator externus can be shown.

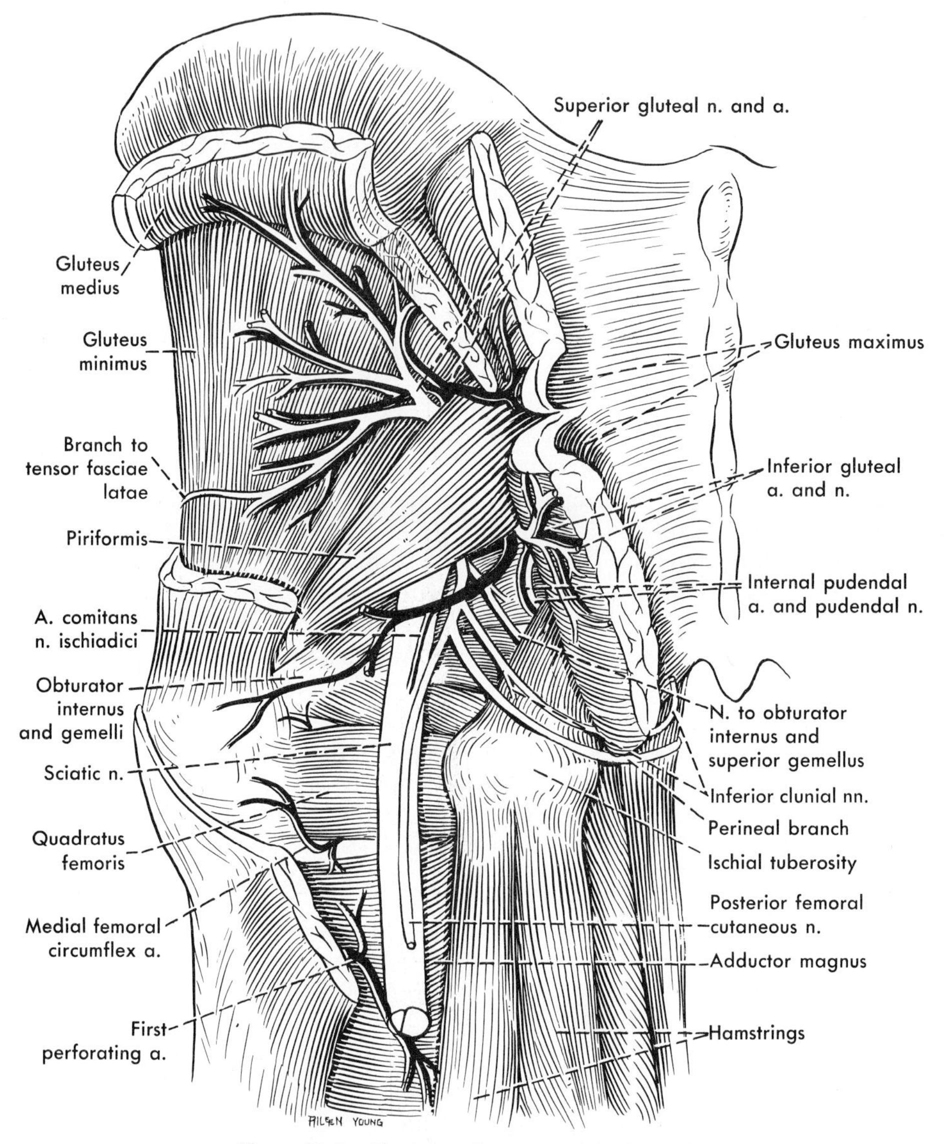

Figure 17–3. Nerves and arteries of the buttock.

Under cover of the gluteus medius is the **gluteus minimus,** arising from the lower part of the lateral surface of the wing of the ilium and inserting on the greater trochanter. This muscle also is supplied by the superior gluteal nerve and vessels as they run between it and the overlying gluteus medius. One branch of the superior gluteal nerve passes forward to supply the **tensor fasciae latae,** an internal rotator and flexor of the hip that is a member of the gluteal group but can best be studied on the front of the thigh (see p. 251).

The gluteus medius and the gluteus minimus have not only the same nerve and blood supply but also a similar action. Both strongly abduct the thigh. When the weight is supported on one limb, the other side of the pelvis tends to sag, a movement equivalent to adduction of the supporting limb. Since the gluteus medius and minimus are abductors, they oppose this movement. In walking, therefore, the muscles of the two sides have to contract alternately, as the weight is shifted from side to side. Stimulation of the anterior part of the gluteus medius also produces flexion and internal rotation of the thigh, stimulation of the posterior part produces extension and external rotation, and it has usually been assumed that both these movements are carried out by both muscles; electromyographic evidence indicates, however, that the gluteus minimus is used in flexion and internal rotation, but not in the opposite movement, whereas the gluteus medius may be used in extension and external rotation, but is not used to flex and internally rotate.

Below the posteroinferior edge of the gluteus medius is a series of small muscles. The **piriformis** (pear-shaped) muscle, the upper one of these, arises from the pelvic surface of the sacrum and passes to an insertion on the inner surface of the upper part of the greater trochanter. The muscle largely fills the greater sciatic foramen as it passes from origin to insertion. Above and below this muscle emerge the important branches of the sacral plexus and the branches of the internal iliac artery (the artery to the pelvis, p. 382) that leave the pelvis through the greater sciatic foramen. The piriformis is supplied by one or two small branches from the ventral rami of the second or first and second sacral nerves, which enter its pelvic surface and are therefore not visible from the buttock. It is a lateral rotator of the thigh, but when the thigh is flexed it becomes an abductor.

Below the piriformis is the tendon of the **obturator internus,** associated with two small muscles on its upper and lower borders, the superior and inferior gemelli. The obturator internus arises on the inner surface of the pelvis, both from the obturator membrane and the edges of the obturator foramen. The muscle passes posteriorly to round the lesser sciatic notch and largely fills the lesser sciatic foramen, becoming tendinous on its deep surface at this point and being provided with a bursa to allow free movement over the bone of the notch. The tendon receives the attachments of the two gemelli and passes to an insertion just above the trochanteric fossa.

The **superior** and **inferior gemelli** are small muscles accessory to the obturator internus, and one or both may be lacking. The superior gemellus arises from the ischial spine to insert on the upper border of the obturator internus tendon, while the inferior gemellus arises from the ischial tuberosity and inserts on the lower border of this tendon. The superior gemellus and obturator internus are supplied by the same nerve, a branch of the sacral plexus (fig. 17-1). It runs across the superficial surface of the gemellus, lateral to the pudendal nerve and vessels, and like these disappears into the lesser sciatic foramen on the surface of the obturator internus. The inferior gemellus is supplied by the nerve to the quadratus femoris, also from the sacral plexus; it runs deep to all the small muscles below the piriformis, against the posterior surface of the capsule of the hip joint. The obturator internus is a lateral rotator of the thigh; when the thigh is flexed it, like the piriformis, may act as an abductor. The gemelli obviously have similar actions.

The **quadratus femoris** is a small quadrangular muscle extending transversely from the ischial tuberosity to the posterior surface of the femur about midway between the lesser and greater trochanters. Its nerve supply is in common with that of the inferior gemellus. It is an external rotator of the thigh and also, because of its position below the head of the femur, an adductor.

The **obturator externus** is an anteromedial muscle of the thigh, and is therefore described on page 257 with other muscles of its group. However, its insertion is visible only from the buttock, and should therefore be mentioned now. It passes backward just below the hip joint and then upward and laterally across the posterior aspect of the joint, deep to the quadratus femoris, to insert into a pit, the trochanteric fossa, immediately below the insertion of the obturator internus.

In addition to these muscles of the buttock the origins of the hamstring muscles from the ischial tuberosity can also be noted at this time. While the hamstring muscles constitute the posterior musculature of the thigh and have an important action on the knee, they also act as extensors at the hip joint, where they cooperate with the gluteus maximus.

NERVES AND VESSELS

The nerves and vessels in the buttock all emerge through the greater sciatic foramen, therefore in close contact with the piriformis muscle (fig. 17-3). Only the superior gluteal nerve and vessels normally pass above the piriformis, however, and all the remaining ones, whether they end in the buttock or merely pass through it, typically first appear at the lower border of the piriformis.

The **inferior gluteal nerve** emerges through the greater sciatic foramen below the piriformis muscle, in company with the inferior gluteal vessels, and passes directly into the gluteus maximus. The larger branches of the inferior gluteal vessels are also distributed to the gluteus maximus, but there are twigs to the adjacent small muscles of the buttock and the upper ends of the posterior muscles of the thigh, and the artery gives off a branch to the posterior surface of the sciatic nerve.

The **superior gluteal nerve** emerges through the greater sciatic foramen above the piriformis muscle, in company with the superior gluteal vessels, and runs laterally between the gluteus medius and minimus, supplying both and being continued beyond these muscles to enter the deep surface of the tensor fasciae latae. The superior gluteal vessels are largely distributed with the nerve, but the artery sends a superficial branch between the piriformis and the gluteus medius to the upper part of the gluteus maximus.

The **nerve to the piriformis** enters the pelvic surface of the muscle and is difficult to demonstrate in a posterior dissection.

The **nerve to the obturator internus and superior gemellus** emerges from the greater sciatic foramen below the piriformis muscle close to the ischial spine and turns around this spine to supply a branch to the superior gemellus. It then enters the lesser sciatic foramen to run on the perineal (internal) surface of the obturator internus. Medial to the nerve to the obturator internus and also passing around the ischial spine from the greater into the lesser sciatic foramen is the pudendal nerve to the perineal or pudendal region (the region of the anus and of the external genitals), and medial to the nerve are the internal pudendal vessels. These have no distribution to the buttock but simply pass through it.

The **nerve to the inferior gemellus and the quadratus femoris** usually arises from the sacral plexus as the main elements converge to form the sciatic nerve. It runs deep to the two gemelli and the obturator internus to enter the deep surface of the quadratus femoris, having previously given off a branch to the inferior gemellus. This nerve supplies also the posterior aspect of the hip joint.

The two remaining nerves appearing in the buttock are the posterior femoral cutaneous and the sciatic. The large **sciatic nerve** appears below the lower edge of the piriformis muscle, thus making its exit from the greater sciatic foramen, and runs down the thigh. It gives off no branches to the buttock. Occasionally the sciatic nerve is split at this level into its two component parts, the tibial nerve and the common peroneal nerve; sometimes when this occurs the common peroneal portion of the nerve may emerge through the piriformis instead of below it; rarely, the entire nerve passes through the piriformis. The abnormal relation of the nerve to the piriformis muscle has been held responsible for some cases of sciatic pain, the concept being that a spastic piriformis muscle may so squeeze the nerve as to produce pain over its distribution.

The small **posterior femoral cutaneous nerve** runs almost exactly in the posterior midline, behind the sciatic, and gives off recurrent branches to the skin of the buttock. It then passes down the posterior surface of the thigh just deep to the deep fascia (fascia lata), gives off a series of branches which pierce the fascia lata to supply the overlying skin, and continues a variable distance down the leg.

MOVEMENTS OF THE BONY PELVIS

The muscles that rotate and tilt the bony pelvis form a heterogeneous group; none of them are, strictly speaking, muscles of the pelvis. Downward rotation of the pelvis is assisted especially by the anterior thigh muscles attaching to the front of the pelvis, and is a concomitant of any increase in the lumbar curvature. This latter can be brought about by the psoas muscles which, taking their fixed points from below, can pull upon the front of the lumbar vertebral column. More commonly it is brought about by gravity and by relaxation of the anterolateral abdominal muscles. Upward rotation of the pelvis is brought about especially by the upward pull on the pubis of the anterolateral abdominal muscles, probably assisted by the downward pull of the hamstrings on the ischial tuberosity. Lateral rotation of the pelvis and trunk as a whole upon one femoral head, as in walking, is brought about mostly by certain of the rotators of the thigh, assisted also by the anterolateral abdominal muscles.

Lateral tilting of the pelvis tends to occur when the weight is put upon one leg, and is opposed by the passive checking action of the fascia lata, particularly the iliotibial tract, and, as already noted, by the active contraction of the gluteus medius and minimus, probably assisted slightly by the tensor fasciae latae.

POSTERIOR MUSCLES OF THE THIGH

The three muscles of the posterior aspect of the thigh (fig. 17-4) are known as the hamstring muscles (or posterior hamstrings when the sartorius and gracilis are also called hamstrings), and consist of the semitendinosus, the semimembranosus and the biceps femoris. The more vertical portion of the adductor magnus, running from the ischial tuberosity to the adductor tubercle, functions with these hamstrings (see the previous chapter) and has been regarded as the remains of a muscle that once continued across the knee joint.

The **semitendinosus** muscle arises from the ischial tuberosity, where it is intimately blended with the long head of the biceps. Diverging from this, it passes down on the medial side of the posterior aspect of the thigh and behind the knee joint, then curves forward and inserts on the tibia medial to and a little below the tibial tuberosity, where it is closely associated with the insertions of the gracilis and sartorius muscles. A bursa intervenes between the tendons of insertion of the semitendinosus and gracilis muscles, on the one hand, and the overlying tendon of insertion of the sartorius. It also extends (or there is a separate bursa) between the two first-mentioned tendons and the tibial collateral ligament. The associated tendons of insertion of these three muscles are sometimes referred to as the pes anserinus (goose's foot), and the bursa is accordingly named the anserine bursa.

The **semimembranosus** arises from the lateral aspect of the ischial tuberosity and crosses deep (anterior) to the semitendinosus and long head of the biceps. The long, wide tendon of origin of the semimembranosus, from which the muscle derives its name, gives rise to a muscular belly which at the knee is succeeded by a thick rounded tendon that passes on the posteromedial side of the knee to be inserted on the posteromedial side of the medial tibial condyle. At the knee the semimembranosus tendon has a bursa between it and the joint capsule. The tendon of insertion of the semimembranosus gives off a heavy band that runs obliquely upward and laterally, blending with the posterior capsule of the knee joint, to form the oblique popliteal ligament (p. 244).

The **biceps femoris** has its long head of origin from the ischial tuberosity in common with the semitendinosus; this head is joined above the knee by the short head, which arises from the linea aspera on the posterior aspect of the femur, and from the lateral intermuscular septum. The tendon derived from the union of the two heads runs on the posterolateral aspect of the knee joint to attach to the head of the fibula. The biceps forms the upper lateral border of the somewhat diamond-shaped popliteal fossa, the semitendinosus and semimembranosus form the upper medial border, and the two heads of the most superficial muscle of the calf, the gastrocnemius, form the lower borders.

All three hamstring muscles give off about the knee expansions to the fascia of the leg. Similarly, all three are good extensors and very weak adductors of the thigh, and good flexors of the leg. Their lengths are such that they can with difficulty be stretched over both joints during combined flexion at the hip and extension at the knee; they therefore act passively to check such movements as high kicking or bending over to touch the floor with the knees straight. The short head of the biceps does not, of course, participate in extension at the hip. In addition to these actions, the semitendinosus and semimembranosus serve as internal rotators at the hip, and as internal rotators of the leg at the knee when the leg is flexed. The long head of the biceps acts as a lateral rotator of the thigh; both heads externally rotate the leg when it is flexed, and both help to flex the leg. However, the long head is said to participate in only the early part of flexion, and to relax as the leg becomes semiflexed.

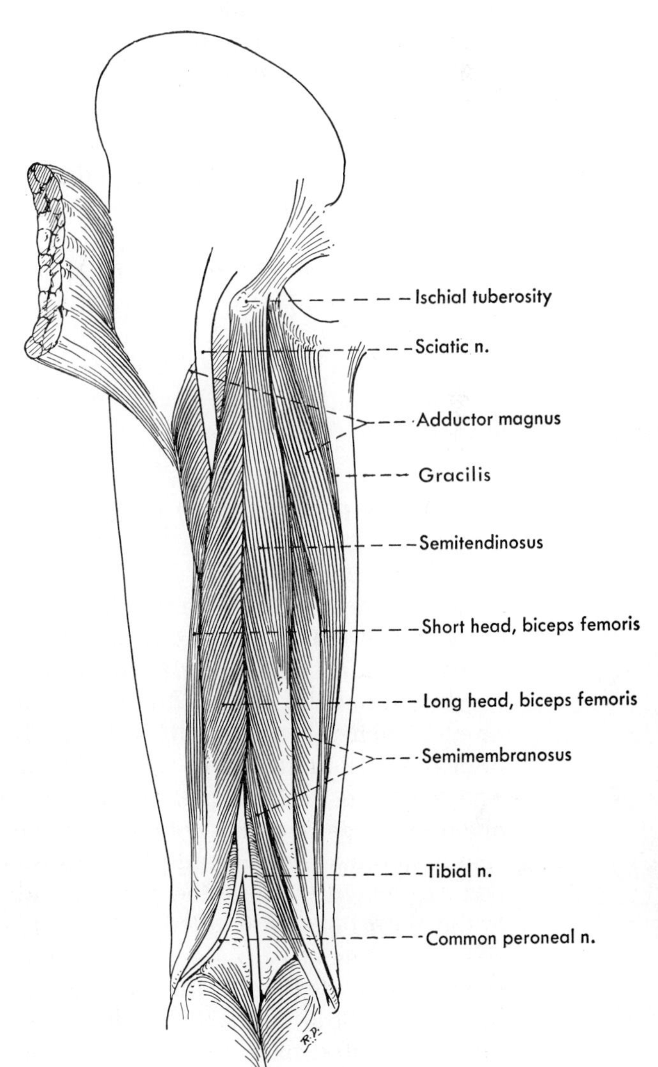

Figure 17–4. The posterior muscles of the thigh.

The hamstrings and the posterior part of the adductor magnus work with the gluteus maximus in extending the thigh (thus in such movements as straightening up from a bending position). It has been reported that they contribute from 31 to 48 per cent to the strength of this movement, their greatest contribution occurring when the hip was flexed to 90°. It might be noted, though, that they cannot contribute much to rising from a sitting position, since this involves simultaneous extension at both hip and knee, and they are flexors at the latter.

Each of these three muscles is supplied by branches (usually multiple) of the sciatic nerve; when, as sometimes occurs, the sciatic nerve is divided into its tibial and common peroneal components at a level higher than normal, it can be seen that the semitendinosus, the semimembranosus and the long head of the biceps are supplied from the tibial portion of the sciatic, while the short head of the biceps is supplied by the common peroneal portion. Most of the branches of the sciatic nerve in the thigh thus arise from the medial, or tibial, portion of the sciatic nerve, and course medially. The surgeon takes advantage of this fact in his operative approach to the posterior aspect of the thigh, as he knows that the lateral side of the sciatic nerve is the side of relative safety.

As has already been noted, the adductor magnus can be considered as two muscles from the standpoint of actions and innervations. The obliquely running portion of the adductor magnus is innervated from the obturator nerve and belongs functionally with the adductor group of muscles. The more vertical posteromedial portion of the adductor magnus arises from the ischial tuberosity as do the hamstrings proper, and is innervated from the tibial portion of the sciatic nerve, as are these muscles. While this portion of the adductor magnus cannot, of course, flex the knee, it functions with the hamstrings in extension at the hip.

POSTERIOR NERVES AND VESSELS

The **sciatic nerve** is the sole posteriorly placed deep nerve of the thigh, although the posterior femoral cutaneous nerve runs for much of its course deep to the fascia lata. The sciatic nerve enters the thigh by passing just lateral to the ischial tuberosity, and runs straight downward in about the midline to its division into common peroneal and tibial branches in the popliteal fossa (fig. 17-5). Besides its terminal branches, its branches in the thigh are to the hamstring muscles. These arise in variable patterns, but the tibial side of the nerve gives off one or more branches into the long head of the biceps and into the semimembranosus, the semitendinosus, and the posterior part of the adductor magnus, while the peroneal side gives rise only to the nerve to the short head of the biceps. The explanation of this distribution is simple. It has already been noted that the posterior midline of the thigh represents the junction of original ventral and dorsal parts of the limb, the ventral part lying medially, the dorsal part laterally. Therefore the ventral or tibial component of the sciatic supplies muscles

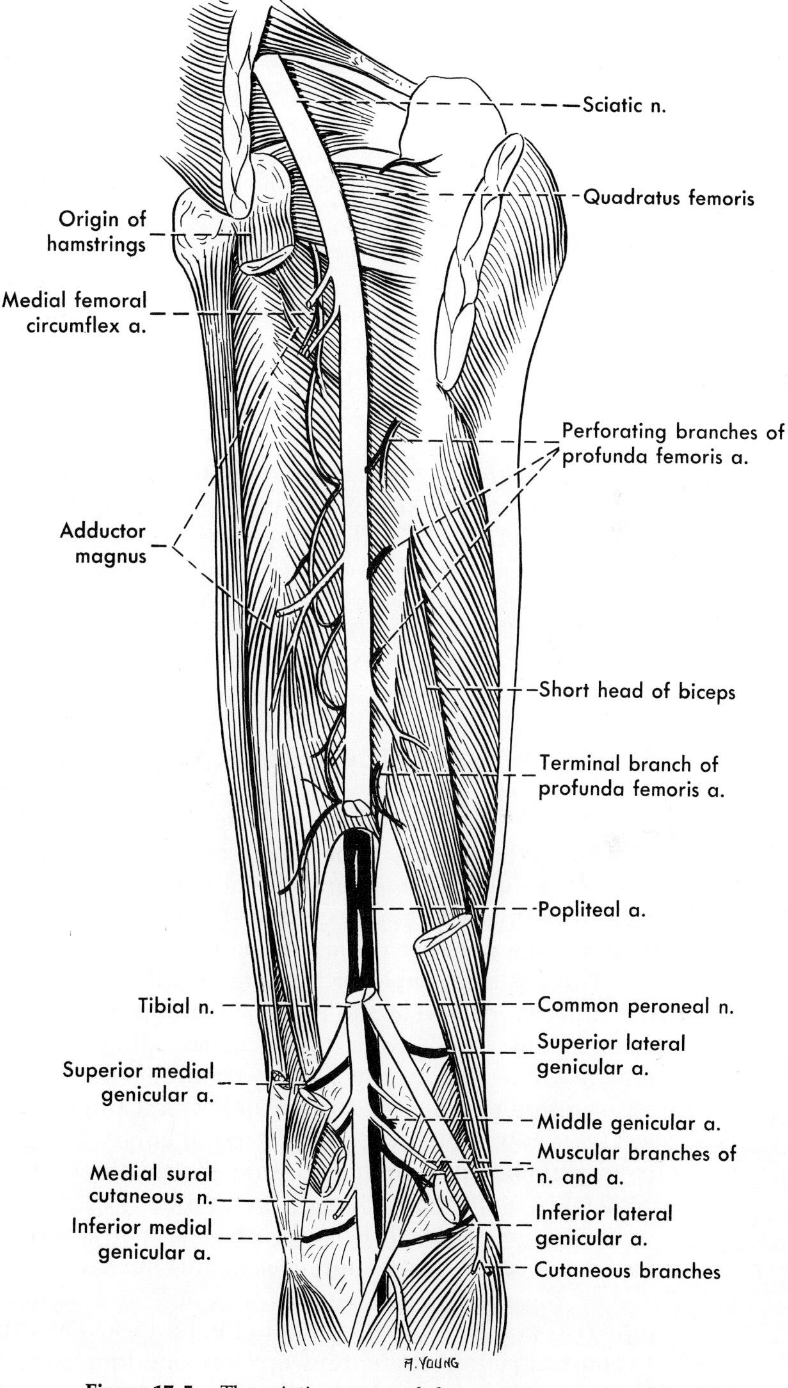

Figure 17–5. The sciatic nerve and the posterior arteries of the thigh.

arising medial to the posterior midline, while the common peroneal component supplies the sole muscle arising laterally.

After they separate, the tibial nerve continues the downward course of the sciatic and disappears deep to the muscles of the calf, while the common peroneal nerve diverges to the lateral side of the leg. Both nerves give off cutaneous branches to the leg as they leave the thigh.

Since there is no longitudinal artery in the posterior side of the thigh above the popliteal fossa, the blood supply to the muscles comes largely from the **perforating branches** of the deep femoral artery that come through the adductor magnus. In addition, there are twigs from the inferior gluteal artery and from branches of the two femoral circumflexes to the upper ends of the muscles, and from the femoral and popliteal arteries to their lower ends.

The **popliteal vessels,** the continuations of the femoral vessels below the adductor hiatus, appear only briefly in the thigh. With the vein behind the artery, they run down through the popliteal fossa and disappear with the tibial nerve deep to the muscles of the calf. Their branches about the knee are small ones primarily to the joint and the adjacent soft tissues. While there are anastomoses between the vessels above and below the knee (including branches from the profunda femoris), their total diameters are so small, in comparison with that of the popliteal artery, that they cannot furnish an adequate blood supply to the leg and foot if the popliteal artery is suddenly occluded.

REVIEW OF THE BUTTOCK AND POSTERIOR THIGH

There is relatively little anatomy of the buttock that can be reviewed on the living person. The crest of the ilium is easily palpable, since it is subcutaneous and the abdominal muscles above it yield readily to pressure. It can be traced forward with no difficulty to the anterior superior iliac spine, which is also prominent.

The posterior superior iliac spine, although it may not be clearly palpable, is located at the posterior end of the iliac crest where this abuts against the sacrum; the permanent dimple of the skin over the region where the crest meets the sacrum marks its location. The posterior inferior spine lies slightly below and anterior to it. The sacrum can be identified between the two coxal bones. The greater trochanter is identifiable on the lateral side of the thigh at its widest part, somewhat more than a hand's breadth below the iliac crest, and the ischial tuberosity is most clearly felt while one is sitting and it is therefore not covered by the gluteus maximus muscle.

Of the **muscles,** the gluteus maximus and medius are about the only two that can be identified in the buttock. The former can be felt to contract as one straightens up from bending over, the latter as one shifts all the weight onto that limb.

In the thigh, the hamstrings can be felt as a unit as they arise from the ischial tuberosity, and at the borders of the popliteal fossa some of their tendons can be identified. The tendon on the lateral side of the popliteal fossa is that of the biceps. The first (most lateral) tendon on

TABLE 17–1. Nerves and Muscles of Buttock and Posterior Part of Thigh

NERVE AND ORIGIN*	MUSCLE *Name*	*Segmental innervation**	*Chief action(s)*
Superior gluteal L4–S1	Gluteus medius	L4–S1	Abduction and external rotation at hip
	Gluteus minimus	L4–S1	Abduction and internal rotation at hip
	Tensor fasciae latae	L4–S1	Flexion and internal rotation at hip
Inferior gluteal L5–S2	Gluteus maximus	L5–S2	Extension and adduction at hip
N. to piriformis S1, 2	Piriformis	S1, 2	External rotation at hip
N. to obt. int. and sup. gemel. L5–S2	Obturator internus	L5–S2	External rotation at hip
	Superior gemellus	L5–S2	" " " "
N. to quadr. fem. and inf. gemel. L4–S1	Inferior gemellus	L4–S1	" " " "
	Quadratus femoris	L4–S1	" " " "
Tibial L4–S3	Semitendinosus	L5, S1	Extension at hip, flexion at knee
	Semi-membranosus	L5, S1	" " " " " "
	Biceps, long head	L5–S2	" " " " " "
	Adductor magnus, post. part	L4, 5	Extension and adduction at hip
Common peroneal L4–S2	Biceps, short head	L5, S1	Flexion at knee

*A common segmental origin or innervation.

the medial side, and the most prominent when the knee is forcefully flexed, is that of the semitendinosus. Medial to it, and not projecting so much upon flexion, is the broader tendon of the semimembranosus. Closely applied to the medial side of this tendon, and distinguishable from it only with difficulty, is the thin flat tendon of the gracilis.

Of the **vessels** and **nerves,** none of those in the buttock can be palpated, and the only one in the posterior aspect of the thigh that can be palpated is the popliteal artery. The pulse of even this large artery may be difficult to obtain because of the depth at which the vessel lies.

Although none of the nerves can be clearly palpated, the courses of the sciatic nerve and its tibial continuation, and that of the posterior femoral cutaneous, can be fairly easily visualized by remembering that the sciatic nerve runs just lateral to the ischial tuberosity and then straight down the midline of the thigh. The tibial nerve therefore lies in the middle of the popliteal space. The posterior femoral cutaneous nerve lies posterior to the sciatic and tibial nerves, therefore also in the posterior midline of the thigh.

Because the sciatic nerve has a much greater distribution to the leg and foot than it does to the thigh, its distribution is illustrated later (figs. 19-14 and 19-15). The innervations and actions of the muscles of the buttock and posterior thigh are summarized in Table 17-1.

18 Movements of the Thigh and Leg

MOVEMENTS AT THE HIP JOINT

The extensors at the hip joint (fig. 18-1) lie in the buttock and the posterior aspect of the thigh, and the chief abductors are also in the buttock. The chief flexors and adductors lie anteriorly and medially, while the rotators, especially the external ones, are found both anteriorly and posteriorly.

On the posterior aspect of the thigh the large gluteus maximus, extending as it does from the sacrum, the sacrotuberous ligament, and the posterior wing of the ilium to the fascia lata and femur, is an especially strong extensor at the hip. In this action it is assisted in particular by the part of the adductor magnus arising from the ischial tuberosity and innervated by the sciatic nerve. Since the posterior hamstring muscles attach above to the ischial tuberosity they also are extensors at the hip, but since they are likewise flexors at the knee they cannot contribute strongly to hip extension unless the knee is kept from flexing. They become active in any forward bending at the hips, and act as the antigravity or postural muscles here. In the movement of bending over to touch the floor with the fingers, the extensors of the hip must contract and then slowly relax (an eccentric contraction) in order to control the movement; the hamstrings are active through the whole movement of bending over and straightening up, but the gluteus maximus contracts most toward the end of flexion and the beginning of extension. The posterior fibers of both the gluteus medius and minimus have been thought to aid in extension, but the medius alone has been shown to do so. Although the piriformis has also been thought to contribute to extension, and the obturator internus may contribute if the movement starts from a sharply flexed position, neither is of any real importance in this movement. Because they pass across both hip and knee joints, the hamstrings are subject to stretching when the knee is kept extended as the hip is flexed; hence the pain behind the knee associated with floor-touching or high-kicking exercises.

The chief function of the abductors of the thigh (fig. 18-2) is to keep the pelvis approximately horizontal when all the weight is put on one leg; normally, they contract enough to raise the unsupported side of the pelvis slightly above the horizontal. If the muscles are weak, however, there will be marked sagging of the pelvis on the opposite side and, in an effort to balance the weight on one limb, a lateral protrusion of the hip that is supporting the weight and a flexion of the trunk toward that side. From the standpoint of supporting the pelvis, there are only two really good abductors of the thigh, namely, the gluteus medius and the gluteus minimus. When these muscles are weakened, the gait is much disturbed by the constant tilting of the pelvis and the consequent side-to-side sway of the trunk. The other

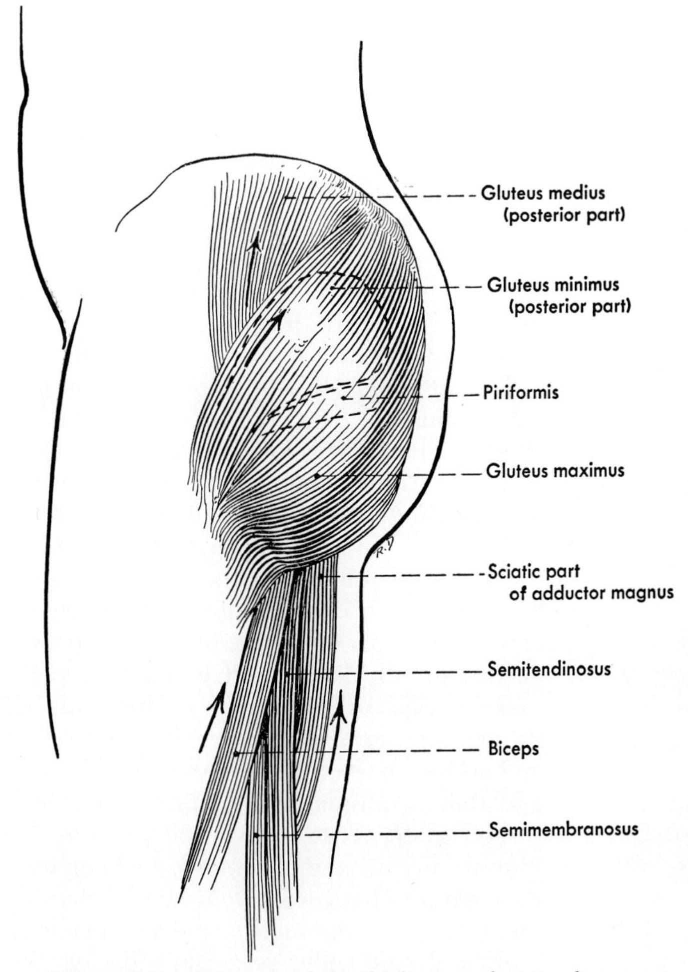

Figure 18–1. Extensors of the thigh. According to electromyographic evidence the posterior part of the gluteus minimus, usually regarded as an extensor, may not participate in extension.

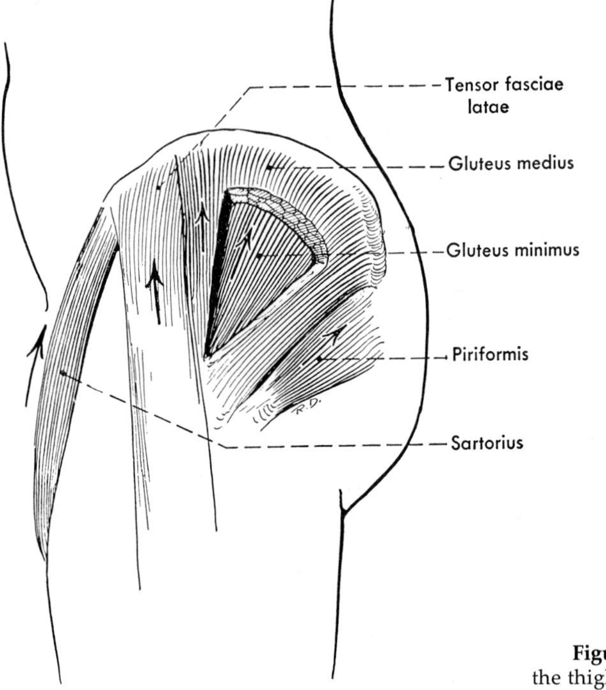

Figure 18–2. The abductors of the thigh.

abductors of the thigh are the tensor fasciae latae, the sartorius, and, to an even less extent, the piriformis, the obturator internus, and the lower fibers of the gluteus maximus (which can function as abductors when the thigh is flexed). While these muscles may assist in abducting the limb when the foot is free from weight-bearing, they are by no means capable of replacing the gluteus medius and minimus in the important weight-bearing function of the abductors. The tensor fasciae latae comes closest to doing that, but it has been estimated that it can exert no more than one fifth of the combined pull of the two glutei, and it contributes almost nothing to abduction of the free limb. Only if the thigh is flexed to a right angle does the gluteus maximus help abduct the limb; in other positions this muscle is an adductor.

External rotation of the thigh is stronger than internal rotation, but beyond the fact that most of the muscles of the buttock may act as external rotators there was for years little agreement as to which muscles externally rotate and which internally rotate. Of the posterior muscles (fig. 18-3), the lateral direction of the fibers of the gluteus maximus makes this a powerful external rotator as well as an extensor, and all the short muscles in the gluteal region—that is, both obturators and the piriformis, the quadratus femoris and the gemelli—are so placed as to assist in external rotation. The posterior fibers of the gluteus medius and minimus have been thought to externally rotate as they extend, because stimulation of the medius produces this movement, but only the medius has been shown electromyographically to participate. Of other posterior muscles, the long head of the biceps probably exerts a very weak external rotatory action.

Except for acknowledgment that the sartorius contributes weakly to external rotation, there has been doubt and confusion as to whether other anterior and anteromedial muscles externally rotate or internally rotate (figs. 18-4 and 18-5). This centers primarily around the function of the iliopsoas and the adductor group of muscles. Stimulation of the iliopsoas and of the various muscles of the adductor group (fig. 18-4) was reported to produce flexion and a variable degree of external rotation (except for the posterior or hamstring portion of the adductor magnus, which was reported to internally rotate when stimulated); subsequently, it was argued on mechanical grounds, and supposedly "proved" in the case of the iliopsoas by strands representing the muscle, that all muscles inserting posteromedially on the upper part of the femur must internally rotate it. More recently, electromyographic evidence has indicated that the iliopsoas is the only member of this group that participates in external rotation; it is said to be an especially important external rotator in the infant. The electromyographic evidence indicates that the pectineus and all the adductor muscles rotate internally. The gracilis, similarly, has been regarded as both an external and an internal rotator; it is now said that it becomes active in internal rotation.

The anterior part of the gluteus medius, like the adductors, has also been reported to internally rotate when it is stimulated, but according to electromyographic evidence it does not contract during internal rotation. The anterior part of the gluteus minimus, however, does internally rotate. The semitendinosus and the semimembranosus have always been regarded as weak internal rotators, and both stimulation and electromyography indicate that the tensor fasciae latae is an internal rotator (fig. 18-5).

Most of the rotators of the thigh rotate only incidentally as they

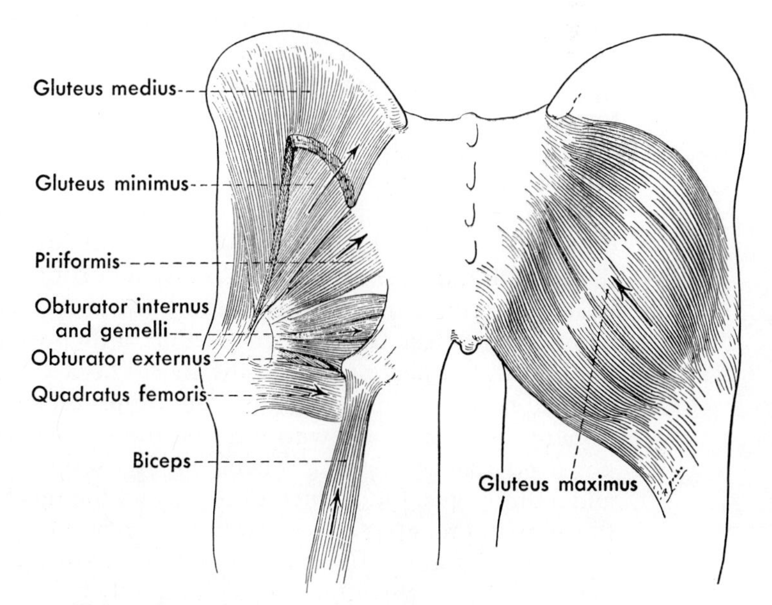

Figure 18–3. The posteriorly placed external rotators of the thigh. The gluteus minimus has long been regarded as aiding external rotation, but its participation has not been shown electromyographically.

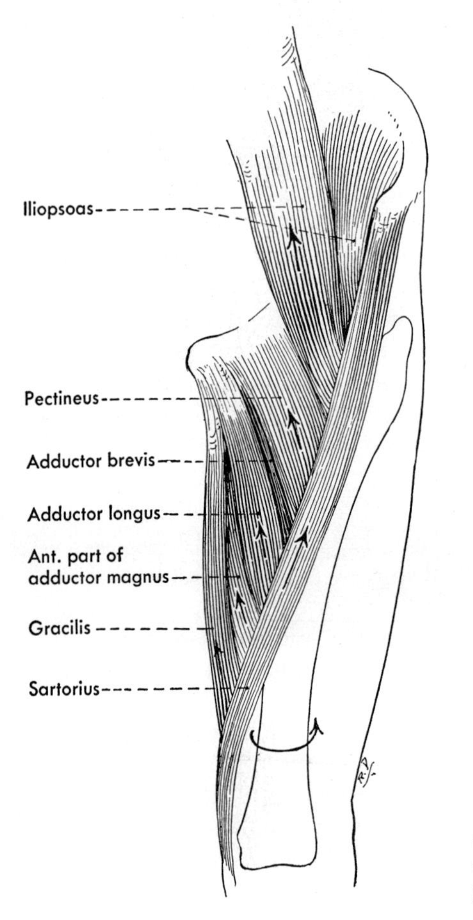

Figure 18–4. External rotators on the front of the thigh as determined by stimulation. According to electromyographic evidence, however, the pectineus, the gracilis, and the adductors rotate internally.

flex or extend the thigh, although the small posteriorly placed external rotators do little except rotate. The gluteus maximus is obviously the most important external rotator; the tensor fasciae latae, the pectineus, and the anterior part of the gluteus minimus are probably the more important internal rotators.

The flexor muscles (fig. 18-6) are those lying mostly on the anterior or anteromedial surface of the hip. The most lateral one of these is the tensor fasciae latae. The only head of the quadriceps that crosses the hip joint, the rectus femoris, is also a flexor at the hip; it can exert little power in flexion, however, until this movement has been started by other muscles. The strongest action of the sartorius is in flexion of the hip; it, the pectineus, and the tensor fasciae latae apparently participate regularly in hip flexion. The iliopsoas, crossing as it does the front of the joint, is a powerful flexor, the strongest of the group, but is not used unless a strong movement is needed. The pectineus, the adductor longus, the adductor brevis, and the more anterior portion of the adductor magnus also assist in flexion of the hip; and while stimulation of the anterior fibers of the gluteus medius produces flexion, it is apparently only the minimus that normally participates in this action.

It may not be amiss to point out that the adductor magnus has been included with both the flexors and the extensors of the hip. This

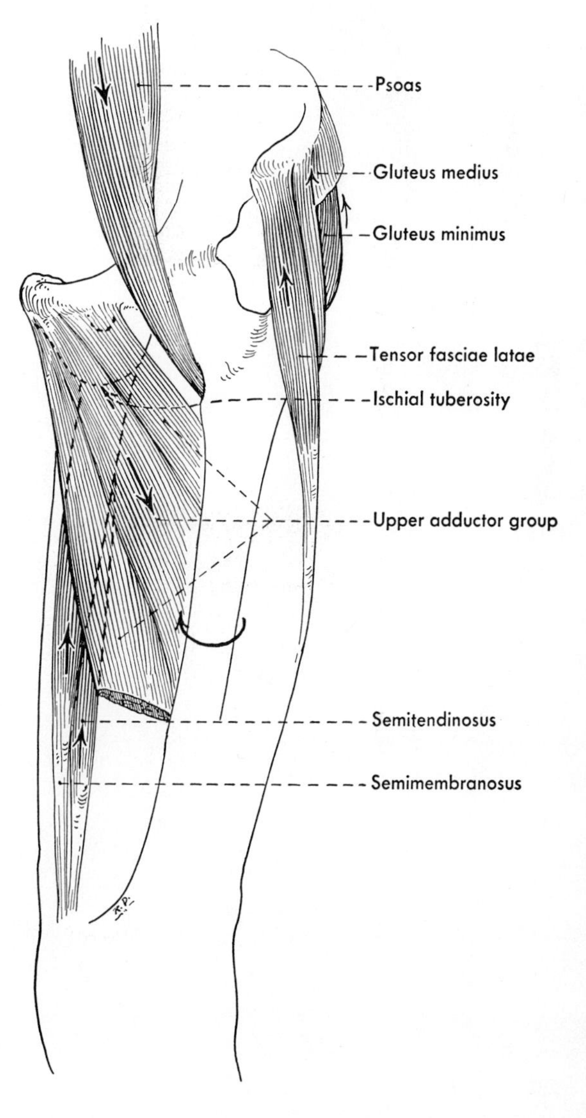

Figure 18–5. Other rotators of the thigh. All those shown have been regarded as internal rotators, but the psoas major apparently rotates externally and the anterior part of the gluteus medius is apparently normally inactive during rotation; the entire adductor group, however, is now regarded as rotating internally.

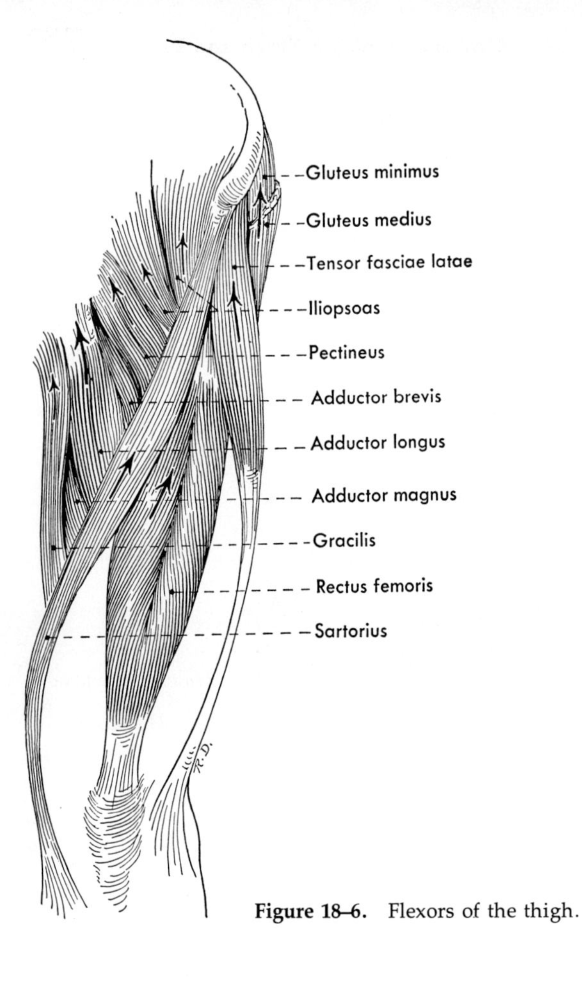

Figure 18–6. Flexors of the thigh.

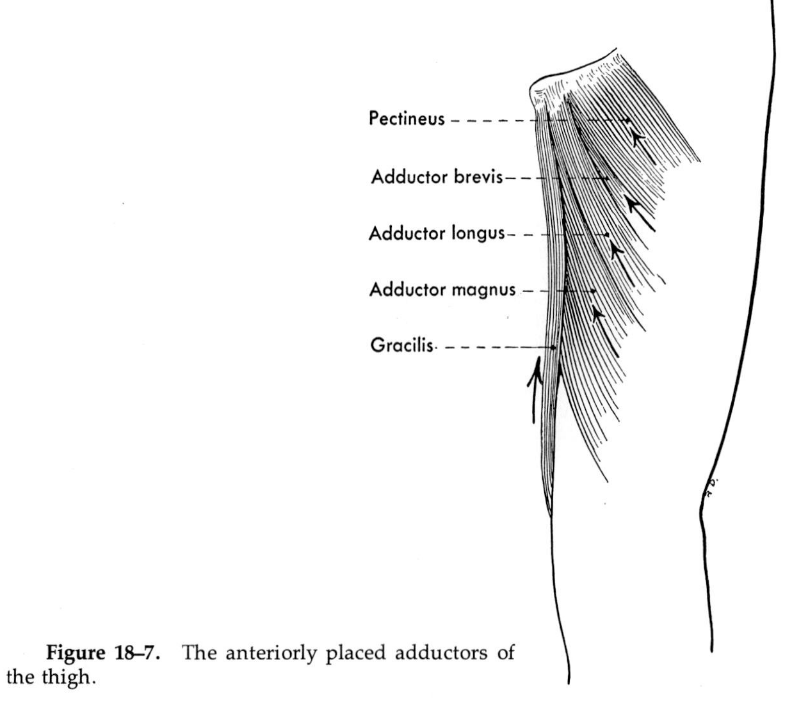

Figure 18–7. The anteriorly placed adductors of the thigh.

is because, as already stated, the muscle actually consists of two portions blended rather well anatomically but nevertheless fairly distinct in function. The portion arising from the ramus of the ischium, sweeping obliquely across to insert on the linea aspera, and innervated by the obturator nerve, acts with other members of the adductor group in flexing, adducting and internally rotating the femur. The second portion of the adductor magnus, arising from the ischial tuberosity, running downward to insert on and a little above the adductor tubercle, and innervated by the tibial portion of the sciatic nerve, is also an adductor and internal rotator, but acts with the hamstrings as an extensor of the thigh.

The adductors of the thigh (figs. 18-7 and 18-8) are the pectineus, the adductors longus, brevis, and obturator portion of the magnus (these are also flexors and, in general, the more posteriorly they arise, the more important they are as adductors, the less important as flexors); the gluteus maximus, quadratus femoris, and obturator externus; the hamstrings, including the sciatic part of the adductor magnus, and the gracilis; and, with the thigh flexed, the iliopsoas.

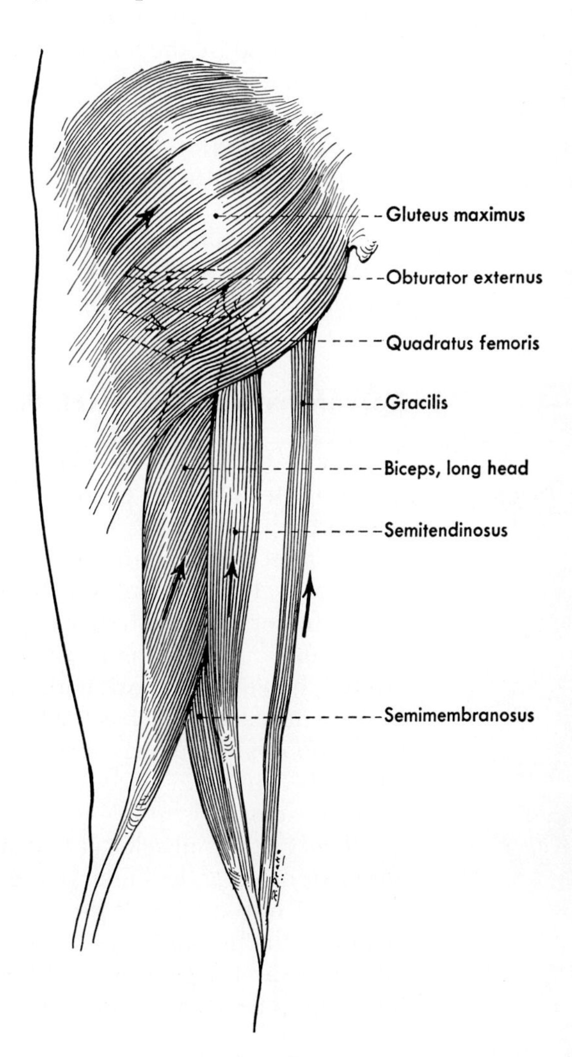

Figure 18-8. The posteriorly placed adductors of the thigh.

The chief abductors, namely, the gluteus medius and minimus and the tensor fasciae latae, are innervated by the same nerve, the superior gluteal. Thus injury to this single nerve will markedly affect stability of the pelvis and both abduction and to some slight extent (since the minimus and tensor are also internal rotators) internal rotation of the femur. The chief extensors, the gluteus maximus and hamstrings, are innervated from the sacral plexus but by separate nerves, the gluteus maximus being supplied by the inferior gluteal nerve while the hamstrings are supplied by the sciatic. The adductor group of muscles is innervated by the obturator nerve (except for the posterior part of the adductor magnus) and injury to this single nerve may much diminish the power of adduction, although the movement may still be carried out by muscles innervated through the sacral plexus.

The external rotators in the buttock are innervated by a number of branches from the sacral plexus; the iliopsoas is innervated by branches directly from the lumbar plexus and also by branches of the femoral nerve. Thus marked disturbance of external rotation by an isolated nerve injury is impossible. The same is true for flexion and internal rotation, for both the flexors and internal rotators of the hip are innervated by the femoral, obturator, and superior gluteal nerves; some of the internal rotators are also supplied by the sciatic nerve.

The segmental innervation of the various muscle groups is so diverse that limited lesions of the lumbar or sacral plexuses affect no one movement in particular. More extensive lesions of the lumbar plexus will affect the muscles supplied by the obturator and femoral nerves, thus especially flexion and adduction at the hip. Similar lesions of the sacral plexus will affect the gluteal and hamstring muscles, therefore especially extension and abduction.

MOVEMENTS AT THE KNEE JOINT

There is only one good extensor at the knee, namely, the quadriceps muscle. The four heads of the muscle insert upon the patella and are continued from this to the tibial tuberosity by the so-called patellar ligament which is, in actuality, the tendon of insertion of the quadriceps muscle. Since all four heads of the quadriceps are innervated by the femoral nerve, injury to this nerve will effectively prevent active extension at the knee against gravity. In walking slowly on level ground, however, the gait with a paralyzed quadriceps may be approximately normal; as long as the forward swing of the affected limb is not great enough to produce flexion at the knee, the limb is stabile, for the weight-bearing extended knee stays extended through the action of other muscles.

These other muscles at first sight would appear to have little to do with extension at the knee. In the weight-bearing limb, however, with the foot fixed, flexion at the knee can occur only when there is also flexion at the hip and dorsiflexion at the ankle. Thus the gluteus maximus, in extending the weight-bearing limb, helps also to extend the leg or keep it extended, and the gastrocnemius and soleus, muscles of the leg that plantar flex the foot, have a similar action as they resist dorsiflexion or promote plantar flexion of the foot.

The main part of the iliotibial tract passes just in front of the center of the knee joint on the lateral side of the knee, and therefore both the tensor fasciae latae and the gluteus maximus (fig. 18-9) have been said to exert an effect upon the knee through this. While the tensor fasciae latae apparently contracts during extension at the knee, stimulation of either it or the gluteus maximus, or pulling upon the iliotibial tract (which, it may be recalled, is anchored to the femur by the lateral intermuscular septum) will not produce extension. It thus seems that neither muscle should be described as acting at the knee through the tract, although why the tensor contracts during extension has not been explained.

Flexion at the knee (fig. 18-10) is brought about largely by the semitendinosus, semimembranosus, biceps femoris, gracilis, and sartorius. The long head of the biceps is said to function as a flexor only until the knee is semiflexed, becoming relaxed as flexion is carried further. In addition to these muscles of the thigh there are also muscles of the calf of the leg that extend across the knee joint and have

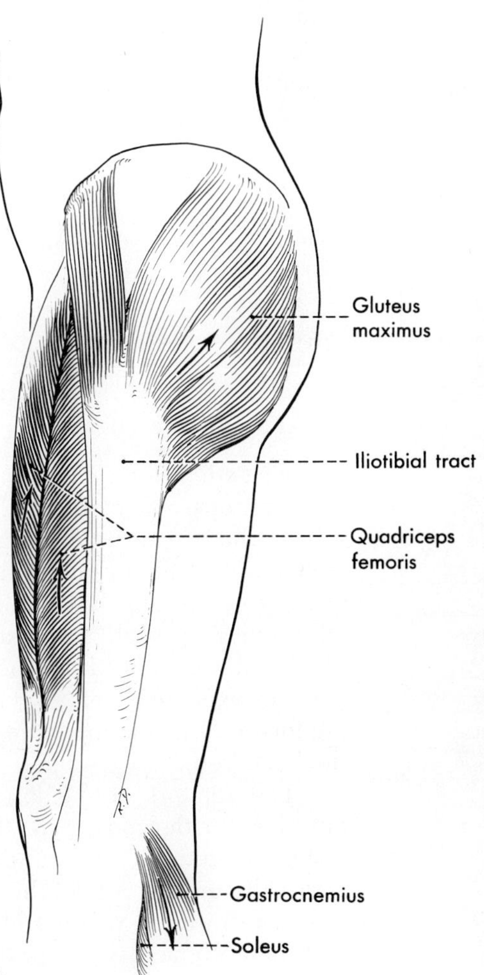

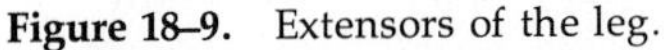
Figure 18–9. Extensors of the leg.

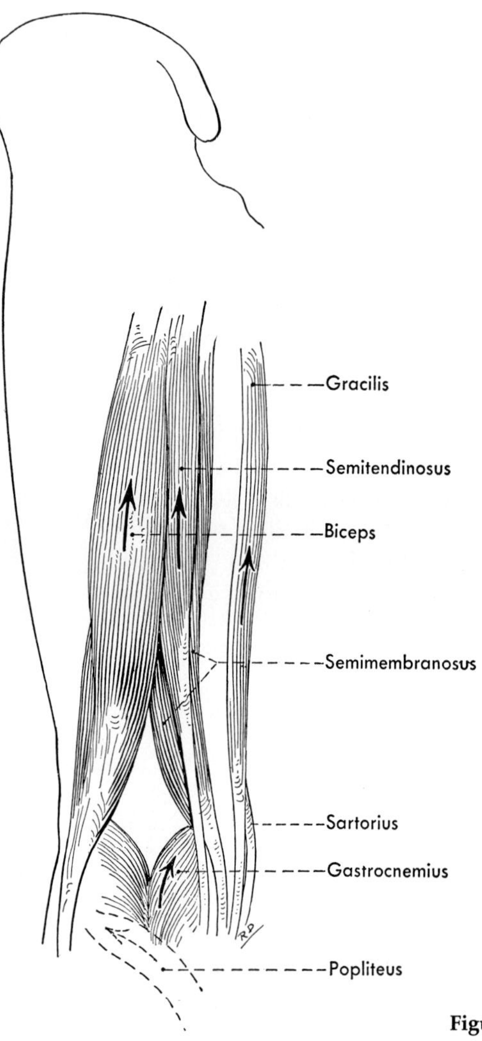

Figure 18–10. The flexors of the leg.

a flexor action here. These are the gastrocnemius, the plantaris, and the popliteus, especially the latter. The gastrocnemius (the plantaris is so tiny that it can be disregarded) is primarily a plantar flexor at the ankle, and as already noted helps to extend the knee when the leg is supporting weight, but when the leg is free it helps to flex it. The popliteus is a weak flexor; although it contracts at the beginning of flexion, its real contribution to this is probably that of rotation at the knee, as explained below. It also helps to stabilize the knee, by resisting forward movement of the femur on the tibia, when one stands with the knee bent.

The rotators at the knee (figs. 18-11 and 18-12) are the same muscles, for the most part, that have just been listed as flexors. The sartorius, gracilis, semitendinosus and semimembranosus all pass across the medial side of the knee joint to insert upon the tibia and are therefore internal rotators of the leg. The popliteus, passing distally and medially across the posterior aspect of the knee joint, is also an internal rotator. The biceps, passing laterally to insert on the fibula, is apparently the only external rotator of the leg.

Rotation of the leg is most free with the knee flexed, in which position the total range of movement, from full external to full internal rotation, may amount to about 40 degrees on the average. Most of the muscles concerned with rotation are at a mechanical disadvantage for this movement when the leg is extended, and even passive rotation in the extended position is very much limited by the tautness of the ligaments of the knee. In complete extension of the knee, however, with the foot implanted on the ground, there occurs during the last of this movement a slight posterior skidding of the medial femoral condyle on the corresponding tibial condyle, resulting in a slight medial rotation of the femur with further tightening of the collateral ligaments. If the leg is free, it is correspondingly rotated laterally or externally during the last part of extension. It is in reversing this terminal lateral rotation of the leg or, in the weight-bearing limb, the terminal medial rotation of the femur, that the popliteus has its chief function during flexion. Taking its fixed point from below, when the foot is firmly on the floor, it will externally rotate the femur on the fixed leg, and thus prepare the knee joint for flexion.

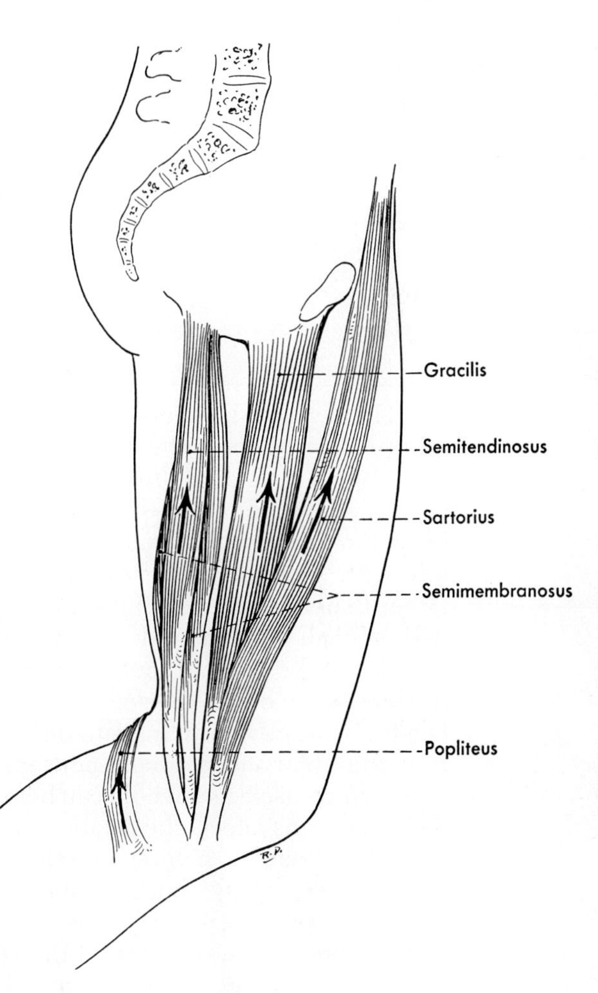

Figure 18–11. Internal rotators of the leg.

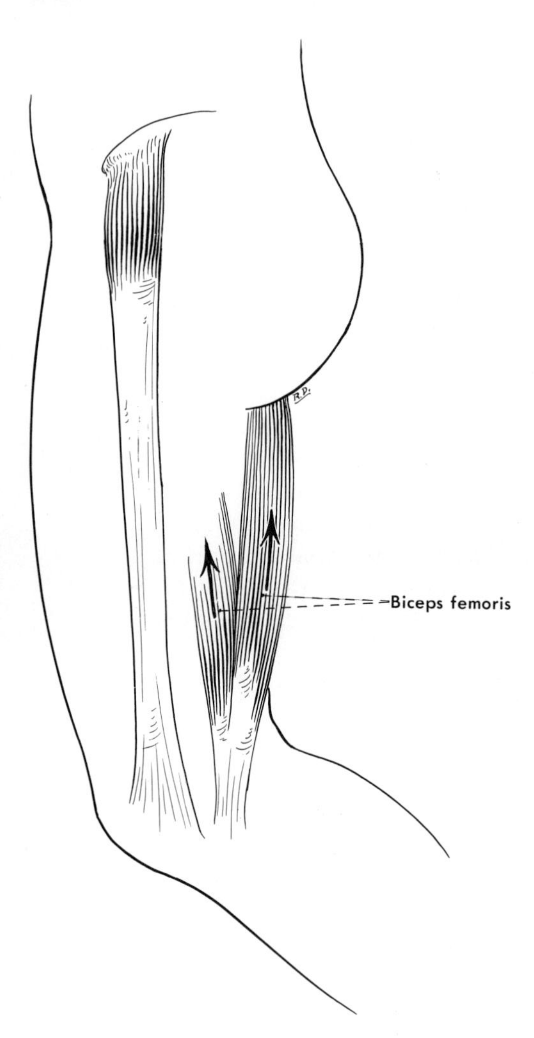

Figure 18–12. The external rotator of the leg. The iliotibial tract, also shown, stabilizes but apparently does not contribute to movement of the knee joint.

The hamstring flexors are innervated by the sciatic nerve; the sartorius by the femoral; the gracilis by the obturator; and the gastrocnemius, plantaris and popliteus by the tibial. As pointed out in a preceding paragraph, injury to the femoral nerve will affect extension at the knee. Similarly, severe injury to the sciatic nerve as a whole will interfere with maintenance of extension at the hip, since the hamstrings rather than the stronger gluteus maximus normally carry this out, and will abolish flexion and rotation at the knee (and also paralyze muscles of the leg and foot), resulting in a flail-like extremity. While the biceps femoris is the sole effective external rotator of the leg, both heads apparently participate in this action. Since separate nerves arise from the tibial and peroneal portions of the sciatic to supply the two heads, the muscle as a whole can be paralyzed by a single nerve lesion only when this affects the sciatic nerve. Other movements of the knee are carried out by several muscles acting together; the diversity of their nerve supply, both segmental and peripheral, is such that limited lesions of either the plexus or peripheral nerves will have no marked effect on movements at the knee.

THE MAINTENANCE OF STABILITY AT HIP AND KNEE JOINTS

The actions of the muscles across the hip and knee have now been considered, and reference has also been made, in the sections on these joints, to the parts played by some of the ligaments. While the following discussion must be then, at least in part, a repetition of what has already been said, the importance of the lower limb in supporting the body is such that the mechanics of the static limb deserve a summary here.

The entire weight of the body, in standing, is transmitted from the pelvis onto the heads of one or both femurs. Since the center of gravity of the erect body falls behind the center of movement of the head of the femur, the weight of the body tends to force the hip joint into hyperextension. The strong iliofemoral particularly, and the pubofemoral and ischiofemoral ligaments to a lesser extent, all resist hyperextension at the hip, therefore the hip can be maintained in the extended weight-bearing position with little or no muscular effort.

When the weight is supported on one limb only, the center of gravity of the body lies to the medial side of the head of the supporting femur, and therefore tends to force the unsupported side of the pelvis downward, a movement equivalent to adduction of the limb. The abductors then resist this movement, normally raising the unsupported side slightly above the horizontal. If there is marked drooping of the unsupported side, the iliotibial tract will exert a passive checking action on this.

With the weight on both limbs, the bracing action of the two femora and the ligamentous checking of hyperextension allow complete relaxation of the muscles about the hip, except for occasional slight contractions of the hamstrings and the iliopsoas that may be necessary to keep the body properly balanced. Thus the hip is equipped to support the weight of the body with little expenditure of energy.

At the knee, also, somewhat the same fundamental conditions hold true. At this joint, the center of weight distribution is anterior to the center of movement through the knee, so that the weight of the body also tends to keep this joint extended, once it has been completely extended by the quadriceps. Mention has already been made of the part that the gluteus maximus and muscles of the calf may play in extension when the limb is supporting weight. The gluteus maximus is relaxed during quiet standing, but since the calf muscles are partly contracted in order to prevent dorsiflexion at the ankle (for the center of weight, being anterior to this joint also, tends to force the foot into dorsiflexion), they tend also to keep the knee extended. Extension of the knee results also in a tightening of the collateral ligaments with further stabilization of this joint. When the weight is rested upon one leg only, the knee is usually further slightly extended; the terminal extension involves a medial rotation of the femur on the tibial condyles with consequent further tightening of the collateral ligaments and so-called locking of the joint. While this "locking" or "screwing home" of the knee in extension is often described as if it were an inevitable concomitant of weight-supporting extension, it might be noted that most persons, standing with the weight equally distributed between the two legs, actually do not completely extend the knee joints. Rather they stand with knees very slightly flexed, therefore avoiding the final locking of the joint, but with the weight so distributed on the condyles of the femur that the joint is easily maintained in this almost completely extended position.

In patients with torn collateral or cruciate ligaments, or both, the stability of the knee joint of necessity depends largely upon muscles. Under these circumstances, as already mentioned, it is especially important that the quadriceps be of good strength, and therapeutic exercises to achieve this end are often prescribed for such patients.

The Leg 19

One muscle of the leg crosses the knee joint only, and two cross both the knee and ankle (talocrural) joints. The remainder originate in the leg and therefore act at the ankle or both here and on a more distal part of the foot, including the toes.

The movements at the knee have already been discussed. Those at the ankle joint (called talocrural joint because it is formed by the bones of the leg—crus—and a single tarsal bone, the talus) are almost entirely limited to flexion and extension. There is no general agreement upon the use of the terms "flexion" and "extension" in regard to movement at the ankle. Some would define flexion as the act of standing upon one's toes, while others would call this extension, since it results in more nearly aligning the longitudinal axis of the foot with that of the leg. Similarly, drawing the toes up toward the leg so that one stands upon one's heel is sometimes described as flexion and sometimes as hyperextension.

While, from an anatomic and embryologic standpoint, the foot may be regarded as being normally in a position of hyperextension, it is preferable to avoid the unqualified terms "flexion" and "extension" in trying to describe movements at the ankle. The sole of the foot is usually referred to as the plantar surface, and the "top" of the foot is the dorsum. By use of these terms, then, it is possible so to qualify the terms of direction that there can be no misunderstanding concerning what movement at the ankle is being described. Thus the movement of rising upon the toes is plantar flexion of the foot; similarly, standing upon the heels is dorsiflexion of the foot.

Just as movement of the hand is not restricted to movement at the radiocarpal joint but occurs also among the carpals, so movement of the foot is not restricted to the talocrural joint but occurs also among the tarsals. Through certain joints among the tarsals the sole of the foot can be turned inward, as if to appose it to the sole of the opposite side, and this movement, occurring distal to the talocrural joint, is known as inversion. The movement in the opposite direction, turning the sole of the foot outward, occurs at the same joints and is known as eversion. Inversion is normally accompanied by adduction (a medial flexion of the anterior part of the foot on the posterior part) and eversion by abduction (lateral flexion); these combined movements are frequently referred to, particularly by clinicians, as supination and pronation respectively. (The terms varus and valgus, already defined as indicating inward and outward bending, respectively, are not used to describe movement, but position; thus a pronated or everted foot is in the valgus position, or is a pes valgus.)

There is not the confusion regarding flexion and extension of the toes that there is in regard to these movements of the foot. Flexion of the toes always means plantar flexion, and extension of the toes means straightening or dorsiflexing them. As is true of the fingers, abduction is spreading the toes apart, adduction bringing them together.

GENERAL ANATOMY

The muscles of the leg are conveniently divided into three groups: posterior muscles or muscles of the calf, lateral muscles, and anterior muscles. The posterior muscles are primarily plantar flexors at the ankle and flexors of the toes, the anterior ones dorsiflexors at the ankle and extensors of the toes, and the lateral ones evertors of the foot (certain of the anterior and posterior muscles invert the foot). Although all three groups of muscles are innervated by the sciatic nerve, they are innervated by different major branches of this nerve. The posterior muscles are innervated by the tibial nerve, the lateral and anterior ones by the superficial and deep peroneal nerves, respectively; the two peroneal nerves are branches of the common peroneal, like the tibial a terminal branch of the sciatic.

The tibia and fibula (fig. 19-1), the bones of the leg, have been studied in part in connection with the thigh and knee, but should be studied again in connection with the leg. The slender **fibula** presents at its upper end a head that bears an articular surface for the synovial joint between it and the tibia, and has a pointed apex. Most of the body is so twisted and so marked by the attachments of muscles that its surfaces and borders are difficult to follow; its sharp edge, however, is its interosseous border, directed in general medially. At its lower end the fibula enlarges to form the lateral malleolus. This bears medially an articular surface through which the fibula enters into the lateral side of the ankle joint.

The upper end of the **tibia,** the much heavier medial bone of the leg, has already been described (p. 242). In summary, the medial and lateral condyles present articular surfaces for the femoral condyles; these articular surfaces are separated from each other by the nonarticular intercondylar areas and intercondylar eminence, and the lateral condyle bears on its posteroinferior aspect an articular surface for the fibula. The body of the tibia is somewhat triangular; its anterior border, marked above by the tibial tuberosity for the attachment of the patellar ligament, is subcutaneous, as is its medial surface. Its interosseous border faces laterally, toward the interosseous border of the fibula. Close to its lower end the tibia expands to form the medial malleolus. On the lateral surface of the downward-projecting malleolus is an articular surface that enters into the medial side of the ankle (talocrural) joint. This is continuous with the inferior articular surface of the tibia, on the lower end of that bone, through which weight is transmitted from the leg to the foot.

The tibia and fibula are united above by the tibiofibular joint (p.

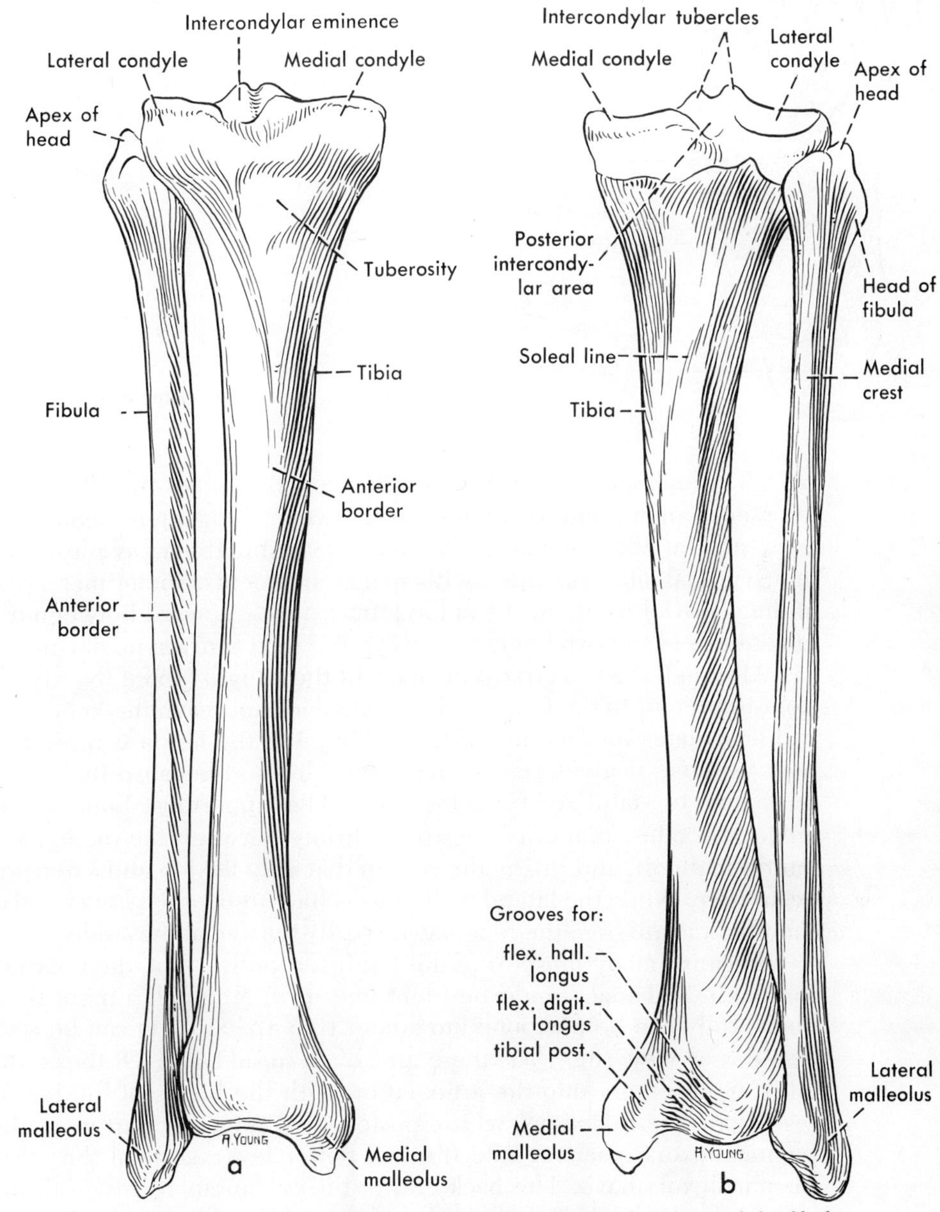

Figure 19–1. Anterior *(a)* and posterior *(b)* views of the tibia and the fibula.

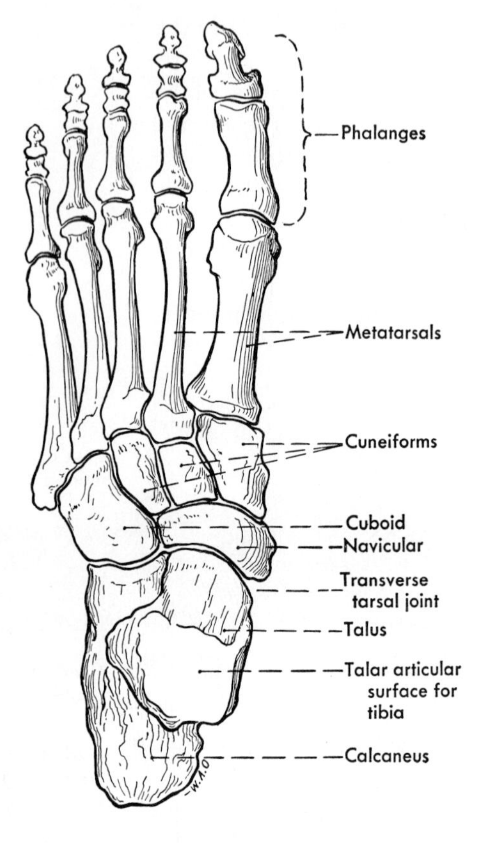

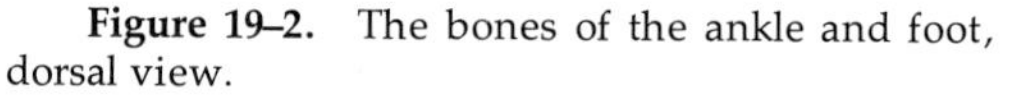
Figure 19–2. The bones of the ankle and foot, dorsal view.

247). Throughout most of their length they are united by a heavy interosseous membrane that stretches between their interosseous borders, and at their lower ends there is a tibiofibular syndesmosis (nonsynovial joint) provided with special ligaments. Sometimes a part of the synovial cavity of the ankle joint extends upward between the lower ends of the two bones to convert the joint into a synovial one.

Although the tibia transmits most of the weight (about five sixths) from the femur to the foot, for the fibula does not reach the knee joint and articulates on the side of the ankle joint, the fibula is important both because it gives rise to many muscles and because its help is necessary to stabilize the ankle joint. The uppermost bone at the ankle, the talus, is normally gripped firmly between the medial and lateral malleoli, and this is the reason that only flexion and extension occur here. When the lateral malleolus is fractured, or the lower end of the fibula is missing, the foot may be badly twisted at the ankle.

Because many muscles of the leg insert on the foot, the bones of the ankle and foot should be identified now, although a more thorough study can be postponed to a later time (p. 316). As can be seen in a dorsal view (fig. 19-2) there are seven **tarsal bones.** Of these, the talus alone enters into the articulation with the bones of the leg. It rests upon the calcaneus, whose posteriorly projecting part gives the plantar flexors inserting into it much better leverage than the other plantar flexors have. The back end of the calcaneus receives all the weight on the heel; its front end, and that of the talus, are higher and normally not in contact with the ground, for they are lateral and medial parts of the arch of the foot. Between the talus and the calcaneus is the important subtalar joint, and at the front ends of both bones, between them and the more distal ones, is the important transverse tarsal joint. These joints are discussed in the chapter on the foot; they need mention here because it is they that allow inversion and adduction (supination), and eversion and abduction (pronation). The transverse tarsal joint also allows additional plantar flexion and dorsiflexion.

Distal to the calcaneus is the cuboid bone. Immediately distal to the talus is the navicular bone, and distal to that are three cuneiforms, medial, intermediate, and lateral. The cuboid and the cuneiforms articulate with the metatarsals, essentially similar to the metacarpals. These in turn articulate with the proximal phalanges of the toes. Except in size, the phalanges of the toes are similar to those of the fingers and thumb. Because of the arch of the foot, the weight in standing is normally transmitted to the ground only through the back end of the calcaneus and the heads (distal ends) of the metatarsals. The toes, especially the big toe (hallux), participate in the thrust in walking, when the weight is shifted forward onto the ball of the foot.

SUPERFICIAL NERVES AND VESSELS; FASCIA

Posteriorly, the skin of the leg is supplied by branches from both the tibial and common peroneal nerves; these branches usually unite to form the **sural nerve** which continues to the lateral side of the foot. Medially, the **saphenous nerve,** from the femoral, supplies the leg and continues onto the foot. Anterolaterally, the upper part of the leg is supplied by a branch of the common peroneal that may arise with or separate from the branch that joins the sural, and the lower part is supplied by the **superficial peroneal nerve,** which also continues onto the foot.

Two large superficial veins are located in the leg. The **great (or greater) saphenous vein,** beginning on the medial border of the foot, runs along the medial side of the leg on up into the thigh where it has already been identified. The **small (or lesser) saphenous vein** begins on the lateral border of the foot and runs up the posterior surface of the leg, communicating with the great saphenous and sometimes joining that vein in the thigh but usually ending by penetrating the deep fascia in the hollow behind the knee (popliteal fossa) and joining the large deep vein here, the popliteal vein.

The deep fascia of the leg (crural fascia) resembles the deep fascia found elsewhere on the limbs; it is a tough fibrous layer that in its upper part gives origin to some of the musculature. It blends with the periosteum of the subcutaneous part of the tibia throughout most of the length of the leg and laterally sends two septa to the fibula; the anterior one of these separates the anterior from the lateral muscles of the leg, the posterior one separates the lateral muscles from the posterior ones. Thus each of the three groups of muscles lies in its own compartment. In addition, the fascia of the calf gives rise to another septum that passes across the calf and separates a superficial group of calf muscles from a deep group. Both the anterior and the deep posterior groups of muscles lie in such tight compartments that any trauma to them that produces swelling interferes very quickly with their circulation; this in turn can lead to their rapid degeneration.

Close to the ankle the crural fascia is thickened by more or less transverse fibers to form retinacula, similar to those at the wrist, that hold the tendons of the muscles of the leg close to the bones as they cross the ankle. Since there are three sets of muscles of the leg, there are three sets of retinacula. The **flexor retinaculum** lies posteromedially (figs. 19-3 and 19-5) and extends between the medial malleolus and the calcaneus. The tendons of the deep muscles of the calf pass deep to it, as do also the nerve and vessels that continue from the calf into the sole of the foot. There are two lateral or **peroneal retinacula** (fig. 19-7; perone refers to fibula); the superior peroneal retinaculum passes between the lateral malleolus and the calcaneus, the inferior is attached at both its ends to the calcaneus. The tendons of the two lateral muscles of the leg (peroneal muscles) pass deep to these retinacula. The anterior or **extensor retinacula** are also two (figs. 19-7 and 19-8); the superior one is a not very distinct transverse thickening that passes between the tibia and fibula above the malleoli. The inferior, once called the "cruciate ligament," resembles a Y lying on its side. The stem of the Y is attached to the lateral and dorsal surfaces of the calcaneus; as it is traced medially across the dorsum of the foot its two limbs diverge, the upper one going to the medial malleolus, the lower one blending with the fascia of the medial side of the foot. The tendons of the anterior muscles of the leg pass deep to the superior and through the inferior retinaculum.

As the posterior tendons pass through the flexor retinaculum, and as the anterior ones pass through the inferior extensor retinaculum, most of them lie in separate compartments, each lined with a synovial membrane. These membranes form tendon sheaths that extend for varying distances above and below the retinacula. The two lateral muscles pass together through a single compartment and have a common sheath deep to the superior peroneal retinaculum, but the sheath divides into a part around each tendon, separated by a septum, deep to the inferior retinaculum.

MUSCLES OF THE CALF OF THE LEG

The muscles of the calf are divisible into a superficial and a deep group, each lying in its own fascial compartment. The muscles of the superficial group are the gastrocnemius, the soleus, and the plantaris. They all insert on the back end of the calcaneus, and the gastrocnemius and soleus share the same tendon of insertion. In addition to being separately named, these two muscles are also grouped together and called the **triceps surae.**

The **gastrocnemius,** the superficial member of the triceps and the most superficial muscle of the calf (fig. 19-3), arises by medial and lateral heads from the corresponding epicondyles of the femur. These two heads quickly unite to form the bulk of the muscle. Approximately half way down the calf the gastrocnemius ends in a flat tendon that receives on its deep or anterior surface the attachment of the next underlying muscle, the soleus, and then, becoming more rounded, proceeds downward as the tendo calcaneus or tendon of Achilles to attach to the projecting back portion of the calcaneus. The tendon attaches to the lower part of the posterior surface, and between it and the upper part there is usually a bursa. The gastrocnemius is a powerful plantar flexor of the foot, and can help flex the knee when the leg is not supporting weight. With the foot fixed in weight-bearing, however, its posterior position at the ankle allows it to resist dorsiflexion here, and since flexion of the weight-bearing limb at the knee cannot occur without concomitant dorsiflexion at the ankle, it helps maintain extension at the knee. It is innervated by the tibial nerve, which passes between its two heads to a deeper position in the calf in company with the large vessels (popliteal vessels) that run from the thigh to the leg.

The **soleus** muscle lies immediately deep to the gastrocnemius (fig. 19-4) and also has two heads. One arises from an oblique line (soleal or popliteal line) across the upper third of the posterior surface of the tibia, and from the middle third of the medial border; the other arises from the upper third of the posterior surface of the fibula. Its

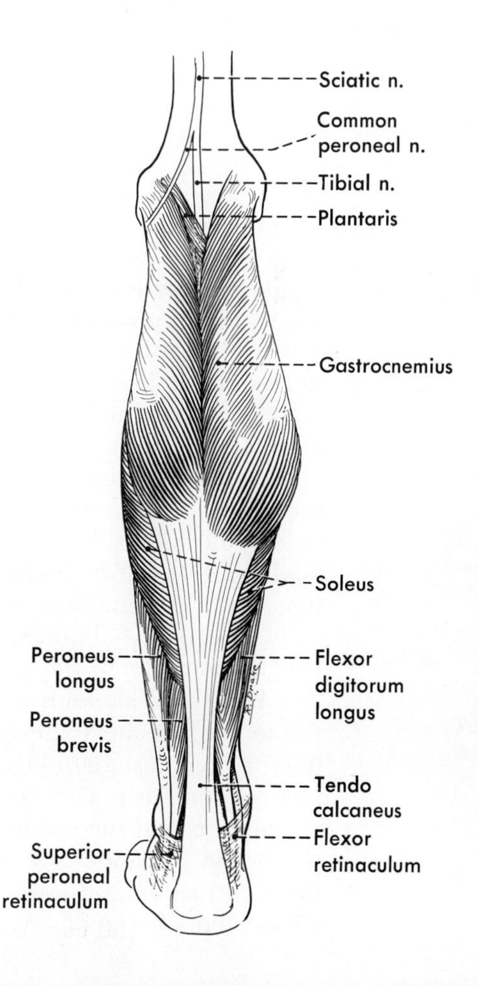

Figure 19–3. Musculature of the calf of the leg.

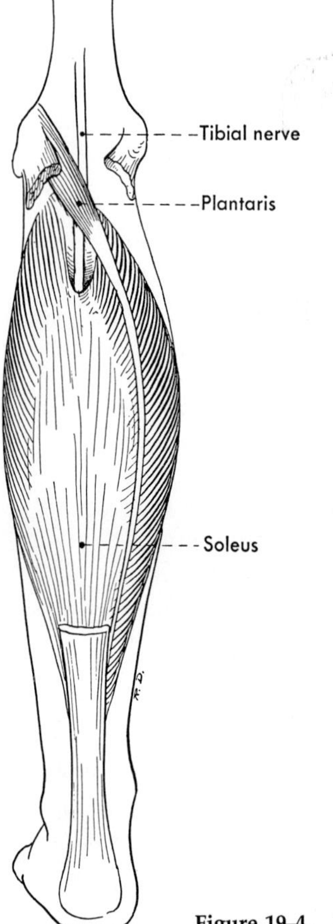

Figure 19-4. Musculature of the calf, second layer.

two heads unite so that the muscle forms a septum across the calf of the leg, but between the heads the tibial nerve and the popliteal vessels pass deeply. The soleus inserts on the deep (anterior) surface of the tendo calcaneus. Since it has no attachment across the knee joint it does not share the function of flexion of the knee with the gastrocnemius. It obviously works with this muscle, however, in plantar flexion of the foot, and completes plantar flexion when the knee is flexed and the gastrocnemius is at a disadvantage. Taking its fixed point from below, the soleus, like the gastrocnemius, prevents dorsiflexion at the ankle and therefore flexion at the knee; it is usually the soleus rather than the gastrocnemius that does this during quiet standing. It typically receives two branches from the tibial nerve, one into its superficial and one into its deep surface.

The small **plantaris** muscle arises from the lateral epicondyle of the femur just above the attachment of the lateral head of the gastrocnemius. Its muscular belly is approximately 2 to 4 inches long, but its slender tendon is very long and passes downward between the gastrocnemius and the soleus to insert on the calcaneus anteromedial to the tendo calcaneus, or sometimes to fuse with that tendon or the flexor retinaculum. In some animals the plantaris is a powerful muscle whose tendon extends over the heel onto the plantar surface of the foot, becoming continuous there with the heavy plantar aponeurosis; thus the plantaris is the equivalent of the palmaris longus of the forearm and hand.

The deep muscles of the calf (fig. 19-5) are the popliteus, a muscle behind the knee, and three muscles that arise from the tibia and fibula and continue deep to the flexor retinaculum into the foot: the flexor hallucis longus, the flexor digitorum longus, and the tibialis posterior. These muscles are separated from the superficial group by a fascial septum that extends across the leg from the deep fascia and is reinforced in its upper part by a slip from the semimembranosus tendon. Below the popliteus this fascia helps give origin to the deep muscles, but lower in the leg it covers them more loosely, and it is this layer that is thickened at the ankle to form the flexor retinaculum. It is pierced in the upper part of the leg by the tibial nerve and the popliteal vessels, the continuations of which therefore lie deep to the fascia and among the deep muscles.

The **popliteus** muscle forms part of the floor of the popliteal fossa; its tendon, variably described as its tendon of insertion or origin, attaches within the fibrous capsule of the knee joint to the lateral epicondyle of the femur, and runs posteriorly and medially between the fibrous and synovial layers of the joint. The muscle also has attachments to the lateral meniscus and to the arcuate popliteal liga-

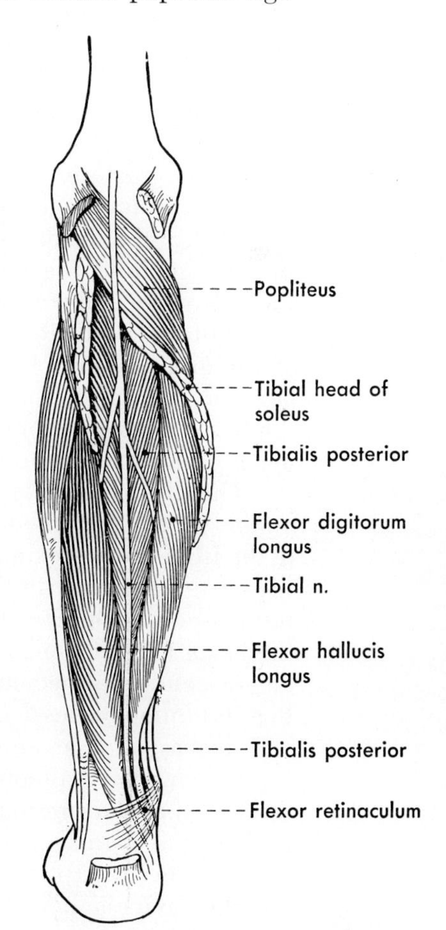

Figure 19–5. The deep muscles of the calf.

ment. It emerges through a gap in the posterior part of the capsule below the arcuate ligament and runs obliquely across the posterior aspect of the knee joint to attach to about the upper third of the posterior surface of the tibia, above and medial to the tibial origin of the soleus. The popliteus is an internal rotator of the leg on the femur or an external rotator of the femur upon the leg, and an unimportant flexor at the knee; however, in a subject standing with the knee partly flexed, it contracts to help prevent forward displacement of the femur on the tibia. It is innervated by a branch of the tibial nerve that typically passes around the lower edge of the muscle to penetrate the muscle on its deep (anterior) surface.

The three remaining muscles of the calf cover the posterior aspect of the tibia, fibula, and interosseous membrane. They are all innervated by the tibial nerve. The **flexor hallucis longus,** the most lateral, arises from the lateral side of about the middle half of the posterior aspect of the fibula; its tendon of insertion begins just above the ankle and passes obliquely downward, medially and forward to enter the foot deep to the flexor retinaculum. The tendon of the flexor hallucis longus is the most posterolateral of the three tendons behind the medial malleolus; in the foot, it runs forward to an insertion upon the distal phalanx of the big toe. The muscle is primarily a flexor of the big toe, as its name implies; it is a very weak plantar flexor of the ankle.

The most medial of the three muscles is the **flexor digitorum longus,** which arises from about the middle third of the posterior aspect of the tibia. As this muscle passes downward it is crossed on its anterior surface by the tibialis posterior, and therefore at the ankle the tendon of the flexor digitorum longus lies between those of the tibialis posterior and flexor hallucis longus. The tendon of the flexor digitorum longus begins somewhat higher above the ankle than does that of the flexor hallucis, passes through the flexor retinaculum and thus behind the medial malleolus, and in the sole of the foot spreads out to attain an attachment to the distal phalanges of the four lateral digits. In the foot these tendons are associated with the origins of the lumbrical muscles, and pass through the divided tendons of the flexor digitorum brevis, a muscle located entirely in the foot. Thus the flexor digitorum longus of the leg corresponds quite exactly to the flexor digitorum profundus of the forearm.

The third muscle of the group, the **tibialis posterior,** arises from the upper two thirds of the lateral side of the posterior surface of the body of the tibia, an approximately corresponding part of the medial part of the posterior surface of the fibula, and from the posterior aspect of the interosseous membrane between these two bones; it thus lies between the flexor hallucis longus and the flexor digitorum longus. As the muscle descends and becomes tendinous it runs medially and forward so that its tendon passes deep to the tendon of the flexor digitorum longus, and thus assumes an anterior rather than an intermediate position behind the medial malleolus. The tendon of the tibialis posterior passes through the flexor retinaculum in its own compartment, grooving the back of the medial malleolus as it does so, and then passes to the medial side of the plantar surface of the foot where it inserts primarily into the navicular bone but extends also to attach, with some variation, to all the other tarsals except the talus, and to the bases of the second, third, and fourth metatarsals. This muscle is an adductor of the fore part of the foot, and aids in inversion and plantar flexion.

NERVES AND VESSELS OF THE CALF

The **tibial nerve** leaves the sciatic as its larger and medial terminal branch in the popliteal fossa. It is, in direction and size, the more direct continuation of the sciatic. It leaves the popliteal fossa by passing between the two heads of the gastrocnemius and then almost immediately passes deep to the soleus to lie between this muscle and the tibialis posterior, deep to the fascia covering the deep group of muscles of the leg. It runs downward on the tibialis posterior in company with the posterior tibial vessels (fig. 19-6); at the ankle it lies between the tendons of the flexor hallucis longus and flexor digitorum longus. It supplies all the muscles in the calf of the leg. As it passes through the flexor retinaculum to reach the plantar surface of the foot it divides into medial and lateral plantar nerves.

The **popliteal artery** is the direct continuation of the femoral artery, which changes its name as it passes from the anteromedial to the posterior aspect of the thigh through the gap in the adductor magnus tendon. The popliteal artery gives off paired superior and inferior genicular branches that anastomose with each other about the knee, and an unpaired middle genicular artery that enters the knee joint; it then divides, usually on the posterior surface of the popliteus muscle, into anterior and posterior tibial arteries.

The **anterior tibial artery** passes forward between the tibia and fibula through a gap at the upper end of the interosseous membrane to attain the anterolateral aspect of the leg; its course there is described later. The **posterior tibial artery** takes the same course down the calf as does the tibial nerve; it gives off muscular branches in the calf and as it passes through the flexor retinaculum in the same compartment as the nerve it divides, as the nerve does, into medial and lateral plantar branches. In addition to its muscular branches and the nutrient artery to the tibia, the posterior tibial gives off one large branch in the leg, the **peroneal artery.** This artery arises high, from the upper end of the posterior tibial. It runs downward on the lateral side of the leg deep to or in the flexor hallucis longus, lying close to the interosseous membrane and the fibula. At the ankle the artery communicates with the posterior tibial, supplies branches about the ankle, and gives rise to a perforating branch that passes forward between the tibia and the fibula to reach the dorsum of the foot.

The tibial and the peroneal arteries are usually each accompanied by two correspondingly named veins, which unite in a variable pattern to form a single popliteal vein. Usually the popliteal vein receives also the small saphenous vein; before the popliteal vein passes through the adductor hiatus to continue as the femoral vein, there is frequently a communication with perforating branches of the profunda femoris vein.

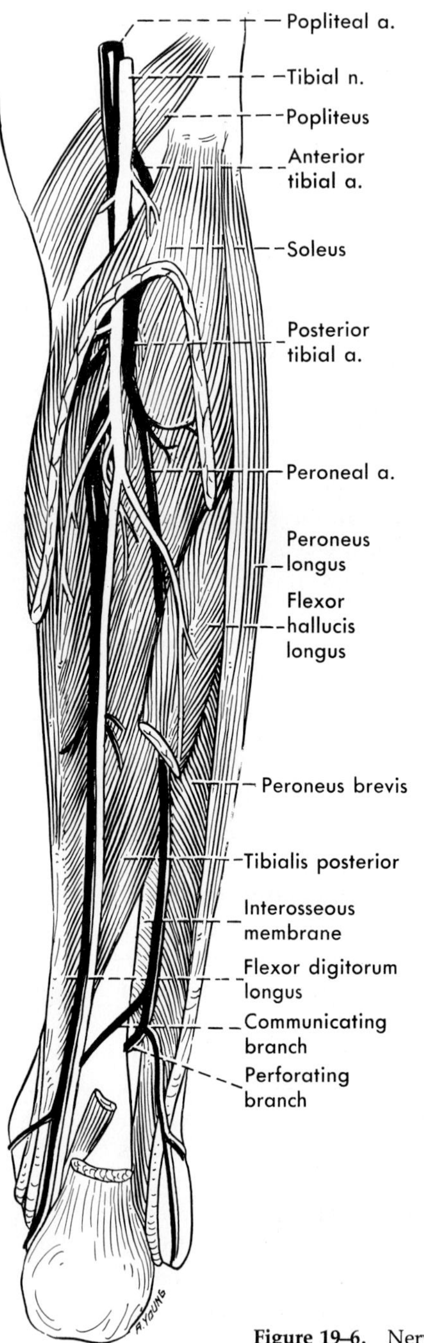

Figure 19-6. Nerves and arteries of the calf.

ANTEROLATERAL MUSCLES OF THE LEG

The lateral muscles are two, the peroneus longus and brevis, and the anterior ones four, the extensor digitorum longus, the closely associated peroneus tertius, the extensor hallucis longus, and the tibialis anterior. The **peroneus longus** and the **peroneus brevis** (fig. 19-7) both arise from the lateral surface of the fibula, and from the fascia about them; the peroneus longus is the more superficial muscle, arising from about the upper half of the bone, while the brevis arises from much of the lower half. The two muscles pass downward in close conjunction; they enter a common synovial tendon sheath above the ankle and pass behind the lateral malleolus and deep to the two peroneal retinacula onto the lateral border of the foot. Here the two tendons diverge. The tendon of the peroneus brevis passes to an attachment on the dorsal part of the base of the fifth metatarsal bone, while the tendon of the peroneus longus rounds the lateral border of the foot to run in a deep position across the sole of the foot and attach to the base of the first metatarsal and the adjacent medial cuneiform bone. Both these muscles are usually innervated by the superficial peroneal nerve, but the peroneus longus often receives also a branch from either the common or the deep peroneal. They are similar in function, being good evertors of the foot and weak plantar flexors. The common peroneal nerve, or the superficial and deep peroneal nerves together, pass deep to the upper part of the peroneus longus between it and the

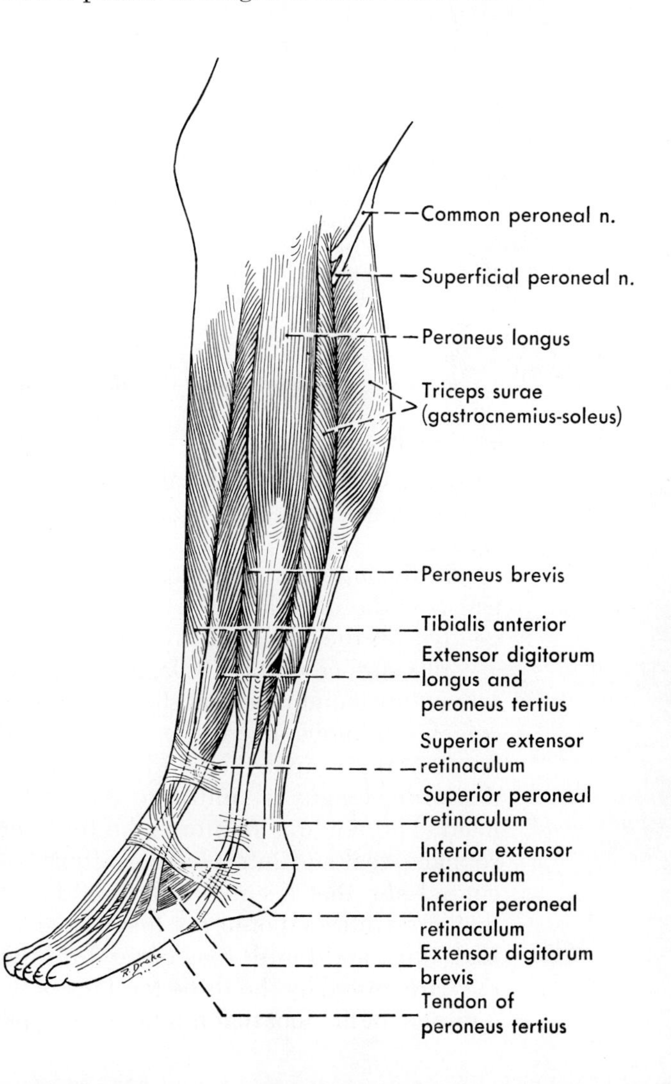

Figure 19–7. The lateral muscles of the leg.

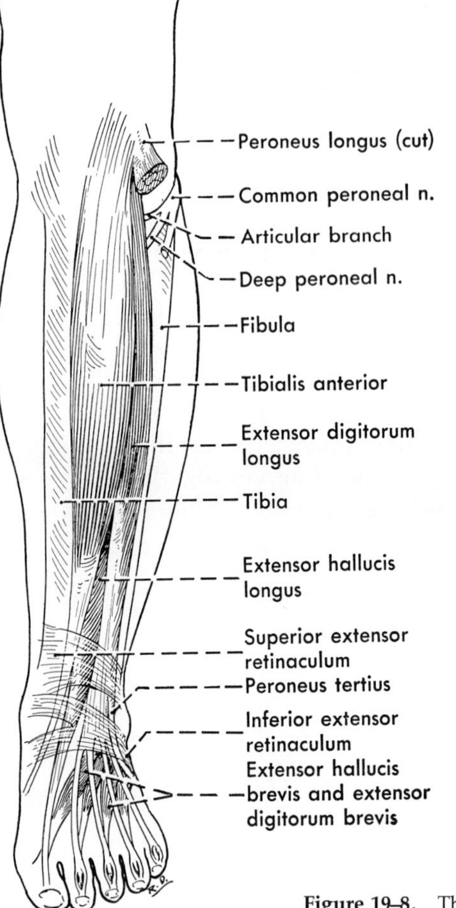

Figure 19–8. The anterior muscles of the leg.

fibula. The deep peroneal continues forward into the anterior muscles, but the superficial peroneal runs downward between the two peroneal muscles to become subcutaneous in the lower third of the leg.

Of the anterior muscles of the leg (fig. 19-8), the **extensor digitorum longus** is the most lateral. It arises from a small portion of the lateral condyle of the tibia and from about the upper three fourths of the anterior surface of the fibula, and has also some attachment to the interosseous membrane and the covering crural fascia. The muscle becomes tendinous above the ankle, and its tendon then divides into four that insert upon the four lateral toes. The insertion of the extensor digitorum longus is similar to that of the extensor digitorum in the hand. The tendons reinforce the thin dorsal capsules of the metatarsophalangeal and interphalangeal joints. Each tendon divides into a central slip that inserts on the middle phalanx and two lateral slips that are continued onto the distal phalanx. As the tendons to the second, third and fourth toes expand on the metatarsophalangeal joints they are joined by the three tendons of the extensor digitorum brevis, a muscle of the foot that has no counterpart in the normal hand.

Closely associated with the extensor digitorum longus, and continuous at its origin from the fibula with the lower fibers of origin of the longus, is the **peroneus tertius.** The fibers of this muscle end in a tendon that seems to be a fifth, most lateral, tendon of the extensor digitorum longus on the dorsum of the foot; however, instead of passing to the toes it attaches to the dorsal surface of the fifth metatarsal bone close to its base. The muscle varies considerably in size and is sometimes absent.

The extensor digitorum longus is of course an extensor of the toes and is a dorsiflexor (extensor) and weak evertor of the foot. The peroneus tertius also dorsiflexes and everts, but is a better evertor than is the extensor. Both muscles are innervated by the deep peroneal nerve (as are the other anterior muscles); this nerve passes deep to the extensor before turning downward in the leg.

The **extensor hallucis longus** is, at its origin, covered by the extensor digitorum longus and a more medial muscle, the tibialis anterior. Arising from approximately the middle third of the anterior surface of the fibula and the adjacent interosseous membrane it appears between the extensor digitorum longus and the tibialis anterior muscles somewhat above the ankle; it runs deep to the superior and through the inferior extensor retinaculum and goes to an insertion on the distal phalanx of the big toe. The extensor hallucis longus is an extensor of the big toe and a weak dorsiflexor at the ankle and invertor of the foot. The deep peroneal nerve supplies it and runs downward with the anterior tibial vessels, between it and the tibialis anterior muscle; nerve and artery pass deep to its lower part in the leg to lie lateral to it on the foot.

The **tibialis anterior** arises from much of the lateral surface of the tibia and from the deep fascia of the leg and the interosseous membrane. After passing deep to the superior and through the inferior extensor retinaculum the tendon of this muscle goes to the medial side of the foot to attach to the medial cuneiform and the first metatarsal, almost over onto the sole of the foot. The tibialis anterior is the strongest invertor and dorsiflexor of the foot. It receives several branches from the deep peroneal nerve.

NERVES AND VESSELS OF THE ANTEROLATERAL ASPECT OF THE LEG

The **common peroneal nerve,** the lateral terminal branch of the sciatic in the popliteal fossa, runs laterally around the leg. It is subcutaneous just below the head of the fibula, in which position it is easily damaged. As it passes between the fibula and the peroneus longus muscle it divides into two branches, or sometimes three—superficial and deep peroneal nerves, and an articular branch to the knee joint that usually arises from the deep peroneal but may arise with both the deep and the superficial as a terminal branch of the common peroneal. The **superficial peroneal nerve** runs downward between the peronei, supplying both, and becomes subcutaneous by

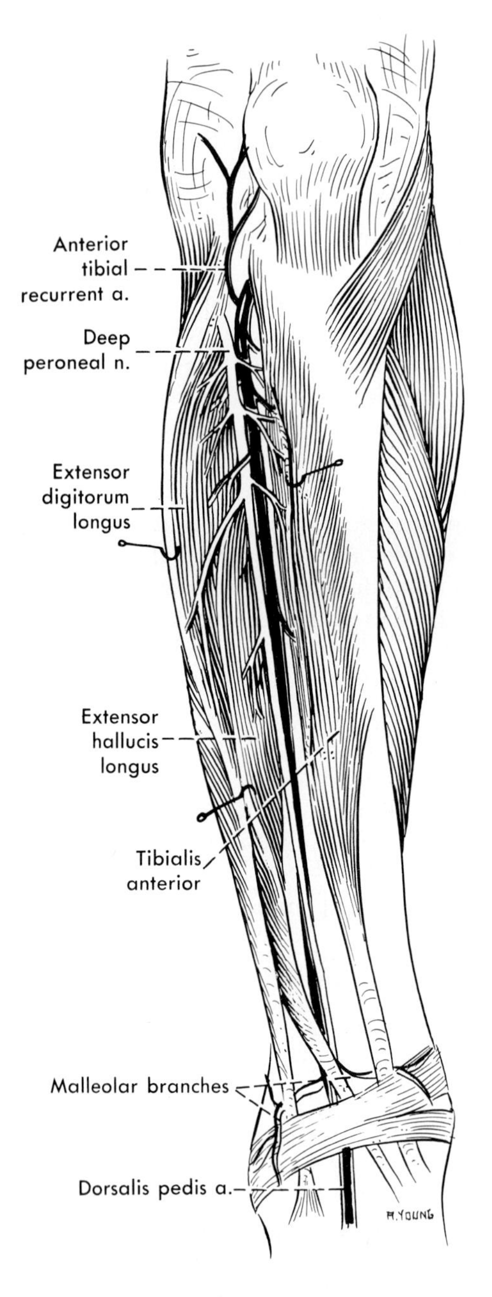

Figure 19–9. The deep peroneal nerve and the anterior tibial artery in the front of the leg.

emerging at the anterior border of these muscles near the middle of the leg. Its distribution upon the foot will be described later.

The **deep peroneal nerve** passes forward deep to the peroneus longus into the anterior muscles of the leg. It gives off branches to all these muscles and courses downward close to the interosseous membrane, in company with the anterior tibial artery (fig. 19-9) to be continued onto the dorsum of the foot.

The **anterior tibial artery** arises in the upper portion of the leg as one of the two terminal branches of the popliteal artery; it passes between the tibia and fibula above the upper edge of the interosseous membrane to attain a position on the anterior surface of this membrane, where it first gives off a recurrent branch to the knee. As it runs downward with the deep peroneal nerve (which crosses from its lateral to its medial side) it supplies the four anterior muscles and, after giving off branches around the ankle, continues onto the dorsum of the foot as the dorsalis pedis artery.

MOVEMENTS OF THE FOOT

Plantar flexion is due (fig. 19-10) primarily to the actions of the powerful gastrocnemius and soleus muscles (triceps surae) on the calcaneus. The other three muscles in the calf of the leg that send their tendons posteriorly around the ankle joint have usually been described as aiding in this plantar flexion, and indeed as playing a very important part therein. Similarly, the peroneus longus and brevis can be plantar flexors. These muscles all have much poorer leverage, however, than does the triceps, for the latter uses the posterior part of the calcaneus as its lever arm while the other muscles pass close to the malleoli. As a result, even the best ones have only about a fifth of the efficiency of the triceps. Because of the shortness of that muscle's

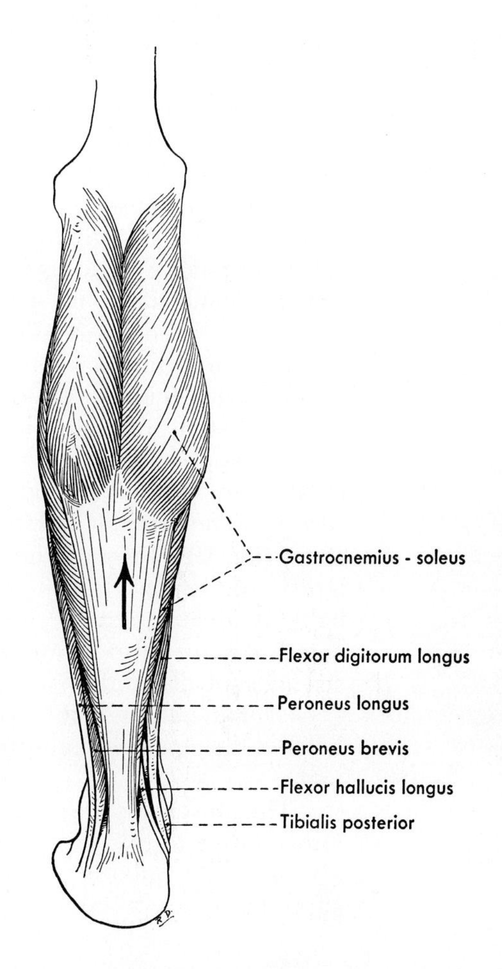

Figure 19–10. The plantar flexors of the foot; all the potential plantar flexors are shown, although some of them (see text) may contribute little or nothing to plantar flexion.

lever arm, it must exert a pull of about 200 lb. in order to plantar flex the foot against a weight of 100 lb. Even the triceps, therefore, is at a disadvantage. Further, these muscles combined are of much less strength than is the triceps, and calculations taking into account both their relative strength and their efficiency indicate that they can produce a pull of only about 5 to perhaps 15 per cent of that of the triceps. With loss of the action of the triceps surae, walking becomes less smooth because of the lack of propulsion during "push-off." In addition, it has been shown that in many individuals the peroneus longus and brevis are not normally used to produce plantar flexion of the foot. Among individuals in whom the tibial nerve had been destroyed, thus paralyzing all potential plantar flexors at the ankle except the peroneus longus and brevis, less than half were able to plantar flex the unopposed foot at all. As the peronei were used when the patients were asked to evert their feet, there was obviously no weakness or disability of these muscles. It seems that some persons are accustomed to using the peroneus longus and brevis primarily as evertors of the foot, and may not have learned to use them as plantar flexors. This can be learned, however, and in irreparable paralysis of the triceps surae the peroneus longus is sometimes attached to the calcaneus to give it better leverage for plantar flexion.

Dorsiflexion is due (fig. 19-11) to the actions of all the muscles crossing the front of the ankle; the tibialis anterior is the most important muscle involved in this action, but the extensor digitorum longus and its associated peroneus tertius assist, and the extensor hallucis longus can contribute weakly. When the tibialis anterior is paralyzed, the other muscles necessarily contract more strongly to dorsiflex the foot, and the extensor hallucis longus therefore dorsiflexes the big toe. The extensor digitorum longus may dorsiflex the other toes, but this is not as noticeable because its action is primarily on the proximal phalanges and these are usually maintained in dorsiflexion anyway. Dorsiflexion with paralysis of the tibialis anterior may be accompanied by eversion of the foot because the peroneus tertius and the extensor digitorum longus, especially the former, may evert more strongly than the extensor hallucis longus inverts.

Inversion and eversion of the foot necessarily involve little movement at the hingelike talocrural joint, almost all of this movement occurring at the subtalar and transverse tarsal joints. Inversion is brought about (fig. 19-12) by all the muscles passing to or around the medial border of the foot. Thus, posteriorly, the tibialis posterior, the flexor hallucis longus, and the flexor digitorum longus all invert the foot; anteriorly, the tibialis anterior is a strong invertor of the foot—the strongest of all—and the extensor hallucis longus is a weak one.

Eversion of the foot is brought about (fig. 19-13) by all three peronei and the extensor digitorum longus, particularly of course its lateral part.

The muscles of the calf are all innervated by the tibial nerve; therefore injuries to this nerve may not only markedly interfere with the "push-off" in walking (for the peronei are the only plantar flexors not supplied by the tibial nerve) but may make it impossible for the limb

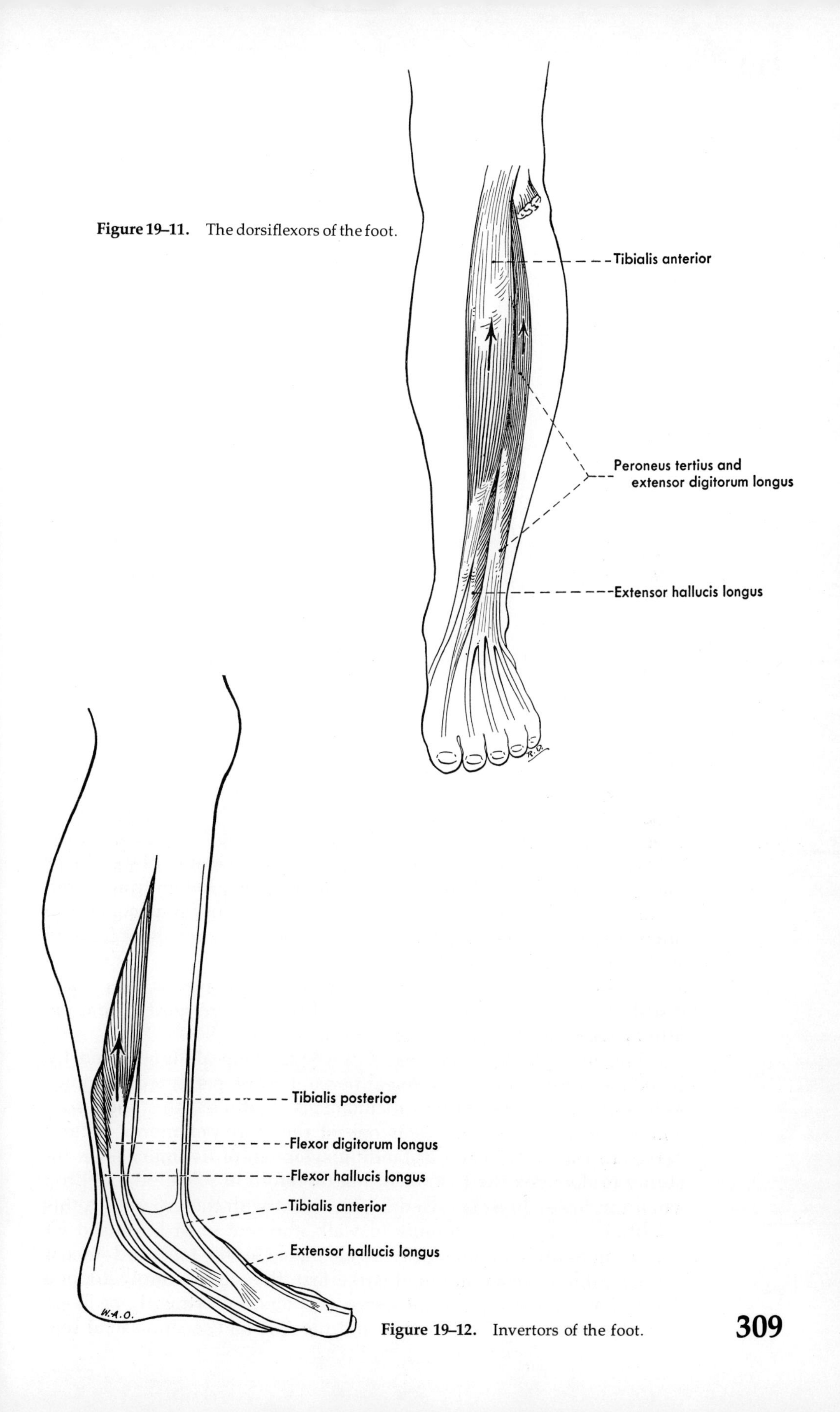

Figure 19–11. The dorsiflexors of the foot.

Figure 19–12. Invertors of the foot.

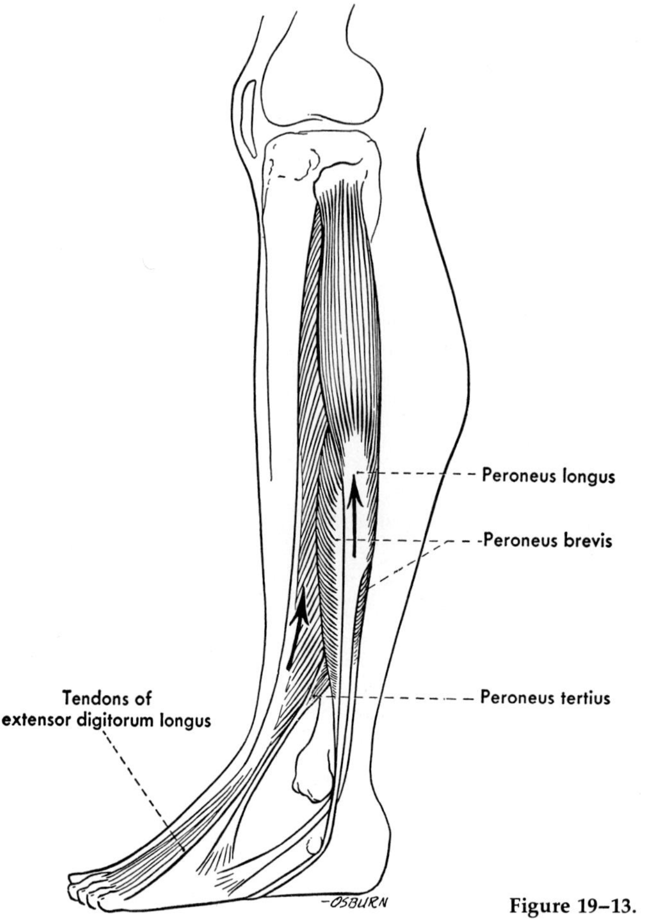

Figure 19–13. Evertors of the foot.

to bear weight unless an ankle brace is worn. The center of weight, or line of gravity, of the body, it should be recalled, lies in front of the ankle joint; therefore, the activity of plantar flexors is necessary to prevent the foot from going into dorsiflexion, with a resultant forward shift of the weight of the body. The tibial nerve receives fibers from almost all the elements entering into the sacral plexus; the fibers distributed through it to the muscles of the calf are mostly from the fifth lumbar and first two sacral nerves.

The anterolateral muscles of the leg, being all innervated by branches of the common peroneal nerve, may be paralyzed by injury to this nerve. Because of its subcutaneous position against the upper end of the fibula this nerve is one of the more commonly injured nerves in the body. The outstanding symptom of its injury is an inability to dorsiflex the foot, resulting therefore in so-called foot drop when the lower limb is raised from contact with the ground. In this condition, when one attempts to walk, the foot must be raised far enough from the ground to provide clearance for the toes, and since it is impossible to make the heel strike first, the foot is simply flopped down. The common peroneal nerve is composed primarily of fibers from the fourth and fifth lumbar and the first and second sacral segments. Most of the anterolateral muscles of the leg are believed to receive fibers from the first three of these segments.

Because the muscles of the leg, not those of the foot, move the foot, imbalance among the muscles of the leg can markedly distort the foot. There are numerous grades and directions of distortion, but a general name to cover all of them is **clubfoot** (talipes). It has long been recognized that postnatal fibrosis or paralysis of a muscle will lead to clubfoot of some type, and that in congenital clubfoot some of the muscles are abnormally short, but there is still disagreement as to whether the muscle imbalance causes or is caused by malformation of the bones of the foot.

To cite a few simple examples of clubfoot: if the triceps surae is spastic or fibrotic, or its tendo calcaneus abnormally short, this powerful plantar flexor will hold the foot in a permanently plantar-flexed position, so that the heel bears no weight. If the tibialis anterior is spastic or fibrotic, it will hold the foot in dorsiflexion and inversion. If the peroneus longus and brevis are spastic, they hold the foot in eversion.

In addition to plantar flexion and dorsiflexion, and inversion and eversion, a clubfoot may also be much abducted or adducted, that is, its longitudinal axis may be bent laterally or medially. The extent of this distortion depends upon the particular muscle or muscles involved, for although the invertors are also adductors and the evertors are also abductors, they are generally much more effective in producing one movement than they are the other. For instance, the tibialis anterior inverts strongly, dorsiflexes strongly, but adducts weakly, so dorsiflexion and inversion are the predominant effects of its overaction or shortness. On the other hand, the tibialis posterior is a relatively weak invertor and a weak plantar flexor, but a strong adductor, so its chief effect is adduction.

Correction of clubfoot is largely a matter of correcting the relations of the bones of the foot to each other, and of restoring muscle balance. This can sometimes be done, especially in congenital clubfoot, by applying a succession of casts. However, operations to realign the bones and to lengthen or transplant tendons may be necessary.

REVIEW OF THE LEG

A part of the tibia is subcutaneous throughout its entire course in the leg, and therefore this bone can be traced without difficulty from the condyles and tuberosity above to the medial malleolus on the medial side of the ankle. The upper end of the fibula is also subcutaneous and can easily be palpated just below the knee; farther down, the fibula is covered by muscles and therefore can be felt only indistinctly; above the ankle it again becomes subcutaneous and can then be palpated down to the prominent lateral malleolus.

The **muscles** of the calf are largely covered by the gastrocnemius. This muscle alone is therefore identifiable in the upper part of the calf. The level at which its muscular part gives way to the tendo calcaneus is plainly visible. The tendo calcaneus can be followed without difficulty to the heel, and below the gastrocnemius the soleus can be palpated deep to and on the sides of the wider upper part of the tendon. The long deep muscles of the calf are not identifiable individually, although their tendons can be identified as a group just behind the medial malleolus, particularly if the foot is inverted or the toes are flexed.

The peroneus longus and brevis can be felt to contract when the foot is everted, and their tendons can be felt together as they emerge from behind the lateral malleolus and pass forward below it.

The anterior muscles of the leg can be felt on the lateral side of the tibia, and at the ankle some of their tendons can be recognized. As the foot is inverted and dorsiflexed the heavy tendon of the tibialis anterior can be seen and felt running across the medial side of the anterior surface of the ankle and dorsum of the foot. If the big toe is dorsiflexed, the sharper tendon of the extensor hallucis longus can be palpated just lateral to the tendon of the tibialis anterior. As the remaining toes are dorsiflexed, the extensor digitorum longus can be felt lateral to the extensor of the big toe and some of the tendons can be seen and felt as they diverge toward the various digits. The tendon of the peroneus brevis may or may not be prominent lateral to these tendons; it is best brought out by dorsiflexion and eversion of the foot. The muscles of the leg are summarized in Table 19-1.

Of the **vessels,** parts of the great and small saphenous veins are frequently visible through the skin. The arteries are for the most part so deep that they cannot be palpated, but it is common practice for the pulse to be obtained from the posterior tibial artery behind the medial malleolus. The peroneal artery lies entirely deep; so does the anterior tibial, although its pulse can be felt in its continuation, the dorsalis

TABLE 19–1. Nerves and Muscles of the Leg

NERVE AND ORIGIN*	MUSCLE		
	Name	*Segmental innervation**	*Chief action(s)*
Tibial L4–S3	Gastrocnemius	S1, 2	Plantar flexion at ankle
	Soleus	S1, 2	" " " "
	Plantaris	L4–S1	" " " "
	Popliteus	L5, S1	Rotation and flexion at knee
	Tibialis posterior	L5, S1	Adduction and inversion of foot
	Flexor digitorum longus	L5, S1	Flexion of lateral four toes
	Flexor hallucis longus	L5–S2	Flexion of big toe
Superficial peroneal L4–S1	Peroneus longus	L4–S1	Eversion of foot
	Peroneus brevis	L4–S1	" " "
Deep peroneal L4–S2	Tibialis anterior	L4–S1	Inversion and dorsiflexion of foot
	Extensor digitorum longus	L4–S1	Extension of four lateral toes, dorsiflexion of foot
	Peroneus tertius	L4–S1	Eversion and dorsiflexion of foot
	Extensor hallucis longus	L4–S1	Extension of big toe

*A common segmental origin or innervation.

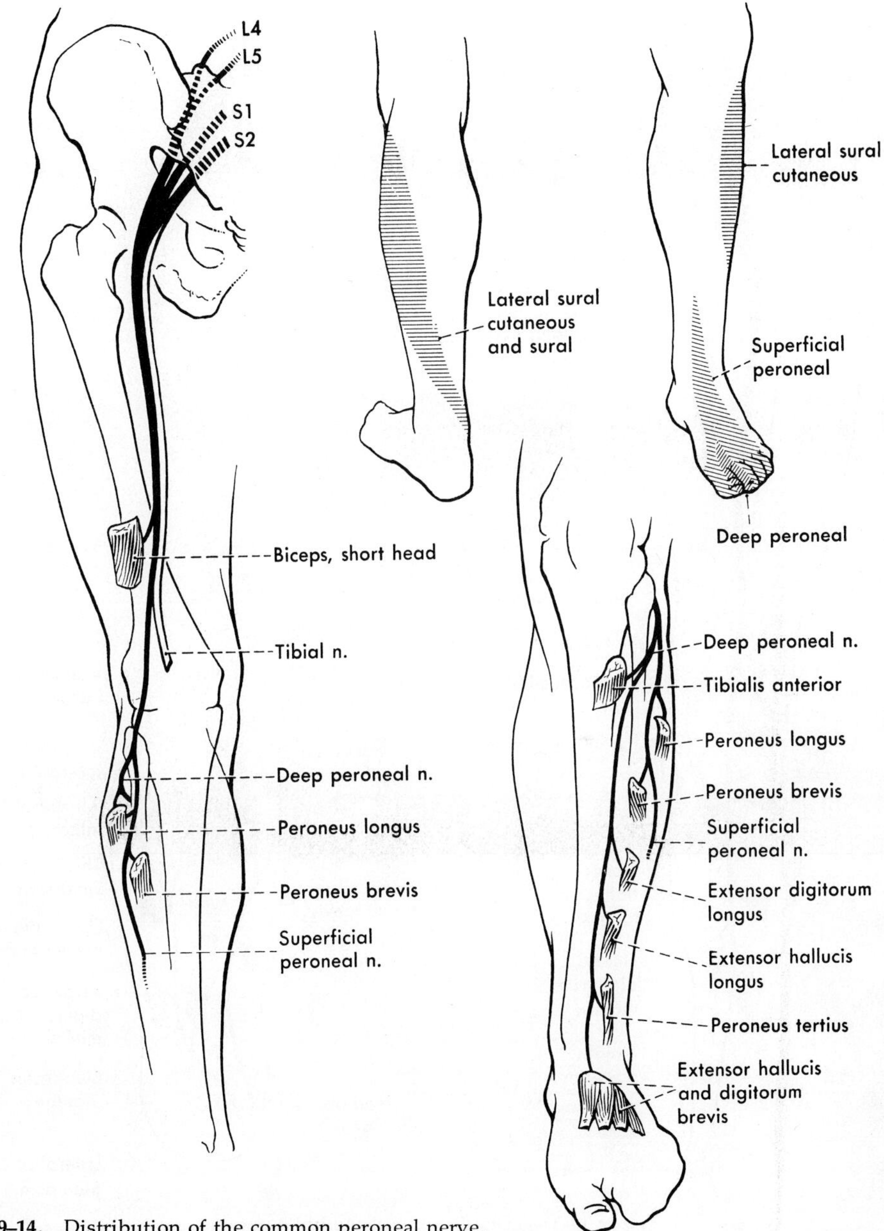

Figure 19–14. Distribution of the common peroneal nerve.

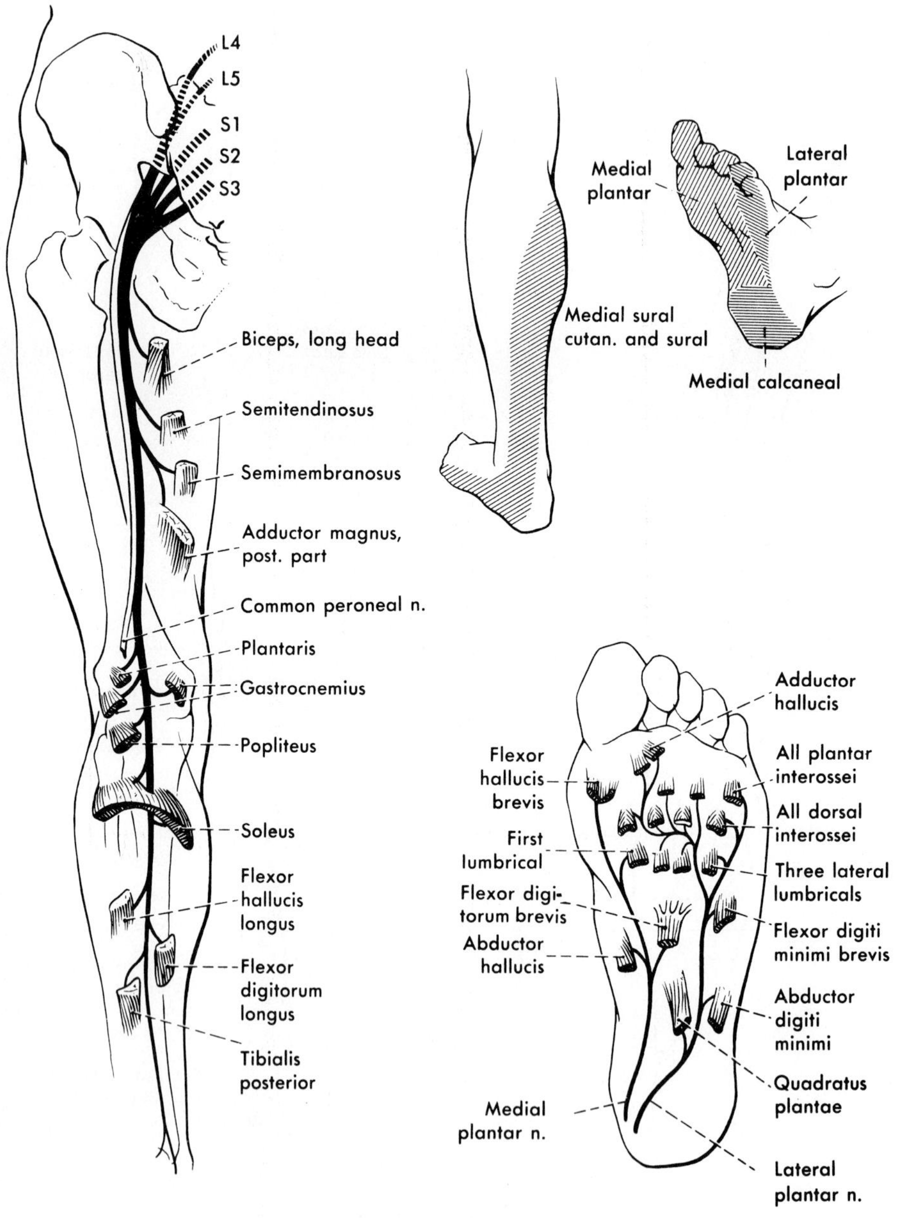

Figure 19–15. Distribution of the tibial nerve.

pedis artery, onto the dorsum of the foot. This artery lies between the tendons of the extensor hallucis longus and extensor digitorum longus muscles, just below the ankle joint. In about 12 per cent of limbs, however, the anterior tibial artery ends largely in the leg so that the dorsalis pedis is not palpable, and in others the dorsalis pedis has an abnormal course or origin and is largely covered by tendons on the foot.

Neither of the two major **nerves,** the tibial and common peroneal, can be satisfactorily located by palpation, although their courses are not particularly hard to visualize. The tibial nerve continues the course of the sciatic through the popliteal fossa and runs down the leg between the superficial and deep muscles of the calf; it diverges just enough medially to pass behind the medial malleolus with the posterior tibial artery. The common peroneal nerve runs laterally and downward, leaving the popliteal fossa to cross the fibula just below the head of that bone; its superficial branch then runs downward between the peronei, and its deep branch continues farther forward and then turns down in company with the anterior tibial artery. The distribution of the branches of the sciatic nerve is shown diagrammatically in figures 19-14 and 19-15.

20 The Foot

The foot (pes) has necessarily been studied in part in connection with the leg. It should be recalled again that its surfaces are referred to as plantar and dorsal ones, and its borders as medial or tibial and lateral or fibular. The big toe is the hallux, the little toe the digitus minimus (littlest toe), and the toes are numbered beginning with the big toe.

SKELETON OF FOOT

The individual bones of the foot (figs. 20-1 and 19-2) have already been identified, but should be reviewed now, and their articulations studied, in preparation for the study of the soft tissues of the foot.

The **calcaneus,** the heel bone, projects backward behind the ankle, thus providing leverage for the triceps surae which inserts upon its posterior end. The lower projecting surface of the posterior end, the tuber, bears rounded medial and lateral processes which support the weight that falls upon the heel. The upper surface has articular facets for the talus, the largest surface being borne by a medially projecting ledge called the sustentaculum tali. The anterior end articulates with the cuboid bone.

The **talus** rests upon the upper surface of the calcaneus. Its upper posterior part is called the trochlea (pulley); it is shaped somewhat like a short transversely placed segment of a rod, and is provided above and on its medial and lateral ends with an articular surface. The upper part of the articular surface articulates with the end of the tibia, the medial part with the medial malleolus of the tibia, and the lateral part with the lateral malleolus of the fibula. The anterior end of the talus is called the head, and articulates anteriorly with the navicular

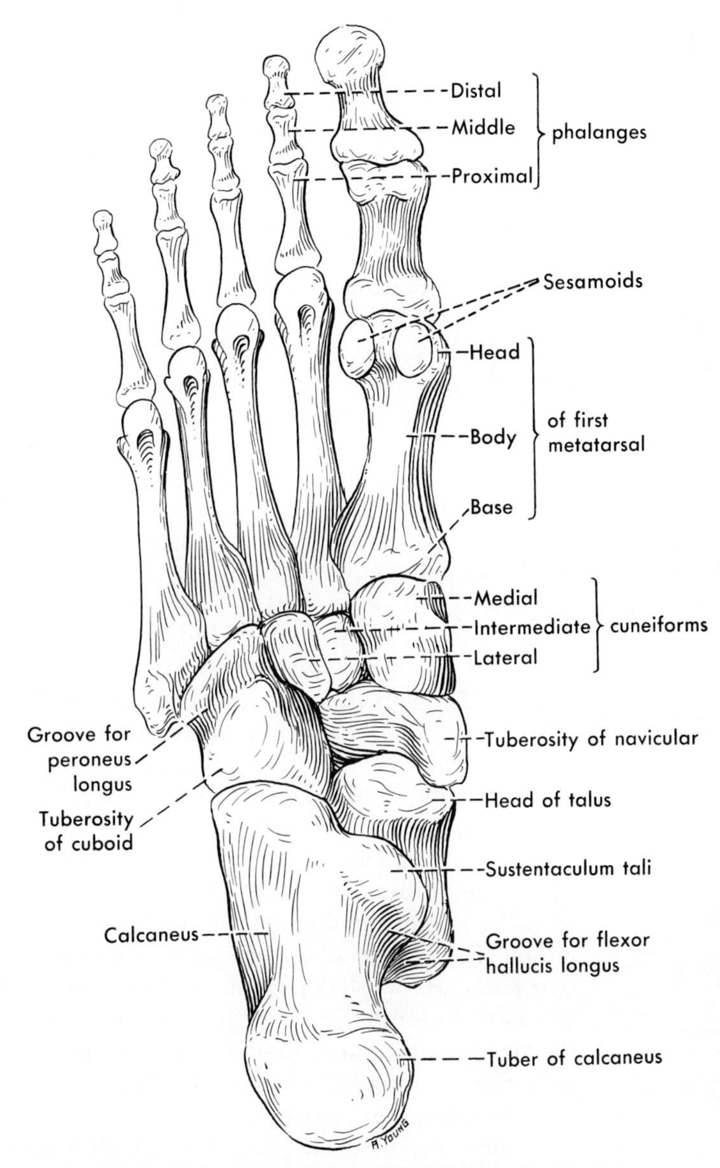

Figure 20–1. Bones of the ankle and foot, plantar view.

bone, below with the calcaneus and a ligament that stretches between the calcaneus and the navicular. The chief articulations between the talus and calcaneus are behind the head.

The remaining bones of the foot are both smaller and simpler. The **navicular** lies anterior to the talus on the medial side of the foot and has a concave proximal articular surface for the head of the talus, and a convex distal articular surface with three impressions for the three cuneiform bones. Its lateral surface is attached to the cuboid by a heavy ligament. The three **cuneiforms,** medial, intermediate, and lateral, lie in front of the navicular, and their proximal ends articulate with that through a synovial joint. They also articulate with each other, and the lateral cuneiform articulates with the cuboid, but the articular surfaces on their sides are small because the articulations are only in part through synovial joints; the rough parts of the adjacent surfaces are united by heavy ligaments. The distal ends are, however, largely smooth for articulation with the bases (proximal ends) of the metatarsals. The **cuboid** has a proximal articular surface for the calcaneus, a distal one for the two lateral metatarsals, and a small medial one for the lateral cuneiform, to which it is also bound by a heavy ligament.

The **metatarsals** resemble in general the metacarpals of the hand. Each consists of a base, a body, and a head. The **phalanges** also have bases, bodies, and heads, although the middle and distal phalanges are such short segments of bone that these parts are not very clear. The bases of the metatarsals articulate with each other and with the cuneiform and cuboid bones; their heads articulate with the bases of the proximal phalanges, whose heads in turn articulate with the bases of the middle phalanges, and so forth. The distal phalanges, of course, end freely. There are rather regularly two sesamoid bones at the metatarsophalangeal joint of the big toe.

The bones of the foot are so held together by ligaments that in a normal foot none of the parts between the posterior end of the calcaneus and the heads of the metatarsal bones transmits weight to the ground. Thus all the weight transmitted to the talus by the leg is in turn transmitted backward and downward to the back end of the calcaneus, or forward and downward to the heads of the metatarsals. In normal standing the weight is divided approximately equally between the calcaneus and the ball of the foot. The curvatures of the plantar surface between these points form the **arches** of the foot. These are two, a longitudinal arch and a transverse arch.

The longitudinal arch is obviously higher on its medial than on its lateral side (fig. 20-2), and because of this is described as consisting of two parts. Its medial part starts posteriorly with the calcaneus, and proceeds through the talus, the navicular, and the three cuneiforms to the heads of the three medial metatarsals. The lateral part also starts with the calcaneus, but proceeds through the cuboid to the heads of the two lateral metatarsals.

The transverse arch is more difficult to describe. The head of the talus and the navicular bone form the highest part of the arch on the medial side, the adjacent ends of the calcaneus and cuboid the highest part on the lateral side. In front of this level it gradually flattens out, and the heads of the metatarsals are all on the same plane, so that they all share in weight bearing. Behind it the arch is directed not only laterally and downward but also posteriorly and downward, so that this part of the arch resembles a segment of a dome, and when the two feet are together the back ends of the two transverse arches form approximately a half dome.

Studies of the way in which weight is transmitted through the foot indicate that this description is at best an anatomic one, and that functionally the foot has a single although admittedly complex arch. Stresses on the arch are apparently not proportioned according to a medial, lateral, and transverse part, but spread out in all directions through the foot. The stress at any particular point is proportional to its

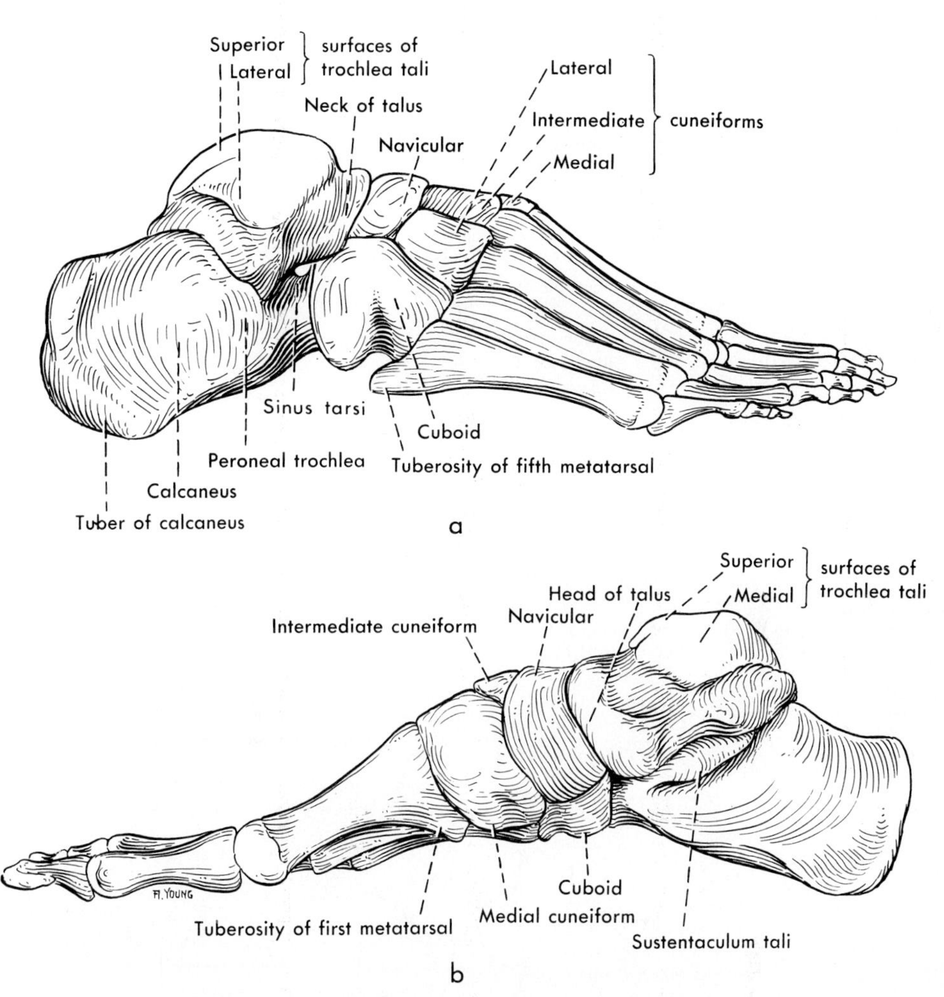

Figure 20–2. Bones of the foot from the lateral and medial sides.

height on the curve, not to what curve it is on, just as is stress on an arch built of stone or concrete.

THE ANKLE JOINT

The lower ends of the tibia and fibula have already been described (p. 292), as has the articular surface of the talus, and it has been noted that the malleoli so grip the trochlea tali between them (fig. 20-3) that the talocrural or ankle joint is primarily a hinge one. The joint is tightened still further when the foot is dorsiflexed, because the trochlea tali is slightly wider in front than it is behind.

As is typical of hinge joints, the capsule of the talocrural joint is thin anteriorly and posteriorly, but is reinforced on its sides by special ligaments. That on the medial side is called the **deltoid ligament** (fig. 20-4) because of the way it fans out from the medial malleolus of the tibia. It is composed of four parts, anterior and posterior tibiotalar, tibiocalcaneal, and tibionavicular. This important ligament resists

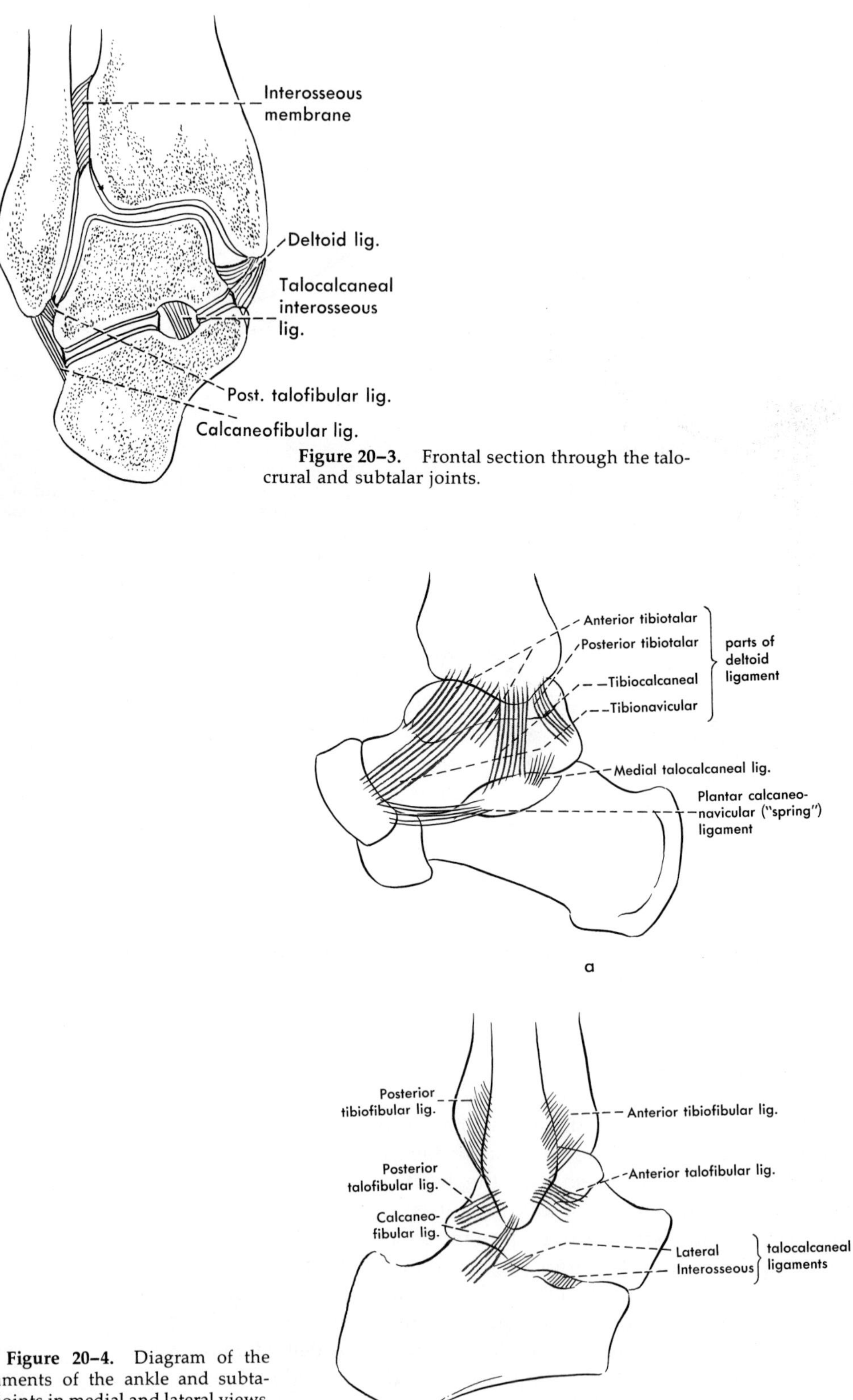

Figure 20–3. Frontal section through the talocrural and subtalar joints.

Figure 20–4. Diagram of the ligaments of the ankle and subtalar joints in medial and lateral views.

eversion of the foot. Weakness of it, allowing eversion and therefore throwing a greater weight than usual on the medial side of the arch, has been thought to be a predisposing cause of flatfoot.

There are three lateral ligaments at the ankle, but they fan out very much as do the parts of the deltoid ligament on the medial side. They are attached above to the lateral malleolus, and are called the **anterior talofibular ligament,** the **posterior talofibular ligament,** and the **calcaneofibular ligament.** These ligaments check inversion of the foot. They and the parts of the deltoid ligament are also so arranged that they check anteroposterior movement at the talocrural joint. ' "Turning" an ankle, or forced eversion or inversion of the foot, may produce a painful strain, in which ligaments are stretched but not torn, or a sprain. In a sprain, however, some of the ligaments are torn, and the resulting swelling and disability are marked. It is some part of the deltoid ligament that is affected by forced eversion at the ankle, one or more of the lateral ligaments—apparently regularly including the anterior talofibular—by forced inversion, which is the more common accident.

JOINTS OF THE FOOT

The tarsal bones are united by three sets of ligaments. Those on the dorsal surface of the foot are called collectively the **dorsal tarsal ligaments,** those on the plantar surface the **plantar tarsal ligaments,** and those that stretch between adjacent surfaces of the bones, and interrupt the synovial cavities, are the **interosseous tarsal ligaments.**

The numerous individual ligaments are usually named according to the bones that they connect. Thus there are dorsal and plantar cuboideonavicular ligaments, dorsal and plantar intercuneiform ligaments, and the like. The dorsal ligaments, since they are on the top of the bony arch, are generally thin, while the plantar ligaments, lying below the arch and therefore acting as tie rods that support it, are very much heavier. Two of the plantar ligaments are particularly important, and deserve special comment. One of these, the **plantar calcaneonavicular ligament** (figs. 20-4 and 20-5), passes from the lower surface of the calcaneus to the lower surface of the navicular and in so doing forms a sling on which the lower surface of the head of the talus rests. This ligament, by resisting downward movement of the head of the talus, thus helps support the highest part of the arch. Because it has been supposed to account for some of the elasticity of the arch, it is frequently called the "spring ligament." The other, the **long plantar ligament** (fig. 20-5), lies more laterally, stretching between the calcaneus posteriorly and the cuboid and lateral three metatarsals anteriorly. This ligament therefore extends for most of the length of the lateral part of the arch, and is the chief support of this side.

Of the interosseous ligaments, the talocalcaneal is particularly strong; one part of it partially fills the grooves on the adjacent surfaces of the talus and calcaneus and divides the joint between these two bones into two synovial cavities, one posterior to the ligament and one

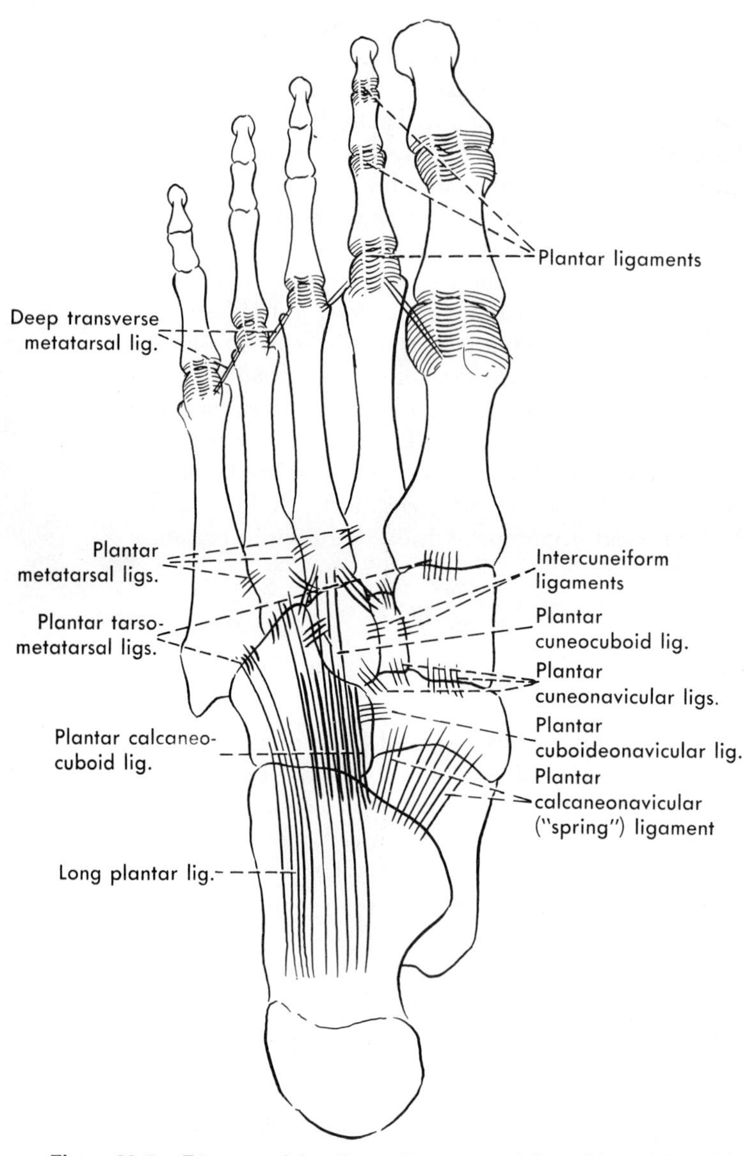

Figure 20–5. Diagram of the plantar ligaments of the ankle and foot. The plantar calcaneocuboid ligament lies deep to the long plantar ligament.

anterior to it (fig. 20-6). There are also interosseous ligaments between the cuboid and the navicular, the cuboid and the lateral cuneiform, and the three cuneiforms.

The **intertarsal joints** (also fig. 20-6) are the subtalar, the talocalcaneonavicular, the calcaneocuboid, the transverse tarsal, and the cuneonavicular. The slight gliding movements between the various tarsals are greatest at the complex subtalar and transverse tarsal joints.

The subtalar joint, between the talus and the calcaneus, has a synovial cavity both posterior and anterior to the talocalcaneal interosseous ligament; the anterior cavity is also part of the talocalcaneonavicular joint. The latter joint, the highest in the foot, is closed inferiorly by the plantar calcaneonavicular ("spring") ligament. Lateral to it is the calcaneocuboid joint. The transverse tarsal joint is a name for the combined calcaneocuboid joint and the talonavicular part of the talocalcaneonavicular joint; the two stretch across the foot almost transversely, and work together and with the subtalar joint in allowing the foot to move in inversion and eversion. The transverse tarsal joint also allows flexion and extension of the fore part of the foot, thus increasing or decreasing the height of the arch. The cuneonavicular joint lies, as its name implies, between the navicular and the cuneiforms, and extends a little bit between the cuneiforms and between the cuboid and the lateral cuneiform.

The **tarsometatarsal joint** of the big toe is a separate cavity; the corresponding joints of the second and third toes are confluent; likewise, the tarsometatarsal joints between the fourth and fifth toes and the cuboid bone also embrace a single synovial cavity. The **intermetatarsal joints** are small extensions from the tarsometatarsal joints. There is typically none between the first and second toes. Dorsal, plantar, and interosseous tarsometatarsal and intermetatarsal ligaments bind the bones together.

The **metatarsophalangeal joints** are condylar ones, usually maintained in a variable degree of dorsiflexion that varies with the height of the heel of the shoe; collateral ligaments, similar to those found in the corresponding joints of the hand, are placed laterally, while dorsally the extensor tendon reinforces the joint capsule. On the plantar surface are heavy **plantar ligaments** (fig. 20-5) that complete the joint capsules and serve as gliding surfaces for the flexor tendons. The foot's plantar ligaments and metatarsal heads are connected to their neighbors by deep transverse metatarsal ligaments, corresponding to the similar ligaments of the hand. While, however, there is no transverse ligament between the thumb and forefinger, thus allowing for greater

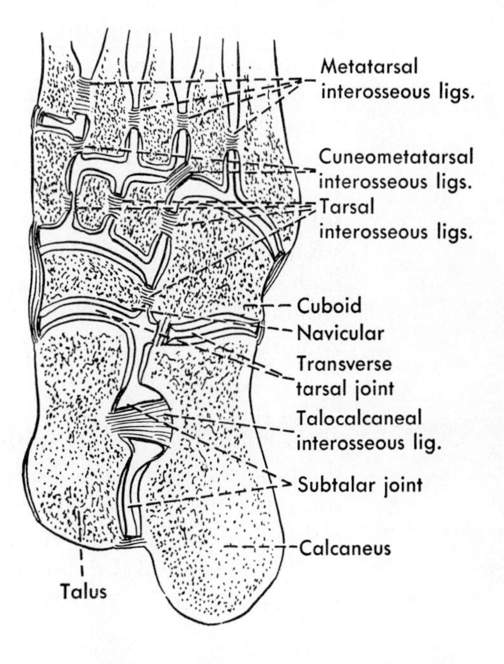

Figure 20–6. An oblique section of the foot to show intertarsal joints.

mobility of the thumb, the big toe is similar to the other toes in being connected to its neighbor by such a ligament.

The **interphalangeal joints** are similar to corresponding joints in the hand. Each has collateral ligaments at the sides, and a plantar ligament on the flexor surface, while the dorsal portion of the capsule is completed by the extensor tendon. The interphalangeal joints are hinge joints.

SUPERFICIAL NERVES AND VESSELS; FASCIA

The cutaneous innervation of the sole of the foot is through branches of the medial and lateral plantar nerves, the two terminal divisions of the tibial nerve. The **medial plantar nerve** is distributed in the foot in a manner similar to the distribution of the median nerve in the hand. It gives off branches to the medial side of the sole of the foot and breaks up into digital branches that supply approximately three and one-half toes. The **lateral plantar nerve** corresponds in its distribution to the ulnar nerve in the hand, except that it has no dorsal cutaneous branch. The cutaneous distribution of the lateral plantar nerve is to the lateral side of the plantar surface of the foot and to the lateral one and one-half toes. Not infrequently a communication exists between the digital branches of the medial and lateral plantar nerves so that the nerve to the adjacent sides of the third and fourth toes is formed from both. This nerve and its branches are particularly subject to a painful tumor (neuroma) that necessitates resection of the portion of nerve involved.

The skin on the medial side of the foot is supplied by the saphenous nerve which continues to about the level of the metatarsophalangeal joint. Similarly, the skin on the lateral border of the foot is supplied by the sural nerve. The skin of the dorsum of the foot is supplied by both branches of the common peroneal nerve, but the superficial peroneal innervates most of the area. This nerve supplies all the skin on the dorsum of the foot and toes except for a variable lateral part supplied by the sural nerve, and the adjacent sides of the first and second toes, which the deep peroneal nerve supplies.

Since superficial veins in the sole of the foot would be constantly subjected to so much pressure that they would be unable to function adequately, there is no large venous network on this aspect of the foot. Rather, the venous drainage passes quickly into the deep veins or around the borders of the foot and between the toes into the dorsal venous network. The veins on the dorsum of the foot are usually fairly obvious, and in addition to the network here they form an arch on the distal part of the foot. This arch is continued along the medial border of the foot as the great saphenous vein, and along the lateral margin of the foot as the small saphenous. The larger veins of the dorsal network unite the two saphenous veins on the foot or drain upward to end in either but primarily in the great saphenous.

The fascia of the dorsum of the foot is thin and merits no particular description here. The plantar fascia markedly resembles the palmar fascia of the hand, but is even stronger and better developed. The **plantar aponeurosis** is essentially a strong, superficially placed ligament that extends in the middle part of the foot from the calcaneus to the toes, and apparently plays an important part in supporting the arch of the foot. Like the palmar aponeurosis, the plantar aponeurosis sends slips to the digits, and these slips help reinforce the flexor digital tendon sheaths, and attach around the tendons to the distal ends of the metatarsals and the bases of the proximal phalanges. The plantar aponeurosis blends laterally and medially with the thinner fascia over the short muscles of the big and little toes, and at these points sends a medial and a lateral intermuscular septum toward the first and fifth metatarsals, respectively. The aponeurosis and its septa form a central compartment in the foot more complicated than, but somewhat comparable to, that in the hand, and containing the flexor digitorum tendons and lumbrical muscles as does that of the hand. Most of the muscles of the big and little toes lie similarly in medial and lateral compartments, as do those of the thumb and little finger.

PLANTAR MUSCULATURE

The muscles of the foot can be more easily dissected in layers than by compartments, and are therefore so described here. They and the long tendons of the toes form four layers.

The **superficial layer** (fig. 20-7) is composed, from medial to lateral sides, of the abductor hallucis, the flexor digitorum brevis, and the abductor digiti minimi. The **abductor hallucis** arises from the medial tuberal process of the calcaneus, the flexor retinaculum, and the medial intermuscular septum, and inserts on the tibial side of the flexor surface of the proximal phalanx. The medial and lateral plantar nerves and vessels, the continuations of the tibial nerve and posterior tibial vessels into the foot, pass deep to its posterior end as they enter the foot, and run forward at first deep to this muscle. This muscle receives its innervation from the medial plantar nerve; it is usually a much better flexor at the metatarsophalangeal joint than it is an abductor.

The **flexor digitorum brevis** is the central one of the superficial muscles. This muscle, the equivalent of the flexor digitorum superficialis in the upper limb, arises from the medial tuberal process of the calcaneus and both intermuscular septa and gives rise to four tendons that run forward to the four lateral toes. (The tendon to the little toe is fairly frequently missing; when it is present it often arises from a separate muscular slip attached to the tendon of the flexor digitorum longus.) Each tendon divides to allow a tendon of the flexor digitorum longus to pass through it, and then unites again to go to an insertion on the middle phalanx. The tendons of the flexor digitorum brevis lie with those of the flexor digitorum longus in digital synovial sheaths; in contrast to those of the hand, none of these sheaths usually join the flexor tendon sheath situated at the ankle. The lateral plantar nerves and vessels run deep to (above) the muscle toward the lateral part of

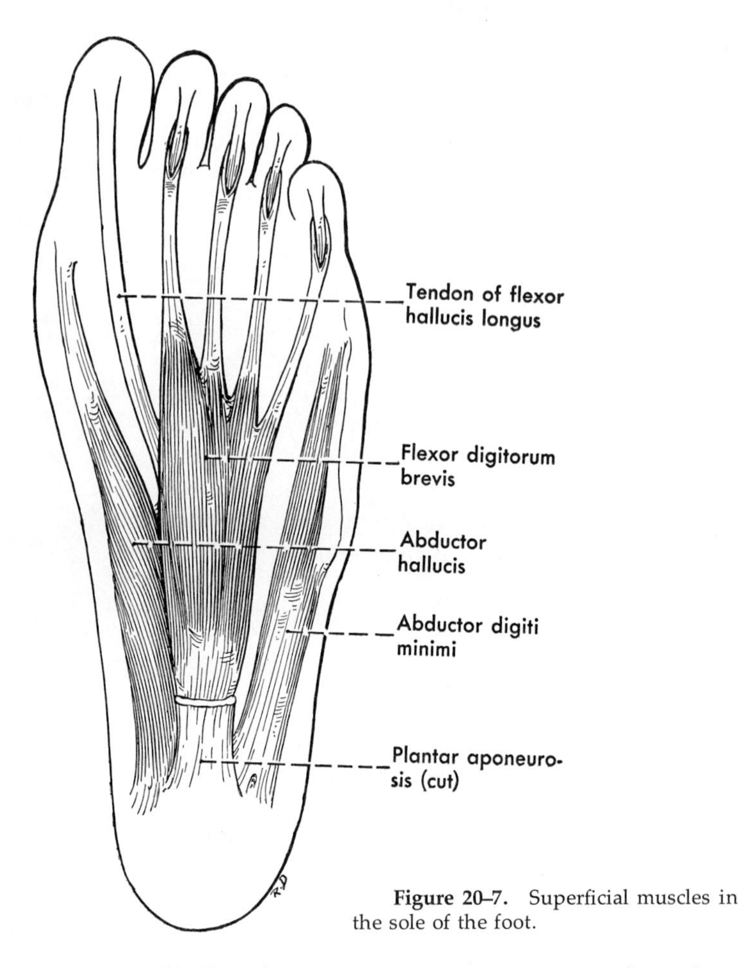

Figure 20–7. Superficial muscles in the sole of the foot.

the foot, but it receives its innervation from the medial plantar nerve. Its action is to flex the middle phalanges of the four lateral toes.

The third muscle of the superficial group, the **abductor digiti minimi,** arises from the lateral process of the tuber calcanei, the bone between the two processes, and a small part of the medial process, and also from the adjacent fascia; it inserts on the lateral side of the proximal phalanx of the little toe. It may send a slip to the extensor tendon of the toe. Occasionally, also, a portion of the muscle attaches to the tuberosity of the fifth metatarsal to form an abductor of this bone (abductor ossis metatarsi quinti). This muscle is supplied by a branch of the lateral plantar nerve. It both flexes and abducts the little toe.

The **second layer** in the foot (fig. 20-8) consists in part of the tendons of the long flexors of the toes, the flexor hallucis longus and the flexor digitorum longus, and in part of muscles, the quadratus plantae and four lumbricals, that are closely associated with the flexor digitorum longus tendon. The **flexor hallucis longus tendon** enters the foot behind the medial malleolus, lying in a tendon sheath which it loses at about the point at which it crosses above the tendon of the flexor digitorum longus. It usually gives off a tendinous slip that joins the flexor digitorum and then runs forward on the lower surface of a muscle of the big toe (flexor hallucis brevis); as it reaches the level of the metatarsophalangeal joint it acquires a digital tendon sheath essentially similar to those of the other toes and to those of the fingers, and within this sheath goes to its insertion on the distal phalanx of the big toe.

The **flexor digitorum longus tendon** also enters the foot by passing deep to the flexor retinaculum and behind the medial malleolus, and usually loses its tendon sheath in the proximal part of the foot. It passes below the tendon of the flexor hallucis longus, and as it nears the center of the foot receives on its lateral border the insertion of the quadratus plantae muscle. As the tendon divides to go to the four lateral toes the lumbrical muscles take origin from the tendons, just as corresponding muscles in the hand take origin from the flexor digitorum profundus. On the toes the tendons of the flexor digitorum longus enter digital tendon sheaths with the flexor digitorum brevis, pass through the split tendons of the brevis, and insert upon the distal phalanges of the toes.

The **quadratus plantae** is a muscle that has no counterpart in the hand. It arises by two heads of origin from the medial and lateral sides of the plantar surface of the calcaneus, distal to the tuber, and inserts on the lateral and posterior margin of the flexor digitorum longus tendon just before this divides into its four terminal slips. It aids the

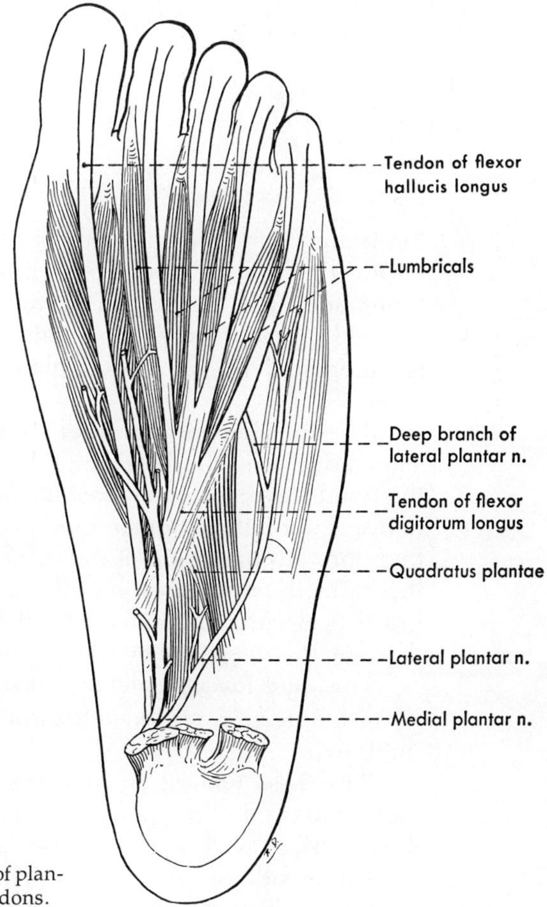

Figure 20–8. The second layer of plantar muscles, and the long flexor tendons.

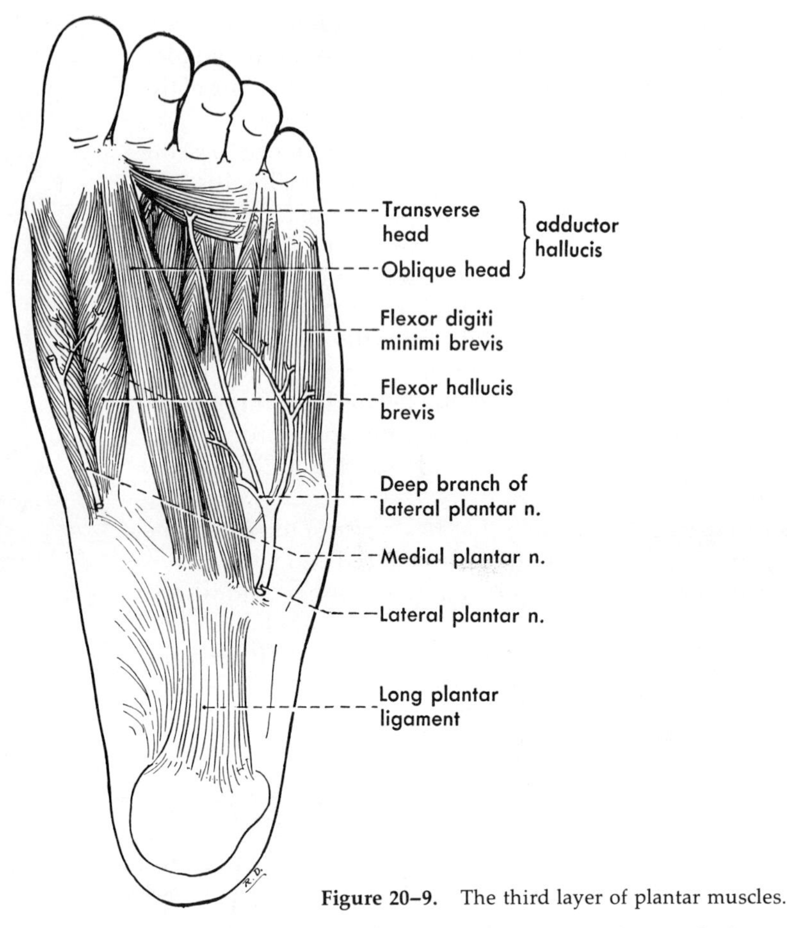

Figure 20–9. The third layer of plantar muscles.

flexor digitorum in flexing the four lateral toes, helping to convert the pull of this tendon from a posteromedial one to a more directly posterior one; and unlike the flexor digitorum longus its action is not affected by the degree of plantar flexion or dorsiflexion of the foot. It is innervated by the lateral plantar nerve as this crosses its lower surface.

The four **lumbrical muscles** are essentially like those in the hand. They arise from the flexor digitorum longus tendons and pass across the tibial side of the metatarsophalangeal joints of the lateral four toes to insert upon the extensor tendons on the dorsum of these toes. They therefore aid in flexion of the metatarsophalangeal joints and, at least theoretically, in extension of the interphalangeal ones. The first lumbrical is usually innervated by the medial plantar nerve, the other three by the deep branch of the lateral plantar nerve.

The **third layer** of plantar muscles (fig. 20-9) is composed of the flexor hallucis brevis, the adductor hallucis, and the flexor digiti minimi brevis.

The **flexor hallucis brevis** arises from the cuneiform bones and divides into two bellies, one of which is inserted on the medial side of the flexor surface of the proximal phalanx of the big toe and the other on the lateral side of this phalanx. The medial part at its insertion fuses with the abductor hallucis; the combined tendons of insertion of the two muscles attach also to the medial sesamoid bone of the metatarsophalangeal joint of the big toe. The lateral part unites with the adductor hallucis, and the combined tendon of insertion here has some attachment to the lateral sesamoid of the metatarsophalangeal joint. The flexor hallucis brevis is a flexor of the proximal phalanx of the big toe; it is innervated by the medial plantar nerve.

The **adductor hallucis** resembles the adductor of the thumb, for it consists of an oblique and a transverse head. The oblique head, the larger of the two, arises from the long plantar ligament, from lateral tarsals, and from the bases of the second and third metatarsals. The small transverse head arises from the capsules of the third, fourth, and fifth metatarsophalangeal joints and the intervening deep transverse metatarsal ligaments. The two heads unite and join the lateral head of the flexor hallucis brevis to insert on the base of the proximal phalanx of the big toe. The muscle is innervated by the deep branch of the lateral plantar nerve. It aids in flexion of the big toe, as well as adducting this digit; like other short muscles of the big toe it may send an expansion to the extensor tendon, and therefore help in extending the distal phalanx.

The **flexor digiti minimi brevis** arises from the cuboid bone and the base of the fifth metatarsal and blends with the abductor digiti minimi to attach, however, more onto the plantar than the lateral surface of the proximal phalanx. It is a flexor of this phalanx, and is innervated by the lateral plantar nerve.

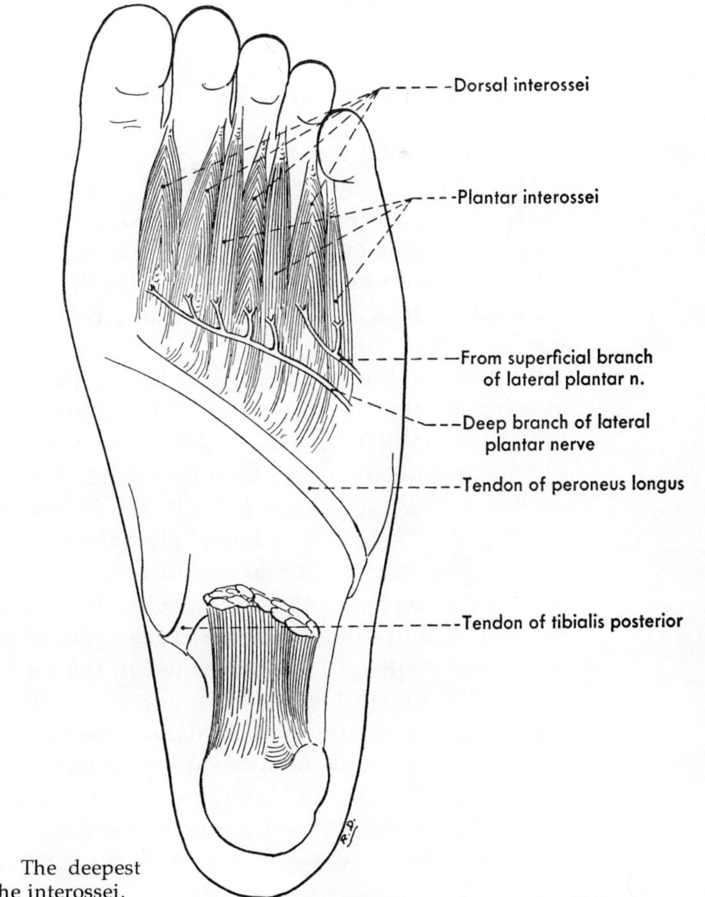

Figure 20–10. The deepest plantar muscles, the interossei.

The fourth or **deepest layer** of plantar muscles (fig. 20-10) consists of seven interossei, three plantar and four dorsal. The interossei of the foot are essentially similar to the corresponding muscles in the hand, but vary in two particulars. In the first place, they are so arranged as to abduct or adduct about the second rather than the middle toe; thus the midline of the foot has to be thought of as passing through the second digit rather than the third digit as is true in the hand. In the second place, all the interossei attach primarily to the proximal phalanges rather than having strong insertions into the extensor tendons as do most of those of the hand.

The three **plantar interossei** arise from the third, fourth, and fifth metatarsals, respectively, and insert on the medial or big toe side of the proximal phalanges of the corresponding digits. The four **dorsal interossei** have two heads of origin each: thus the first arises from the adjacent surfaces of the first and second metatarsals and inserts on the medial side of the proximal phalanx of the second toe; the second arises from the second and third metatarsals and inserts on the lateral side of the proximal phalanx of the second toe; the third arises from the third and fourth metatarsals, the fourth from the fourth and fifth metatarsals, respectively, and they insert on the lateral sides of the proximal phalanges of the third and fourth toes.

As in the hand, the dorsal interossei abduct the digits and the plantar interossei adduct. All these muscles are innervated by the lateral plantar nerve.

PLANTAR NERVES AND VESSELS

As the tibial nerve passes through the flexor retinaculum at the ankle with the posterior tibial artery, lying between the compartments of the flexor hallucis longus and the flexor digitorum longus, it divides into medial and lateral plantar branches which pass deep to the origin of the abductor hallucis to enter the foot. These nerves are distributed in the same manner that the median and ulnar nerves are in the hand (figs. 20-8 to 20-10). The **medial plantar nerve,** comparable to the median nerve, supplies skin of about three and one-half digits, and also innervates the flexor digitorum brevis, the abductor hallucis and the flexor hallucis brevis, and the first lumbrical, while the median nerve supplies the equivalent flexor digitorum superficialis, abductor pollicis brevis, and flexor pollicis brevis (but the first two lumbricals instead of only one). The **lateral plantar nerve,** comparable to the ulnar nerve, runs laterally deep to the flexor digitorum brevis, turns forward between this and the abductor digiti minimi, and divides into superficial and deep branches. In its course this nerve supplies the quadratus plantae (not represented in the hand) and the musculature of the little toe, just as the ulnar nerve supplies the musculature of the little finger. The superficial branch of the nerve supplies approximately one and one-half or more digits, and may give off the muscular branches to the third plantar and fourth dorsal interosseous muscles. The deep branch runs transversely from the lateral to the medial side of the foot in close association with the plantar arterial arch. It supplies the two or three lateral lumbricals, all the interossei not supplied by the superficial, and both heads of the adductor hallucis. It thus corresponds very nicely to the deep branch of the ulnar nerve in the hand, for it innervates lumbricals and interossei, and the equivalent of the adductor pollicis. The minor differences are only that some interossei may be supplied by the nerve's superficial branch, and that the nerve usually supplies three lumbricals.

The posterior tibial artery also divides into medial and lateral plantar branches deep to the flexor retinaculum. The **medial plantar artery** is small and is distributed largely to the muscles of the big toe. Its superficial branch, however, helps supply the skin of the medial side of the sole and sends tiny twigs distally, along the digital branches of the medial plantar nerve, toward the toes. These join the metatarsal branches of the plantar arch, but contribute little to the circulation to the toes.

The **lateral plantar artery** accompanies the nerve of the same name, thus running laterally and forward at first deep to the abductor hallucis and then deep to the flexor digitorum brevis. It gives off a branch to the lateral side of the little toe, and may give a twig that joins the metatarsal artery between this and the fourth toe, and then arches medially across the foot, with the deep branch of the lateral plantar nerve, as the plantar arch (fig. 20-11). The plantar arch is completed on the medial side of the foot by the deep plantar artery, a branch of the dorsalis pedis (see the following section) that reaches the plantar surface by passing between the two heads of the first dorsal interosseous muscle (as the radial artery does in the hand).

From the plantar arch are given off a metatarsal artery for the big toe and adjacent part of the second toe, and three other metatarsal arteries that run forward to form digital branches for the remaining toes. Between the heads of the interossei the plantar arch is connected by perforating branches to the dorsal metatarsal arteries; other perforating branches pass from the plantar to the dorsal vessels between the heads of the metatarsals or in the webs of the toes.

THE DORSUM OF THE FOOT

The cutaneous innervation of the dorsum of the foot has already been considered.

In contrast to the hand, the dorsum of the foot contains two muscles. They are closely associated at their origin; the lateral and larger one, the **extensor digitorum brevis** (fig. 19-8), arises laterally from the upper surface of the calcaneus and divides into three tendons that run toward the second, third and fourth toes. Near the distal ends of the metatarsals these short extensor tendons unite with the long extensor tendons of the corresponding toes; sometimes there is also a tendon to the fifth toe. The second muscle, the **extensor hallucis brevis,** may seem to be simply a larger medial part of the extensor digitorum brevis, and has been so described; its single tendon attaches to the base of the proximal phalanx of the big toe. It may send a slip to the tendon of the long extensor. The extensor digitorum brevis (including

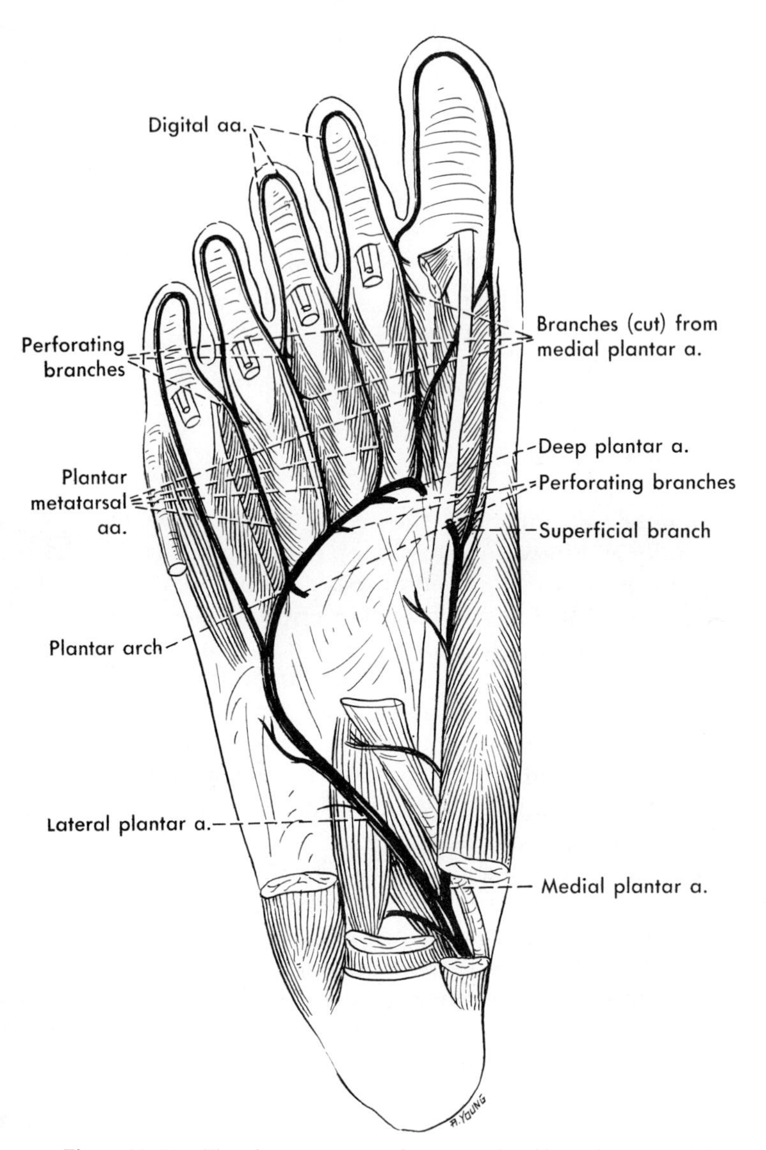

Figure 20–11. The plantar arteries; their superficial branches, except for the digital ones, have been cut away.

the extensor hallucis) assists the long extensors in extension of the toes and is innervated by the deep peroneal nerve.

The **deep peroneal nerve,** after supplying muscles of the leg, passes onto the dorsum of the foot where it supplies the short extensors, sends branches to the intertarsal joints, and ends as the cutaneous branch, already mentioned, to the adjacent surfaces of the big and second toes.

The **anterior tibial artery** (figs. 19-9 and 14-3) gives off branches about the ankle and is continued onto the foot as the **dorsalis pedis artery.** This artery supplies small branches to the region of the tarsus, gives off the arcuate artery running transversely across the foot, and then ends by dividing into the deep plantar and the first dorsal metatarsal arteries. The **deep plantar artery** passes between the two heads of the first dorsal interosseous muscle to form, with the lateral plantar artery, the plantar arch. The **first dorsal metatarsal artery** divides into branches that supply both sides of the big toe and the medial side of the second toe. The **arcuate artery,** as it runs across the foot, gives off three more dorsal metatarsal branches that run forward to supply the skin of the digits. The dorsal metatarsal arteries are connected to the plantar arch and to the plantar metatarsal arteries by perforating vessels.

The arcuate artery is often very small and the perforating branches may be the chief source of blood to the dorsal metatarsal arteries. In about 3.5 per cent of feet the dorsalis pedis is not the continuation of the anterior tibial artery but is formed by the perforating branch of the peroneal artery.

MOVEMENTS OF THE TOES

Movements of the toes are generally of little importance. Extension of the distal phalanx of the big toe is brought about by the extensor hallucis longus, and extension of the proximal phalanx by the extensor hallucis brevis. The short muscles that sometimes attach in part to the tendon of the extensor hallucis longus—the extensor brevis, for instance—can help extend the distal phalanx when they so attach. Extension of the other toes is carried out by the extensor digitorum longus and the parts of the extensor digitorum brevis associated with each of these except the little toe. These muscles, although they insert on the middle and distal phalanges, act primarily at the metatarsophalangeal joints which are normally hyperextended and can be hyperextended still more. The distal phalanges are usually kept flexed by the pull of the flexors. Since the interossei of the foot, in contrast to those of the hand, have little insertion on the extensor tendons, they are largely ineffective in extension of the interphalangeal joints. Although the lumbricals do insert into the extensor tendons, it is obvious that they also have little effect.

Flexion of the big toe is carried out by the short and long flexors, the abductor, and the adductor. Flexion of the remaining toes is carried out by the flexor digitorum longus and brevis, assisted by the lumbricals, both sets of interossei, and the quadratus plantae. The fifth digit is flexed also by its own short flexor and its abductor. All these muscles act at the metatarsophalangeal joints, directly or indirectly. The interphalangeal joint of the big toe is flexed by the flexor hallucis longus. The flexor digitorum brevis acts upon the proximal interphalangeal joints of the remaining toes, while the flexor digitorum longus, assisted by the quadratus plantae, is the flexor of the distal phalanges. Because of the shortness of the toes, in contrast to the fingers, and the relatively slight importance of their action at specific joints, much less is known regarding the roles played by the various muscles working on the toes.

Abduction of the toes is brought about by the dorsal interossei and the abductors of the big and little toes. Adduction is brought about by the plantar interossei and the adductor hallucis.

Since all the muscles in the sole of the foot are innervated by the tibial nerve, tibial nerve injury may result in paralysis of them all. Their segmental innervation is from the fifth lumbar and first and second sacral nerves. The extensor digitorum brevis, innervated by the deep peroneal, probably receives fibers from the fifth lumbar and first sacral nerves, just as do most of the anterolateral muscles of the leg.

THE ANKLE AND FOOT IN SUPPORTING WEIGHT

It has already been noted (p. 289) that the center of gravity of the body passes behind the hip joint and in front of the knee joint, so that the weight borne on the extended limb helps keep these joints extended. Thus, since hyperextension is resisted by ligaments, no sustained muscular effort is required to hold the pelvis, thigh, and leg together as a supporting pillar. The situation is different at the ankle, however, for here the center of gravity passes in front of the normally dorsiflexed joint, so that the weight of the body tends to dorsiflex it even farther. Thus even quiet standing requires contraction of the plantar flexors of the foot, a duty that normally falls upon the soleus. If there is any difficulty in keeping the balance, as there is in standing upon one foot, practically all the muscles of the leg contract in order to stabilize the intertarsal joints as well as the talocrural one and prevent any inversion or eversion of the foot.

Since all weight is transmitted to the rest of the foot through the talus, the position of this bone on the arch of the foot is important. For one thing, the talus is not centered over the longitudinal midline of the foot but somewhat to the medial side of this line. This means that weight is first transmitted to the medial side of the arch, which is also the highest side. Indeed, although it thereafter spreads out in all directions through the arch, measurements indicate that the medial side of the anterior end of the arch, represented by the head of the first metatarsal, bears approximately twice the weight of any of the other metatarsals. Further, the medial position of the talus means that weight bearing produces a tendency toward eversion (pronation) of the foot, with therefore greater weight than normal being thrown upon the medial side, and possibly some flattening of the medial side of the arch. If eversion does occur, it can set up a vicious cycle: the subtalar joint is normally slightly tilted downward and forward so that there is a tendency for the talus to slip in this direction; eversion both increases the weight on the medial side of the foot and increases the tendency of the talus to be displaced medially and downward, and this in turn distributes more weight to the medial side and encourages more eversion and flattening.

The fact that the talus is situated posterior to the middle of the longitudinal arch of the foot has less obvious consequences, but does increase the strain thrown upon the arch. Because of the position of the talus it would normally transmit to the calcaneus, through the shorter back end of the arch, 80 per cent of the weight it bears, and only 20 per cent forward to the heads of the five metatarsals. However, even in quiet standing the contraction of the posterior leg muscles pulls upward upon the calcaneus (sufficiently to redistribute the weight equally, rather than in an 80:20 ratio, between the calcaneus and the metatarsals), and the arch is thus subject not only to the weight bearing upon it but also to the pull exerted by the posterior muscles. This pull obviously becomes much greater when all the weight is shifted forward onto the ball of the foot and the heel is lifted from the ground. In fact, because of the short lever arm upon which the triceps surae has to work, the pull must be sufficient to support twice the weight that is on the ball of the foot. Thus the strain upon the arch of a plantar-flexed weight-bearing foot is very great, approximately three times the actual weight that the foot is bearing.

In view of the above considerations, it is obvious that an arch that remains normal must be very strongly constructed. Probably everyone would agree that a fundamental consideration is the bony conformation—whether or not, for instance, the subtalar joint is slanted slightly more downward and medially than usual—but beyond this there have been varied opinions as to which soft tissues contribute to the support of the arch, and to what degree. Extremes have been (1) that the heavy plantar ligaments normally contribute no support whatever, and (2) that they are the primary support of the arch.

Numerous experiments, now backed by electromyographic evidence, have shown that the plantar ligaments are indeed the primary support of the arch; in quiet standing the normal foot needs no other support, for the muscles of the leg and foot show only slight intermittent activity associated with small shifts in balance. When additional support is needed, as in standing on the toes or in walking, the short muscles of the foot become active. The support of the arch, therefore, comes from the plantar ligaments and, when necessary, the plantar muscles which stretch along the arch like tie rods.

The long muscles seem to contribute to the preservation or destruction of the arch in one way only: they are responsible for keeping the foot properly balanced between eversion and inversion, therefore for the normal distribution of weight over the arch. If they fail to do this, excess weight is thrown upon either the medial or the lateral side of the arch, and this may be more than the ligaments, even with the aid of the short muscles, can withstand without stretching. Thus while imbalance of the long muscles may lead to deformation of and pain from a normal arch, strengthening them by exercise cannot be expected to increase the support of the arch.

GAIT

While little muscular activity is required during quiet standing, walking obviously requires the co-operation of a number of muscles. The following brief account relates to the lower limb only, omitting such concomitant activity as the usual rhythmic swinging of the arms, and support of the trunk by the appropriate muscles.

The muscles involved in walking vary according to what the limb is doing, and this can be conveniently described as consisting of two phases, each initiated by a particular momentary activity: the swing phase is initiated by push-off (or toe-off) and lasts until heel-strike. The stance phase begins with heel-strike and ends with push-off. The swing phase occupies about one third of the cycle, the stance phase two thirds; thus both limbs are simultaneously in the stance phase about a third of the time.

Push-off is normally produced by plantar flexion of the foot; if the triceps surae is paralyzed, or the tendo calcaneus has been severed, it can be accomplished to a lesser extent by the use of the gluteus maximus and the posterior hamstrings to extend the hip. The swing phase involves almost simultaneous flexion at the hip and flexion at the knee, followed by dorsiflexion at the ankle. There is usually also a concomitant forward swing of the hip (rotation of the pelvis to the opposite side) of varying degree. There is some uncertainty as to which of the several hip flexors actually participate in the swing, but the tensor fasciae latae, the pectineus, and the sartorius are usually regarded as being primarily responsible. It has been said that neither the iliopsoas, the most powerful hip flexor, nor the adductors participate; the iliopsoas and the upper part of the adductor magnus, however, have been shown by electromyography to be active throughout the swing phase (although there is some question as to whether the iliopsoas is acting on the limb or the vertebral column). Flexion at the knee is presumably aided by the contraction of the sartorius, but gravity and the passive pull of the posterior hamstrings are adequate to produce this movement.

As the forward swing continues, the foot is dorsiflexed by the action of the anterior crural muscles, and the quadriceps usually contracts slightly to initiate extension at the knee. Much of this extension, however, is passive, brought about by the forward swing. Thus if one walks on level ground at an appropriate speed the limb can be extended at heel-strike even if the quadriceps is paralyzed; and as the weight is shifted forward during the stance phase the knee stays extended as long as the weight is centered anterior to the knee joint.

At heel-strike the gluteus maximus contracts to promote extension at the hip; the posterior hamstrings also contract, but this is usually interpreted as being primarily in order to prevent hyperextension of the knee. (If the hamstrings are paralyzed, a hyperextension deformity develops at the knee.) A person with a weak gluteus maximus may walk normally on level ground, or may lurch backward at heel-strike to stop the forward movement of his trunk and produce a passive extension at the hip.

During the first part of the stance phase, until the foot is flat on the ground, the triceps surae contracts in order to control the forward shift of the body over the foot. Passive dorsiflexion of the foot then occurs as the body is carried farther forward, and the plantar flexors then contract again to initiate the push-off of the next cycle. The short muscles of the sole of the foot also contract as the weight is shifted onto the ball. During the part of the stance phase that the other foot is clear of the ground, the gluteus medius and minimus contract to prevent undue drooping of the unsupported side of the pelvis. Thus, if they are weak, the person will lurch to that side, putting the weight

TABLE 20–1. Nerves and Muscles of the Foot

NERVE AND ORIGIN*	MUSCLE Name	Segmental innervation*	Chief action(s)
Medial plantar L5, S1	Abductor hallucis	L5, S1	Abduction-flexion of big toe
	Flexor hallucis brevis	L5, S1	Flexion of big toe
	Flexor digitorum brevis	L5, S1	Flexion of four lateral toes
	First lumbrical	L5, S1	Flexion of second toe
Lateral plantar S1, 2	Three lateral lumbricals	S1, 2	Flexion of lateral three toes
	Quadratus plantae	S1, 2	Assist flexion of four lateral toes
	Flexor digiti minimi brevis	S1, 2	Flexion of little toe
	Abductor digiti minimi	S1, 2	Abduction of little toe
	Adductor hallucis	S1, 2	Adduction of big toe
	Plantar interossei	S1, 2	Adduction and flexion of three lateral toes
	Dorsal interossei	S1, 2	Abduction and flexion of toes II, III, and IV
Deep peroneal L4–S2	Extensor digitorum and hallucis brevis	L5, S1	Extension of toes

*A common segmental origin or innervation.

more over the joint; since that lessens the length of the resistance arm, it also reduces the effort required of the abductors.

Hip-swing that may occur at the first part of the swing phase can be due to the anterior part of the gluteus minimus and to the adductors of the supporting side (internal rotators when acting on the free limb), to the abdominal muscles, or to all combined.

In running, extension at the hip and knee is very powerful, and the triceps surae contracts strongly before the foot touches the ground, thus preventing heel-strike by transferring all the weight onto the ball of the foot (and in consequence, subjecting the arch to enormous stress).

REVIEW OF FOOT

There is little living anatomy that can be reviewed on the foot.

Of the **bones,** the malleoli, already identified, are useful landmarks but do not really belong to the foot. The back end of the calcaneus can be felt with no difficulty where it forms the prominence of the heel. Below and anterior to the medial malleolus the talus and the sustentaculum tali can be felt on the medial border of the foot, but otherwise the tarsals are not identifiable. The metatarsals are of course palpable from the dorsum, especially toward their heads.

Although the actions of some of the **muscles** of the foot can be observed, none of the muscles are identifiable: the short extensors because they lie deep to the long tendons on the dorsum of the foot (reviewed on page 312), and the plantar muscles because of the heavy padding and fascia between them and the skin. The muscles of the foot are summarized in Table 20-1.

The situation is not much better concerning the **vessels** and **nerves.** The pattern of superficial veins on the dorsum may be visible, and when the dorsalis pedis is in its proper position and of normal size it can be palpated on the foot where it lies between the long extensor of the big toe and that of the other digits. No nerves are identifiable. The distribution of the medial and lateral plantar nerves (fig. 19-15) can perhaps be most easily recalled by remembering that the medial plantar has a distribution almost exactly comparable to that of the median nerve in the hand, the lateral plantar one comparable to that of the palmar portion of the ulnar nerve.

The Head and Neck 21

THE SKULL

The skull (figs. 21-1 and 21-2) may be divided roughly into two portions, the cranium proper (cerebral cranium) which supports, surrounds and protects the brain, and the facial skeleton. The bones of the cranium are tightly knit together by sutures (seams) so that movement between them is impossible, and with the exception of the lower jaw the facial bones are similarly united with each other and with the cranial bones.

The roof of the skull (calvaria) is formed anteriorly by the unpaired frontal bone; the paired parietal bones succeed this, and the unpaired occipital bone forms the most posterior portion of the skull. The occipital bone surrounds the large foramen magnum for the transmission of the lower part of the brain stem as it joins the spinal cord. The floor of the cranial cavity is formed in part by the occipital bone, in part by the sphenoid, which is an unpaired bone lying anterior to the occipital, and in part by a horizontal portion of the frontal bone, this portion also representing the roof of the orbit, the cavity that contains the eyeball and its associated muscles, nerves, and vessels. The floor of the skull between cranial cavity and nose is formed by the ethmoid bone. Finally, a part of the temporal bone forms a lateral part of the floor between the occipital and sphenoid bones. Most of the bones of the floor and roof of the cranial cavity also extend laterally to form its lateral walls, and so does the temporal bone, appearing on the side of the skull in the region of the external ear. The temporal bone contains the middle ear cavity; the part of it behind the external ear, the mastoid process, contains mastoid air cells that communicate with the middle ear cavity; and a particularly hard part of the bone also contains the complicated inner ear.

In the newborn infant the bones of the roof of the skull, instead of being tightly bound together by sutures, are united by membranes

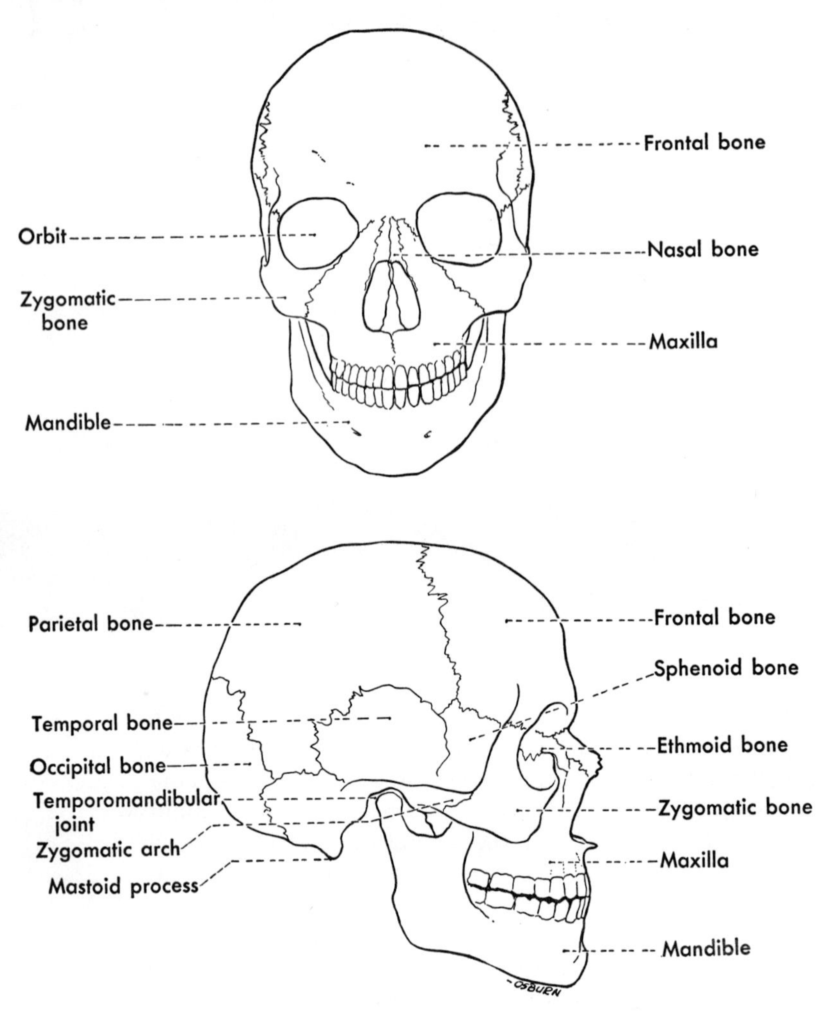

Figure 21–1. Anterior and lateral views of the skull.

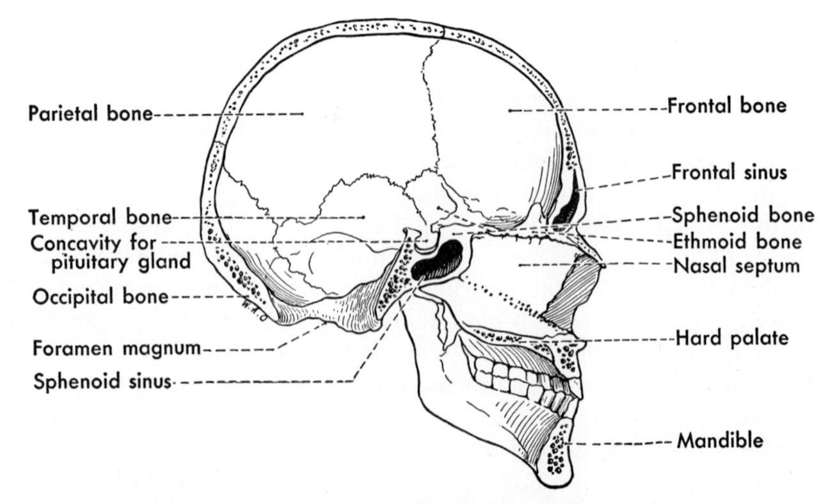

Figure 21–2. Medial view of the hemisected skull.

which will gradually be converted into bone. The membranes allow for considerable deformation of the infant's head during birth. In locations where more than two bones come together the membranes are particularly extensive, and constitute fontanels or fonticuli, soft spots in the baby's head. The two fontanels on each side of the skull are covered by muscle, but the anterior fontanel, where the originally paired frontal bones and the two parietal bones approach each other in the midline, is easily palpable; the posterior fontanel, between the two parietal bones and the occipital bone is also palpable, but for a much shorter period.

The most important paired facial bones include the zygomatic or cheek bone, forming the prominence of the cheek and having a projecting process that unites with a similar process of the temporal bone to form the zygomatic arch; the nasal bone or bone of the bridge of the nose; and the maxilla, the tooth-bearing bone of the upper jaw. Processes from the maxilla aid in the formation of the hard palate, separating the anterior part of the nasal cavity from the corresponding part of the oral cavity. Of the remaining small bones in the skull the ethmoid is perhaps the most important; this unpaired bone forms a part of the medial wall of each orbit, and thus of the lateral walls of the nose; it also helps to form the septum separating the two nasal passages from each other; and the upper part of the bone, connecting these downward-projecting parts, forms the roof of the nasal cavities and transmits to the cranial cavity the nerves concerned with olfaction (the sense of smell). It is a fragile bone containing large cavities filled with air, the ethmoidal air cells or ethmoidal sinuses, connecting with the nasal cavity. Other paranasal sinuses lie in the frontal, sphenoid and maxillary bones and receive their names from these bones. All of the paranasal sinuses communicate with the nasal cavity and are therefore subject to infection from this cavity.

The mandible is the unpaired bone of the lower jaw; it consists of paired rami that articulate with the temporal bone to form the temporomandibular joint, and a heavy tooth-bearing body. The synovial cavity of the temporomandibular joint is divided into two parts by a fibrocartilaginous articular disk; the hinge action at this joint takes place between mandible and articular disk, while protraction and retraction of the jaw occur between the articular disk and the temporal bone.

THE BRAIN, MENINGES AND CRANIAL NERVES

The brain lies within the cranial cavity and like the spinal cord (p. 216) is supplied with three meninges, the dura mater, the arachnoid, and the pia mater. An outer part of the dura mater (fig. 21-3) in contact with the skull is actually the periosteum of the inner surface of the cranium. Folds of the dura (not including the periosteal layer) also pass between the two cerebral hemispheres (as the falx cerebri) and between cerebral hemispheres and cerebellum (the tentorium cerebelli). In certain locations, the dura contains venous sinuses that receive the blood from the brain.

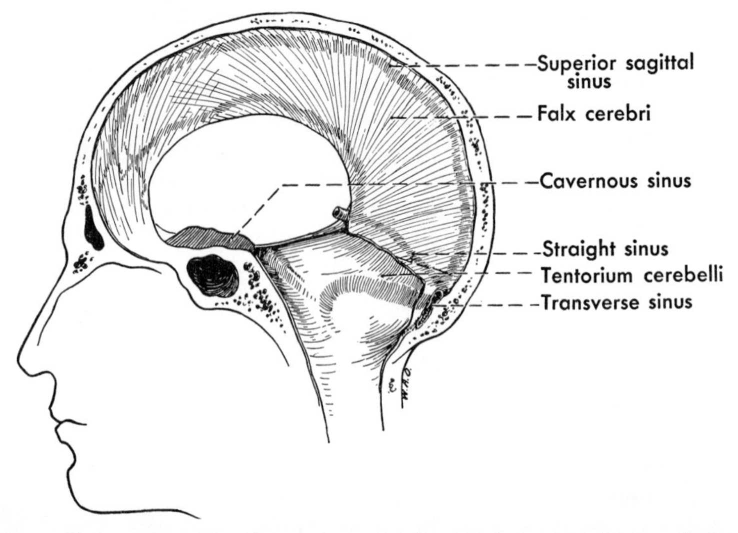

Figure 21–3. The dura mater and the chief venous sinuses of the cranium, as seen in a medial view of the hemisected head.

The larger and more important **cranial venous sinuses** consist of the unpaired superior sagittal sinus, situated over the convexity of the cerebral hemispheres in the midline (where in the infant a needle can be easily inserted into it through the anterior fontanel); the paired transverse and sigmoid sinuses which form continuous channels that first run laterally and then curve downward to leave the skull as the internal jugular veins; and the paired cavernous sinuses situated on the floor of the skull, on the sides of the central portion of the sphenoid bone. The superior sagittal sinus receives veins from the brain and also the drainage of the cerebrospinal fluid, which enters the venous system through the mediation of arachnoid villi projecting into this sinus. The transverse sinuses receive the blood from the superior sagittal sinus and also veins from much of the rest of the brain, and they and the sigmoid sinuses form the chief venous pathways leaving the skull. The cavernous sinuses are smaller than those described above, and are chiefly of importance because they lie adjacent to a very important part of the brain, the hypothalamus, and because through these sinuses run the internal carotid artery and a number of nerves.

The dura mater of the brain resembles that of the spinal cord in being a tough fibrous membrane. Where venous sinuses occur, the walls of these sinuses consist largely of the dura, that is, the vessels have no special walls except for lining endothelium. The dura and arachnoid are separated from each other by a potential space containing a film of fluid. The arachnoid is more closely applied to the dura than to the brain substance (fig. 21-4), and runs smoothly from one high point to another; thus it fails to dip into the folds of the cerebral hemispheres (see the following paragraph) or to dip in any farther than the dura between the cerebral hemispheres and between those and the cerebellum. The pia, on the other hand, follows very exactly the outer surface of the brain, dipping in at every fold of this organ; thus a considerable amount of space is left between the arachnoid and pia. This space is occupied by the cerebrospinal fluid, a fluid that helps to cushion the brain and at the same time takes the place of the lymphatic system, which is lacking in the brain. In some locations, such as around certain parts of the base of the brain, the subarachnoid space is particularly large and considerable accumulations of cerebrospinal fluid exist there. These larger subarachnoid spaces are known as cisterns; the largest of these, the cerebellomedullary cistern or cisterna magna, lies between the lower surface of the cerebellum and the posterior surface of the medulla. It is possible to insert needles into this cistern for the purpose of obtaining cerebrospinal fluid or measuring the pressure of this fluid without injuring delicate nerve structures that lie immediately adjacent to it.

The **brain** (fig. 21-5) consists of the large cerebral hemispheres, with their surfaces thrown into numerous convolutions or gyri; the smaller cerebellum which also presents numerous but smaller folds of its outer layers; and the brain stem representing an upper, much modified continuation of the spinal cord. The cerebral hemispheres and the cerebellum, instead of having all their nerve cell bodies buried close to their centers, have layers of cell bodies, gray matter, on their surfaces also. This outer layer of gray matter has traditionally been called "cortex" (bark), although the more rarely used "pallium" (cloak) is now the official name for the cerebral cortex. In the brain stem, however, the arrangement of cells and fibers is essentially like that in the spinal cord (p. 48)—the outside is composed of white matter, and the gray matter lies deeper.

Since the brain is derived from a hollow tube, its various parts contain cavities within them. The largest cavities lie within the cerebral hemispheres and are known as the lateral ventricles of the brain; vascular folds, called choroid plexuses, project into the lateral ventricles and are the chief source of cerebrospinal fluid. This fluid then circulates downward through other cavities or ventricles to escape into the subarachnoid space at about the junction of the brain and spinal cord.

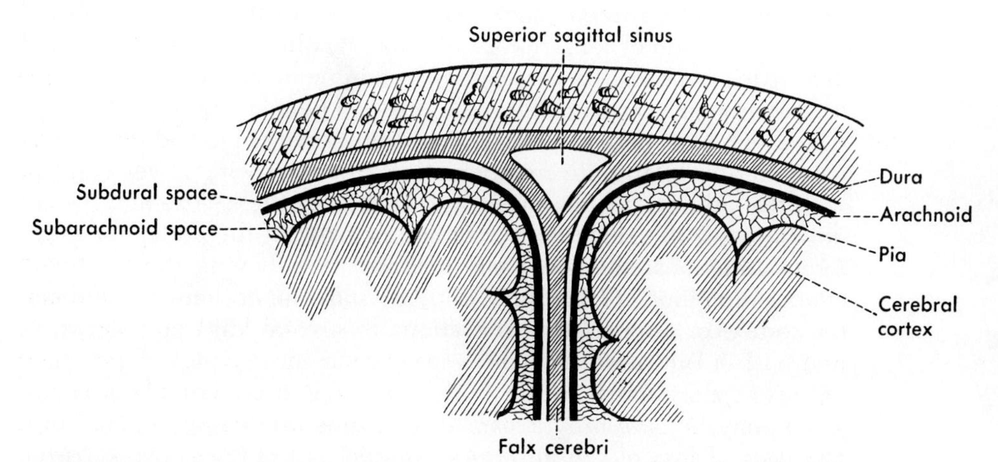

Figure 21–4. Diagram of the cranial meninges close to the midline of the roof of the skull. The subdural space is a potential space only; it is created by separation of the closely adherent arachnoid and dura (as by a subdural hemorrhage).

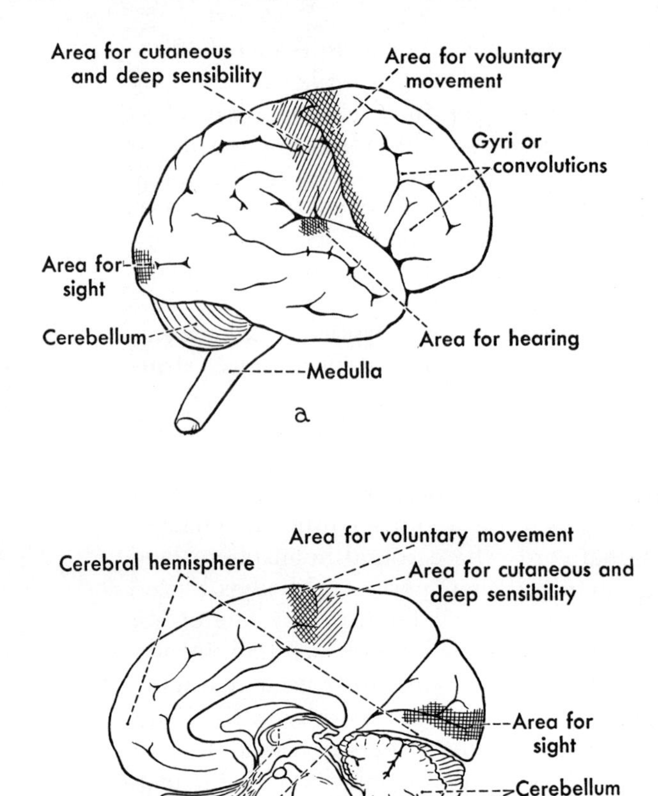
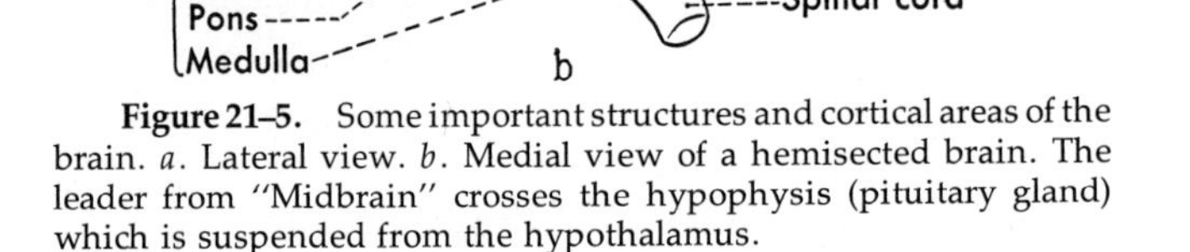

Figure 21–5. Some important structures and cortical areas of the brain. *a*. Lateral view. *b*. Medial view of a hemisected brain. The leader from "Midbrain" crosses the hypophysis (pituitary gland) which is suspended from the hypothalamus.

The cerebral hemispheres are the parts of the brain concerned with recognition of sensations, initiation of voluntary movements, and the intricate mental processes involved in memory, judgment and interpretation. The cortex of a single cerebral hemisphere has been estimated to contain approximately 7 billion nerve cells, and these cells have an almost infinite number of connections with lower centers, with the other cerebral cortex, and with both their close and distant neighbors in the same cortex. Different parts of the cortex have different functions; the locations of those parts having the relatively simple functions of initiating voluntary movement (hence called motor cortex) or appreciating sensations of several kinds are shown in figure 21-5. Injuries to the cortex may or may not involve one or more of these cortical areas; further, since the areas are relatively large, injuries may involve only a part of one, thus producing, for instance, paralysis or loss of sensation of a limited part of the body, or partial loss of sight.

Certain masses of gray matter that lie deeper in the cerebral hemispheres are sometimes grouped together as the basal ganglia. The largest and most important of these, the **corpus striatum,** consists of the caudate nucleus, a major part of which bulges into the floor of the lateral ventricle (fig. 21-6), and a more laterally lying lenticular or lentiform nucleus, which can in turn be subdivided into an outer putamen and an inner globus pallidus.

The corpus striatum is an important part of the extrapyramidal motor system (see p. 221); although it does not itself send fibers into the spinal cord, it sends nerve impulses to various centers in the brain stem that do give rise to extrapyramidal fibers, and it also sends impulses to the cerebral cortex. The exact part it plays in helping to govern voluntary movements is not known; however, lesions of it or of other extrapyramidal centers with which it is closely connected are typically associated with an increased contraction of all muscle (rigidity), making movements difficult, and with some type of abnormal, unwilled, movement. Thus paralysis agitans, in which muscular rigidity and rhythmic involuntary movements are combined, is one of the better known syndromes associated with disease of the extrapyramidal system.

On the inner surface of the lenticular nucleus, separating this nucleus posteriorly from the thalamus and anteriorly from the caudate nucleus, is a very heavy bundle of fibers known as the **internal capsule** (fig. 21-6). The internal capsule contains almost all the fibers that are either going to the cerebral cortex or leaving it, except those that go from one cortex to the other through the corpus callosum. Because sensory and motor fibers are so closely packed together in the internal capsule, a lesion of it, unlike one of the cortex, is likely to cause both paralysis and loss of sensation over one entire side of the body.

The diencephalon, the uppermost part of the brain stem, consists of two important portions: the large paired thalami which receive incoming sensory impulses, integrate them in complex patterns and for-

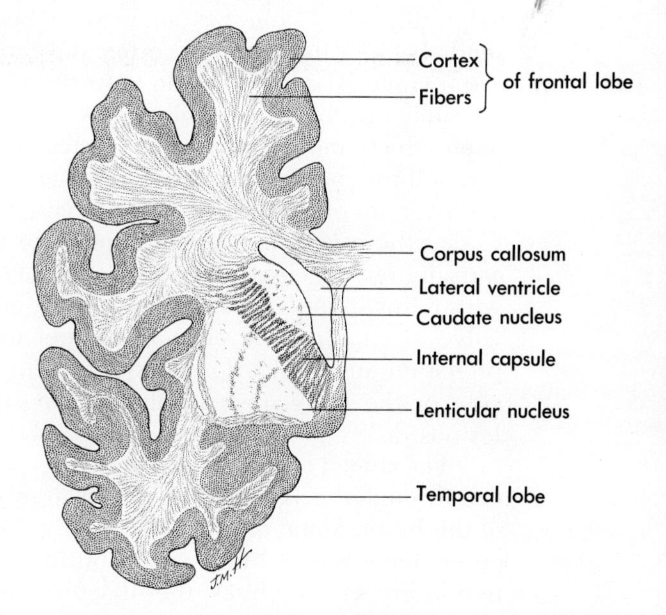

Figure 21–6. A frontal section through the front part of an entire cerebral hemisphere of a child.

ward them to the cerebral hemispheres; and the hypothalamus which is an important center of reflex action, being especially concerned with reflexes involving the autonomic nervous system, and correlations between the autonomic and the voluntary nervous system such as the maintenance of body temperature. It also governs the release of hormones from the hypophysis (pituitary gland). The midbrain serves as a center for reflexes involving the eye and ear, but much of it consists of fibers passing to or from higher centers. The pons is the segment of brain stem succeeding the midbrain. It gets its name, which means "bridge," from the heavy bundle of transverse fibers on its ventral surface that go to the cerebellum, and thus seem to connect the two sides of the cerebellum (they are really a connection from the cerebral cortex to the cerebellum). The pons, and the medulla that succeeds it, both transmit large ascending and descending tracts, and contain the nuclei of various cranial nerves; they are also important sources of extrapyramidal fibers to the spinal cord. The various centers giving rise to these latter fibers (reticulospinal, vestibulospinal) are in turn under the influence of higher extrapyrămidal centers. Besides those features just mentioned, the medulla (medulla oblongata) contains various cell groups or combinations of groups, usually called centers, that have particularly to do with the control of certain vital functions such as blood pressure, cardiac rate, and respiration. On each side of the ventral midline of the medulla the corticospinal fibers form a prominent projection known as a pyramid; this gives the alternative name, the pyramidal tracts, to the corticospinal tracts, and is the anatomic basis for distinguishing between pyramidal and extrapyramidal motor fibers.

The cerebellum, lying dorsal to the pons and an upper portion of the medulla, is primarily concerned with helping to control voluntary movement. It receives impulses from many sources, but chiefly from muscles and from the cerebral cortex; unlike the cerebral cortex, the cerebellar cortex has nothing to do with sensation. The cerebellum helps to control voluntary movement both through impulses that it sends back to the cerebral cortex and through others that it sends to extrapyramidal centers in the brain stem. Damage to the cerebellum may affect primarily the ability to keep the trunk balanced, but more often affects movements of the limbs: without the influence of the cerebellum, the muscles of the limb lack the coordination in sequence and strength of contraction necessary to produce a smooth movement, so that there is tremor on attempting to move the limb, difficulty in carrying out precise movements such as touching the tip of the nose with a finger, and difficulty in carrying out rapid movements that involve alternating use of two different muscle groups—for instance, pronation and supination. Lesions of the cerebellum affect the same side of the body, instead of the opposite side as cerebral cortical lesions do.

The **cranial nerves** consist of twelve pairs of nerves that make their exit through the skull and therefore largely arise from various portions of the brain. Some of these nerves (fig. 21-13) are widely distributed; others have a very limited distribution. The first and second cranial nerves are sensory ones, the olfactory nerve for smell and the optic nerve for sight. The fibers of the olfactory nerves originate from cells that lie in the nasal mucosa and join small olfactory bulbs that lie just above the nose on the under surface of the cerebral hemispheres, while those of the optic nerves originate from cells in the retina of the eyeball and run to the diencephalon. With the exception of these two nerves, the sensory fibers of the cranial nerves arise from ganglia similar to those of the dorsal root ganglia of spinal nerves; and the motor fibers originate from cell groups in the brain stem known as motor nuclei, the equivalent of the ventral horn cells in the spinal cord, except that instead of being a continuous mass most of the motor nuclei supply fibers to only one nerve. The incoming sensory fibers also end largely in a discrete collection of nerve cells, so that there are both motor and sensory nuclei for the cranial nerves. Those nerves that contain parasympathetic preganglionic fibers also have nuclei composed of the cell bodies of these fibers.

The third (oculomotor), fourth (trochlear), and sixth (abducens) nerves are motor to the muscles of the eyeball; the third and fourth nerves arise from the midbrain, the latter being the only nerve to leave the brain dorsally, and the sixth nerve arises from the upper end of the medulla. The large fifth, or trigeminal, nerve contains both motor and sensory fibers; the motor fibers supply the muscles of the lower jaw, the sensory fibers supply all the scalp in front of a line between the ears and practically the entire face, including not only skin but also the teeth, the anterior part of the tongue, and the membranes of the mouth, nose, and paranasal sinuses. This nerve possesses a large sensory ganglion which lies within the skull; the three branches from which the nerve derives its name make their exits through separate openings in the skull. The fifth nerve attaches to the brain stem at the pontine level. Its long sensory root between the trigeminal ganglion and the brain stem is sometimes cut in order to alleviate attacks of severe pain in the face (trigeminal neuralgia).

The seventh or facial nerve has motor fibers distributed to all the muscles of the face and to a few other, deeper, muscles. In addition to this it sends sensory fibers of taste to the anterior two thirds of the tongue, and parasympathetic motor fibers to be distributed to some of the salivary glands (which open into the mouth, p. 62) and to the mucous membrane of the nose. In its course the facial nerve passes very close to the internal ear and then runs through the mastoid portion of the temporal bone. In this latter location it may be affected by infection of the mastoid air cells, or may be injured in operations upon this bone. Injury to the facial nerve in its peripheral course causes paralysis of the facial muscles on that side, and this condition is referred to as facial or Bell's palsy. In such paralysis the muscles of the sound side, by their contractions, pull the mouth toward that side, producing an asymmetry of the face even at rest. The facial asymmetry is more marked if an individual so affected attempts to smile, frown or close the eyes tightly.

The eighth or vestibulocochlear nerve is distributed to the internal ear, and conveys impulses from the parts of the ear that are concerned both with hearing and with balance; this nerve does not leave the skull, as the internal ear is entirely enclosed within the temporal

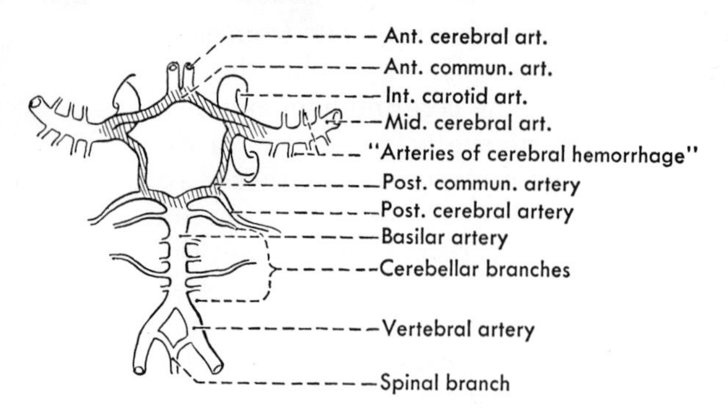

Figure 21–7. Schema of the arteries at the base of the brain and their connections through the arterial circle; the latter is shaded.

bone. The ninth (glossopharyngeal) and tenth (vagus) nerves arise as a series of rootlets from the lateral side of the medulla and leave the skull with the accessory or eleventh cranial nerve through the anterior part of the foramen (jugular foramen) by which the sigmoid sinus passes to the internal jugular vein. The ninth nerve contains both sensory and motor fibers, the motor fibers being distributed to a single small muscle connected with the pharynx, the sensory fibers being distributed to the pharynx and also to the posterior third of the tongue, in which location they subserve taste and general sensations. This nerve also contains parasympathetic fibers for the parotid salivary gland.

The tenth cranial nerve sends voluntary motor fibers to the muscles of the pharynx and larynx, sensory fibers to the pharynx, larynx, and trachea, and involuntary motor fibers to most of the thoracic and abdominal viscera. Its wide distribution has earned for this nerve the name of vagus or wanderer. The accessory or eleventh cranial nerve arises in part by rootlets in line with the ninth and tenth nerves and in part by rootlets that arise from the upper four or five segments of the spinal cord and run upward to join the medullary rootlets. As the accessory nerve passes through the jugular foramen with the vagus nerve the fibers in the cranial rootlets of the accessory join the vagus and are thereafter distributed with it; they are really, therefore, much more logically regarded as vagal than as accessory nerve fibers. The fibers of the spinal rootlets alone form the accessory nerve of gross anatomy, which carries voluntary motor fibers to the sternocleidomastoid and trapezius muscles (pp. 354 and 91). The twelfth or hypoglossal nerve arises from the floor of the medulla; it is a purely motor nerve, carrying fibers to the voluntary muscles of the tongue.

Blood Supply of the Brain

The brain receives its blood supply from the paired vertebral and internal carotid arteries (fig. 21-7). The vertebral arteries, branches of the subclavian arteries at the base of the neck, run up in a deep position in the neck (in part through the holes in the transverse processes of cervical vertebrae) and penetrate the posterior atlantooccipital membrane (p. 206) to enter the vertebral canal, pass upward through the foramen magnum, and unite upon the anterior surface of the medulla to form the basilar artery. From the vertebrals and the basilar are given off vessels to the cerebellum, medulla and spinal cord, and to the pontine region of the brain. The basilar artery ends in the lower diencephalic region by dividing into two posterior cerebral arteries; these arteries are distributed to the lower part of the medial surfaces of the cerebral hemispheres and to the inferior surfaces of the hemispheres. Close to its origin from the basilar each vessel receives a posterior communicating branch that unites it to the internal carotids.

The internal carotid arteries have a devious course through the skull and come in contact with the brain at about the front end of the diencephalon. They give off large middle cerebral arteries that are distributed to most of the lateral surfaces of the cerebral hemispheres. Smaller anterior cerebral arteries also arise from the internal carotids and are distributed to the upper portions of the medial surfaces of the hemispheres. The internal carotids are connected by the posterior communicating arteries to the posterior cerebrals and hence to the basilar, and the anterior cerebrals are connected to each other by an anterior communicating artery; thus a complete "circle" of arteries (really a hexagon), called the arterial circle of the cerebrum (circle of Willis), is formed at the base of the brain. A knowledge of the detailed distribution of the various vessels just mentioned, and of other small but important branches, is of great importance to the neurologist and neurosurgeon but cannot be entered into here. It might be pointed out, however, that certain small branches from the middle cerebral artery and the arterial circle supply the internal capsule with its ascending and descending tracts connecting to the cerebral hemispheres; since these vessels have no free anastomoses with other vessels their occlusion leads to damage to the function of some or most of these pathways, resulting in the more common type of "stroke." Because of the relative frequency with which these arteries are involved when vascular accidents occur in the brain, these vessels are known as the arteries of cerebral hemorrhage. This should not be interpreted, however, to mean that all cerebral hemorrhages or all strokes are due to damage to these specific vessels.

The venous drainage of the brain is by numerous, mostly unnamed, veins that in general run into the nearest cranial venous sinuses. Thus the veins from the convexity of the cerebral hemispheres pass into the superior sagittal sinus, those from posterior parts of the inferior surfaces of the hemispheres and from the upper surface of the cerebellum pass into the transverse sinuses, and so forth.

THE FACIAL MUSCLES

The facial muscles differ from most voluntary muscles in that, instead of moving one bone upon another, they move primarily skin. While many have some attachment to the bones of the face their insertions are chiefly into the skin; because of their effect upon facial expression they are sometimes known as the **mimetic muscles.**

The facial muscles (fig. 21-8) may be divided into four groups, one group centering about the mouth, a second about the nose, a third about the eye and forehead, and a fourth about the ear. A single muscle, the platysma, forms a fifth group which, although it lies

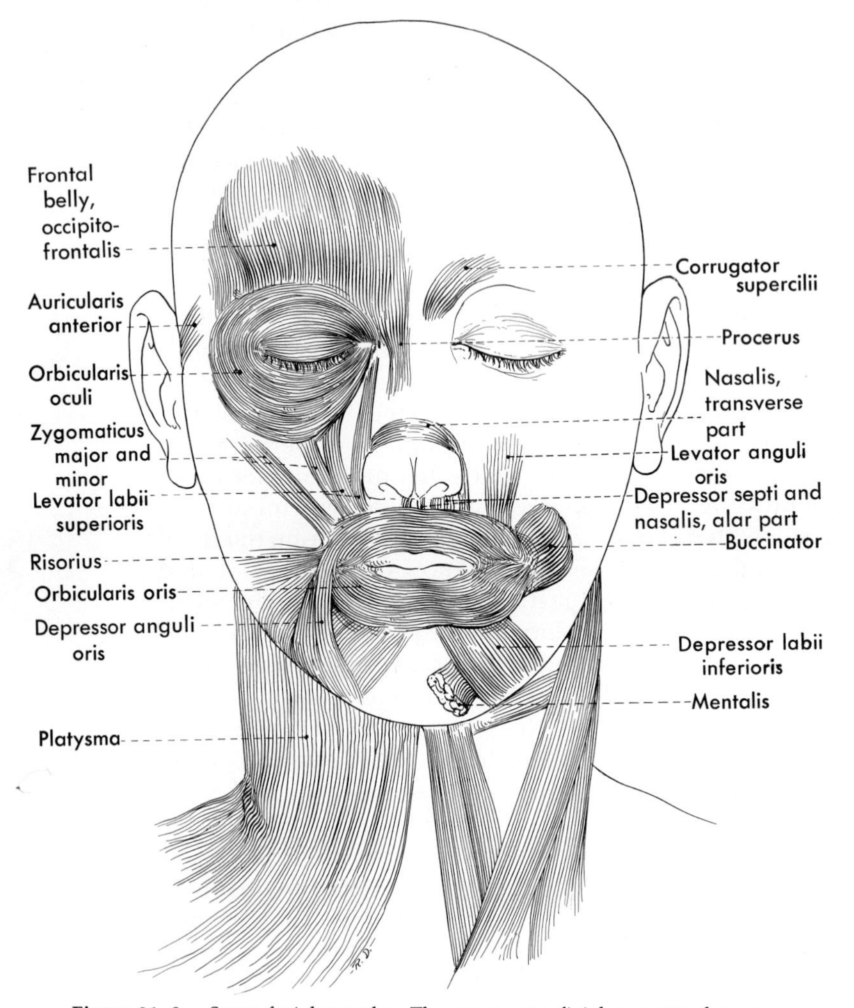

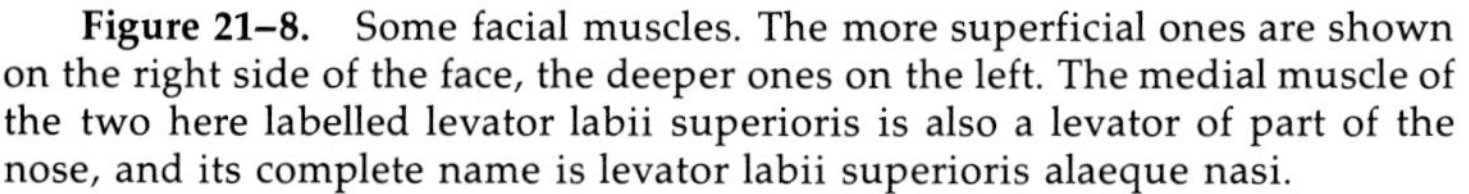

Figure 21–8. Some facial muscles. The more superficial ones are shown on the right side of the face, the deeper ones on the left. The medial muscle of the two here labelled levator labii superioris is also a levator of part of the nose, and its complete name is levator labii superioris alaeque nasi.

chiefly in the neck rather than in the face, really belongs to the facial group.

The paired muscles about the mouth include the **depressor anguli oris** and the **depressor labii inferioris,** which pull the lower lip downward; the **risorius** and the two **zygomaticus muscles** which pull the corners of the mouth laterally, and laterally and upward; and the **levator labii superioris** and the **levator anguli oris** which lift the upper lip. The **orbicularis oris** encircles the mouth and therefore closes it or puckers the lips as for a kiss.

The **buccinator** (trumpeter) muscle forms the fleshy wall of the cheek, and cooperates with the tongue in moving the food about the mouth during chewing; by its contraction it decreases the size of the mouth cavity, and this action is made use of in swallowing, blowing wind instruments, and the like.

The muscles connected with the nose are primarily compressors, depressors or dilators of the nares or external openings of the nose. They are small and vary considerably in their development.

The important facial muscle connected with the eye is the **orbicularis oculi,** a broad muscle that surrounds the orbit and extends into both upper and lower lids; this muscle is responsible for closing the lids. The upper lid is raised by a muscle lying within the orbit.

The musculature of the forehead consists of the **frontalis** (the frontal belly of the occipitofrontalis) which runs downward from the scalp to the skin above the orbit, and wrinkles the forehead transversely, and the **corrugator supercilii** which produces the small vertical folds between the eyebrows that we associate with a worried look. Since the frontalis muscle attaches to the scalp, it and a corresponding scalp muscle in the occipital region, the **occipitalis** (the occipital belly of the occipitofrontalis), may move the scalp backward and forward. The occipitofrontalis is the major part of the musculature of the scalp, the epicranial muscle.

There are several small muscles connected with the ears, so placed that they may move the ears forward, upward or backward. These muscles are rudimentary in man, and are not subject to voluntary control in many individuals.

The **platysma** (flattest) muscle lies in the superficial fascia of the neck, extending from the upper part of the thorax to the lower jaw, where its fibers interlace with the musculature about the mouth. This muscle aids in depressing the lower jaw, thus in opening the mouth, or taking its fixed point from above makes taut the superficial fascia of the neck.

The entire group of facial muscles, including also the platysma, is innervated by branches of the seventh cranial, or facial, nerve. The **facial nerve** rounds the posterior aspect of the ramus of the mandible, passing through the parotid gland, and branches out to reach all the muscles of the face. Interruption of this nerve as a whole, therefore, produces a unilateral facial paralysis (p. 347).

THE ORBIT

The eyeball is moved by a number of muscles (fig. 21-9) all of which lie within the orbit. They are divided into four rectus (straight) muscles and two obliques according to the way in which they insert on the eyeball. The **superior, inferior, medial** and **lateral recti** direct the pupil upward, downward, medially and laterally respectively. The **superior oblique** directs the pupil downward and laterally, the **inferior oblique** directs it upward and laterally, and the two muscles together tend to overcome the slight medial pull exerted by the superior and inferior recti. All these muscles, with the exception of the inferior oblique, arise from or close to a tendinous ring placed at the apex (back end) of the orbit; the inferior oblique arises forward, from the floor of the orbit. In addition to these muscles of the eyeball, the orbit contains also the levator of the upper lid, termed the **levator palpebrae superioris.** This muscle works in conjunction with the superior rectus, so that as the pupil is turned upward the lid is further raised. It is

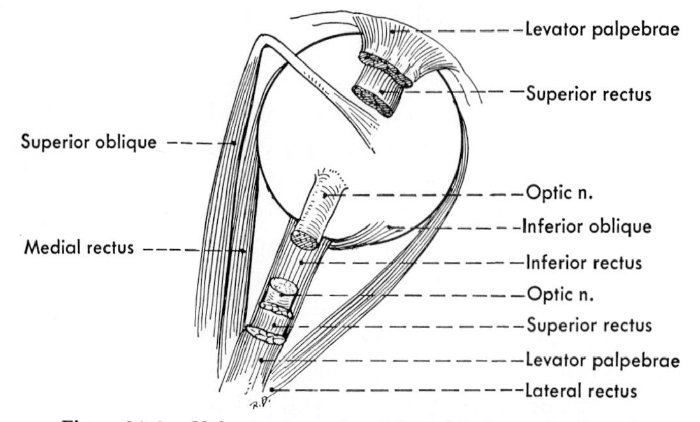

Figure 21–9. Voluntary muscles of the orbit. Segments have been removed from the levator palpebrae, superior rectus, and optic nerve, and the back of the eyeball with its attached portion of optic nerve has been rotated up and forward.

attached to the lid both through tendinous fibers and through a smooth muscle that by its contraction aids in maintaining the open position of the lid.

Three cranial nerves supply the voluntary muscles of the orbit; the fourth cranial, or trochlear, nerve supplies the superior oblique muscle, the sixth or abducens supplies the lateral rectus, and the remaining muscles of the orbit are supplied by the third or oculomotor nerve. The oculomotor nerve also brings into the orbit fibers for the supply of the smooth muscles regulating the size of the pupil and the shape of the lens. These nerve fibers end in the ciliary ganglion, a ganglion about the size of the head of an ordinary pin located between the optic nerve and the lateral rectus muscle. The activity of the parasympathetic fibers in the oculomotor nerve and the branches from the ciliary ganglion produces constriction of the pupil of the eye and accommodation of the lens for near vision. Sympathetic fibers derived from the plexus on the internal carotid artery also supply the eye, to produce dilation of the pupil.

In addition to the three nerves supplying ocular muscles the optic nerve, the large nerve of sight, occupies a prominent position in the orbit. The branches of the ophthalmic division of the fifth nerve traverse the orbit in their course toward the face, and the maxillary division of the fifth nerve runs for some distance in the floor of the orbit.

THE JAW AND TONGUE

Two muscles of the lower jaw are largely superficial (fig. 21-10), others lie deeper. The **masseter** muscle arises from the zygomatic arch and passes almost straight downward to cover and insert into the ramus (vertical part) and angle of the mandible. The **temporal** muscle arises from much of the side of the skull in front of and above the ear, and inserts into the anterior of the two projections on the upper border of the ramus (coronoid process), and into the inner surface of the anterior part of the ramus. The two pterygoids lie deeper, and arise from the base of the skull. The **lateral** (external) **pterygoid** arises from the outer surface of a projection, the lateral lamina of the pterygoid process, and from part of the base of the skull lateral to that; instead of running downward, its fibers run largely backward to insert into the neck of the mandible (below the condyle) and into an articular disk that lies in the temporomandibular joint. The **medial** (internal) **pterygoid** arises mostly from the inner surface of the lateral pterygoid lamina and inserts into the inner surface of the angle of the mandible.

The lateral pterygoid draws its side of the jaw forward; both sides working together thus protract the jaw, a movement necessary to wide opening of the mouth. The masseter, temporal, and medial pterygoids are powerful closers of the jaw, and also move the mandible from side to side to produce the grinding movement between upper and lower teeth. These masticator (chewing) muscles are all innervated by the third (lowest) or mandibular branch of the fifth cranial (trigeminal) nerve.

Most of the other muscles attached to the jaw form the floor of the mouth or aid in movement of the tongue. The tongue muscles attach in part to the mandible and in part to the hyoid bone, which may be considered the bone of the tongue. The tongue itself (fig. 21-12) is a fleshy mass composed of interlacing voluntary muscle fibers arranged longitudinally, horizontally, and transversely. The muscles passing into the tongue from the hyoid bone and mandible act with the intrin-

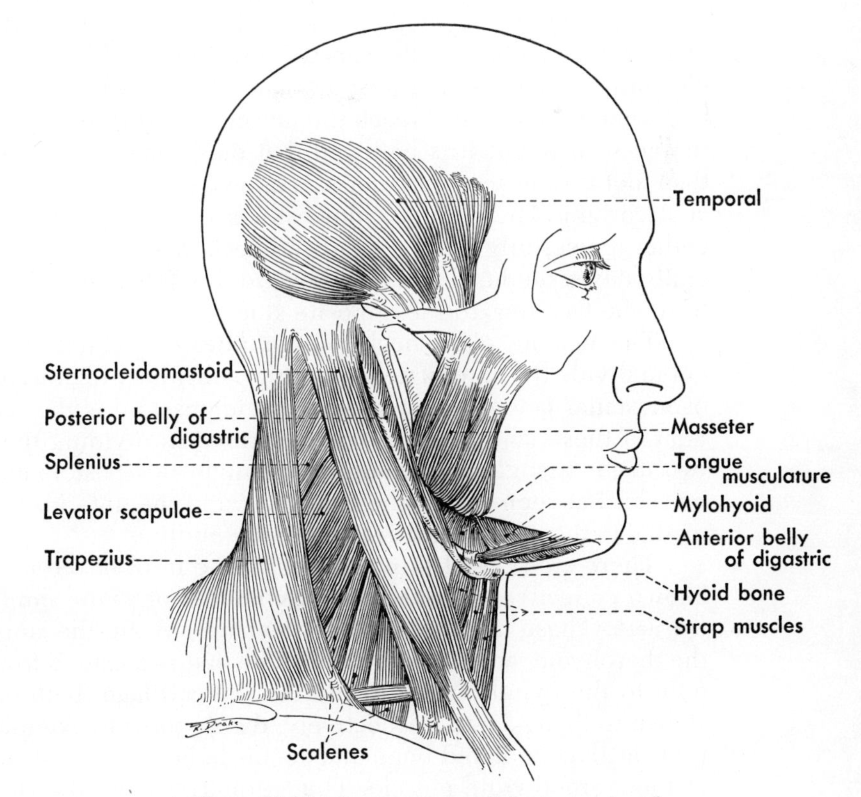

Figure 21–10. Some muscles of the jaw and neck.

sic muscles of the tongue to carry out the various movements of protraction, retraction, changes in shape, and so forth, which this very mobile organ undergoes in speaking, chewing, swallowing, and the like. The muscles of the tongue are innervated by the hypoglossal, or twelfth cranial, nerve.

The most prominent muscle of the floor of the mouth, the **mylohyoid,** stretches across from one side of the mandible to the other, and is innervated by the mandibular branch of the fifth nerve. By its contraction it raises the floor of the mouth and the base of the tongue; since it is attached to the hyoid bone, it may also assist the digastric in opening the mouth. The **digastric** (two-bellied) muscle extends from the mastoid process of the skull to the back of the chin, the tendon connecting the two bellies being attached by a sling of fascia to the hyoid bone. The muscle aids in opening the mouth or raising the hyoid bone; its posterior belly is innervated by the facial nerve, its anterior belly by the trigeminal.

THE MUSCULATURE OF THE NECK

The heaviest musculature in the neck is the upward continuation of the muscles of the back, which we have already considered. In addition to these muscles and the trapezius, the **sternocleidomastoid** is a prominent component of the muscles of the neck. This muscle (fig. 21-10) arises by a tendinous head from the manubrium sterni and by fleshy fibers from the medial third of the clavicle; it inserts upon the mastoid process behind the ear. Its innervation is by the accessory, or eleventh cranial, nerve. Fibers from upper cervical nerves, usually C2 and sometimes C3, also reach the muscle either directly or by joining the accessory, but it is believed that these fibers are sensory rather than motor—the accessory nerve is known to contain no sensory fibers at its origin. The sternocleidomastoids of the two sides, acting together, powerfully flex the head and neck; the muscle of one side laterally flexes the head and neck toward this side and at the same time turns the face toward the opposite side.

The anatomy of the neck is so complicated that it is often convenient to divide the anterolateral part into subsidiary regions on the basis of muscular landmarks. The sternocleidomastoid is the most prominent of these landmarks, and is described as dividing the neck into two major triangles: the posterior triangle lies posterolateral to the muscle, between it and the trapezius, and the anterior triangle lies medial to the muscle and extends to the midline.

There are several thin muscles in the front of the neck that are known collectively as the **infrahyoid muscles** or as the **strap muscles of the neck.** These muscles include the sternohyoid, the sternothyroid, the thyrohyoid, and the omohyoid. The first two extend from the sternum to the hyoid bone and the thyroid cartilage (both of these are shown in figure 21-12) respectively; the thyrohyoid extends from thyroid cartilage to hyoid bone and seems to be an upward continuation of the sternothyroid muscle. The sternohyoid and the sternothyroid muscles cover the anterior and lateral surfaces of the thyroid gland.

The omohyoid muscle arises from the scapula, hence the name "omo" (shoulder) and inserts upon the hyoid bone. It consists of two bellies, a superior and an inferior, united by a tendon that is anchored to the clavicle by a sling of fascia. The superior belly of the omohyoid muscle runs almost parallel but lateral to the sternohyoid; the inferior belly runs almost transversely across the base of the neck from its attachment on the shoulder.

The strap or infrahyoid muscles are innervated by a nerve loop, the ansa cervicalis, derived usually from branches of the first three cervical nerves. The suprahyoid and infrahyoid muscles cooperate in moving the hyoid bone and thyroid cartilage up and down as in swallowing or singing up or down the scale.

The remaining muscles in the neck (those of the pharynx will be mentioned in connection with this organ) are connected either with the larynx (voice box) or with the ribs and vertebral column. The musculature of the larynx is complex and there is no necessity of describing it here. Most of the laryngeal musculature is supplied by the recurrent branch of the vagus nerve, which leaves the main vagus stem in the lower part of the neck or in the chest (differing on the two sides) and ascends alongside the trachea and esophagus to reach the level of the larynx. The recurrent nerve is subject to injury during operations upon the thyroid gland, and such injury, if bilateral, may have serious consequences. Paralysis of the musculature of the larynx may lead to gradual closure of the air passage, so that breathing becomes more and more difficult.

Most of the muscles connected with the vertebral column in the neck have already been described in connection with the back muscles; anterolaterally in the neck, however, there are three that arise from the vertebral column and insert upon the ribs. These are the three **scalene muscles,** anterior, middle, and posterior, respectively (fig. 21-11). The anterior scalene muscle arises from anterior tubercles of the transverse processes of approximately the third to sixth cervical vertebrae and inserts on the upper surface of the first rib toward its sternal attachment. The scalenus medius has a similar origin from the

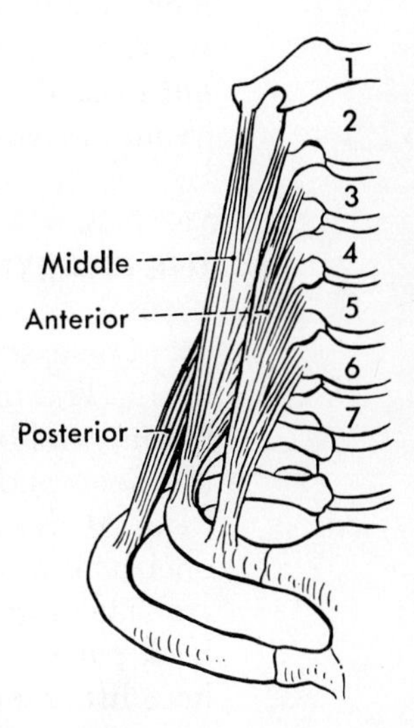

Figure 21–11. The scalene muscles.

lower five or six or all the cervical vertebrae and inserts also into the first rib. The scalenus anterior and the scalenus medius are separated from each other by the roots of the brachial plexus and by the subclavian artery. The scalenus posterior lies behind the scalenus medius and arises from the transverse processes of about the fifth and sixth vertebrae, and passes down to insert onto the second rib.

The scalenes are innervated by short branches from the ventral rami of the cervical nerves corresponding to the levels of their origins. Taking their fixed points from below, they flex the neck laterally or turn it slightly toward the opposite side. Taking their fixed points from above they are respiratory muscles, serving to fix the first two ribs in quiet inspiration, and to raise these ribs in forced inspiration. Because of its position between the anterior and middle scalenes the brachial plexus is subject to pressure from the contraction of these muscles, or to being stretched over a first rib raised by their spastic contraction. Spasm of the anterior scalene, in particular, has been thought to be responsible for certain cases of injury to the plexus at the base of the neck, and the operation of scalenotomy or section of the anterior scalene has been reported to relieve selected cases. The "scalenus anticus syndrome," like other signs of compression of the brachial plexus, is often precipitated or exaggerated by a lower position of the shoulder, and is therefore often amenable to physical therapy (p. 88).

The deepest anterior muscles of the neck are two short ones that extend between the atlas and the skull, and thus are similar to the muscles lying deep to the long muscles of the back (p. 215); and two muscles, the **longus colli** and **longus capitis,** that largely cover the fronts of the bodies of the cervical vertebrae (fig. 21-14). The longus muscles arise from the bodies and transverse processes of cervical and upper thoracic vertebrae, and insert upon upper cervical vertebral bodies and the skull. They are supplied by twigs from the adjacent spinal nerves, and primarily flex the neck and head.

An abnormal position of the head and neck (*torticollis* or wryneck) may be maintained by various muscles of the neck. In the simplest type the sternocleidomastoid is responsible, but in other types there seems to be a contraction of many muscles—the large muscles such as the sternocleidomastoid, trapezius, scalenes, and vertebral muscles, the short muscles associated with the skull, and even the hyoid musculature.

THE PHARYNX, LARYNX, TRACHEA AND ESOPHAGUS

The essential anatomy of these structures is diagrammed in figure 21-12. The pharynx is formed by the union of nasal and oral cavities behind and below the soft palate, and is therefore the common tube in the head and upper part of the neck for the passage of both air and food. At the level of the thyroid cartilage, airway and food passages once again separate.

The soft palate is essentially a muscular partition between the back parts of the oral and nasal cavities. The most important muscles here lift or tense the soft palate; these attach to the skull, but others extend from the soft palate to the walls of the pharynx or to the tongue. Movements of the soft palate are essential for proper phonation and for swallowing. Defects in the soft palate, especially as represented in cleft palate but as also found in paralysis of the musculature of the palate, lead to difficulty in swallowing and sometimes to the outflow of liquids through the nose.

Also essential to swallowing is the action of the voluntary muscles in the walls of the pharynx; these muscles are so arranged as to compress the pharynx and are thus termed the constrictors of the pharynx. The upper constrictor is continuous with the buccinator muscle, the facial muscle bridging the gap between the two jaws and thus forming the substance of the cheek. The superior, middle, and inferior pharyngeal constrictors so overlap each other that by their contraction, beginning above, they pass food or liquids down into the esophagus; their malfunction thus leads to difficulty in swallowing. The pharyngeal constrictors and most of the muscles of the soft palate are innervated by pharyngeal branches of the vagus nerves; the sensory supply to the posterior wall of the pharynx is partly through the ninth, or glossopharyngeal, nerves, partly through the vagi.

When the pharynx ends by dividing into a respiratory and a digestive tube, the air passage is continued ventrally as the larynx and trachea and the food passage is continued dorsally as the esophagus. The walls of the air passage are held open by special cartilages in the region of the larynx and by a series of C-shaped cartilaginous rings in the trachea; the opening into the larynx can, however, be closed, as in holding the breath, by the action of special muscles acting upon certain laryngeal folds. One pair of these folds is termed the vocal folds or vocal cords, which not only can be brought together in the midline but also can be lengthened or shortened; these folds are thrown into vibration by the passage of air between them to produce sound, and

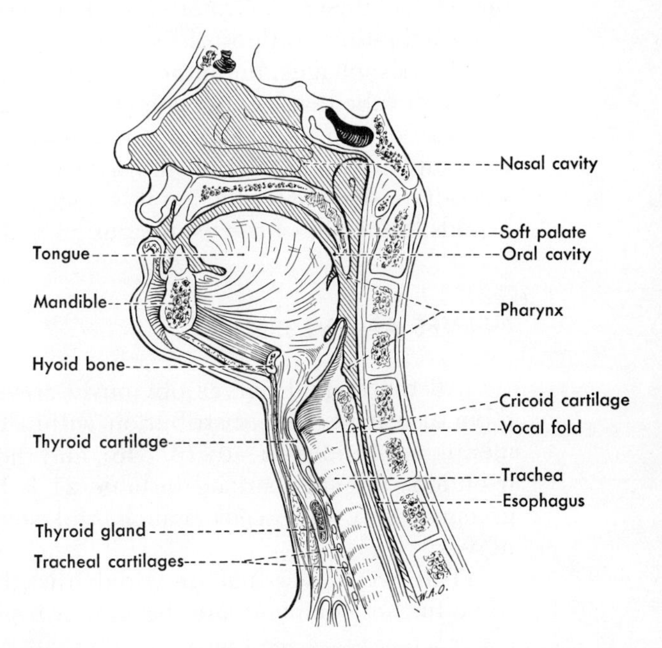

Figure 21–12. The digestive and respiratory tracts in the head and neck, as seen in sagittal section.

TABLE 21–1. Chief Functions and Distributions of the Cranial Nerves

NERVE	AFFERENT	EFFERENT
I. Olfactory	Smell: nose	
II. Optic	Sight: eye	
III. Oculomotor		Vol. motor: levator of eyelid, sup., med., and inf. recti, inf. oblique of eyeball Autonomic: smooth muscle of eyeball
IV. Trochlear		Vol. motor: sup. oblique of eyeball
V. Trigeminal	Touch, pain: skin of face, mucous membranes of nose, sinuses, mouth, anterior tongue	Vol. motor: muscles of mastication
VI. Abducens		Vol. motor: lat. rectus of eyeball
VII. Facial	Taste: anterior tongue	Vol. motor: facial muscles Autonomic: lacrimal, submandibular and sublingual glands
VIII. Vestibulocochlear	Hearing: ear Balance: ear	
IX. Glossopharyngeal	Touch, pain: posterior tongue, pharynx Taste: posterior tongue	Vol. motor: unimportant muscle of pharynx Autonomic: parotid gland
X. Vagus	Touch, pain: pharynx, larynx, bronchi Taste: tongue, epiglottis	Vol. motor: muscles of palate, pharynx, and larynx Autonomic: thoracic and abdominal viscera
XI. Accessory		Vol. motor: sternocleidomastoid and trapezius
XII. Hypoglossal		Vol. motor: muscles of tongue

the pitch of the voice is altered by changes in the length and tenseness of the vibrating portions of the cords.

The esophagus, the more direct continuation of the pharynx, is a muscular tube that lies between the trachea and the vertebral column. The part of it in the neck has voluntary muscle in its walls, but before the esophagus reaches the stomach this muscle has been replaced by smooth muscle. Except when it is distended by the passage of food or liquids, the walls of the esophagus are collapsed.

NERVES AND VESSELS

All the cranial nerves obviously arise in the head, and most of them also have their distribution within the head. They have been mentioned briefly already (p. 346), and their chief functions and distributions are summarized in table 21-1. Figure 21-13 presents diagrammatically the major branches of several of the larger cranial nerves.

The two nerves that are widely distributed in the head, and deserve further comment, are the fifth or trigeminal and the seventh or facial nerves. The **trigeminal nerve** consists of three branches, of which the upper two contain only sensory fibers while the third is a mixed nerve. The upper, first, or ophthalmic division of the trigeminal nerve is distributed to the skin of the forehead and much of the scalp as far back as the ears, the upper eyelid, and the bridge of the nose; it also supplies certain paranasal sinuses.

The second, middle, or maxillary division of the trigeminal nerve supplies skin over the prominence of the cheek, the side of the nose, and the upper lip. It also supplies much of the interior of the nose, most of the paranasal sinuses, the roof of the mouth, and the upper jaw and its teeth. This is the nerve the dentist blocks for work on the upper jaw.

The lowermost, third, or mandibular branch of the fifth nerve supplies most of the muscles of mastication, and is also distributed to the lower jaw and its teeth, to the anterior part of the tongue, and to the skin over the lower jaw including an area extending upward in front of the ear.

The cutaneous branches of these three divisions of the trigeminal nerve appear in various positions upon the face. The main branch of the ophthalmic nerve becomes subcutaneous by rounding the upper border of the orbit not far from the bridge of the nose. The chief cutaneous branch of the maxillary nerve appears on the cheek by passing through a foramen below the orbit. The cutaneous branches of the mandibular are rather widely separated as they approach the skin, one appearing just in front of the ear, one appearing in the cheek, and one

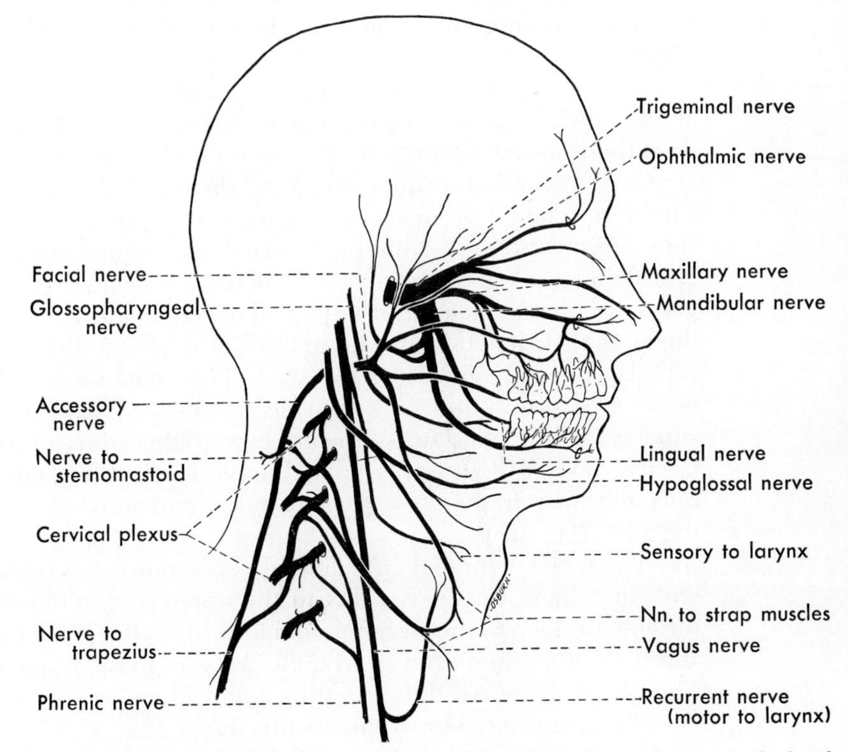

Figure 21–13. Major branches of some of the principal cranial nerves in the head and neck, and some connections and branches of the cervical plexus.

presenting itself on the chin through the prominent mental foramen after having run within the lower jaw and supplied the teeth.

The **facial nerve** is likewise widely distributed on the face; in contrast to the trigeminal nerve the facial is almost entirely voluntary motor. The major portion of the facial nerve rounds the posterior border of the ramus of the mandible, traversing the parotid gland, and divides within the gland into a number of branches that spread out to supply the musculature of the face. In addition, the facial nerve contains sensory fibers for taste that are carried to the tongue by a branch of the trigeminal nerve, and contains also parasympathetic (secretory and vasomotor) fibers that accompany various branches of the trigeminal nerve to some of the salivary glands and blood vessels of the head.

In the neck, the upper cervical nerves, with the frequent exception of the first, enter into the formation of the **cervical plexus** (fig. 21-13). This is formed chiefly by the union of the ventral rami of the second to fourth cervical nerves. From this plexus arise a number of cutaneous nerves that supply skin of the neck and of the head, and muscular branches into adjacent muscles, especially the longus muscles and the levator scapulae. A portion of the ansa cervicalis, the major supply to the strap muscles, is derived directly from the cervical plexus; the other portion is derived indirectly from the cervical plexus by the union of fibers of this plexus with the hypoglossal nerve. The **phrenic nerve,** the nerve to the diaphragm, is also derived largely from the cervical plexus since it usually contains fibers from C3, C4 and C5. The phrenic nerve traverses the lower part of the neck almost vertically, lying on the anterior surface of the anterior scalene muscle. Leaving this muscle, it runs the length of the thorax to reach the diaphragm.

In addition to the cervical plexus and its branches, mentioned above, and the brachial plexus and its branches, which have been described in connection with the upper limb, the nerves in the neck are the tenth and eleventh cranial and the cervical sympathetic trunk. The tenth cranial or **vagus nerve** after making its exit from the skull runs downward just behind the internal and common carotid arteries. It supplies most of the musculature of the soft palate and pharynx, and gives off a branch (superior laryngeal nerve) that is largely sensory to the larynx. Most of the muscles of the larynx are supplied by the recurrent laryngeal branches of the vagi. The right recurrent laryngeal nerve arises from the right vagus as this nerve passes in front of the subclavian artery and turns upward behind this artery to ascend to the larynx; the left recurrent laryngeal nerve leaves the vagus in the thorax rather than in the neck, passing below and then behind the arch of the aorta (fig. 22-4).

The **accessory** or eleventh cranial nerve at first lies in close contact with the vagus, but leaves this in the upper part of the neck to pass through or deep to the sternocleidomastoid muscle, which it supplies, and then cross the posterior triangle and end in the trapezius muscle. The accessory nerve thus supplies motor fibers to only two muscles.

The paired **cervical sympathetic trunks** (fig. 21-14)—also called chains—lie deeply, on the front of the longus muscles. Each is an upward continuation of the thoracic sympathetic trunk (see page 56 for a summary of the sympathetic nervous system), and like that is composed of ganglia and strands of nerve fibers that connect the ganglia. The ganglia are not regularly placed, however, and instead of the expected eight, to correspond with the eight cervical nerves, there are typically only three or four: a superior cervical ganglion, placed high in the neck, an inferior or cervicothoracic (stellate) ganglion at the base of the neck, a small "middle" ganglion lying closer to the inferior than to the superior one, and a vertebral ganglion lying still lower; either of the latter two ganglia may be missing. These ganglia give off fibers, gray rami communicantes, to all the cervical nerves (the inferior or stellate being therefore the chief sympathetic supply to the blood vessels and sweat glands of the upper limb). They also give rise to descending nerves that help supply the heart, increasing the rate and strength of the heart beat. Finally, the superior cervical ganglion sends a great number of fibers upward along the external and particularly the internal carotid arteries (see the following paragraph). These fibers supply blood vessels, sweat glands, and other structures of the face and head including a muscle that dilates the pupil of the eye. Because all the preganglionic fibers to the cervical sympathetic ganglia come through thoracic nerves (cervical nerves have no white rami communicantes) interruption of the cervical sympathetic trunk anywhere will cut off the sympathetic innervation to the face and head.

At the base of the neck on the right side the **brachiocephalic trunk** (previously called innominate artery), the first large branch from the aorta, divides into right subclavian and right common carotid vessels (fig. 21-15). On the left side the left common carotid and left subclavian arteries appear as separate branches from the aorta (fig. 22-4). The **subclavian arteries** arch across the base of the neck to enter the axillae, while the **common carotid arteries** run upward lateral to the trachea and esophagus. The pulse of these latter vessels may easily be felt in the neck. At about the level of the hyoid bone the common carotids

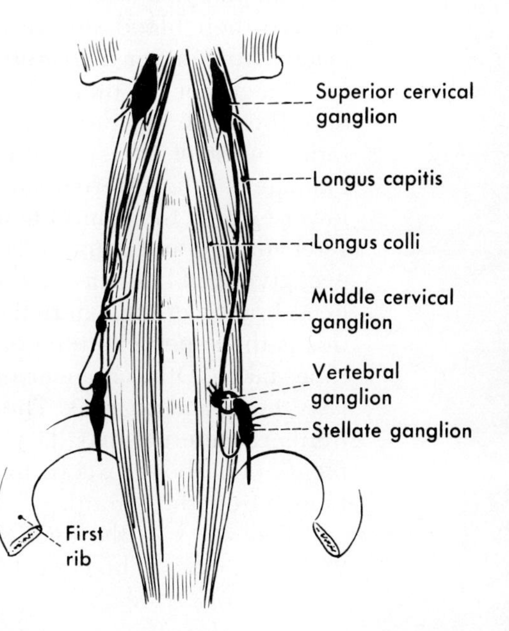

Figure 21–14. The cervical sympathetic trunks; the two trunks are not necessarily symmetrical, and there may be both a middle cervical and a vertebral ganglion on one side.

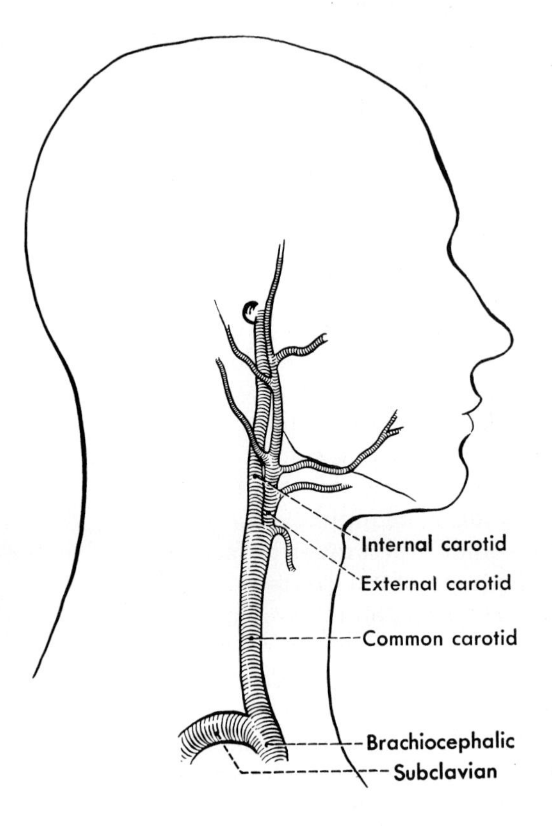

Figure 21–15. Diagram of the carotid arteries.

branch into external and internal carotids. The internal carotids continue the upward course of the common carotids and end by supplying the brain, while the external carotids branch to supply numerous vessels to structures in the region of the head. The thyroid gland (p. 64) is supplied by branches from both external carotids and both subclavians; the pharynx, larynx and upper parts of the trachea and esophagus receive their blood supplies from numerous branches derived ultimately from either the carotid or subclavian arteries.

Except for the brain, the blood supply of which has already been described as being derived from the internal carotid and vertebral arteries, most of the structures of the head are supplied by various branches of the external carotid artery. The facial artery crosses the lower edge of the mandible just in front of the attachment of the masseter muscle and runs upward toward the angle between nose and eye, giving off branches to the face in its course. A second large superficial branch of the external carotid ascends close in front of the ear; this is the superficial temporal artery, from which the pulse is sometimes taken. Other branches of the external carotid supply the tongue, jaws, nose, and pharynx. The largest of these branches is the maxillary (formerly internal maxillary) artery, supplying most of the structures mentioned above except for the tongue which has its own special branch from the carotid.

The veins in the neck are numerous, and connect rather freely with each other. The largest from the head and upper part of the neck are the internal and external jugular veins, the former accompanying the common and internal carotid arteries and the latter lying largely subcutaneously. These veins join the subclavian veins at the base of the neck.

REVIEW OF HEAD AND NECK

The surface anatomy of the head is largely the anatomy of the skull. Thus we speak of frontal, parietal, temporal, and occipital regions of the head, obviously named from the corresponding bones of the cranium. Below the external occipital protuberance, the prominent bump on the back of the head close to the midline, the occipital bone is largely covered by the muscles attaching to it. The part of the temporal bone on the side of the skull in front of the ear is also covered by muscle, but this is thin and the bone can be felt through it. The mastoid process of the temporal bone is easily felt below and behind the external ear; the zygomatic process of the temporal forms the back part of the zygomatic arch, stretching from in front of the ear to the prominence of the cheek. The parietal and frontal bones are largely subcutaneous.

On the face, the rim of the orbit (the upper part formed by the frontal bone) is easily palpable, and the bony and cartilaginous parts of the external nose can be distinguished without difficulty. The prominence of the cheek is formed by the maxillary and zygomatic bones; the tooth-bearing alveolar process of the maxilla can be palpated through or examined by retracting the lips and cheek. Much of the hard palate is formed by horizontal processes from the two maxillae. Most of the mandible can be palpated; its lower border and the back end of its angle are particularly prominent, but its ramus is largely covered by muscle.

The skeletal anatomy that can be examined in the intact neck (except for parts of the vertebral column) is limited to the hyoid bone and the larynx and trachea. The hyoid bone is easily palpated in the anterior midline at about the level of junction of the neck and lower surface of the jaw. Its greater horns curve posteriorly on the sides of the pharynx, and can be followed to their tips. Below the hyoid bone the thyroid cartilage is palpable (and its projection often visible), especially close to the midline; its two sides meet in an anterior ridge, and its upper border is notched above this ridge. Of the several other cartilages of the larynx, only a part of the cricoid cartilage can be felt; this is the rounded bar of cartilage passing across the front of the larynx immediately below the thyroid cartilage. The rings of tracheal cartilage that lie below the cricoid cartilage are palpable with more difficulty, but give the trachea its rough feeling when a finger is drawn along its length (trachea means rough).

Of the **muscles**, only a few can be recognized easily. The masseter and the temporal can both be felt to contract when the jaws are clenched, the former where it covers the angle of the jaw and the latter in front of the ear above the zygomatic arch. Muscles of the floor of the mouth can be felt to contract when swallowing occurs or the tongue is moved vigorously. The only prominent muscle of the neck is the sternocleidomastoid, best felt and seen when the neck is flexed and the face is at the same time turned to the opposite side.

Of the **nerves** and **vessels,** none of the former can be palpated, and neither can any of the large veins. The external jugular vein may however be visible through the skin, particularly where it crosses the anterior surface of the sternocleidomastoid muscle. The pulse of the common carotid artery can be felt lateral to the larynx, and that of the superficial temporal artery (from the external carotid) just in front of the upper part of the ear. Similarly, the facial artery (another branch of the external carotid) can be felt as it crosses the lower border of the mandible immediately in front of the masseter muscle.

The Thorax 22

THE THORACIC WALL

The thoracic wall (fig. 5–1) consists of the sternum anteriorly, the vertebral column posteriorly, and the ribs (costae) and their connecting muscles between the two. Internally, the thorax is separated from the abdomen by the diaphragm, but there is no similar separation between neck and thorax. The sternum consists of three portions: an upper part, the manubrium; a larger part, the body; and a small xiphoid process. It is the manubrium and the body of the sternum that meet to form the sternal angle, and at this location the second rib attaches. To the sternum are attached the clavicle and the costal cartilages of the upper seven ribs.

The thoracic vertebrae form the posterior medial aspect of the wall of the thoracic cavity. Infections—for example, tuberculosis—of this portion of the vertebral column may therefore present themselves within the thoracic cavity.

The greater portion of the thoracic wall consists of the twelve ribs and their cartilages, and the muscles connecting them. The majority of the ribs articulate with both the body and a transverse process of their corresponding vertebrae (for more details see the discussion of the vertebral column), and freely movable joints exist between vertebral and costal elements. The last five ribs, which do not directly reach the sternum, are termed false ribs. The upper three false ribs attach by their costal cartilages to each other and to that of the seventh rib, and thus help to form the costal arch, but the costal cartilages of the last two ribs end freely in the musculature to which they give attachment. The last two ribs are sometimes referred to as floating ribs.

The muscles passing between two adjacent ribs are known as intercostals, and since they are arranged in two distinct layers, there are external and internal intercostals (shown in their relations to the abdominal muscles in figure 23-1). The **external intercostals** slant downward and forward from one rib to the next below, while the **internal intercostals** slant downward and backward from one rib to the next below. Connected with the vertebral ends of the ribs are several small muscles; **innermost intercostals,** having the same slant as the internal intercostals, lie deep to parts of the latter; and ventrally the **transversus thoracis** radiates upward and laterally from the posterior surface of the sternum to attach to the inner aspects of the ribs. Intercostal nerves (the ventral rami of thoracic nerves, kept apart by the ribs so that they do not form a plexus) run along the lower borders of the ribs, accompanied by intercostal arteries that are for the most part branches of the aorta (p. 370). The muscles connected with the ribs are innervated through the intercostal nerves; they serve to adjust and maintain the relations of the ribs to each other; to what extent they actively participate in respiration is not agreed upon.

Between the thorax and abdomen is the curved muscular mass, convex above, known as the **diaphragm** (fig. 23-5); this completely separates the thoracic from the abdominal cavity but is of course penetrated by structures that pass from one to the other. The diaphragm consists of voluntary muscle that arises from the inner surfaces of the sternum and the lower ribs, and from the bodies of the upper lumbar vertebrae. These muscle fibers insert upon a central tendon that completes the partition between thoracic and abdominal cavities. The diaphragm's motor innervation is from the phrenic nerve which contains fibers from C3, C4 and C5. Since the phrenic also contains sensory nerve fibers, pain from the diaphragm is sometimes felt in the neck or shoulder, where other fibers of the three cervical nerves are distributed—this is an example of "referred pain." Contraction of the diaphragm lowers its dome so that the length of the thoracic cavity is increased; upon relaxation of the diaphragm the elasticity of the abdominal wall pushes the viscera and the diaphragm upward in expiration.

Breathing is usually brought about by movements of both diaphragm and ribs. In inspiration, the ribs are somewhat lifted so that their downward inclination is lessened, and their upper borders are turned outward at the same time, thus resulting in an increase in both anteroposterior and lateral diameters of the chest. During quiet respiration inspiration is brought about primarily through the action of the diaphragm and the scalene muscles. During forced inspiration many other muscles attaching to the ribs may assist in raising them or fixing the lower ribs against the pull of the diaphragm, and active contraction of the abdominal muscles is called into play during forced expiration.

SURVEY OF THE THORACIC VISCERA

The thoracic cavity contains three large serous sacs (fig. 22-1) lined with mesothelium, which enclose the two lungs and the heart, respectively. Each is a closed sac, with one wall carried inward around the viscus with which it is related, much as the side of a balloon may be pushed in with one's fist. A pleural sac surrounds each lung and the pericardial sac surrounds the heart. The outer wall of each pleural sac is attached to the thoracic wall, to the diaphragm below, and to the pericardial sac medially; this wall is known as the **parietal pleura.** The **pulmonary** or **visceral pleura** (pulmonary is the adjective derived from pulmo, the Latin name for the lung) is that layer of the wall of the sac which has been pushed inward by the growth of the lung and forms the smooth outer surface of this organ. Visceral and parietal pleura are normally in contact with themselves and each other, and the pleural cavity consists only of a potential space between the immediately adjacent surfaces of the pleura; it is occupied by a very thin layer of fluid which allows the visceral pleura to slide freely on the parietal pleura but resists separation of the two. Elevation of the ribs and contraction with consequent depression of the diaphragm result in an enlargement of the chest, and the parietal pleura follows the movements of the diaphragm and thoracic wall, to both of which it is firmly attached. The outward movement of the parietal pleura naturally results in an attempt to enlarge the pleural cavity by separating parietal from visceral pleura, but the cohesive force of the fluid makes the visceral pleura, and therefore the lung, move with the parietal pleura. As the lungs are expanded, the atmospheric pressure, which has free access to them through the air passages, moves air into them. This movement is inspiration.

Although the lungs are highly elastic, the cohesive film thus normally keeps them in contact with the parietal pleura. If, however, the pleura has been so injured that air attains access to the pleural cavity, either from a wound in the thoracic wall or through the lung itself, the air dries the fluid and the lung therefore collapses. The presence of air in the pleural cavity is termed "pneumothorax." It may occur spontaneously through damage to the thoracic wall or the lung, or air may be introduced purposely into the pleural cavity to collapse a lung and thus prevent it from following the respiratory movements. Such resting of the lung may allow healing that would not go on so well when the lung was constantly being expanded and contracted.

The **pericardial sac** is built upon the same fundamental plan as are

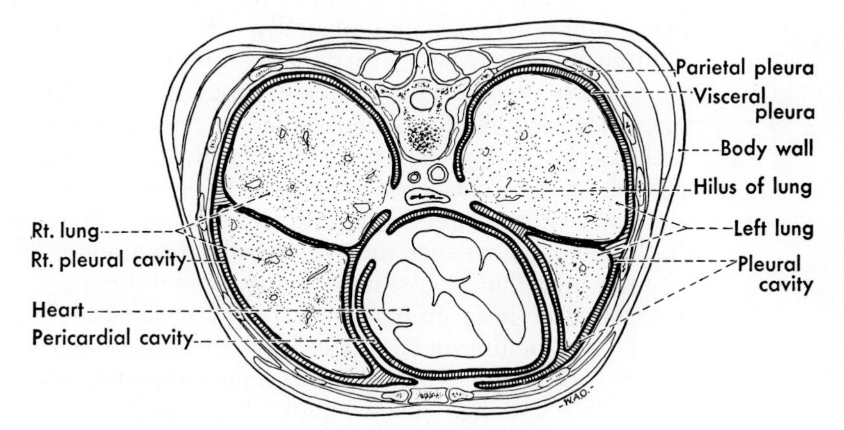

Figure 22–1. Schema of the thorax in cross section, to show relationships of pleural and pericardial cavities.

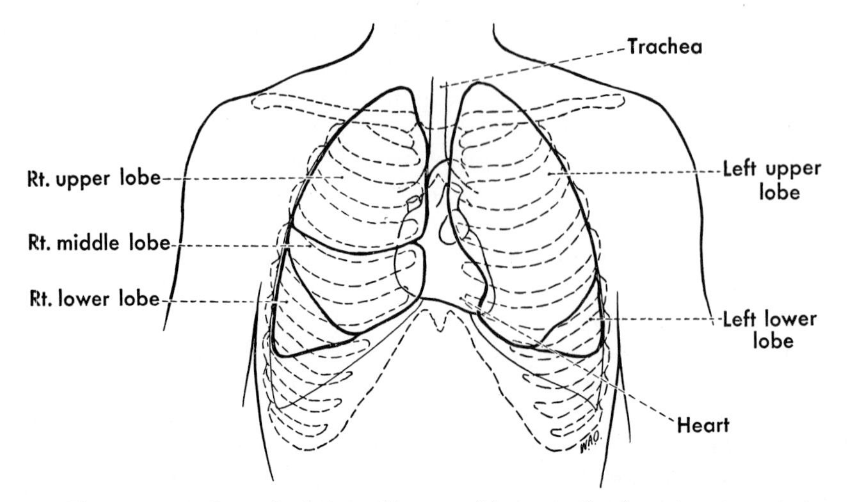

Figure 22–2. General relation of heart and lungs to the thoracic cage, anterior view.

the pleural sacs. The outer layer of the pericardial sac, the **pericardium,** is a tough fibrous membrane lined internally with mesothelium; where the pericardium is attached to the great vessels as they leave the heart the mesothelium is reflected down over these vessels and over the heart to form the epicardium. The pericardial cavity consists of the potential space between epicardium and pericardium. While the pericardial sac is fairly loose in order to allow for the rhythmic changes in heart volume necessary for the pump action of the heart, the pericardium is so tough that it can be expanded suddenly only very slightly. Thus accumulations of fluid within the pericardial cavity may markedly interfere with the ability of the heart to receive incoming blood.

The **lungs** (figs. 22-1 and 22-2) receive air by way of the trachea and the bronchi; the trachea ends in the upper part of the chest by branching into two principal bronchi, one for each lung. The right principal bronchus in turn branches into three lobar bronchi, as the right lung contains three fairly separate but overlapping lobes; there are only two lobes in the left lung and thus two lobar bronchi on this side. The bronchi branch repeatedly within the lungs. Their smallest subdivisions, known as bronchioles, finally end in connection with small thin-walled air sacs, or alveoli (p. 63), through the walls of which takes place the essential interchange of gases between air and blood stream. The lungs receive blood from the pulmonary trunk, which arises from the right ventricle (fig. 22-3); the pulmonary trunk divides into right and left pulmonary arteries, each of which tends to follow the bronchus of its own side but gives off more numerous branches into the lung than does the bronchus. The pulmonary veins, usually two from each lung, empty into the left atrium of the heart.

The branching of the lobar bronchi is particularly complicated, and their direction of branching becomes important when it is desired to facilitate drainage of some particular bronchus by gravity. Each lobar bronchus gives off within a lobe from two to five subsidiary bronchi, each with a descriptive name but known collectively as segmental bronchi, and these usually run in different directions within a lobe; only in the case of the small middle lobe of the right lung do all the segmental bronchi (there are only two here) run parallel enough with each other to allow one position to facilitate drainage of the entire lobe. For the other lobes, the patient may have to be erect in order to drain one segmental bronchus, lying upon his back in order to drain another, or lying on one side, but face down, in order to drain another, and the positions differ from lobe to lobe.

The **heart** (cor) is a muscular pump for propelling the blood. It contains four chambers (fig. 22-3), of which two, known as the atria, are thin-walled and receive incoming blood, while the other two chambers, the ventricles, are heavy-walled and propel the blood into the arterial systems. As is true in all pump systems, the heart contains valves to prevent backflow of fluid and allow a propulsive pressure to be built up within the system. The right atrium receives blood returning from the head and neck, the upper extremities and the thorax by the superior vena cava, and from the abdomen and lower extremities by the inferior vena cava. This blood then passes into the right ventricle and is pumped by the right ventricle out the pulmonary trunk and arteries to the lungs, where the blood is aerated. The tricuspid valve, at the opening between atrium and ventricle, prevents backflow from the ventricle into the atrium. The pulmonary valve, with three valvules somewhat resembling pockets on a vest, protects the mouth of the pulmonary artery and prevents blood in this vessel from running back into the ventricle as the latter relaxes.

Blood returning from the lungs by way of the pulmonary veins enters the left atrium, is then passed into the left ventricle and from the left ventricle is pumped out into the aorta to be distributed to the body in general. The bicuspid or mitral valve protects the atrioventricular opening; an aortic valve, similar to the pulmonary valve, protects

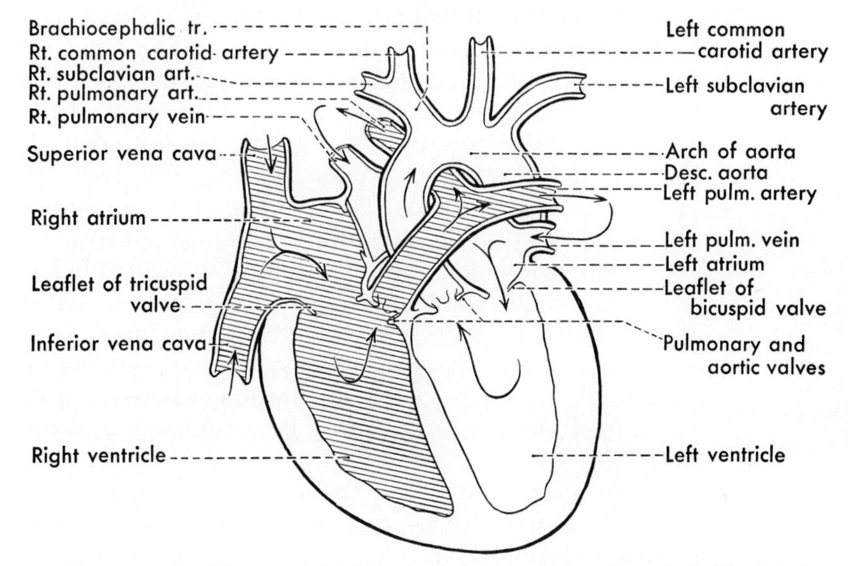

Figure 22–3. Schema of the heart and the circulation through it. The interior of the right side of the heart and of the great vessels connected with it is shaded; on both sides, the arrows indicate the direction of flow of the blood.

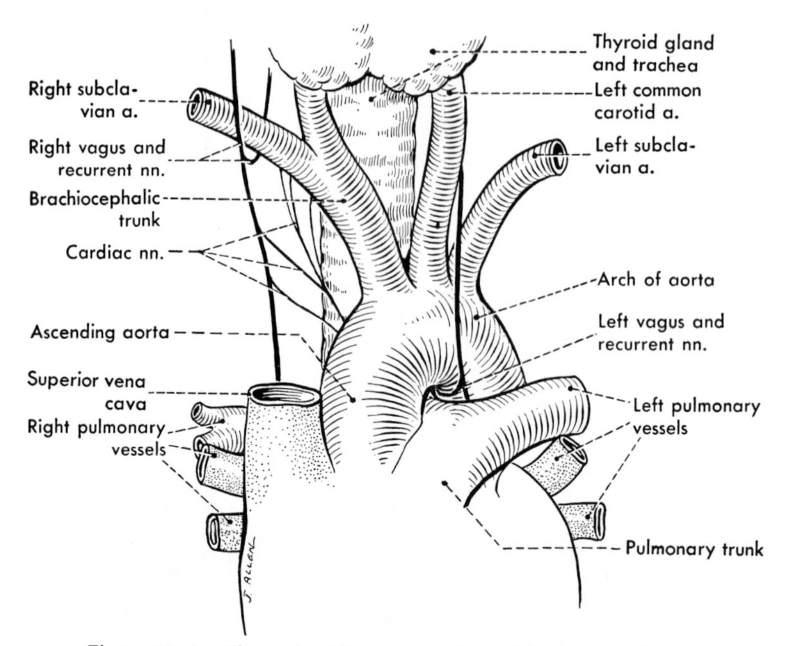

Figure 22–4. The arch of the aorta and neighboring structures.

the orifice of the aorta. Any of the various valves of the heart may on occasion be thickened or otherwise defective and fail to close properly, thus allowing blood under high pressure to flow back into the region of less pressure; this is referred to as a leaking or "insufficient" valve and produces a sound known as a heart murmur. Since the left ventricle must pump blood all over the body, while the right ventricle pumps blood only to the lungs, the left ventricle does considerably more work than does the right; correspondingly, the wall of the left ventricle is very much thicker than is that of the right.

After giving off its branches to the head and neck and the upper limbs, the **aorta,** the great artery leaving the heart, descends slightly to the left of the midline and passes through the diaphragm to become the abdominal aorta. The thoracic aorta consists of three parts: the ascending aorta, running upward from the left ventricle; the arch of the aorta (fig. 22-4), that curves to the left, posteriorly, and downward; and the descending aorta. The coronary arteries, supplying the heart muscle, arise from the ascending aorta just above the left ventricle. The brachiocephalic trunk (dividing into right subclavian and right common carotid), the left common carotid, and the left subclavian are given off in that order from the arch. Other branches of the thoracic aorta are small; the largest and most numerous are the paired intercostal arteries which run beneath the lower borders of the ribs. Small branches from the aorta supply the esophagus and the walls of the bronchi.

The **great veins** in the thorax consist of paired brachiocephalic (formerly called innominate) veins, formed by the union of internal jugular and subclavian veins and therefore returning blood from head, neck and upper limbs; the superior vena cava, formed by the union of the brachiocephalic veins in the upper part of the chest; the inferior vena cava which penetrates the diaphragm to end in the immediately adjacent heart (having, therefore, a very limited thoracic course); and the azygos and hemiazygos system of veins which receive blood mostly from the thoracic wall by means of the intercostal veins, and empty into the superior vena cava.

The pulmonary circulation has already been briefly described in connection with the heart and lungs and need not be repeated here.

The important nerves connected with the thoracic viscera are the paired vagus nerves and the sympathetic trunks, both parts of the autonomic system (p. 56). The **vagus nerves,** descending from the neck, pass behind the roots of the lungs to lie upon the esophagus with which they travel through the thorax and enter the abdomen. These nerves supply involuntary motor nerve fibers to the heart (which decrease the strength and rate of the heart beat) and also involuntary motor fibers to the lungs and esophagus. The **thoracic sympathetic trunk** consists of some eleven or twelve ganglia, or accumulations of nerve cells, connected together by fibers which run up or down the trunk. The trunk is continuous above with the cervical and below with the lumbar trunk; the individual ganglia are connected to spinal nerves by rami communicantes (p. 224), through which they receive fibers from the spinal cord and send fibers back into the spinal nerves to be distributed with these latter. The heart receives its sympathetic innervation largely by branches which descend into the thorax from the cervical trunk but additional fibers reach the heart from the thoracic trunk. Vagal and sympathetic fibers unite to form plexuses in connection with the heart. Other fibers from the thoracic trunk enter into plexuses supplying the lungs and the esophagus. In general, stimulation of sympathetic fibers produces effects opposite to those produced by stimulation of the vagus; thus sympathetic stimulation increases the rate and strength of the heart beat. The thoracic sympathetic trunks also send large nerves, the splanchnic nerves, to plexuses in the abdomen.

It might be noted that the heart normally generates its own impulse to beat, and that the sympathetic and parasympathetic fibers to it merely exert a limited control over this beat. Thus a transplanted heart, although deprived of its nerve supply, can continue to function. The heart muscle will also contract in response to an appropriate electric shock; a heart that stops beating can sometimes be started again by this method, or a ventricle that is contracting irregularly can have a steady, properly timed, beat if regularly spaced electric impulses are delivered to it through an electrode implanted in the muscle.

In addition to these two sets of nerve fibers, concerned primarily with the viscera, the thorax is traversed by the important **phrenic nerves.** These run downward anterior to the hilus of the lungs between pleura and pericardium, to reach and supply the diaphragm.

23 The Abdomen

The abdomen is so large that it is convenient to subdivide it for purposes of description. One method of subdivision consists in the erection of imaginary lines, according to a definite plan, that divide it into nine regions. The terminology employed for these regions is somewhat cumbersome, and much of it is rarely used in actual practice; the most commonly used term is "epigastrium" or "epigastric region," referring to the area below the sternum and between the two costal arches. The simplest and most convenient method of subdividing the abdomen is to think of it as consisting of quadrants divided from each other by the anterior midline and a line passing horizontally through the umbilicus. Thus we can speak of upper right and lower right quadrants, and of upper left and lower left quadrants.

THE ABDOMINAL WALL

The abdominal wall consists functionally not only of the anterolateral abdominal muscles (fig. 23-1) and of the lumbar portion of the vertebral column, but also of the lower ribs and the diaphragm. Because of the domelike shape of the diaphragm, the abdominal viscera extend upward beneath it and are therefore in part protected by the lower ribs. The diaphragm intervenes between these ribs and the abdominal viscera. Posteriorly in the midline are the bodies of the lumbar vertebrae, flanked on each side by the psoas major muscle, a muscle of the lower limb.

The anterolateral abdominal muscles obviously serve as an elastic corset to retain and support the abdominal viscera, and also through their attachments to the ribs and sternum play an important part in movements of the trunk. The lateral abdominal muscles are arranged in three layers; two of these run obliquely and are therefore known as the external and internal obliques, while the third runs almost transversely and is named the transversus abdominis. The **external oblique** muscles arise from about the lower six ribs and form broad sheets whose fibers run downward and medially to attach to the anterior part of the iliac crest, the pubis, and the linea alba. (The linea alba—white line—lies deep to the skin in the anterior midline. It is formed by the union of the aponeuroses of all three lateral abdominal muscles of one side with those of the other side.) Most of the insertion of the external oblique is tendinous, the tendon of insertion being known as the aponeurosis of the external oblique. The aponeurosis in its extension to the anterior midline passes in front of the rectus abdominis muscle, and the lower edge of the aponeurosis as it passes from the anterior superior iliac spine to the pubis forms the inguinal ligament. The lateral part of the inguinal ligament is attached to the fascia of the iliopsoas muscle, but medially it has a free edge behind which the external iliac vessels become continuous with the femoral vessels.

The **internal oblique** muscle corresponds in its direction to the internal intercostal muscles, just as the external oblique corresponds to the external intercostals. The internal oblique arises from the iliopsoas fascia adjacent to the lateral half of the inguinal ligament, the more anterior portion of the iliac crest, and by an aponeurotic layer that extends backward to split around the muscles of the back and attach to both the spinous and transverse processes of lumbar vertebrae as the thoracolumbar fascia. The insertion of the internal oblique is by an aponeurosis into the linea alba, a portion of this aponeurosis passing in front of the rectus muscle to blend with the aponeurosis of the external oblique, and a second portion passing behind the rectus abdominis and blending with the aponeurosis of the transversus.

The **transversus** (transversus abdominis) muscle, the deepest of the lateral abdominal muscles, arises likewise by thoracolumbar fascia from the lumbar vertebrae, and also from the tips of the lower six ribs,

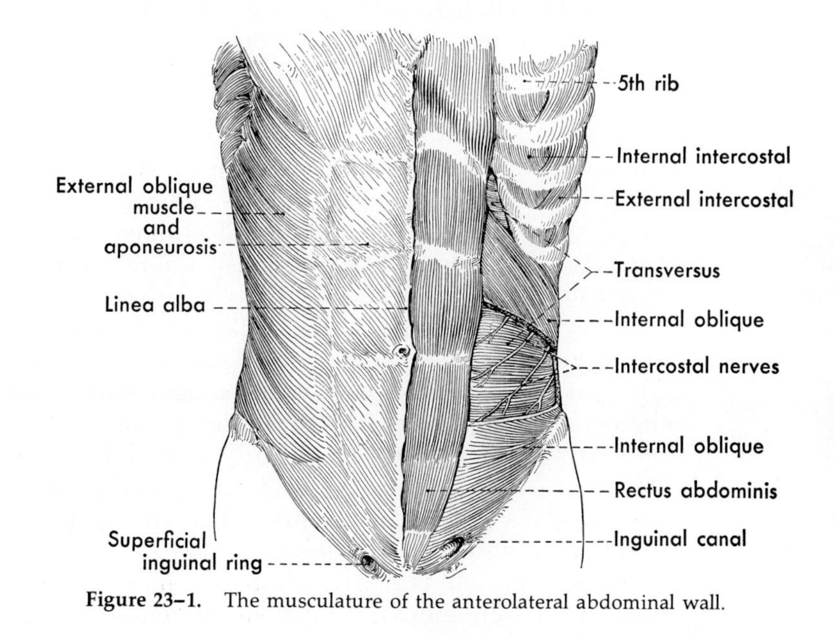

Figure 23–1. The musculature of the anterolateral abdominal wall.

from the anterior portion of the iliac crest, and from the iliopsoas fascia adjacent to a lateral part of the inguinal ligament. This muscle also ends in an aponeurosis, the fibers of which pass for the most part behind the rectus abdominis to attain an attachment to the midline. For a variable distance above the pubis, however, the aponeuroses of all three muscles pass in front of the rectus.

The external and internal oblique muscles in the male are split in their lower portions to allow the structures of the spermatic cord (the duct and vessels of the testis) to make their exit from the abdominal cavity; the transversus may be involved in this, but usually lies above the level of exit. The oblique passageway of the spermatic cord through the abdominal muscles is known as the inguinal canal, and obviously forms a weak place in the anterior abdominal wall. The abdominal end of the canal is known as the deep inguinal ring, its external opening (the split in the external oblique aponeurosis) as the superficial inguinal ring. Approximately 97 per cent of abdominal hernias, that is, protrusions of abdominal viscera through the abdominal wall, in the male involve all or a part of the inguinal canal. An inguinal canal is also present in the female, although it is small because it transmits only a small ligament and a few tiny blood vessels. Inguinal hernia is therefore less predominant in the female than in the male, although even in the female it is the most frequent type of hernia (50 per cent). Other particularly weak places in the abdominal wall are the femoral canal (medial to the femoral vessels behind the inguinal ligament, p. 258) and the umbilicus. Due presumably to the small inguinal canal and the less common exposure of the abdominal viscera of the female to the excessive increased pressure produced by heavy lifting or other exertion, abdominal hernias of all types are only about one-sixth as common in females as in males.

The obliques and the transversus, working as a group, compress the abdomen and therefore increase the pressure upon the abdominal viscera. In so doing, they may cooperate with the diaphragm to fix the thorax, a fact that is taken advantage of when especially delicate or powerful movements of the upper limb are to be carried out, at which times we ordinarily hold our breath. The external and internal obliques of both sides, acting together, aid in flexion of the trunk. The external oblique of one side is usually described as working with the internal oblique of the opposite side in flexing and rotating the trunk to the side of the internal oblique. Electromyography indicates, however, that this is carried out mostly by the internal oblique, although both external obliques become slightly active.

The **rectus abdominis** is a straplike muscle of the anterior abdominal wall; one of the pair is situated just on each side of the midline. The two muscles correspond, in the abdomen, to the strap muscles in the neck. The aponeuroses of the more lateral abdominal muscles as they pass partly in front of and partly behind the rectus muscles form sheaths for the two muscles. Each rectus has its own sheath, but the two sheaths are united at the linea alba. The rectus is attached above to the sternum and to the lower costal cartilages, and below to the pubis. The muscle is partly subdivided into segments by several fibrous bands (tendinous intersections) that cross it. In muscular individuals the rectus muscles, even their segments, can be plainly recognized. Their curved lateral borders are called the **semilunar lines,** and may be a prominent feature of the surface anatomy of the abdomen.

The rectus abdominis is an important flexor of the trunk or, taking its fixed point from above, an upward rotator of the pelvis. Unlike the obliques and the transversus, the rectus does not aid in compressing the abdomen except incidentally as it flexes the trunk or depresses the thoracic wall.

A posterior muscle of the abdominal wall, situated in the lumbar region, is the **quadratus lumborum** muscle. It is a complex muscle extending between the medial part of the iliac crest and the last rib, just lateral to the vertebral column; some of its fibers also insert upon and some arise from the lumbar transverse processes. It is primarily a lateral flexor of the vertebral column; taking its fixed point from above, it will therefore tilt upward the side of the pelvis to which it is attached. It is innervated by direct branches from several of the upper lumbar nerves.

The lower intercostal nerves and vessels, continuing the downward direction of the ribs, run into the lower part of the abdomen. Thus the tenth intercostal nerve ends at about the level of the umbilicus, and the twelfth ends only a short distance above the pubic symphysis. The lower intercostal nerves, and branches (iliohypogastric and ilioinguinal) from the first lumbar nerve, supply the musculature and skin of the abdomen. The main branches of the nerves run roughly parallel to each other between internal oblique and transversus muscles; they exchange enough branches so that they form a loose plexus in this position. The external oblique receives fibers from all the intercostal nerves corresponding to the ribs from which it takes origin, that is, from the last six or seven; the internal oblique and transversus receive fibers from approximately the seventh to twelfth intercostal nerves and from the branches of the first lumbar; the rectus is supplied by lower intercostal nerves that pierce the lateral wall of the rectus sheath to reach the muscle.

Small intercostal arteries accompany the abdominal portions of the intercostal nerves, and anastomose here with ascending branches from the external iliac vessels (fig. 23-5) and with lateral branches of the inferior epigastric artery, also derived from the external iliac. In the substance of the rectus muscle the inferior epigastric anastomoses with the superior epigastric artery, a continuation of an artery (internal thoracic) that arises from the subclavian artery at the base of the neck and runs downward on the inner surface of the thoracic wall a little lateral to the sternum.

An important relation of the inferior epigastric artery close to its origin is the deep inguinal ring or abdominal end of the inguinal canal. The inferior epigastric artery lies just medial to this opening; therefore a hernia that traverses the whole of the inguinal canal (an indirect inguinal hernia) must have the neck of its hernial sac situated lateral to this vessel. The second chief type of inguinal hernia, direct inguinal hernia, bulges directly toward or through the superficial inguinal ring rather than starting at the deep ring, and therefore lies medial to the artery.

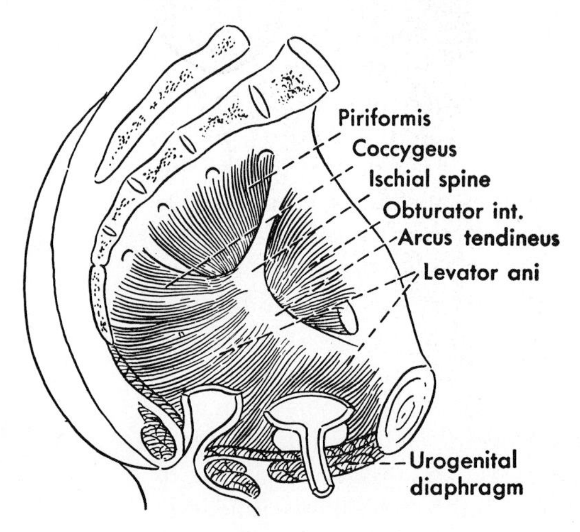

Figure 23–2. Muscles of the walls and floor of the pelvis.

THE PELVIC FLOOR AND THE PERINEUM

The walls of the abdomen have already been briefly considered. The diaphragm forms the "roof" of the abdominal cavity, and since abdominal and pelvic cavities are continuous with each other the floor of the pelvis is also the floor of the abdominal cavity.

The gap between the coccyx, the pubic symphysis, the ischia, and the inferior rami of the pubes is known as the pelvic outlet, and the muscles bridging the pelvic outlet constitute the pelvic floor, or **pelvic diaphragm.** The chief muscle of the pelvic diaphragm is the **levator ani** (fig. 23-2), which is attached posteriorly to the coccyx, anteriorly to the pubis, and laterally to the lateral pelvic wall. This muscle is somewhat funnel-shaped, converging toward the rectum and anal canal (p. 380) at its apex. About the anal canal the fibers of the levator ani blend with the musculature of the canal, thus affording firm attachment to this terminal portion of the digestive tract as it leaves the pelvis. Anteriorly, the levator ani divides to allow the urethra to pass through, and in the female it is also perforated by the vagina (fig. 23-4). Its attachments to the vaginal wall give support to this viscus. The lower portion of the rectum and the urinary bladder or, in the male, the prostate rest upon the upper surface of the levator ani. By its contraction the levator ani resists downward movement of the pelvic viscera incident to increased abdominal pressure, and pulls the anal canal upward during defecation. The muscle is innervated, usually on its pelvic surface, by branches from S3 and S4. Modern concepts of its function are that it largely controls voluntary emptying of the urinary bladder and is extremely important in the support of the uterus. Weakness of it (or of a part of it called the pubococcygeus) predisposes to urinary incontinence, and exercise of it and adjacent muscles may restore continence.

Behind the levator ani is the **coccygeus,** the other muscle of the pelvic diaphragm. This extends from the ischial spine to the lower part of the sacrum and upper part of the coccyx. It adds little to the pelvic diaphragm, but spasm of it and the part of the levator ani that attaches to the coccyx has been thought to be responsible for certain cases of painful coccyx (coccygodynia or coccydynia).

External to (below) the levator ani, passing transversely from one inferior ramus of the pubis and an associated portion of the ramus of the ischium across to the corresponding bony elements on the other side, is a structure known as the **urogenital diaphragm.** This consists in part of muscle and in part of fascia; since it bridges the more anterior part of the pelvic outlet, between the diverging inferior pubic rami, it affords additional support to the pelvic viscera. The urogenital diaphragm is perforated by the urethra, and in the female by the vagina also.

The external aspect of the pelvic outlet, that is, the region between the thighs including both the area around the anus and the external genital organs, is known as the perineum. The musculature of the perineum includes the muscle in the urogenital diaphragm, special muscles in connection with the penis or the vagina and clitoris, and an external, voluntary, sphincter muscle of the anus. These muscles are all innervated by the pudendal nerve from the lower portion (S2, S3, and S4) of the sacral plexus. The external surface of the levator ani forms the roof of the perineum; between this muscle, as it converges on the anal canal, and the ischial tuberosity on each side is a considerable gap filled with fat and tough strands of connective tissue and known as the **ischiorectal fossa.** The levator ani can be massaged through the ischiorectal fossa.

ABDOMINAL VISCERA

The abdomen is lined by a serous membrane that is, for the most part, in intimate contact with the abdominal wall and is known as the **parietal peritoneum.** The viscera are also covered by a peritoneal layer, the **serosa** or **visceral peritoneum;** the peritoneal cavity lies between visceral and parietal peritoneum, and, although more complicated in form, it is built upon exactly the same plan as the pericardial and pleural cavities. If we compare the walls of the peritoneal cavity to a balloon into one side of which most of the viscera have sunk down, carrying the wall of the balloon reflected over themselves, it will be obvious that visceral and parietal peritoneum are continuous with each other. Where the viscera have deeply invaginated the peritoneal sac the visceral peritoneum covering them is attached to the parietal peritoneum of the body wall by a double layer of peritoneum known as a **mesentery.** The vessels and nerves to the viscera run between the two layers of mesentery.

The peritoneal cavity in the male is a completely closed sac; in the female the uterine tubes open into the pelvic portion of the peritoneal cavity. Peritonitis, or infection of the peritoneal cavity, involves grave danger to the patient since the peritoneal surface is warm, moist, and very extensive, offering almost ideal conditions for the growth of bacteria. Infections of the peritoneal cavity in the male usually result from a ruptured viscus or penetration of the abdominal

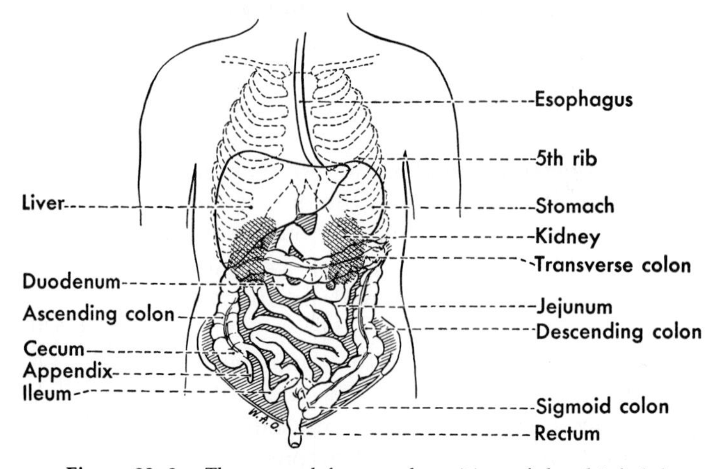

Figure 23–3. The general form and position of the chief abdominal viscera. The coils of the jejunum and ileum, which in reality very largely fill the abdominal cavity, are shown here in a simplified, very diagrammatic manner in order that the other viscera may be more clearly visible.

wall. Those in the female may occur through similar causes, but peritonitis may also arise here through infections of the genital tract with subsequent spread through the uterine tubes.

The pelvic portion of the peritoneal cavity is directly continuous with the abdominal portion, and the division between abdomen and pelvis is therefore largely an artificial one. The portion between the flared wings of the ilia is usually described as the major or false pelvis and considered as part of the abdominal cavity proper. The portion below a plane passing from the sacral promontory to the upper border of the pubic symphysis is described as the pelvis minor or true pelvis (or simply as *the* pelvis).

In brief summary of the abdominal viscera (fig. 23-3), the **liver** (hepar, hence the adjective "hepatic") occupies the upper portion of the abdomen on the right side, and extends over to the left. It is attached to the curved dome of the diaphragm and moves with this; it is almost completely covered by the lower ribs. The liver is by far the largest gland in the body; as a gland, it secretes bile which is stored in a blind sac, the gallbladder, that is attached to the posteroinferior surface of the liver. The liver and the gallbladder are connected to each other and to the intestine (duodenum) by ducts; contraction of the gallbladder therefore discharges bile, necessary to proper digestion of fat, into the intestine. Gallstones, formed from bile, may obstruct a duct and cause serious symptoms. The liver is also an important "chemical factory," and as such is essential to life. It stores both glycogen and fat. A part of its essential activity is the removal from the blood stream of certain poisonous products of digestion; the blood from the abdominal part of the digestive tract runs through the liver before reaching the heart.

The **stomach** (gaster, hence "gastric") lies mostly to the left in the upper part of the abdomen and extends over toward the right; since it is a hollow viscus it varies greatly in size and position according to whether it is empty or full, according to the degree of fullness of other parts of the digestive tract, and according to the position of the individual at any particular time. The stomach not only churns the food and helps to liquefy it but adds to it hydrochloric acid and an enzyme that digests proteins. Oversecretion of acid is an important cause of ulcers (peptic ulcers) which usually occur in the distal end of the stomach or the adjacent first part of the small intestine.

The **pancreas** (not shown in figure 23-3) lies behind most of the abdominal viscera, across the front of the vertebral column at the level of the kidneys. The majority of its cells secrete digestive enzymes that act upon all three basic foodstuffs – proteins, carbohydrates, and fats – and its duct opens with that of the liver and gallbladder into the duodenum. In addition to its digestive function, the pancreas also has endocrine functions, the most important being the production of insulin.

The **spleen** is an organ of the vascular system rather than the digestive system (serving as a storage place for red blood cells and as a place for formation of certain types of white blood cells), and lies to the left of the stomach and against the diaphragm and ribs.

The coils of the **small intestine** occupy most of the abdominal cavity. The duodenum, or first part of the small intestine, lies against the posterior abdominal wall behind the peritoneum, but the remainder of the small intestine is suspended by a fan-shaped mesentery that allows it considerable freedom of movement. Besides the duodenum, the small intestine consists of jejunum and ileum in that order. Some coils of the ileum usually lie within the true pelvis, and the terminal portion of the ileum then ascends into the lower right quadrant to end in the cecum.

The small intestine joins the large intestine not at the end of the latter but rather on its side, and leaves, therefore, a short blind end projecting below the junction of these two parts. The **cecum** is this blind end of the large intestine, and the appendix is attached to it. The **appendix** was originally the end of the cecum, but in the adult rarely retains this position; most commonly the appendix, originating behind the cecum, projects down below it, but it may lie entirely behind the cecum or in some other position. The **large intestine** begins with the cecum; above the ileocecal junction the large intestine is known as the **ascending colon** and runs upward to come in contact with the lower posterior surface of the liver. Here, in the upper right quadrant, the large intestine makes a sharp bend to the left to become the **transverse colon** which then crosses the abdominal cavity to the upper left quadrant in the region of the spleen. The transverse colon may run almost transversely or may droop very markedly in its course across the abdomen. On the left side a second flexure occurs and the **descending colon** then passes downward. Ascending and descending colons lie close against the posterior body wall and are covered with peritoneum only on their front and sides; they have no mesenteries. The transverse colon, however, is attached by a mesentery to the posterior abdominal wall and is also attached to a redundant mesentery of the stomach, the greater omentum. The descending colon is succeeded by a short section of colon having a mesentery, and known from its shape as the **sigmoid colon.** The sigmoid colon crosses the brim of the pelvis, loses its mesentery, and becomes the rectum.

The **kidneys** (renes, hence "renal") are also situated in the abdomen, against the posterior abdominal wall on each side of the vertebral column, and behind the peritoneum. The right kidney usually but not always lies at a slightly lower level than the left; the upper pole of the left kidney usually extends up to the level of the eleventh rib as it attaches to the vertebral column, while that of the right kidney often lies at the level of the twelfth rib. The ureters run downward approximately parallel to the vertebral column, and pass along the lateral pelvic walls to reach the bladder.

The **suprarenal glands** (not shown in figure 23-3) lie on the upper poles of the kidneys. They belong to the endocrine system rather than to the digestive or urogenital ones, and are discussed briefly on page 64.

The **pelvic viscera** in the male (fig. 23-4*a*), in addition to such coils of small intestine and sigmoid colon as may be present in the pelvis, consist of the rectum, urinary bladder, prostate, and seminal vesicles. The upper portion of the rectum is covered anteriorly and on its sides by peritoneum, and lies against the posterior pelvic wall; the **bladder** (vesica) lies against the pubis and anterior abdominal wall and is covered above and posteriorly by peritoneum. The pelvic portion of the peritoneal cavity extends downward between bladder and rectum, and ends blindly some distance above the pelvic floor, leaving lower portions of rectum and bladder without peritoneal contact. As the bowel penetrates the levator ani it turns backward, and this lower portion is the **anal canal.** The **seminal vesicles** and the **prostate,** connected with the male genital tract, lie in close connection with the base of the bladder and urethra below the level of the peritoneal cavity. Both are glands, and together secrete most of the fluid in which the male germ cells (spermatozoa) are suspended. Since the prostate almost completely surrounds the urethra, prostatic enlargement may markedly interfere with the emptying of the bladder.

In the female pelvis (fig. 23-4*b*) the bladder and the rectum have essentially the same peritoneal relations as in the male; the space between the two is occupied by the **uterus** and by the **broad ligaments** that extend from the sides of the uterus to the pelvic walls. The **uterine tubes,** in the upper border of the broad ligaments, open at their ovarian ends into the peritoneal cavity and at their uterine ends into the uterine cavity. The uterus and the broad ligaments divide the lower part of the peritoneal cavity in the female into two portions, one lying between bladder and uterus, the other between uterus and rectum. The **ovaries** lie on the lateral pelvic walls just behind the broad ligaments; the **vagina** extends downward from the uterus, lying mostly below the level to which the peritoneum reaches.

The great artery of the abdomen, the **abdominal aorta** (fig. 23-5), is the direct continuation of the thoracic aorta and ends below at about the level of the fourth lumbar vertebra by dividing into the two common iliac arteries. In its course the abdominal aorta gives off lumbar vessels to the abdominal wall, renal and testicular or ovarian branches to the kidneys and gonads, and three vessels to the digestive tract. The uppermost of these last-mentioned vessels, the **celiac trunk** or artery, arises from the aorta as it lies between the crura of the diaphragm and supplies primarily the stomach, liver, spleen, pancreas, and duodenum. The second branch, the **superior mesenteric artery,** arises from the front of the aorta directly below the celiac and supplies branches to all the small intestine and to the ascending and transverse portions of the large intestine, including the appendix and cecum. The third branch to the digestive system, the **inferior mesenteric artery,** arises somewhat

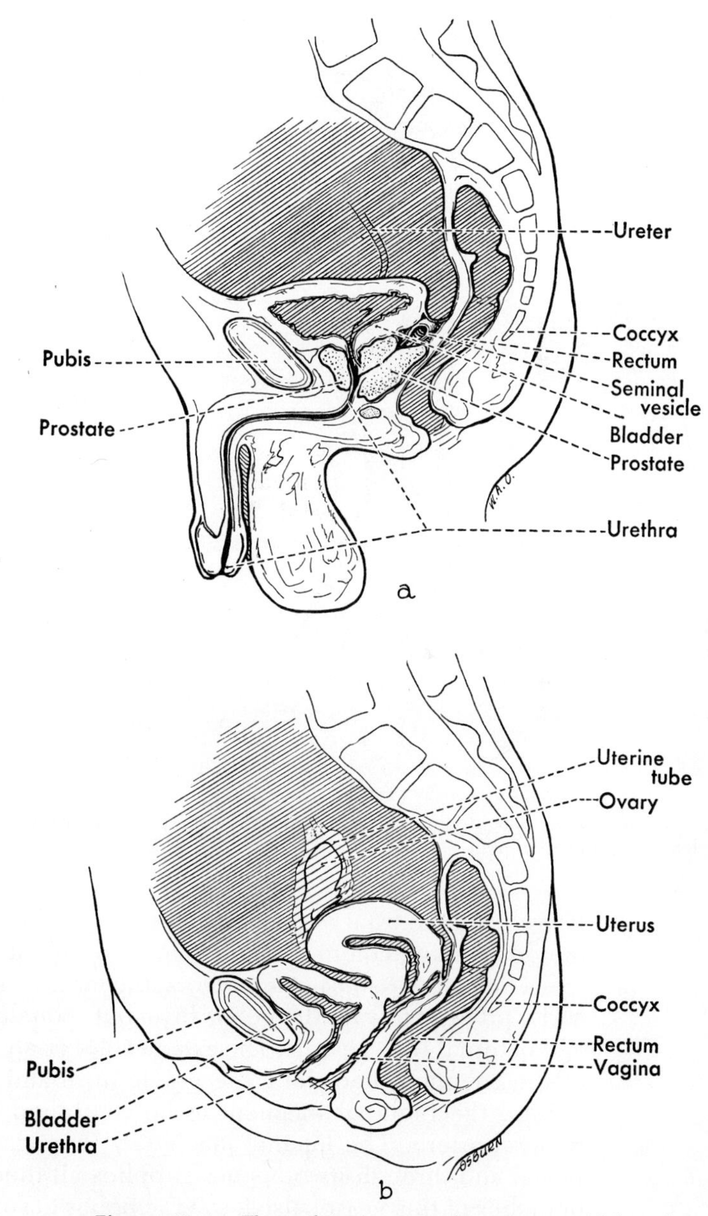

Figure 23–4. The pelvis in sagittal section. *a*. Male pelvis. *b*. Female pelvis.

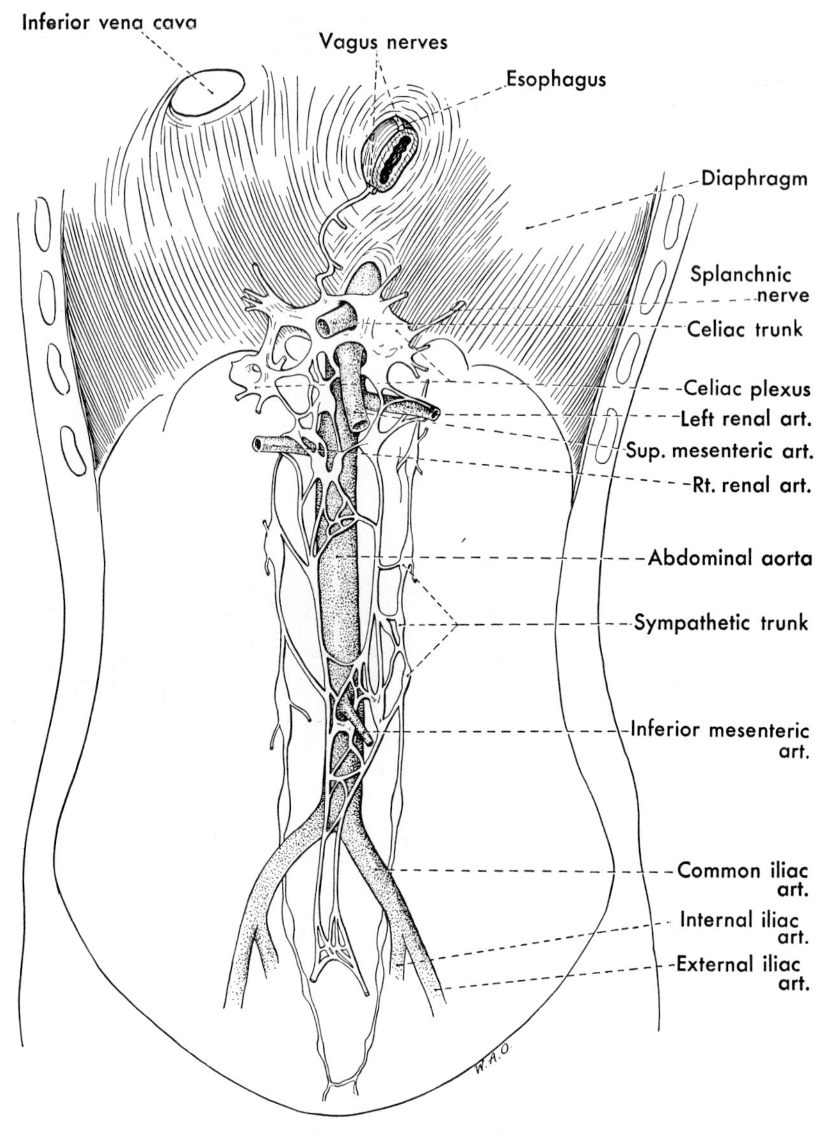

Figure 23–5. The abdominal aorta and the larger autonomic ganglia and plexuses of the abdomen.

lower from the front of the aorta and runs to the left where it supplies the descending colon, sigmoid colon, and rectum.

The paired **common iliac arteries,** the large terminal branches of the aorta, proceed toward the pelvic brim but soon divide into the internal and external iliac vessels. The **external iliac artery** continues the course of the common iliac along the pelvic brim and leaves the abdomen behind the inguinal ligament to be continued into the thigh as the femoral artery. The **internal iliac artery** (hypogastric) passes into the pelvis and through its branches supplies all the pelvic viscera; some branches of this vessel also leave the pelvis to supply the musculature of the buttocks and the perineal region.

There are two great venous systems within the abdomen, the inferior vena cava and the portal vein with its tributaries. The **inferior vena cava** parallels the abdominal aorta, lying to its right, and receives vessels corresponding to the branches of the abdominal aorta with the exception of veins from the digestive tract. The veins from the digestive tract parallel the branches of the celiac, superior mesenteric and inferior mesenteric arteries, but instead of joining the inferior vena cava they join each other to form the **portal vein.** This vein runs upward into the liver, through which organ all its blood must pass in intimate contact with the hepatic cells. Blood from the liver, whether brought there by the celiac artery or by the portal vein, is carried into the inferior vena cava by the hepatic veins just before the former penetrates the diaphragm.

The **nerves** concerned with the abdominal viscera (fig. 23-5) are a continuation of those that are concerned with the thoracic viscera. The sympathetic trunks in the abdomen contribute fibers to the lumbar nerves and also help form a plexus, the aortic plexus, across the front of the abdominal aorta. The upper portion of this plexus, in the region of the celiac and superior mesenteric arteries, is especially well developed and contains several ganglia, the celiac ganglia being the largest of these. The fibers to this upper portion of the plexus (celiac or solar plexus) are derived only in small part from the lumbar trunks; large nerves come into it from the thoracic trunks. These nerves from the thorax are the splanchnic nerves.

From the celiac and aortic plexuses numerous nerve fibers pass along the various arteries to the gut and supply both the blood vessels to it and the smooth muscle of its wall. The aortic plexus is also continued into the pelvis, where it aids in the innervation of the organs there.

The vagus nerves also are distributed to the digestive tract, or at least to the major portion of it. They give off branches to the stomach while they lie on the lower end of the esophagus, and parts of them then leave the stomach to reach the celiac plexus, through the branches of which they are distributed to most of the abdominal viscera, and at least as far down the digestive tract as the descending colon.

The pelvic viscera are innervated in part through the sympathetic fibers that are a continuation downward from the aortic plexus; in addition, they receive fibers from approximately the second, third and fourth sacral nerves. These latter fibers form the sacral parasympathetic nerves (pelvic splanchnic nerves, nervi erigentes), which are the chief nerves concerned with the emptying of the bladder and the rectum, and the most important innervation of the pelvis.

In general, stimulation of the sympathetic nerves of the abdomen and pelvis produces constriction of blood vessels and cessation of movement in the gut. Parasympathetic stimulation, whether of the vagi or of the sacral parasympathetic nerves, increases the peristaltic activity of the digestive tract.

Synonyms

A complete list of synonyms would be far too long for this text, since many structures have in the past been known by several names. The following list, therefore, includes only commonly used terms that the student is likely to encounter in reading older texts or in conversation with people trained in a different terminology. The older term here comes first, followed by the presently accepted one.

Angle of Louis = sternal angle
Ansa hypoglossi = ansa cervicalis
Artery or arteries
  external maxillary = facial
  hypogastric = internal iliac
  innominate = brachiocephalic
  internal maxillary = maxillary
  interosseous, dorsal and volar = posterior and anterior
  superficial cervical = transverse cervical
  transverse scapular = suprascapular
  volar interosseous = anterior interosseous
Astragalus = talus
Band, iliotibial = tract, iliotibial
Bone or bones
  innominate = coxal
  multangular, greater and lesser = trapezium and trapezoid
  navicular (of hand) = scaphoid
Canal, Hunter's = adductor canal
Cartilages, semilunar, of knee = menisci
Centrum (of vertebra) = body
Chain, sympathetic = trunk
Circle of Willis = circulus arteriosus cerebri
Cisterna magna = cisterna cerebellomedullaris
Dorsal (for forearm) = posterior
Epistropheus = axis

Fascia, Camper's and Scarpa's = superficial fascia of lower abdominal wall
coracoclavicular = clavipectoral
Fossa, infra- and supraspinous = infra- and supraspinatous
ischio-anal = ischiorectal
ovalis, of thigh = saphenous hiatus
Ganglion, Gasserian or semilunar = trigeminal
Gland, lymph = lymph node
submaxillary = submandibular
Joint, transverse carpal = midcarpal
Lacertus fibrosus = bicipital aponeurosis
Ligament, carpal
dorsal = extensor retinaculum of wrist
transverse = flexor retinaculum of wrist
volar = no longer named
collateral, medial and lateral = ulnar and radial collateral at elbow, tibial and fibular at knee
costocoracoid = no longer named
cruciate crural = inferior extensor retinaculum of ankle
intercapitular = deep transverse metacarpal or metatarsal
laciniate = flexor retinaculum of ankle
plantar, short = calcaneocuboid
round, of hip = lig. of femoral head
teres, of hip = lig. of femoral head
transverse carpal = flexor retinaculum of wrist
transverse crural = superior extensor retinaculum
triangular = urogenital diaphragm
vaginal, of digits = fibrous, or annular and cruciform parts, of tendon sheaths
volar accessory = palmar or plantar of digits
Muscle or muscles
abductor digiti quinti = abd. digiti minimi
caninus = levator anguli oris
extensor digiti quinti = ext. digiti minimi
digitorum communis = ext. digitorum
indicis proprius = ext. indicis
flexor digiti minimi proprius = flex. digiti minimi
digiti quinti proprius = flex. digiti minimi
digitorum sublimis = flex. digit. superficialis
opponens digiti quinti = opp. digiti minimi
quadratus labii inferioris = depressor labii inferioris
labii superioris = zygomaticus minor, levator labii superioris, and levator labii superioris alaeque nasi
sacrospinalis = erector spinae
triangularis = depressor anguli oris
zygomaticus = zygomaticus major
Nerve, acoustic or stato-acoustic = vestibulocochlear
anterior thoracic = pectoral
dorsal interosseous = post. interosseous
spinal accessory = accessory
volar interosseous = ant. interosseous
Posterior (for spinal nerves) = dorsal
Process, odontoid = dens
Quinti = minimi
Triangle, Scarpa's = femoral triangle

Tubc, Eustachian = auditory tubc
  Fallopian = uterine tube
Tuberosity, of humerus = tubercle
  bicipital = radial
Tunnel, carpal = carpal canal
Vas deferens = ductus deferens
Vein, innominate = brachiocephalic
Volar = anterior in forearm, palmar in hand
Xiphisternum = xiphoid process

Index